T0141846

Smart Innovation, Systems and Technologies

Volume 98

The Smart Innovation, Systems and Technologies book series encompasses the topics of knowledge, intelligence, innovation and sustainability. The aim of the series is to make available a platform for the publication of books on all aspects of single and multi-disciplinary research on these themes in order to make the latest results available in a readily-accessible form. Volumes on interdisciplinary research combining two or more of these areas is particularly sought.

The series covers systems and paradigms that employ knowledge and intelligence in a broad sense. Its scope is systems having embedded knowledge and intelligence, which may be applied to the solution of world problems in industry, the environment and the community. It also focusses on the knowledge-transfer methodologies and innovation strategies employed to make this happen effectively. The combination of intelligent systems tools and a broad range of applications introduces a need for a synergy of disciplines from science, technology, business and the humanities. The series will include conference proceedings, edited collections, monographs, handbooks, reference books, and other relevant types of book in areas of science and technology where smart systems and technologies can offer innovative solutions.

High quality content is an essential feature for all book proposals accepted for the series. It is expected that editors of all accepted volumes will ensure that contributions are subjected to an appropriate level of reviewing process and adhere to KES quality principles.

More information about this series at http://www.springer.com/series/8767

Giuseppe De Pietro · Luigi Gallo
Robert J. Howlett · Lakhmi C. Jain
Ljubo Vlacic
Editors

Intelligent Interactive Multimedia Systems and Services

Proceedings of 2018 Conference

Editors
Giuseppe De Pietro
Istituto Di Calcolo E Reti Ad Alte Prestazioni (Icar)
National Research Council
Rome, Italy

Luigi Gallo
Istituto Di Calcolo E Reti Ad Alte Prestazioni (Icar)
National Research Council
Rome, Italy

Robert J. Howlett
Bournemouth University
Poole, UK

and

KES International
Shoreham-by-Sea, UK

Lakhmi C. Jain
Centre for Artificial Intelligence, Faculty of Engineering and Information Technology
University of Technology Sydney
Sydney, NSW, Australia

and

Faculty of Science, Technology and Mathematics
University of Canberra
Canberra, ACT, Australia

and

KES International
Shoreham-by-Sea, UK

Ljubo Vlacic
Griffith Sciences - Centres and Institutes
Griffith University
South Brisbane, QLD, Australia

ISSN 2190-3018 ISSN 2190-3026 (electronic)
Smart Innovation, Systems and Technologies
ISBN 978-3-030-06388-7 ISBN 978-3-319-92231-7 (eBook)
https://doi.org/10.1007/978-3-319-92231-7

Printed on acid-free paper

This Springer imprint is published by the registered company Springer International Publishing AG part of Springer Nature
The registered company address is: Gewerbestrasse 11, 6330 Cham, Switzerland

Preface

This volume presents a series of carefully selected papers on the theme of Intelligent Interactive Multimedia Systems and Services (IIMSS-18), but also includes contributions on Innovation in Medicine and Healthcare (InMed-18) and Smart Transportation Systems (STS-18).

The papers were presented at the Smart Digital Futures 2018 multi-theme conference, which grouped conferences on Agent and Multi-Agent Systems: Technologies and Applications (AMSTA-18), Intelligent Decision Technologies (IDT-18), Innovation in Medicine and Healthcare (InMed-18), Smart Education and E-Learning (SEEL-18), Smart Transportation Systems (STS-18) together with IIMSS-18 in one venue in Gold Coast, Australia, during 20–22 June 2018.

IIMSS-18 included sessions on 'Cognitive Systems and Big Data Analytics', 'Data Processing and Secure Systems', 'Innovative Information Services for Advanced Knowledge Activity', 'Autonomous System' and 'Image Processing'. InMed-18 papers cover major areas of 'Digital Architecture for Internet of Things, Big data, Cloud and Mobile IT in Healthcare' and 'Advanced ICT for Medical and Healthcare'. STS-18 papers provide a comprehensive overview of various aspects of current research into intelligent transportation technology.

We would like to acknowledge and thank all those who made the conference possible through their hard work. We are grateful to the Programme Co-Chairs, the General Track Chairs, the International Programme Committee members and reviewers for their valuable efforts in the review process, thereby helping us to guarantee the highest quality possible for the conference. We would also like to thank the organisers and chairs of the special sessions which make an essential contribution to the success of the conference.

Lastly, we would like to thank all the authors, presenters and delegates for their valuable contribution in making this an extraordinary event.

We hope and intend that this volume will make a significant contribution to international research in the leading edge topics of the conferences.

R. J. Howlett
L. Gallo
Y.-W. Chen
X. Qu
L. C. Jain

Contents

Intelligent Interactive Multimedia: Systems and Services (KES-IIMSS-18) Introduction

Intelligent Interactive Multimedia: Systems and Services (KES-IIMSS-18) Introduction

We introduce to you a series of carefully selected papers presented during the 11th KES International Conference on Intelligent Interactive Multimedia Systems and Services (IIMSS-18).

At a time when computers are more widespread than ever, and computer users range from highly qualified scientists to non-computer expert professionals, Intelligent Interactive Systems are becoming a necessity in modern computer systems. The solution of "one-fits-all" is no longer applicable to wide ranges of users of various backgrounds and needs. Therefore, one important goal of many intelligent interactive systems is dynamic personalization and adaptivity to users. Multimedia Systems refer to the coordinated storage, processing, transmission and retrieval of multiple forms of information, such as audio, image, video, animation, graphics, and text. The growth rate of multimedia services has become explosive, as technological progress matches consumer needs for content.

The conference took place as part of the Smart Digital Futures 2018 multi-theme conference, which groups AMSTA, IDT, InMed, SEEL, STS with IIMSS in one venue. It was a forum for researchers and scientists to share work and experiences on intelligent interactive systems and multimedia systems and services. It included a general track and four invited sessions.

The invited session "Cognitive Systems and Big Data Analytics" (Chaps. 1–3) specifically focuses on Big Data security, word representations in a machine-readable way, and interactive virtual environments. The invited session "Data Processing and Secure Systems" (Chaps. 4–6) focuses on models, techniques, and algorithms capable of analysing, mining and processing both critical and social data. Differently, the invited session "Innovative Information Services for Advanced Knowledge Activity" (Chaps. 7–13) discusses models, techniques and algorithms for the detection and recognition of human activities, as well as sensing architectures for control systems. The invited session "Autonomous System" (Chaps. 14–16) considers theoretical and practical issues in the design of intelligent and autonomous systems. Finally, the general track (Chapter 17) focuses on image processing, specifically on novel approaches to saliency detection.

Our gratitude goes to many people who have greatly contributed to putting together a fine scientific program and exciting social events for IIMSS 2018. We acknowledge the commitment and hard work of the program chairs and the invited session organizers. They have kept the scientific program in focus and made the discussions interesting and valuable. We recognize the excellent job done by the program committee members and the extra reviewers. They evaluated all the papers on a very tight schedule. We are grateful for their dedication and contributions. We could not have done it without them. More importantly, we thank the authors for submitting and trusting their work to the IIMSS conference.

We hope that readers will find in this book an interesting source of knowledge in fundamental and applied facets of intelligent interactive multimedia and, maybe, even some motivation for further research.

Giuseppe De Pietro
Luigi Gallo
Robert J. Howlett
Lakhmi C. Jain
Ljubo Vlacic

Organization

IIMSS-18 International Programme Committee

Awais Ahmad	Yeungnam University, Korea
Flora Amato	Università degli Studi di Napoli Federico II, Italy
Marco Anisetti	University of Milan, Italy
Koichi Asakura	Daido University, Japan
Monica Bianchini	Dipartimento di Ingegneria dell'Informazione e Scienze Matematiche, Università degli Studi di Siena, Italy
Francesco Bianconi	University of Perugia, Italy
Alfredo Cuzzocrea	University of Trieste and ICAR-CNR, Italy
Ernesto Damiani	Khalifa University, Abu Dhabi, UAE
Dinu Dragan	University of Novi Sad, Faculty of Technical Sciences, Serbia
Massimo Esposito	Institute for High Performance Computing and Networking (ICAR) of the National Research Council of Italy (CNR)
Margarita Favorskaya	Reshetnev Siberian State University of Science and Technology, Russia
Christos Grecos	Central Washington University, USA
Katsuhiro Honda	Osaka Prefecture University, Japan
Ignazio Infantino	ICAR-CNR, Italy
Gwanggil Jeon	Incheon National University, Korea
Dimitris Kanellopoulos	University of Patras, Greece
Chengjun Liu	New Jersey Institute of Technology, USA
Antonino Mazzeo	University of Naples Federico II, Italy
Giovanni Luca Christian Masala	Plymouth University, UK
Marian Cristian Mihăescu	University of Craiova, Romania
Aniello Minutolo	National Research Council of Italy, Italy
Vincenzo Moscato	University of Naples Federico II, Italy
Francesco Piccialli	University of Naples Federico II, Italy
Radu-Emil Precup	Politehnica University of Timisoara, Romania
Antonio M. Rinaldi	Università degli Studi di Napoli Federico II, Italy
Milan Simic	RMIT University, School of Engineering, Australia
Maria Spichkova	RMIT University, School of Science, Australia
Toyohide Watanabe	Nagoya Industrial Science Research Institute, Japan

Big Data Security on Cloud Servers Using Data Fragmentation Technique and NoSQL Database

Nelson Santos(✉) and Giovanni L. Masala

Big Data Group, School of Computing, Electronics and Mathematics, Plymouth University, Plymouth PL4 8AA, UK
nelson.santos@students.plymouth.ac.uk,
giovanni.masala@plymouth.ac.uk

Abstract. Cloud computing has become so popular that most sensitive data are hosted on the cloud. This fast-growing paradigm has brought along many problems, including the security and integrity of the data, where users rely entirely on the providers to secure their data. This paper investigates the use of the pattern fragmentation to split data into chunks before storing it in the cloud, by comparing the performance on two different cloud providers. In addition, it proposes a novel approach combining a pattern fragmentation technique with a NoSQL database, to organize and manage the chunks. Our research has indicated that there is a trade-off on the performance when using a database. Any slight difference on a big data environment is always important, however, this cost is compensated by having the data organized and managed. The use of random pattern fragmentation has great potential, as it adds a layer of protection on the data without using as much resources, contrary to using encryption.

Keywords: Cloud security · Data fragmentation · NoSQL database
Big data

1 Introduction

Cloud computing can be considered one of the most promising technology for IT applications. It is defined by NIST [1] as the model that enables on-demand access to a pool of resources (e.g., networks, storage, applications, and services) that can be rapidly provisioned with minimal effort from the service provider. This technology is growing in such a way that most modern applications are delivered as hosted services. Such services are divided into Infrastructure-as-a-Service (IaaS), Platform-as-a-Service (PaaS) and Software-as-a-Service (SaaS). This scenario has two main cornerstones: virtualization and distributed computing. They provide many benefits including terms of flexibility, elasticity and resource management. Big data is a big adept of this technology, as customers take advantage of the features offered to utilize and pay the resources needed to accommodate the business model and extend such resources when required [2]. This allows the customers to reduce the cost of the storage and computing clusters, as well deviate from the maintenance of the infrastructure and shift all the focus to the development [3].

G. De Pietro et al. (Eds.): KES-IIMSS-18 2018, SIST 98, pp. 5–13, 2019.
https://doi.org/10.1007/978-3-319-92231-7_1

Despite its benefits, cloud computing also brings many challenges. Among them, is the protection of the data and the privacy of the user. In cloud computing, the user's information is handed to the cloud provider and they are responsible for the storage and safekeeping of the data, often without disclosing their procedures to the end-user [2–5]. Furthermore, storing all the data with a single provider, along with the large number of mining algorithms available, leaves users susceptible to mining attacks from attackers with unauthorized access to the cloud and escalated privileges [6].

This paper investigates the use of random pattern fragmentation [7, 8] on different cloud providers, to add a layer of security on the data, by measuring the performance to fragment, send and retrieve the data. In addition, a novel approach of managing the fragmented information on a NoSQL (Not Only SQL) database is proposed, with its performance also measured and compared. It will start by investigating the state of the art (Sect. 2), followed by the methodology in Sect. 3. Afterwards, in Sect. 4 the results will be displayed and discussed and compared to similar approaches, to provide a better evaluation of the performance, as well as a better understanding of the benefits and disadvantages of data protection by means of random pattern fragmentation.

2 State of the Art

Encryption schemes present a satisfactory solution to the data privacy problem, however, they are very complex and computationally expensive [9, 10]. Therefore, research has been shifting towards other alternatives. Kapusta et al. [11] attempted to avoid encryption by splitting information on two distinct groups and provide different protection, according to the sensitivity of the data. Dev et al. [6], approached the problem by categorizing and fragmenting data into chunks and store them in different providers, to avoid mining from providers, as well as attackers. Bahramim et al. [9] proposed a lightweight modality for mobile phones, where random pattern fragmentation, based on chaos system, is used to split a JPEG file and store in multi cloud systems. Bahramim et al. [10] investigates the use of databases to store and manage chunks created with the same method and adding a layer of encryption to the database. Lentini et al. [12] measured the performance of different fragmentation techniques on Amazon Web Services and compared them with the AES cryptography.

However, to improve the organization and overall management of the data in the server, it is imperative to use a database. Rafique et al. [13] proposed a mapping strategy that leverages columnar NoSQL databases to perform data encryption at various levels of granularity dynamically. Alsirhanni et al. [14] proposed a technique that stores data in different providers, by splitting into a master cloud that contains indexes of the fragments, and various slave clouds that store the data encrypted in columnar databases. Masala et al. [15] proposed data fragmentation on the cloud environment using a NoSQL approach, based on MongoDB [16] to take advantage of the highly scalable distributed architecture, which is the main characteristic of NoSQL.

The aim of this paper is the comparison of a novel approach (RPFNoSQLDB), having a mixed solution between a random pattern fragmentation approach and a NoSQL database, with a random pattern fragmentation approach (RPF). The NoSQL solution adds a management layer on the scrambled data, offering therefore better scalability.

3 Methodology

3.1 Random Pattern Fragmentation

In the random pattern fragmentation (RPF), originally proposed by [9, 10], but referencing the version implemented in [12], the original file is divided into N chunks and the pattern indexes are created with a random function, in other words, a random permutation of N elements before being stored in split files. The split files, are then saved on a cloud instance. The pattern indexes get stored in the client's machine, to reconstruct the original file when needed. With this technique, the attacker does not possess the knowledge of the random order and therefore cannot reconstruct the file. In the Fig. 1 the method is shown.

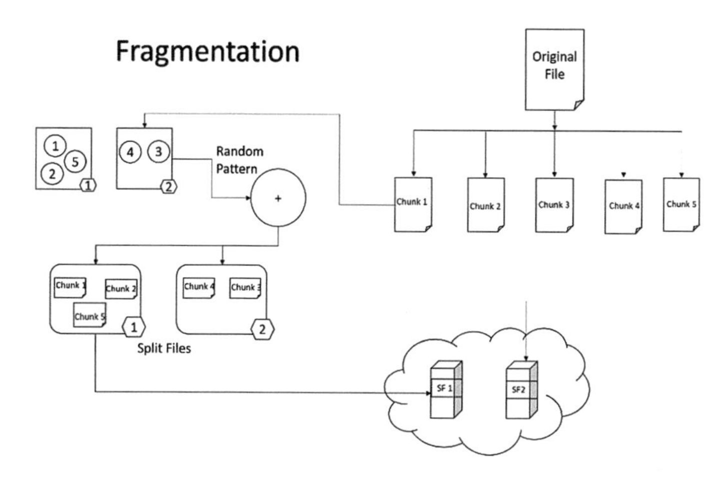

Fig. 1. The process of random pattern fragmentation. The original file gets split into chunks that are stored in split files. The split files are then saved on the cloud server.

In the reconstruction phase (Fig. 2), the split files get downloaded from the cloud and reconstructed using the dictionary format, by combining the stored indexes on the client machine to the different chunks inside the split file. The chunks are then reshuffled back into the original order before being stored back into the client's device.

3.2 The Use of a NoSQL Database Combined with the Random Pattern Fragmentation

We propose a novel approach (Fig. 3) where we combine the use of the combination of the random pattern fragmentation with a NoSQL database (RPFNoSQLDB), where the original file gets split into chunks and those chunks are then inserted to split files. The chosen database management system was CouchDB, version 2.1.1 [17].

The split files are then stored inside the NoSQL database that resides inside an instance on a cloud provider. The data is secured in transit with the use of the virtual

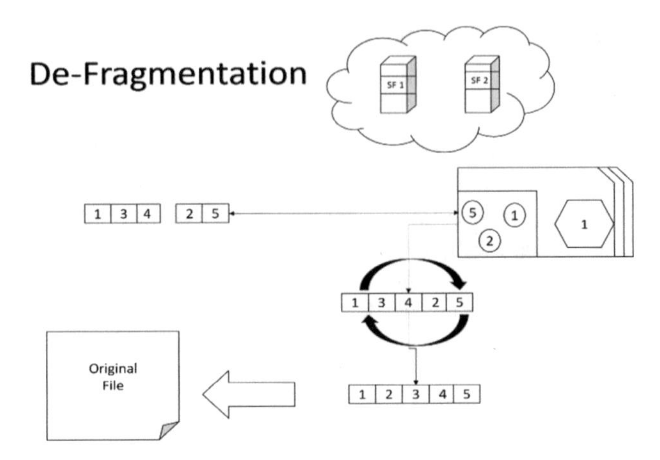

Fig. 2. The process of reconstructing the chunks back to the original file. The file is downloaded from the server and reconstructing using the indexes on the client's machine via a dictionary data structure. After the reconstruction, the file is stored on the client's machine.

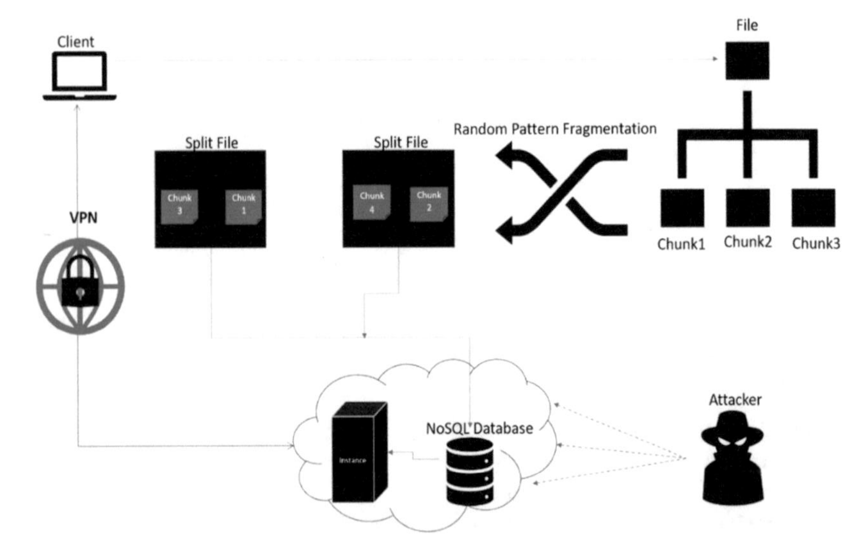

Fig. 3. Proposed model that uses random pattern fragmentation and stores the random chunks in split files, which are then stored on a NoSQL database.

private network (VPN) [18], and in case an attacker accesses the database, the chunks are in a random order, discouraging therefore any attempts to reconstruct the data. The details of the patterns are stored in the client's machine, which are then used to reconstruct the original file.

Using NoSQL to presents an advantage over relational databases, as the files are not structured, making the process of analyzing and retrieving the files faster. In the

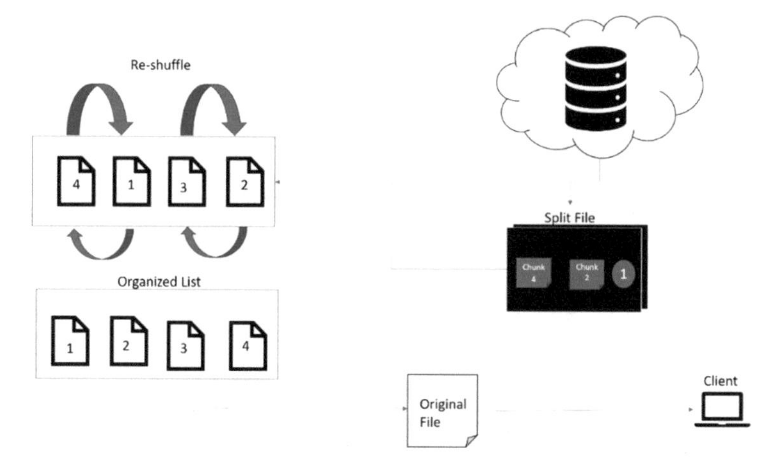

Fig. 4. Process of retrieving and reconstructing the original file. The chunks are sent from the database to the client via a VPN. In the client program, the chunks get re-arranged in a dictionary manner, where the client machine holds the indexes of all the chunks in the correct order.

reconstruction phase, a method based on a dictionary is used, where the client machine uses the stored indexes, combined with the downloaded split files, to re-shuffle the chunks into the correct order, as shown in the Fig. 4.

4 Results and Discussion

In the first part of this paper we are aiming to analyze the performance of using data fragmentation on different cloud providers, as well as the performance of the connection type. This work investigates the performance of the most promising pattern fragmentation technique [12] in a virtual machine hosted by Amazon Web Services (AWS) [19], in comparison with the cloud offered by Microsoft Azure [20]. During the investigation, we always consider sending the files to a single provider via a secure connection. The single provider is the worst-case scenario, as the entire data is available, providing a single point of attack for attackers to mine the data. Nevertheless, we are considering the typical scenario, related to the public cloud.

We are presenting different experiments, using the same algorithm and database in [12], with three different file types (.docx, .jpg, and .pdf), all with 100 KB of size. The result presented in [12] determines that the random pattern fragmentation is faster than the traditional AES encryption [21]. As a result, we are exploring the use of the random pattern fragmentation in the cloud environment.

In the first experiment, we test the random pattern fragmentation approach on a virtual machine in AWS [19] and Azure [20]. The time of splitting a file, storing in a virtual machine, retrieving and reconstructing back to the original file is compared between both providers, in Fig. 5. The communication between the client and the instance is done via tunnel-SSH. In addition, the time of sending a single .docx file, without fragmentation, is highlighted to compare the performance of using the

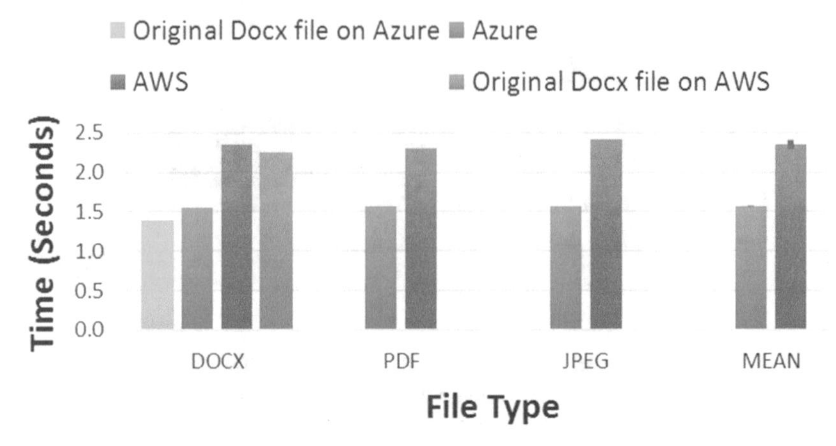

Fig. 5. Performance using tunnel-SSH on two different cloud providers. In the docx file is shown also the difference between sending the original file (called original DOCX) without fragmentation in both providers. On the mean bar is indicated also the standard deviation.

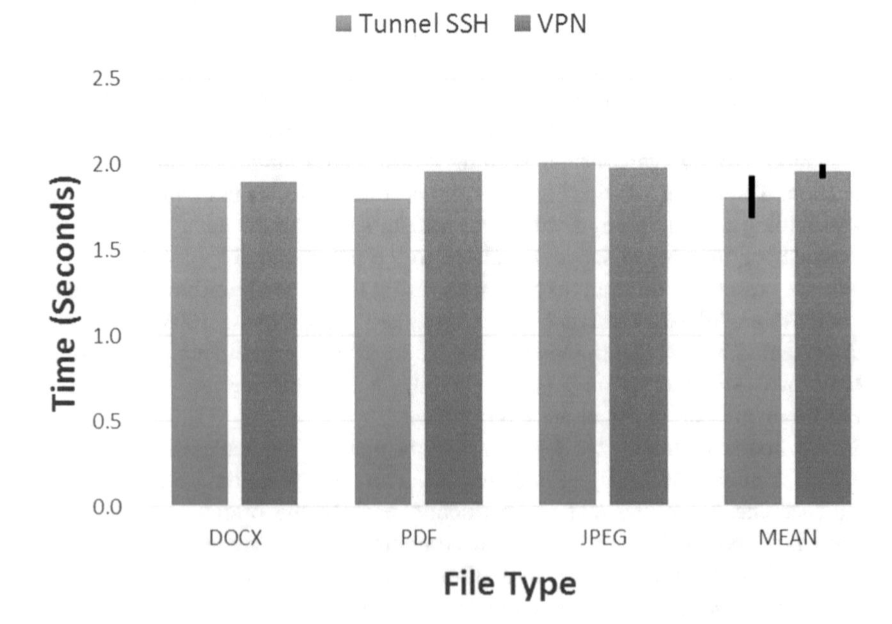

Fig. 6. Performance of the RPFNoSQLDB connecting Azure cloud with tunnel SSH vs VPN.

fragmentation. It is visible, in Fig. 5, that Azure performs better than AWS, with an average of just above 1.5 s (i.e. considering also the sending of the original file without fragmentation).

In the second experiment, we tested the proposed approach RPFNoSQLDB on two different scenarios, regarding the connection between the cloud and the client application. The chosen cloud environment to test the use of the database was Azure. On one hand the program connected to the database using tunnel-SSH, and on the other hand the program interacted with the database using an encrypted Point-to-Site VPN. The results are displayed in Fig. 6. Tunnel-SSH displays slightly better results than its counterpart, however, given the standard deviation calculated in the mean, the difference can be considered neglectable. Nevertheless, using a VPN allows a clear communication channel between the cloud and the client, whereas with the SSH tunnel the client is opening a single connection to the host, complicating the process of transferring multiple files, as well as having multiple users on the application. In addition, with SSH the files are sent sequentially or with multiple connections from the same client, consuming therefore more resources from the server.

In the last experiment, using Azure cloud, in Fig. 7, our proposed method RPFNoSQLDB was compared with the RPF, which does not contain a database. Further details are also published on Table 1.

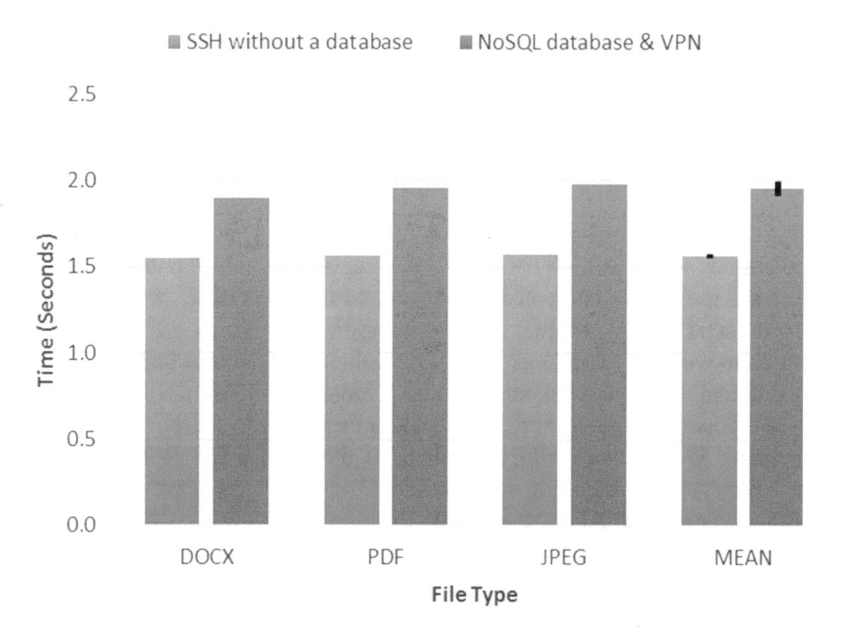

Fig. 7. Comparison of the proposed method RPFNoSQLDB (random pattern fragmentation + NoSQL database), which uses a VPN, with respect to the RPF (random pattern fragmentation without database), with uses a SSH connection.

Table 1. Evaluation of the performance of using RPFNoSQLDB with VPN over sending the files to the instance with respect to the RPF using a SSH connection. It encompasses the time to fragment the file, upload it, download and reconstructing the original file.

File Type	RPFNoSQLDB with VPN	RPF with SSH	Length chunks
100 KB	Time (Seconds)	Time (Seconds)	Bytes
DOCX	1.90	1.55	**1000**
PDF	1.95	1.57	**1000**
JPEG	1.98	1.57	**1000**
MEAN	1.96	1.57	**1000**
ST. DEV	± 0.04	± 0.01	**1000**

It can be derived from the Fig. 7 and Table 1 that using a database to manage the fragments affects the performance. On the base of the first two experiments the results don't depend by the connection used (SSH or VPN). Such performance costs are relevant on a big data environment; however, this tradeoff compensates by having the data organized and structured, facilitating the management of the data.

5 Conclusion

Cloud computing offers many advantages in terms of flexibility, scalability and reliability. Nevertheless, it also brings new challenges on security, data privacy and protection. We compared the use of splitting files and shuffling chunks on different cloud environments.

We also proposed a novel method of combining random pattern fragmentation and a NoSQL database (RPFNoSQLDB), to facilitate the organization and management of the data. When applying RPFNoSQLDB, trough the database structure, there is a trade-off on the performance, and the difference is compensated by having the data stored in an organized manner.

Furthermore, the use of a VPN creates a direct channel of communication between the client and the server, encrypted with IPsec, compared to SSH, where the different connections need to be created, to send the fragments without affecting the performance. Future work would include the use of columnar databases and storing the split files in different environments, and in binary large object formats, instead of using document-oriented databases, which store the information in JSON. These techniques show potential to the data security problem, as they add a further layer of security, without using many computing resources, which is not the case when traditional encryption methods like AES are applied.

References

1. NIST, Definition of Cloud Computing. National Institute of Standards and Technology (2011)
2. Cloud Security Alliance: Top threats to cloud computing. Version 1.0 (2010)

3. Bahrami, M., Singhal, M.: The role of cloud computing architecture in Big Data. In: Pedrycz, W., Chen, S.-M. (eds.) Information Granularity, Big Data, and Computational Intelligence. SBD, vol. 8, pp. 275–295. Springer, Cham (2015)
4. Kumar, P., Raj, H., Jelciana, P.: Exploring data security issues and solutions in cloud computing. Procedia Comput. Sci. **125**, 691–697 (2018)
5. Hegarty, R., Haggerty, J.: Extrusion detection of illegal files in cloud-based systems. Int. J. Space Based Situ. Comput. **5**(3), 150–158 (2015)
6. Dev, H., Sen, T., Basak, M., Ali, M.: An approach to protect the privacy of cloud data from data mining based attacks. In: Companion: High Performance Computing, Networking Storage and Analysis, pp. 1006–1115. IEEE, Salt Lake City (2012)
7. Chakraborty, D., Sarkar, P.: A new mode of encryption providing a tweakable strong pseudo-random permutation. In: Robshaw, M. (ed.) FSE 2006. LNCS, vol. 4047, pp. 293–309. Springer, Heidelberg (2006)
8. Gharajedaghi, J.: Systems Thinking: Managing Chaos and Complexity: A Platform for Designing Business Architecture. Elsevier, Boston (2011)
9. Bahramim, M., Singhal, M.: A light-weight permutation based method for data privacy in Mobile Cloud Computing. In: 3rd IEEE International Conference on Mobile Cloud Computing, Services, and Engineering, pp. 189–198. IEEE, San Francisco (2015)
10. Bahrami, M., Singhal, M.: CloudPDB: a light-weight data privacy schema for cloud-based databases. In: 2016 International Conference on Computing, Networking and Communications (ICNC), pp. 1–5, Kauai (2016)
11. Kapusta, K., Memmi, G.: Data protection by means of fragmentation in distributed storage systems. In: 2015 International Conference on Protocol Engineering (ICPE) and International Conference on New Technologies of Distributed Systems (NTDS), pp. 1–8. IEEE, Paris (2015)
12. Lentini, S., Grosso, E., Masala, G.: A comparison of data fragmentation techniques in cloud servers. In: Proceedings of International Conference on Emerging Internet, Data and Web Technologies (EIDWT 2018), Tirana (2018)
13. Rafique, A., Van Landuyt, D., Reniers, V., Joosen, W.: Leveraging NoSQL for scalable and dynamic data encryption in multi-tenant SaaS. In: 2017 IEEE Trustcom/BigDataSE/ICESS, pp. 885–892, Sydney (2017)
14. Alsirhani, A., Bodorik, P., Sampalli, S.: Improving database security in cloud computing by fragmentation of data. In: International Conference on Computer and Applications, pp. 43–49. IEEE, Dubai (2017)
15. Masala, G.L., Ruiu, P., Grosso, E.: Biometric authentication and data security in cloud computing. In: Daimi, K. (eds.) Computer and Network Security Essentials. Springer, Heidelberg (2018)
16. MongoDB Homepage. https://www.mongodb.com/. Accessed 3 Mar 2018
17. CouchDB Homepage. http://couchdb.apache.org/. Accessed 3 Mar 2018
18. SANS Institute: Extending your business network through a virtual private network (VPN). SANS Infosec Reading room (2016)
19. AWS Amazon Homepage. https://amazon.com. Accessed 2 Mar 2018
20. Microsoft Azure Homepage. https://azure.microsoft.com/en-gb/. Accessed 2 Mar 2018
21. Federal Information. Announcing the Advanced Encryption Standard (AES). Federal Information Processing Standards Publication 197 (2001)

A Comparison of Character and Word Embeddings in Bidirectional LSTMs for POS Tagging in Italian

Fiammetta Marulli(✉), Marco Pota, and Massimo Esposito

Institute for High Performance Computing and Networking - National Research Council of Italy, Via Pietro Castellino 111, 80131 Naples, Italy
{fiammetta.marulli,marco.pota,
massimo.esposito}@icar.cnr.it

Abstract. Word representations are mathematical items capturing a word's meaning and its grammatical properties in a machine-readable way. They map each word into equivalence classes including words sharing similar properties. Word representations can be obtained automatically by using unsupervised learning algorithms that rely on the distributional hypothesis, stating that the meaning of a word is strictly connected to its context in terms of surrounding words. This assessed notion of context has been recently reconsidered in order to include both distributional and morphological features of a word in terms of characters co-occurrence. This approach has evidenced very promising results, especially in NLP tasks, e.g, POS Tagging, where the representation of the so-called Out of Vocabulary (OOV) words represents a partially solved issue. This work is intended to face the problem of representing OOV words for a POS Tagging task, contextualized to the Italian language. Potential benefits and drawbacks of adopting a Bidirectional Long Short Term Memory (bi-LSTM) fed with a joint character and word embeddings representation to perform POS Tagging also considering OOV words have been investigated. Furthermore, experiments have been performed and discussed by estimating qualitative and quantitative indicators, and, thus, suggesting some possible future direction of the investigation.

Keywords: Deep neural network · Natural Language Processing
POS tagging · Character and word embeddings

1 Introduction

Inspired by deep hierarchical structures of human speech perception and analysis systems, the concept of deep learning algorithms was introduced and widely applied to Natural Language Processing (NLP), currently reaching very surprising and promising results. Among the variety of NLP tasks, text classification and Part of Speech (POS) Tagging took much advantage from adopting deep neural network schemes and methods, such as convolutional neural networks (CNNs), recurrent neural networks (RNNs), and recursive neural networks. As very recent text classification accuracy-augmenting strategies, current research is exploring the possibilities enabled

G. De Pietro et al. (Eds.): KES-IIMSS-18 2018, SIST 98, pp. 14–23, 2019.
https://doi.org/10.1007/978-3-319-92231-7_2

by joining the exploitation of a Bidirectional Long Short Term Memory (bi-LSTM) fed with both distributional semantic information and morphological features of the processed words.

Recent works and results obtained investigating in this direction are suggesting that this kind of joint approach could reveal particularly effective and useful in the analysis of morphological rich languages, among which the Italian. This kind of hybrid approach could respond to the effective need to fill and reduce the gap in the accuracy when textual snippets and corpus under a real domain analysis include a relevant amount of unfrequently occurring and very specific words. The problem of accounting these words, better known as Out of Vocabulary words (OOV) is relevant in NLP tasks. Typically, a deep neural networks exploit a distributional word representation model built over large but generic language corpora. However, they are trained for text classification tasks on very specific domains, such as cultural heritage, medicine or law and justice, with a notable amount of words for which no word representation exists, implying a worsening in the classification performance.

In this perspective, the main contribution of this work is to perform an accurate evaluation over the potential deep neural network based solutions addressing the general problem of POS tagging accuracy and the particular problem of improving the POS tagging performance when OOV words are present. By extending the experiences documented in the state of the art for POS Tagging, this exploration is addressed to the case of study of the Italian Language.

To this aim, two different strategies for calculating distributional word representations, typically named word embeddings, have been adopted, namely word2vec [1, 2] and FastText [3], both retrained on a textual snippet of Wikipedia Dump dataset for Italian including a collection of about 500.000 words. Moreover, representations of words at a character level, also called character embeddings, have been calculated by adopting a further Bi-LSTM network. The two typologies of word embeddings and the character embeddings have been combined by concatenation to jointly exploit both distributional semantic and morphological features in the POS tagging task. A set of experiments has been arranged and performed aimed at evaluating the effectiveness of applying a bi-LSTM network for POS tagging on the Universal Dependencies v 2.1 (UD) [4] Dataset for Italian, by considering different possible configurations combining word and character embeddings.

The structure of the paper is as follows: Sect. 2 describes the most recent work related to the topic discussed. Section 3 introduces the most significant concepts (the fundamentals) concerning distributed representations for words and characters, simply known as word and character embeddings. Section 4 introduces the system architecture and the POS tagging model adopted for the aim of comparing the joint exploitation of distributed representations at word and character level in a Bidirectional LSTM network. Sections 5 and 6 describe, respectively, the experiments performed and the results obtained.

2 Related Work

Deep learning methods employ multiple processing layers to learn hierarchical representations of data, and have produced state-of-the-art results in many domains. Recently, a variety of model designs and methods have blossomed in the context of NLP. An interesting and comprehensive overview concerning current methods and strategies exploiting deep learning and deep neural networks applied to numerous NLP tasks has been provided in [5]. In this work, authors also provides a walk-through of their evolution.

In [6], a simple deep learning framework is demonstrated to outperform most state-of-the-art approaches in several NLP tasks such as Named Entity Recognition (NER), Semantic Role Labeling (SRL), and POS tagging. Since then, numerous complex deep learning based algorithms have been proposed to solve difficult NLP tasks.

In particular, bi-LSTM networks have been recently employed for many NLP tasks, reaching very promising results in terms of accuracy improvement in POS tagging [7, 8] and transition-based dependency parsing [9, 10]. LSTMs are a particular kind of multilayer RNNs designed for preventing the vanishing gradients problem. bi-LSTMs, indeed, implement a bidirectional process consisting in a backward and forward pass through the sequence before passing on to the next layer, as described in [11].

Another recent trend that has proven successful in POS tagging and word level classification modeling is represented by the proposal discussed in [12] of adopting a multi-level bi-LSTM network: in a first step, word embeddings are built according a compositional criteria employing an inner bi-LSTM working on the character level and, in a second step, the tagging is performed by an outer bi-LSTM over words.

In [13], authors describe several experiments performed in order to evaluate bi-LSTMs with word, characters and Unicode byte embeddings for a POS tagging task, reaching very high accuracy levels over a set of 22 different languages, including the Italian. Moreover, authors adopted a dynamic modeling approach in designing their bi-LSTM, based on the exploitation of DyNet Libraries [14], introducing the possibility of switching the POS tagging loss function with an auxiliary loss function for managing OOV words.

As regard to the Italian Language, TINT [15] can be currently considered as the state of the art tool for POS tagging and dependency parsing tasks. Indeed, the POS tagging model adopted in TINT is based on the Stanford NLP Core, proposed in [16], showing a quite different but performing approach if compared to other current neural network based solutions. Anyway, no mention is made about how to handle OOV or rare words.

In [17], some experiments aimed to compare TINT and Google SyntaxNet [18] approach for Italian, evidenced that, for a particular category of words, that is verbal forms including enclitics pronouns (very usual in the Italian language), TINT failed much more times than SyntaxNet in assigning the correct POS tag to this type of words. Furthermore, the latest version of SyntaxNet, released as ParseySaurus [19], adopts a bi-LSTM network, both for the POS tagger and the dependency parser.

With respect to the problem of handling OOV words in NLP tasks, in [20] was provided an interesting approach aiming to aimed to build, by adopting a miming inverse function, at training time, word embeddings for unknown words. Furthermore, adapting the basic idea to consider sub-word level information as proposed in [12, 13] a dynamic character-based bi-LSTM model is proposed. This allowed learning a compositional mapping from character to word embeddings, thus tackling the OOV problem.

In [13, 20], some potential solutions are discussed to explicitly address efficiently the problem of rare or OOV words, but the common idea to all the current approaches is to take advantage from joining character level and word level information in order assign a weighted representation in the word space distribution to these words instead of a random representation.

These observations have been source of inspiration for this work to further investigate in this direction for Italian language that is currently missing. More in particular, to the best of our knowledge, no approach exists in literature, where word embeddings including distributional semantic and morphological features and character embeddings have been jointly exploited in a bi-LSTM for POS tagging, also effectively handling the problem of OOV words in Italian.

3 Distributed Representations

3.1 Word Embeddings

Word embeddings [5, 21] essentially express the distributional hypothesis, according to which words with similar meanings tend to occur in similar context. The main advantage of distributional vectors is that they capture similarity between the neighbors of a word. Similarity between vectors can be measured in more than one way; the cosine similarity represents one of the most effectively employed measure. Furthermore, word embeddings are often used as the first data processing layer in a deep learning model.

Typically, word embeddings are pre-trained by optimizing an auxiliary objective in a large unlabeled corpus [21, 22] and the learned word vectors can capture general syntactical and semantic information. These embeddings have proven to be proficient in capturing context similarity, analogies and due to its smaller dimensionality, are characterized by fastness and efficiency in computing core NLP tasks.

The first work showing the utility of pre-trained word embeddings is described in [7]. The authors proposed a neural network architecture that forms the foundation to many current approaches. The work also establishes word embeddings as a useful tool for NLP tasks. However, the immense popularization of word embeddings was arguably due to [1, 2] who proposed the continuous bag-of-words (CBOW) and skipgram models to efficiently construct high-quality distributed vector representations.

The most popular word embedding methods were represented, until the few past years, by the Mikolov's word2vec [1, 2] and by Pennington's Glove [23], which is essentially a "count-based" model.

3.2 Character Embeddings

Character embeddings [5] allow to represent each word as no more than a composition of individual letters. Differently from word embeddings, which are able to capture syntactic and semantic information, character embeddings can capture intra-word morphological and shape information. Generally speaking, building NLP systems at the character level has attracted certain research attention [12, 24]. Better results on morphologically rich languages are reported in certain NLP tasks.

In [9], authors exhibit positive results on building a neural language model using only character embeddings. Bojanowski et al. [3] also tried to improve the representation of words by using character-level information in morphologically rich languages. They approached the skipgram method by representing words as bag-of-characters ngrams. Their work thus had the effectiveness of the skip-gram model along with addressing some persistent issues of word embeddings. The method was also fast, which allowed training models on large corpora quickly. Popularly known as FastText, such a method stands out over previous methods in terms of speed, scalability, and effectiveness.

4 System Architecture

The deep neural network architecture here proposed for the POS tagging task resembles the one proposed in [13, 20]. It is essentially composed by two bi-LSTM networks, used in a cascade configuration, as shown in Fig. 1. In particular, a first bi-LSTM network is responsible of calculating the character embedding for each word in input whereas a second bi-LSTM network is charge of receiving, for each word in input, both the calculated character embedding and the word embedding, combining them and performing the whole POS tagging process.

In more detail, the first bi-LSTM network calculates the character embedding for each input word by employing the compositional model proposed in [12]. More formally, the input of this network is a single word w, and the output is a d-dimensional vector representing w. Additionally, an alphabet of characters C is defined, containing an entry for each uppercase and lowercase letter as well as numbers and punctuation.

The input word w is decomposed into a sequence of characters $c_1, \ldots. c_m$, where m is the length of w. Each c_i is defined as a one hot vector 1_{ci}, with one on the index of c_i in vocabulary C. A projection layer $P_C \in \mathbb{R}^{dC\,x\,|C|}$ is defined, where d_C is the number of parameters for each character in the character set C.

This is, of course, just a character lookup table, and is used to capture similarities between characters in a language (e.g., vowels vs. consonants). For this reason, the projection of each input character c_i can be written as $e_{ci} = P_c \cdot 1_{ci}$.

Thus, given a sequence of character representations $e^C_{c1}, \ldots . e^C_{cm}$ as input, the forward LSTM, yields the state sequence $s^f_0, \ldots . s^f_m$, while the backward LSTM receives as input the reverse sequence, and yields states $s^b_m, \ldots . s^b_0$. Both LSTMs use a different set of

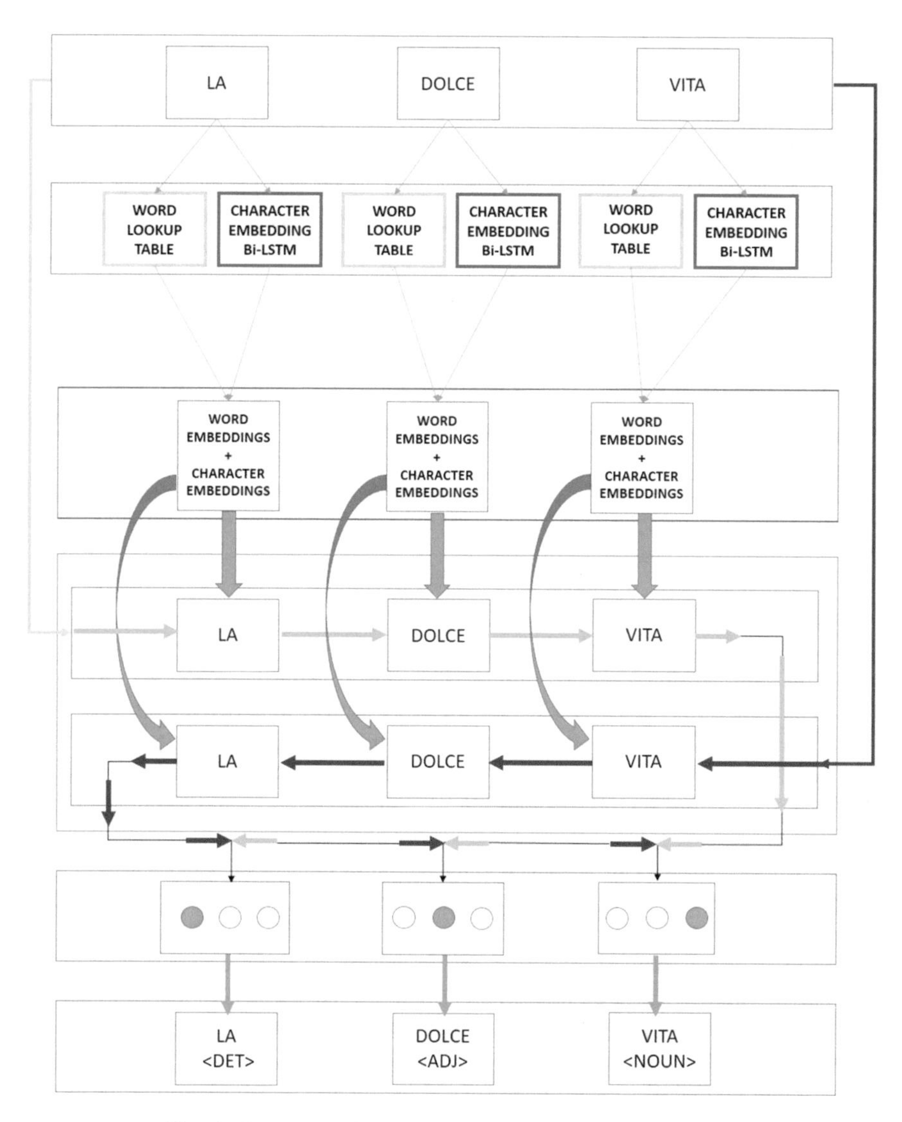

Fig. 1. Deep neural network architecture for POS Tagging.

parameters W^f and W^b. The representation of the word w is obtained by combining the forward and backward states:

$$e_W^C = \mathbf{D}^f s_m^f + \mathbf{D}^b s_0^b + \mathbf{b}_d$$

where D^f, D^b and b_d are parameters that determine how the states are combined.

On the other hand, the second bi-LSTM network operates on a sequence of embeddings $f(w_1), \ldots f(w_n)$, one for each word in input. Each embedding is equal to the one calculated by the previous bi-LSTM, i.e. $f(w_i) = e_{wi}^C$, in case of OOV words, or to the concatenation of the word vector gathered from a pre-trained lookup table with the one calculated by the previous bi-LSTM, i.e $f(w_i) = e^W_{wi} + e^C_{wi}$.

In particular, sequential features $f(w_1), \ldots f(w_n)$ are fed into the bi-LSTM network, obtaining the forward states $s^f_0, \ldots s^f_n$, and the backward states $s^b_n, \ldots s^b_0$.

Thus, state s^f_i contains the information of all words from 0 to i and s^b_i from n to i. The forward and backward states are combined, for each index from *1* to n, as follows:

$$l_i = tanh\left(L^f s^f_i + L^b s^b_i + b_t\right)$$

where L^f, L^b and b_t are parameters defining how the forward and backward states are combined. The size of the forward s^f and backward states s^b and the combined state l are hyperparameters of the network, denoted as d^f_{ws}, d^b_{ws} and d_{ws}, respectively.

Finally, the output labels for index i are obtained as a softmax over the POS tagset, by projecting the combined state l_i.

5 Experimental Evaluation

The experimental evaluation of the proposed architecture has been arranged to estimate the influence of a joint character and word level representation over the accuracy in POS tagging on an Italian dataset. As evaluation metrics, the overall accuracy reached in the POS tagging and the local accuracy obtained by estimating the fraction of OOV words properly tagged have been considered. For all the tests, the implementation of the proposed architecture in DyNet [14] has been used.

Two different algorithms for calculating the word embeddings have been adopted, namely word2vec and FastText. They have been trained on the same Treebank and under the same tuning. In particular, a textual corpus has been extracted in a standard XML format from Italian Wikipedia data[1], using the Wikipedia export tool[2]. A normalization process has been performed on the raw data in order to extract only the plain text from the XML articles by adopting a python script provided as part of the Wikipedia extractor tool[3], just discarding noisy information as images, tables, references. Finally, textual corpus has been refined by deleting a list of stopwords, by using the Lucene Project[4]. The final data set for training word embeddings was composed by about 500 thousands (500.718) token. The word2vec and FastText algorithms for calculating the word embeddings were trained adopting the skipgram model with the same hyperparameters settings (*vectors size = 300; window size = 5; negative samples = 5; loss function = Hierarchical Softmax; learning rate 0.1; number of epochs = 20*).

[1] https://dumps.wikimedia.org/itwiki.

[2] https://it.wikipedia.org/wiki/Speciale:Esporta.

[3] http://medialab.di.unipi.it/wiki/Wikipedia_Extractor.

[4] https://lucene.apache.org/core/4_4_0/analyzers-common/org/apache/lucene/analysis/it/ItalianAnalyzer.html.

For training the proposed architecture for POS tagging, the Italian fragment from the Universal Dependencies Project v.2.1 [25, 26] Treebank has been considered, including 17 POS labels with the canonical data split. This dataset has consisted of a training, a development and a test set, including 12838 instances and 252631 tokens, 564 instances and 11133 tokens, 482 instances and 10417 tokens, respectively.

Four different models have been trained to accomplish the POS tagging task in accordance with the four different experimental configurations. In particular, the proposed architecture has been fed with word2vec and FastText word embeddings, turning off and on the supplementary bi-LSTM for calculating character embeddings and concatenating them to the provided word embeddings. The precise tuning for each of the four experiments is summarized in Table 1.

Table 1. Configurations of hyperpameters for POS tagging experiments.

Parameters name	Test C1	Test C2	Test C3	Test C4
Reference language	IT	IT	IT	IT
POS tagging training – Dev- Test Set	UD 2.1	UD 2.1	UD 2.1	UD 2.1
Word embedding model	Word2vec	Word2vec	FastText	FastText
Number of training epochs	30	30	30	30
LSTM layers number	2	2	2	
Hidden layers number	128	128	128	128
Word embeddings dimension	300	300	300	300
Learning rate	0.01	0.01	0.01	0.01
Character embedding concatenation	OFF	ON	OFF	ON
Character embeddings dimension	20	20	20	20

6 Results and Discussion

Experimental results are given in Table 2. First of all, it is possible to observe that the two POS tagging models (C3 and C4) trained with FastText embeddings show a better accuracy both using the character embeddings and not.

Table 2. Experimental results.

Parameters Name	Test C1	Test C2	Test C3	Test C4
Number of test tokens	10417	10417	10417	10417
Number of OOV tokens	377	377	377	377
%OOV	3.62	3.62	3.62	3.62
Number of tokens positively tagged	10116	10138	10136	10211
Number of OOV tokens positively tagged	319	338	335	362
POS tagging global accuracy	0.971104	0.973216	0.973024	0.980224
Pos tagging OOV local accuracy	0.846153	0.896551	0.888594	0.960212

Furthermore, even if the improvement in the global POS tagging accuracy with or without the character embeddings is about of 1% both by using word2vec and FastText embeddings, more remarkable is the improvements in the accuracy reached in the labeling of OOV words, thus suggesting that the best configuration is represented by the model using the FastText word embeddings jointly with the character embeddings.

7 Conclusions and Future Work

In this study, an empirical evaluation has been performed aiming to outline the behavior held by a well-assessed model of neural network for POS tagging when it is trained with a word representation model including both character and word level features. The results of the experimental tests performed have confirmed the most recent research trend in the NLP, showing that morphology of the words provides evidence about its meaning and such an evidence is useful when it enriches the distributional information. This study has provided the further evidence that by using internal morphological structure of a word is useful to improve accuracy in text classification task as the POS tagging.

The proposed architecture consisting into a bi-LSTM network enabled to concatenate sub-word level representations (character embeddings) to word level representations (word embeddings) has been able to obtain a global accuracy on a POS tagging task of about 98% and a local accuracy over the subset of OOV words of about 96%.

Even if these results are related to the POS tagging for a specific language, the Italian, the applied methodology and tools are quite language-independent and could be used for all those languages showing a wide morphological richness.

As the proceeding of this work, this methodology will be applied to the study of typical expressions in the Italian language, more frequently occurring in the spoken language than in the written language. As a further future goal, the improvements obtained by this POS tagging solution will be used in order to increase natural language understanding capabilities of conversational systems.

References

1. Mikolov, T., Sutskever, I., Chen, K., Corrado, G.S., Dean, J.: Distributed representations of words and phrases and their compositionality. In: Advances in Neural Information Processing Systems, pp. 3111–3119 (2013)
2. Mikolov, T., Chen, K., Corrado, G., Dean, J.: Efficient estimation of word representations in vector space. arXiv preprint arXiv:1301.3781 (2013)
3. Bojanowski, P., Grave, E., Joulin, A., Mikolov, T.: Enriching word vectors with subword information. arXiv preprint arXiv:1607.04606 (2016)
4. Nivre, J., de Marneffe, M.C., Ginter, F., Goldberg, Y., Hajic, J., Manning, C.D., Tsarfaty, R.: Universal Dependencies v1: A Multilingual Treebank Collection. In: LREC (2016)
5. Young, T., et al.: Recent trends in deep learning based natural language processing. arXiv preprint arXiv:1708.02709 (2017)
6. Goldberg, Y.: A primer on neural network models for natural language processing. ArXiv, arXiv:1510.00726 (2015)

7. Collobert, R., Weston, J.: A unified architecture for natural language processing: deep neural networks with multitask learning. In: Proceedings of the 25th International Conference on Machine Learning, pp. 160–167. ACM (2008)
8. Collobert, R., Weston, J., Bottou, L., Karlen, M., Kavukcuoglu, K., Kuksa, P.: Natural language processing (almost) from scratch. J. Mach. Learn. Res. **12**, 2493–2537 (2011)
9. Ballesteros, M., Dyer, C., Smith, N.A.: Improved transition-based parsing by modeling characters instead of words with LSTMs. In: EMNLP (2015)
10. Kiperwasser, E., Goldberg, Y.: Simple and accurate dependency parsing using bidirectional LSTM feature representations. ArXiv e-prints (2016)
11. Wang, P., Qian, Y., Soong, F.K., He, L., Zhao, H.: Part-of-speech tagging with bidirectional long short-term memory recurrent neural network. Pre-print, abs/1510.06168 (2015)
12. Ling, W., Dyer, C., Black, A.W., Trancoso, I., Fermandez, R., Amir, S., Marujo, L., Luis, T.: Finding function in form: compositional character models for open vocabulary word representation. In: EMNLP (2015)
13. Plank, B., Søgaard, A., Goldberg, Y.: Multilingual part-of-speech tagging with bidirectional long short-term memory models and auxiliary loss. arXiv preprint arXiv:1604.05529 (2016)
14. Neubig, G., Dyer, C., Goldberg, Y., Matthews, A., Ammar, W., Anastasopoulos, A., Duh, K.: Dynet: the dynamic neural network toolkit. arXiv preprint arXiv:1701.03980 (2017)
15. Aprosio, A.P., Moretti, G.: Italy goes to Stanford: a collection of CoreNLP modules for Italian. CoRR abs/1609.06204 (2016)
16. Toutanova, K., Klein, D., Manning, C.D., Singer, Y.: Feature-rich part-of-speech tagging with a cyclic dependency network. In: Proceedings of the 2003 Conference of the North American Chapter of the Association for Computational Linguistics on Human Language Technology, NAACL 2003, vol. 1, pp. 173–180. Association for Computational Linguistics, Stroudsburg (2003)
17. Marulli, F., Pota, M., Esposito, M., Maisto, A., Guarasci, R.: Tuning SyntaxNet for POS tagging Italian sentences. In: Xhafa, F., Caballé, S. (eds.) Advances on P2P, Parallel, Grid, Cloud and Internet Computing, 3PGCIC 2017. LNDECT, vol. 13, pp. 314–324. Springer, Cham (2018). https://doi.org/10.1007/978-3-319-69835-9_30
18. SYNTAXNET: Announcing. The Worlds Most Accurate Parser Goes Open Source (2016)
19. Alberti, C., Andor, D., Bogatyy, I., Collins, M., Gillick, D., Kong, L., Thanapirom, C., et al.: SyntaxNet models for the CoNLL 2017 shared task. arXiv preprint arXiv:1703.04929 (2017)
20. Pinter, Y., Guthrie, R., Eisenstein, J.: Mimicking word embeddings using subword RNNs. arXiv preprint arXiv:1707.06961 (2017)
21. Rong, X.: word2vec parameter learning explained. arXiv preprint arXiv:1411.2738 (2014)
22. Cho, K.: Natural language understanding with distributed representation. ArXiv, arXiv: 1511.07916 (2015)
23. Pennington, J., Socher, R., Manning, C.D.: Glove: global vectors for word representation. In: EMNLP, vol. 14, pp. 1532–1543 (2014)
24. Santos, C.D., Zadrozny, B.: Learning character-level representations for part-of-speech tagging. In: Proceedings of the 31st International Conference on Machine Learning (ICML 2014), pp. 1818–1826 (2014)
25. Bosco, C., Dell'Orletta, F., Montemagni, S., Sanguinetti, M., Simi, M.: The Evalita 2014 dependency parsing task. In: CLiC-it 2014 and EVALITA 2014 Proceedings, pp. 1–8. Pisa University Press (2014). ISBN/EAN: 978-886741-472-7
26. Bosco, C., Montemagni, S., Simi, M.: Converting Italian treebanks: towards an Italian stanford dependency treebank. In: Proceedings of the 7th Linguistic Annotation Workshop & Interoperability with Discourse (LAW VII & ID at ACL-2013), Sofia, Bulgaria, 8–9 August, pp. 61–69 (2013)

The Vive Controllers vs. Leap Motion for Interactions in Virtual Environments: A Comparative Evaluation

Giuseppe Caggianese(✉), Luigi Gallo, and Pietro Neroni

Institute for High Performance Computing and Networking,
National Research Council of Italy, Naples, Italy
{giuseppe.caggianese,luigi.gallo,pietro.neroni}@icar.cnr.it

Abstract. In recent years, virtual reality technologies have been improving in terms of resolution, convenience and portability, fostering their adoption in real life applications. The Vive Controllers and Leap Motion are two of the most commonly used low-cost input devices for interactions in virtual environments. This paper discusses their differences in terms of interaction design, and presents the results of a user study focusing on manipulation tasks, namely Walking box and blocks, Block tower and Numbered cubes tasks, taking into account both quantitative and qualitative observations. The experimental findings show a general preference for the Vive Controllers, but also highlight that further work is needed to simplify complex tasks.

Keywords: Virtual reality · Human computer interaction
Head-mounted displays · Input devices · User study

1 Introduction

In recent years, the availability of low cost Virtual Reality (VR) technologies and devices has contributed to extend their use into a wide range of different domains, such as entertainment, marketing, education, training and tourism. These solutions enable a user to dive into an immersive virtual environment (VE) allowing her/him to encounter many different experiences, in any place and at any time, without having to wear cumbersome equipment. However, immersive environments require a suitable solution to manage the natural interaction of the user with the surrounding environment. This requirement has attracted the attention of the research community, which has started to investigate intuitive interfaces able to increase the usability of VR applications while preserving their mobility and comfort.

In this context, many hardware solutions have been introduced, such as data gloves or haptic devices, that generally prove to be too expensive requiring an substantial set up time. An interesting possibility arises from the availability of new low cost and ready to use devices that in different ways allow an improvement in the sense of presence perceived by the user during the interaction.

G. De Pietro et al. (Eds.): KES-IIMSS-18 2018, SIST 98, pp. 24–33, 2019.
https://doi.org/10.1007/978-3-319-92231-7_3

Such devices are divisible into two main categories: advanced controllers and hand-tracking sensors. The advanced controllers are characterized by the presence of buttons and tactile surfaces allowing an indirect tracking of the position and orientation of the hand that is holding them. Alternatively, the hand-tracking sensors are small sensing devices that by exploiting egocentric vision track the user's hands, transforming gestures into interactions with synthetic objects. While both these devices allow the user to see the movement of her/his hands while being immersed in a VE, the interaction proves to be subject to the hardware characteristics of the device.

This paper is mainly focused on user interactions with virtual objects. Our contribution consists in a preliminary investigation of the best performing input devices applicable to VEs. We have designed three simple manipulation tasks in which volunteers were asked to select, position and rotate synthetic objects. We have used a pair of Vive Controller and a Leap Motion device to determine which one of the two types of devices proves to be the more suitable solution to be employed in manipulation tasks, both considering the performances and perceived difficulty experienced by the subjects involved.

2 Related Work

Several studies investigate different aspects of interaction in immersive VEs focussing on input devices, interaction techniques and application domains.

Dias et al. [1] evaluate in a user study uni-manual vs. bi-manual freehand interactions to object manipulation on a large display, showing that the bi-manual interaction provides a more efficient method for rotation and scaling. In [2] the study evaluates the efficiency of spatial interaction techniques to manipulate digital content within a holographic system in the cultural heritage domain. In a similar context, Dangeti et al. [3] compare hand-in-air and object-in-hand interaction for the navigation of 3D objects from the user experience perspective. Meanwhile, an evaluation of the parameters for the design of comfortable bare-hand object manipulation gestures in immersive VR systems is presented in [4], and a controller-based interaction to improve the understanding of a software system into an ad hoc VE is showed in [5].

By exploiting a similar configuration, with a sensor placed in egocentric vision, in [6–8] touchless interactions, different target selection techniques and the degrees of freedom in touchless interaction are investigated. Meanwhile, in [9] a comparison between two 3D manipulation techniques is performed showing a user preference for direct manipulation in a positioning task and for constrained manipulation in a rotational task. A free-hand gesture interface is also explored in [10] where it is compared with a multimodal interaction technique in terms of its usability for translation, rotation and scaling tasks. The user study reports a user preference for gesture-based techniques in translation and rotation tasks, while multimodality perform better in scaling. Similar investigations have been performed also in collaborative VEs, which in last years have been increasingly made such as composition of cloud services [11,12] also considering data privacy issues [13]. For instance, in [14] controller-based and sensor-based interactions are compared, with the authors concluding that controllers offer the best performance in term of accuracy and usability.

Tscharn et al. [15] compared bare-hand and touch interaction for navigation tasks, showing a better efficiency of the bare-hand interactions with respect to a manipulation task because of the lower degree of precision required. Finally, the utility of VR as distraction during medical procedures is evaluated in [16] highlighting the need to design hardware/software systems tailored to the specific needs of patients.

3 Methods

3.1 Motivations and Goals

Both the input devices selected for this evaluation present advantages and disadvantages in terms of interaction design, a fact which explains why users tend not to express any preference for one device rather than the other.

- The advantages of the Vive Controllers include:
  - The ability to track even when out of the user's view; and
  - The ability to give feedback to the users during the interaction with the VE.
- The disadvantages include:
  - Interaction based on the use of buttons and triggers; and
  - The requirement to hold the controller even when the user is not interacting with any synthetic object.
- The advantages of Leap Motion include:
  - Interaction based on the use of metaphors inspired by natural behaviors; and
  - The tracking of mid-air user gestures with a degree of accuracy that permits the identification of individual finger movements.
- The disadvantages include:
  - The requirement for the user's hand actions to be performed in the sensor's FOV; and
  - The fact that, with the sensor placed in egocentric vision, each time the user moves her/his head the FOV of the tracking sensor rotates accordingly.

In this evaluation, the standard features made available by the libraries of both devices have been exploited to manage the interactions of users with the VEs. However, in order to offer a comparable experience with respect to both the devices, we decided to mitigate one of the disadvantages of the Leap Motion. In fact, each time the system loses the tracking of the user's hand during an interaction, the interaction state is stored allowing the user to continue with the interaction left pending by placing her/his hand again in the sensor's FOV. Moreover, with both interaction devices, the user's hand position in the VE has been reproduced by using a synthetic hand. However, the user's fingers positions tracked with the Leap Motion device allow a more natural behavior of the virtual hand in comparison with the limited number of possible states representable by the trigger of the Vive Controller.

The object of this investigation was to compare these two devices in VEs in order to identify which proves to be more suitable in order to address the problem of the manipulation of virtual objects taking into account both objective measurements and user preferences. We were also interested in verifying the interaction between the input devices and the sub-tasks included in the manipulation, namely selection, positioning and rotation. Our prediction was that the Vive Controllers, thanks to the physical feedback provided, could prove to be advantaged in the selection and user comfort while, due to the sensation of interact by using a proxy object, it could be disadvantaged in the rotation. On the contrary, the natural movement achievable by using the Leap Motion device could represent an advantage during a rotation but the involvement of a large muscle group in the mid air interaction could be a disadvantage in terms of precision [17] and, in that it could result in a Gorilla arm effect [18], user comfort.

3.2 Participants

The investigation involved 8 unpaid volunteers (6 males and 2 females) all recruited from the institute we belong to, ICAR, the Institute for High Performance Computing and Networking. Their ages ranged from 30 to 40 years old ($M \approx 33.62$, $sd \approx 3.54$). All the participants were right-handed and had normal or corrected-to-normal vision. In a background questionnaire, three declared a previous experience with immersive VEs.

3.3 Apparatus

The controllers supplied by HTC were two wireless and ergonomic controllers that can be easily handled by a user in one hand. These controllers are fully tracked in the three-dimensional space and allow an indirect tracking of the user's hand movements in a room-scale environment ensuring a stable experience. Moreover, the controllers include also accelerometers so that the hand tracking is the result of a fusion of positional and tracking data.

The Leap Motion sensor has proved to be one of the lightest and best performing devices available off-the-shelf. Thanks to its low weight it can be mounted it in front of the Head-Mounted Display (HMD) so that a user can wear both together. Leap Motion works by using two infrared cameras arranged so that their FOV intersects three infrared red (IR) light-emitting diodes (LEDs) positioned alternately to the cameras.

The VEs realized, in which the manipulation tasks were performed, were proposed to the subjects by using the HTC Vive HMD. The Vive includes two lighthouse base stations aspects of the room-scale VR technology, allowing a tracking of the user's head and controllers in a rectangular area of almost

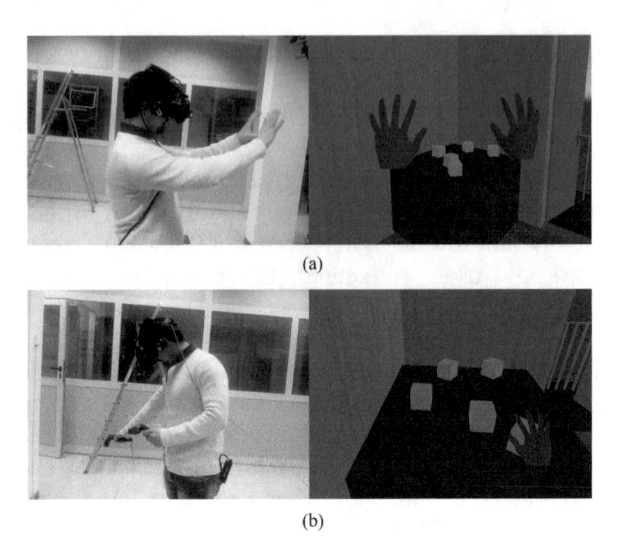

Fig. 1. A subject during the experimentation using the Leap Motion sensor in (a) and the Vive Controllers in (b).

3.5 by 3.5 m. The HMD consists of two screens with a resolution of 1080×1200, one for each eye, a refresh rate of 90 Hz, and a FOV of about 110°. Finally, the VEs were rendered on a VR Ready computer (running a 64 bit Windows 10, with an Intel Core i5-7400 8 GB and a Nvidia Geforce GTX 1060).

Concerning the software used, the tasks were realized entirely using the Unity 3D render engine, while the interaction was implemented in the C# language exploiting Steam VR and Orion SDK respectively for the Vive Controllers and Leap Motion.

3.4 Procedure

The experiments were performed by using the apparatus previously described guaranteeing the same set-up and conditions for each tester. Before starting with the tests, facilitator explained the entire procedure to the subjects presenting the system and the interaction modalities. Moreover, the facilitator asked each volunteer to complete a background questionnaire with questions about their profile and previous experience with immersive VEs. After, the subjects were left free to try both the interaction modalities without any limitations in order to improve their confidence with the system and, at the same time, to make them feel more comfortable (see Fig. 1).

Next, the test session started with the subject receiving precise instructions from the facilitator about the tasks to perform. No time limit was imposed on the participants during the trial and each subject's interaction was observed to collect all the possible impressions of her/his experience. The participants were also asked to think aloud, describing their intentions and possible difficulties.

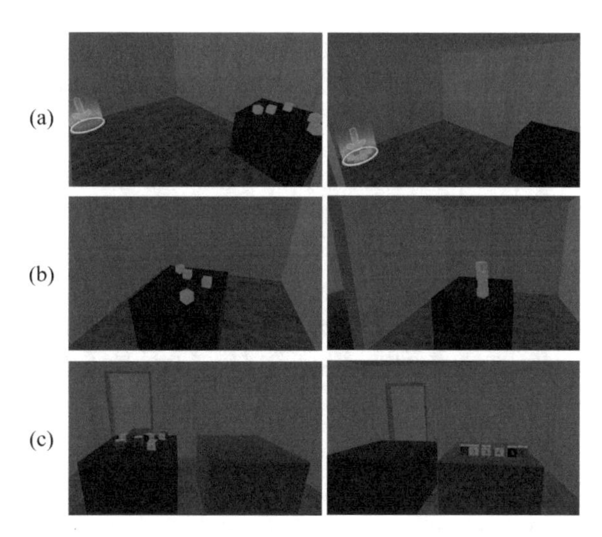

Fig. 2. In (a) the Walking box and blocks task. In (b) the Block tower task. In (c) the Numbered cubes task

Before the execution of each trial, the facilitator revealed the task and the modality to use to the subject. The facilitator was also responsible for helping the subject, if necessary, to report errors that prevented her/him from declaring the task accomplished, e.g. a forgotten block or a block positioned incorrectly and for reporting the execution time. Each subject performed three manipulation tasks characterized by an increasing complexity in order to address the problem of canonical manipulation including, besides the selection, also the positioning and rotation as suggested in the main design guidelines described in Bowman et al. [19]. In the following section, the three tasks are presented in detail, starting from the least complex:

- **the Walking box and blocks task** in which the subject is asked to move five blocks from one box to the another on the other side of the room. Grabbing one block at a time, the subject walks towards the other box to place there the block carried before returning to the previous box to start again until all the blocks have been moved. The manipulation required in this task involves the selection of an object and its positioning in the destination box (see Fig. 2-a).
- **the Block tower task** where the subject has to realize a tower by using the four blocks made available on the table. Interacting with one block at a time, the subjects places them correctly one on top of the other until all the blocks have been used. Compared to the previous task, in this case the manipulation requires more accuracy in the arrangement of the objects to avoid the tower collapsing, so forcing the subject to start from scratch (see Fig. 2-b).
- **the Numbered cubes task** that presents to a subject a set of six blocks with numbered faces and a number sequence that has to be replicated using

the blocks. All the blocks are identical so each time the subject grabs a block she/he has to rotate it in order to identify the number and then place it on the table making the entire number sequence reproduced with the numbers readable. This task forces the subject to rotate the blocks and this rotation can be performed either by rotating the block on the table using a single hand or by rotating the object in mid-air using both hands (see Fig. 2-c).

Finally, at the end of each task the facilitator presented the subject with a task-level questionnaire in order to perform an evaluation of the perceived difficulty. In more detail, the questionnaire was in the Single Ease Question (SEQ) format, with a rating scale ranging from 1 (very easy) to 7 (very difficult) [20].

3.5 Design

The experiment was of a 2×3 within-subjects design, in which the dependent variables are the two interaction modalities and the three manipulation tasks while the independent variable is the time to completion needed by the subjects to accomplish the required tasks. Each of the participants completed the three tasks for each of the two interaction modalities proposed, repeating each task three times, for a total of $8 \times 3 \times 2 \times 3 = 144$ trials. Each of the six trials was executed three times consecutively in order to log an average of the subject's performances. The order of presentation of the combination of the dependent variables to each subject was counterbalanced by using a balanced Latin square to offset any learning effects.

4 Results and Discussion

The grand mean for the tasks completion time was 57.83 s. The interactions with the Vive Controllers were the fastest at 35.33 s, while the interactions with the Leap Motion required on average 80.33 s. In fact, the main effect of the used device on the tasks completion time was statistically significant ($F_{1,7} = 34.933, p < 0.01$). This expected result was confirmed by the views expressed by the subjects at the end of the experiment, in which they revealed that by using the Leap Motion device they were tempted to grab the objects by using only the fingers and not by closing the hand in a fist. In addition, they also complained of some difficulties in performing the selection of the object when using Leap Motion, which could be attributable to the inexperience of users.

In the same way, also for the second within-subjects variable of the task, the analysis yields an effect statistically significant ($F_{2,14} = 11.362, p < 0.01$). The task completion time was 58.41 s for the Walking box and blocks task (task 1). The Block tower task (task 2) was definitely the fastest to complete at 36.56 s, while the Numbered cubes task (task 3) was the slowest at 78.52 s. This latter execution time reflected the increasing task difficulty with the slowest task that forced the subject to perform a manipulation involving selection, positioning and rotation. Actually, further analysis reveals a significant effect in the comparisons

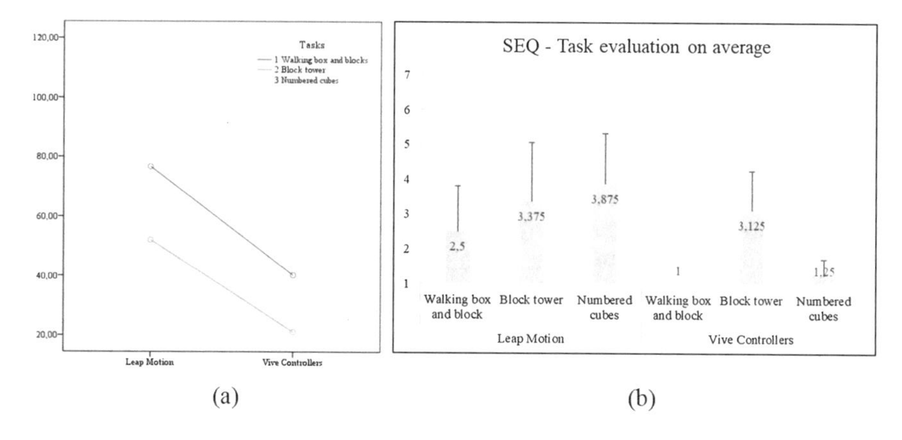

Fig. 3. In (a) the estimated marginal means of execution time of the three tasks performed with each devices. In (b) the subjects' perceived difficulty averaged over the three tasks when using the two devices (the scale ranges from 1 to 7, with higher scores representing the greater difficulty).

of task 2 vs. task 3 ($F_{1,7} = 32.026, p < 0.01$) which can be attributed to the presence in task 3, compared to the former, of the rotation of the objects, an aspect that made the achievement of the result more difficult.

However, the interaction effect between the two variables was not statistically significant ($F_{2,14} = 2.872, ns$). The results by device and task are shown in Fig. 3-a. These findings do not support the hypothesis that subjects would feel more confident in performing selection (task 1) and precision positioning (task 2) when using the Vive Controllers and rotation (task 3) when using the Leap Motion, instead suggesting that the devices work similarly in all three tasks proposed, presenting the same gaps when the different aspects of the manipulation have to be controlled at the same time.

In short, the results show how the presence of tactile feedback in the Vive Controllers, combined with a continuous tracking not interrupted by the exit of the hands from the FOV of the sensor, becomes decisive in terms of performances. The improvement is evident in all the tasks, but is particularly marked in the tasks that include rotations characterized by repeated grab and release actions that become more complex in the absence of buttons.

Finally, the evaluation performed by means of the SEQ questionnaires (see Fig. 3-b) shows a slightly higher perceived difficulty for the Leap Motion device ($M = 2.687$, $sd = 1.539$) compared to the Vive Controllers ($M = 1.125$, $sd = 0.337$). Furthermore, with respect to the task difficulty the investigation highlights that the first task seemed easier ($M = 1.750$, $sd = 1.18$), while the second and the third tasks were considered more or less at the same level of difficulty, ($M = 2.250$, $sd = 1.653$) and ($M = 2.562$, $sd = 1.711$), respectively.

5 Conclusions

The rapid development and dissemination of VR technologies and devices has prompted the investigation of new devices characterized by interaction modalities able to improve user immersion by reducing the required physical effort. However, the design of new devices and interaction paradigms is still in its infancy. In this work we have detailed a comparison between the two currently most commonly used devices to manipulate objects in VEs: the Vive Controllers and the Leap Motion sensor. We have developed three manipulation tasks and performed a user study in which both devices have been evaluated both quantitatively and qualitatively. The experimental findings have revealed a better performance and a lesser perceived difficulty of the Vive Controllers thanks to its stability, accuracy, and lower learning curve compared to that required by the Leap Motion sensor. However, with both devices, general difficulties have been identified in accomplishing tasks in which there has been a requirement to manage simultaneously different aspects of manipulation (e.g. selection, positioning and rotation), an aspect that highlights that neither of the two devices can be considered as the definitive solution to the problems relating to effective interactions in VEs.

References

1. Dias, P., Cardoso, J., Ferreira, B.Q., Ferreira, C., Santos, B.S.: Freehand gesture-based 3D manipulation methods for interaction with large displays. In: International Conference on Distributed, Ambient, and Pervasive Interactions, pp. 145–158. Springer (2017)
2. Caggianese, G., Gallo, L., Neroni, P.: Evaluation of spatial interaction techniques for virtual heritage applications: a case study of an interactive holographic projection. Future Gener. Comput. Syst. **81**, 516–527 (2018)
3. Dangeti, S., Chen, Y.V., Zheng, C.: Comparing bare-hand-in-air gesture and object-in-hand tangible user interaction for navigation of 3D objects in modeling. In: Proceedings of the TEI 2016: Tenth International Conference on Tangible, Embedded, and Embodied Interaction, pp. 417–421. ACM (2016)
4. Lin, W., Du, L., Harris-Adamson, C., Barr, A., Rempel, D.: Design of hand gestures for manipulating objects in virtual reality. In: International Conference on Human-Computer Interaction, pp. 584–592. Springer (2017)
5. Capece, N., Erra, U., Romano, S., Scanniello, G.: Visualising a software system as a city through virtual reality. In: International Conference on Augmented Reality, Virtual Reality and Computer Graphics, pp. 319–327. Springer (2017)
6. Brancati, N., Caggianese, G., Frucci, M., Gallo, L., Neroni, P.: Touchless target selection techniques for wearable augmented reality systems. In: Intelligent Interactive Multimedia Systems and Services, pp. 1–9. Springer (2015)
7. Brancati, N., Caggianese, G., Frucci, M., Gallo, L., Neroni, P.: Experiencing touchless interaction with augmented content on wearable head-mounted displays in cultural heritage applications. Pers. Ubiquit. Comput. **21**(2), 203–217 (2017)
8. Gallo, L.: A study on the degrees of freedom in touchless interaction. In: SIGGRAPH Asia 2013 Technical Briefs, p. 28. ACM (2013)

9. Caggianese, G., Gallo, L., Neroni, P.: An investigation of leap motion based 3D manipulation techniques for use in egocentric viewpoint. In: International Conference on Augmented Reality, Virtual Reality and Computer Graphics, pp. 318–330. Springer (2016)
10. Piumsomboon, T., Altimira, D., Kim, H., Clark, A., Lee, G., Billinghurst, M.: Grasp-shell vs gesture-speech: a comparison of direct and indirect natural interaction techniques in augmented reality. In: 2014 IEEE International Symposium on Mixed and Augmented Reality (ISMAR), pp. 73–82. IEEE (2014)
11. Amato, F., Moscato, F.: Pattern-based orchestration and automatic verification of composite cloud services. Comput. Electr. Eng. **56**, 842–853 (2016). Cited By 13
12. Amato, F., Moscato, F.: Exploiting cloud and workflow patterns for the analysis of composite cloud services. Future Gener. Comput. Syst. **67**, 255–265 (2017)
13. Amato, F., Moscato, F.: A model driven approach to data privacy verification in e-health systems. Trans. Data Priv. **8**(3), 273–296 (2015)
14. Gusai, E., Bassano, C., Solari, F., Chessa, M.: Interaction in an immersive collaborative virtual reality environment: a comparison between leap motion and HTC controllers. In: International Conference on Image Analysis and Processing, pp. 290–300. Springer (2017)
15. Tscharn, R., Schaper, P., Sauerstein, J., Steinke, S., Stiersdorfer, S., Scheller, C., Huynh, H.T.: User experience of 3D map navigation–bare-hand interaction or touchable device? Mensch und Computer 2016-Tagungsband (2016)
16. Indovina, P., Barone, D., Gallo, L., Chirico, A., De Pietro, G., Antonio, G.: Virtual reality as a distraction intervention to relieve pain and distress during medical procedures: a comprehensive literature review. Clin. J. Pain (2018)
17. Zhai, S.: Human Performance in Six Degree of Freedom Input Control. University of Toronto (1996)
18. Boring, S., Jurmu, M., Butz, A.: Scroll, tilt or move it: using mobile phones to continuously control pointers on large public displays. In: Proceedings of the 21st Annual Conference of the Australian Computer-Human Interaction Special Interest Group: Design: Open 24/7, pp. 161–168. ACM (2009)
19. Bowman, D., Kruijff, E., LaViola Jr., J.J., Poupyrev, I.P.: 3D User Interfaces: Theory and Practice, CourseSmart eTextbook. Addison-Wesley (2004)
20. Sauro, J., Dumas, J.S.: Comparison of three one-question, post-task usability questionnaires. In: Proceedings of the SIGCHI Conference on Human Factors in Computing Systems, pp. 1599–1608. ACM (2009)

A MAS Model for Reaching Goals in Critical Systems

Flora Amato[1(✉)], Giovanni Cozzolino[1], Antonino Mazzeo[1], and Francesco Moscato[2]

[1] Dipartimento di Ingegneria Elettrica e delle Tecnologie dell'Informazione DIETI, University of Naples "Federico II", Naples, Italy
{flora.amato,giovanni.cozzolino}@unina.it
[2] Dipartimento di Scienze Politiche. DiSciPol, University of Campania "Luigi Vanvitelli", Caserta, Italy
francesco.moscato@unicampania.it

Abstract. The exploitation of Cloud infrastructure in Big Data management is appealing because of costs reductions and potentiality of storage, network and computing resources. The Cloud can consistently reduce the cost of analysis of data from different sources, opening analytics to big storages in a multi-cloud environment. Anyway, creating and executing this kind of service is very complex since different resources have to be provisioned and coordinated depending on users' needs. Orchestration is a solution to this problem, but it requires proper languages and methodologies for automatic composition and execution. In this work we propose a methodology for composition of services used for analyses of different Big Data sources: in particular an Orchestration language is reported able to describe composite services and resources in a multi-cloud environment.

1 Introduction and Related Works

Framework for Big Data Management, like Hadoop [1] supported distributed file systems inside clusters architectures. On the other hand, Cloud Architecture has emerged as a de facto standard for achieving best performances considerably reducing costs: there are no good reasons (except the ones related to security and data ownership) to do not use Cloud services in Big Data management. In particular, when Big-Data Source is distributed with different and heterogeneous sources, cloud infrastructure results very appealing both from performance and economic points of view.

One of the main problem here is to overcome problems related to the creation of composite services that uses multi-cloud resources like several virtual storages containing data sources.

Let us imagine a scenario where an organization wants to perform some analytics on data from social networks, from weather and other geographical data and from other economical issues. Obviously, because of their heterogeneity,

G. De Pietro et al. (Eds.): KES-IIMSS-18 2018, SIST 98, pp. 34–42, 2019.
https://doi.org/10.1007/978-3-319-92231-7_4

data are distributed and maintained in different data storage. The goal of the analysis may be the identification of co-relation among information drilled-up from the different sources. For example, this can be used by car sellers in order to optimize their production depending on past weather conditions, year period, geographic region etc.

If data are stored in different virtual storage, the execution of proper analyses on data will need the interaction of different Cloud Resources and services, belonging to different cloud providers (both private or public): this is a multi-cloud scenario where a complex service (the analysis of the whole data) needs the execution of component (sub)services that executes on different and heterogeneous resources. The execution of the complex service needs both interoperability among resources and something able to define and execute a workflow process calling proper sub-service when needed.

The National Institute of Standards and Technology (NIST) defined this kind of execution *Orchestration* [2]. Following the NIST definition, orchestration refers to the composition of system components to support arrangement, coordination and management of resources in order to provide (composite) cloud services to cloud consumers.

Some important issues about are: (a) providers are responsible of the management and enactment of activities in orchestrated services; (b) orchestration involves services and resources in all levels of Cloud stack; (c) orchestration must face Quality of Service (QoS) of both component and composite services.

In this context, it is clear that a methodology able to compose cloud Services needs:

- a language like BPEL4WS [3] able to describe not only services composition at Service-as-a-Service (SaaS) level, but Cloud Resources too,
- A framework able to orchestrate at all Cloud Layers [2] services and resources, and also able to analyse and manage the composite service as whole, answering to questions like: are results and Input-Output resources compliant with cloud components services?

We think that compliance cannot be evaluated only by means of syntactical checks or by type checking. In fact, enabling technologies for automatic composition, Cloud architects are now going to use Semantics-based methodologies. Anyway, cloud resources are not like web services messages: they include some complex elements like virtual infrastructures. In addition, in order to achieve automatic Cloud Service composition, resources should carry out a semantic description of their functionality, as well as a semantic description of their parameters.

Actually, several semantics-based approaches for *simple* web services composition exist (a survey is in [4]). Some of them ([5]) exploit BPEL4WS orchestration language and OWL-based ontologies for services description.

In this work we present an architecture and a language able to define, analyse and manage multi-cloud Orchestration and we will show how these can be used for the analysis of a distributed Big-Data sources [6,7]. For what analysis of the semantics and the soundness of the composition concerns, we do not describe here the Ontology-based description of resources: it is based on the work described in [3] and on the Ontology for the Cloud defined in [8,9]. In brief, we use OWL-S with IOPE grounding in order to analyse interoperability among services, but composition is managed as a workflow process and formal operational semantics [10–12] allow for analysis of composite processes. Similarly, the analysis of the Quality of Service of composite services is out of the scope of this work and it is introduced in [13,14].

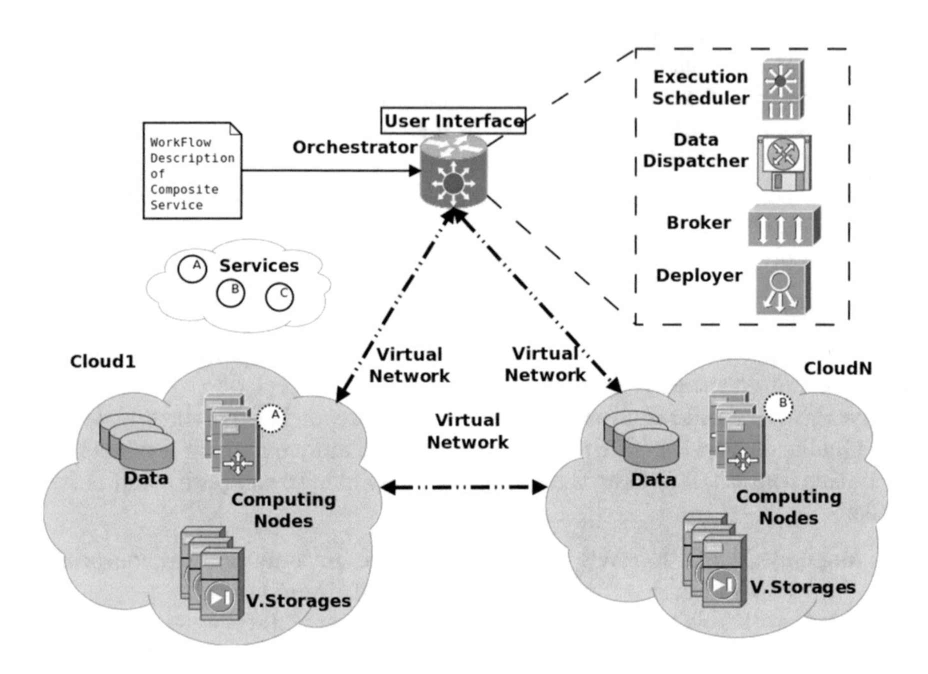

Fig. 1. System architecture

The paper is organized as follows: Sect. 2 describes the overall architecture of our system and its methodology; Sect. 3 introduces the workflow-based language. Section 4 reports an example of the application of the methodology to a Big-Data analysis scenario.Finally, Sect. 5 contains some concluding remarks.

2 Methodology and Architecture

Figure 1 show the overall architecture of the framework we propose for the composition and orchestration of Cloud services.

We work upon the existence of several, eventually heterogeneous, Cloud Providers (**CloudA,**$\cdots$**, CloudN**). Each providers is able to instantiate common Cloud Resources like Computing Nodes, Virtual Storages or Virtual Network (inter and intra-clouds). In addition, Virtual Storages can of course maintain different sets of Data.

The architecture of the **Orchestrator** we are describing consists of:

- an **Execution Scheduler**: the scheduler reads the description of a composite service and executes the proper services when needed, eventually scheduling data migration too from a virtual storage to another if needed.
- a **Data Dispatcher**: It executes physical data migration and maintains information about data produced during the execution of the composite service.
- a **Broker**: it enacts common service brokering actions: if the resource is not yet acquired on a provider, it provides for acquisition and management. It is also responsible for the provisioning of the resources and their configuration.
- a **Deployer**: this component deploys needed services at SaaS (from a pool of available services) on proper resource in the Cloud.
- the **Resources Orchestrator Manager** interface with existent resources orchestrators [15]. At the moment this module supports COPE [16] and OpenStack HEAT [17] Orchestrators.

The workflow-based language we use for description of the whole service is called Operational Flow Language (OFL). The language is complex enough to describe several patterns, as well as simple enough to be defined by means of clear operational semantics [18]. Compositional rules enable patterns descriptioncite.

OFL is able to describe simple workflow graphs that are expressive enough to catch many control-flow and data-flow workflow patterns as explained in the next section. In addition workflow graphs described in OFL allow for the creation of analysis models by using Model Transformation techniques (see [19,20] for more details).

3 Workflow and Orchestration

In this section we briefly introduce the basic elements and operational semantics of OFL language. Than, we show how patterns can be defined by using OFL.

OFL is a workflow-based language. According to the definition provided by the Workflow Management Coalition[1] we consider a workflow process definition as a network of activities and their relationships. Each activity represents one logical step within a process, i.e. the smallest unit of work to be performed. The completion of an activity and the starting of another activity is a Transition point in the workflow execution. A Transition may be unconditional, but the sequence of the activity execution may also be decided at run time according to the value assumed by one or more logical expressions, in this case the sequence of operations depends on Transition Conditions that are evaluated after an activity has started or ended. The activation of an execution thread may be affected by the evaluation of the associated Transition Conditions.

Some points are defined within the workflow that allow the flow of the activities to be controlled: AND-split is a point in where a single thread of control splits into two or more threads which are executed in parallel. AND-join is a

[1] http://www.wfmc.org/.

point where two or more parallel activities converge into a single thread of execution. XOR-split is a decision point where only one of alternative branches is executed. XOR-join is a point in the workflow where two or more parallel activities converge in a single thread of execution without synchronization. OR-split is a decision point between several alternative workflow branches. OR-join is a point in which several alternative branches re-converge into a single thread. In the following XOR, AND and OR are called split or join Conditions. Figure 2 illustrates these elements.

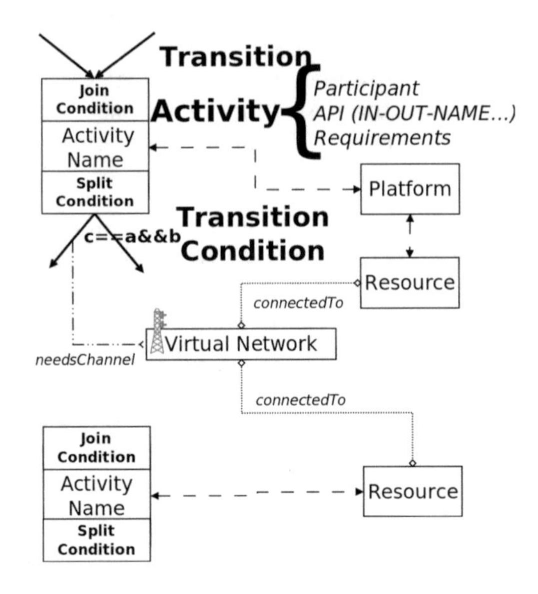

Fig. 2. Workflow elements

Therefore in the OFL language, workflow processes consist of a network of Activity nodes and edges (Transitions) identified by a pair of nodes $(FromActivity, ToActivity)$.

Activities represents atomic Cloud Service or resource invocation, as well as composite activities (i.e. sub workflow processes).

OFL skeletons are graphs defining preconditions and postconditions for cloud services execution and Cloud resources usage.

Activities definitions include effective information about Cloud Services to execute. The main components of this part of the language are:

- *Participant*: it specifies the Cloud Provider where the Orchestrator will execute the activity;
- *API*: this include the description of the service, the REST API used to invoke the service (with Inputs, Outputs, name etc.)
- *Requirements*: the QoS requested by the activity and eventually the SLAs with the Participant

This describes the SaaS level of the Composite Service. In addition, the Activity is linked to further description at PaaS and IaaS levels as depicted in Fig. 2. Platform and Resource elements shares the same main components of SaaS level (i.e. Participant, API and Requirement description of the Resource). The only main difference is that at resource level, the API may refer to an existing resource, or to a Resource Orchestrator.

In addition, Activities or Transitions may require the existence of proper virtual-network resources. They connects other Resources and may be required by transitions in order to create channel for data routing or by Activities if services need to exchange data or events with other resources during their execution.

Proper edges will define the aforementioned dependencies: *needsChannel* when declaring the need to exchange data and events; *connectedTo* in order to state which resources (usually at IaaS level) are connected by the channel.

All this is resumed in Fig. 2

OFL is an XML based language, but here we report only its graphical representation for simplicity's sake.

4 A Case Study

The example we want describe is inspired by a recent commercial study[2] reporting the importance of performing analytics over Weather, Geo-referenced data.

The main problem here is that analytics must cope with proper data about commercial activities in a given field and all drilled-up results both from weather and commercial fields, have to be collected and further analysed in order to achieve good results.

This obviously leads to the need of automatize a composite analytics service by exploiting Cloud platforms and resources.

Figure 3 resumes this scenario and the OFL representation of the composite process.

The top and the bottom of the figure depicts the two Cloud providers (here called CloudWeather and CloudPrivate) where some resources (four computational nodes and a Virtual Storage in CloudWeather; two computational nodes and a VirtualStorage in CloudPrivate) are deployed. On the CloudWeather provider, a PaaS Apache Spark Service is available, while an Hadoop service is available on the other provider.

The Workflow process describing composition is in the middle of the figure. It consists in two Activities executing in parallel after the beginning of the process: the first (W) requests the execution of analytics on the Spark PaaS, while the second (BS) enacts the execution of analytics on Hadoop. When both terminates, the Orchestrator controls the existence of the Virtual Network VN1. If it is not allocated, the Broker provides (if necessary resources exist) the necessary resource. Then, the Collect service is ready to start. The Deployer deploys the

[2] http://advertising.weather.com/big-data-weather-data-enhanced-business-strategy/.

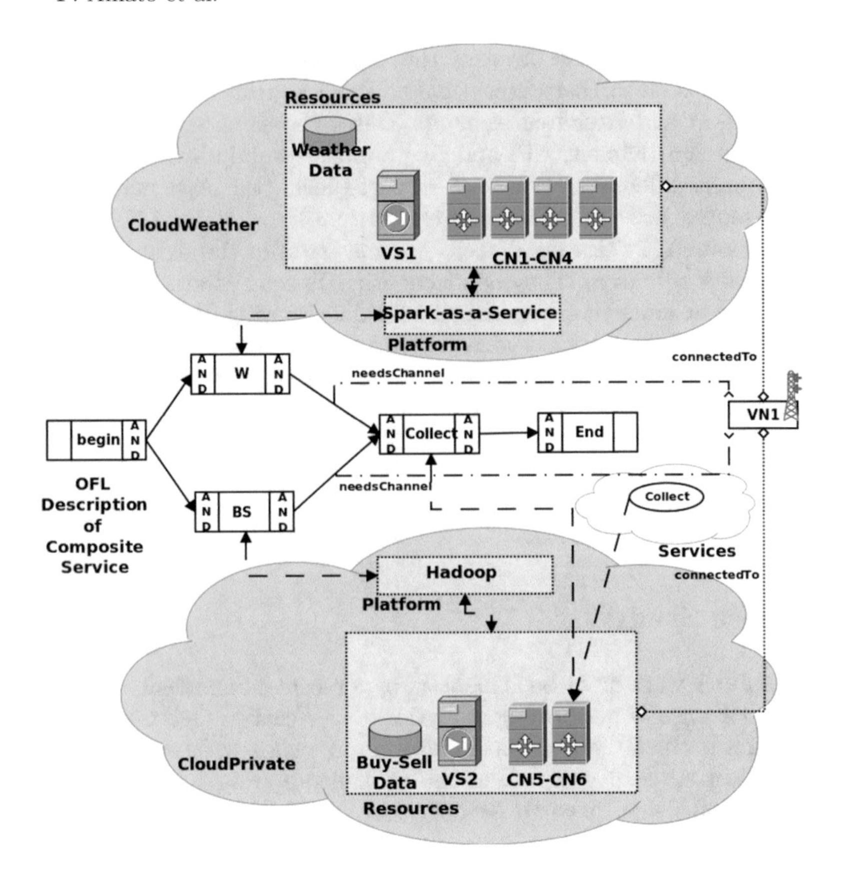

Fig. 3. Case study

service on CN6: the activity can execute on proper data that are collected by this very service. The process can finally terminate leaving results on CloudPrivate Virtual Storage.

Data routing and some minor details are not reported for brevity.

5 Conclusions and Future Works

We have described a language and a framework for Cloud service composition. The language, OFL, is able to describe complex composition patterns, navigating all Cloud architectural layers: SaaS, PaaS, IaaS.

We an example showing how the language and the framework can be exploited in order to describe Composite Big-Data processes based on the use of Multi-Cloud platforms [21,22].

Future work will integrate ontology-based reasoning into the methodology in order to automatically build composite Cloud Services with given semantics and QoS.

References

1. White, T.: Hadoop: The definitive guide. O'Reilly Media, Inc., Sebastopol (2012)
2. VV.AA. Us government cloud computing technology roadmap release 1.0 (draft). In: Special Publication 500-293, vol. 2, pp. 1–85. NIST (2011)
3. Lorenzo, G.D., Mazzocca, N., Moscato, F., Vittorini, V.: Towards semantics driven generation of executable web services compositions. Int. J. Softw. JSW **2**(5), 1–15 (2007)
4. Dustdar, S., Schreiner, W.: A survey on web services composition. Int. J. Web Grid Serv. **1**(1), 1–30 (2005)
5. Traverso, P., Pistore, M.: Automated composition of semantic web services into executable processes. In: The Semantic Web-ISWC 2004, pp. 380–394. Springer (2004)
6. Amato, F., Moscato, F.: Pattern-based orchestration and automatic verification of composite cloud services. Comput. Electr. Eng. **56**, 842–853 (2016)
7. Amato, F., Barbareschi, M., Casola, V., Mazzeo, A., Romano, S: Towards automatic generation of hardware classifiers. In: Lecture Notes in Computer Science (Including Subseries Lecture Notes in Artificial Intelligence and Lecture Notes in Bioinformatics). LNCS, PART 2, vol. 8286, pp. 125–132 (2013)
8. Moscato, F., Martino, B.D., Aversa, R.: Enabling model driven engineering of cloud services by using mosaic ontology. Scalable Comput. Pract. Exp. **13**(1), 29–44 (2012)
9. Moscato, F., Aversa, R., Martino, B.D., Fortis, T.-F., Munteanu, V.I.: An analysis of mosaic ontology for cloud resources annotation. In: IEEE Proceedings of the FedCSIS 2011 Conference, pp. 973–980 (2011)
10. Della Vecchia, G., Gallo, L., Esposito, M., Coronato, A.: An infrastructure for smart hospitals. Multimed. Tools Appl. **59**(1), 341–362 (2012)
11. Essmaeel, K., Gallo, L., Damiani, E., De Pietro, G., Dipanda, A.: Comparative evaluation of methods for filtering kinect depth data. Multimed. Tools Appl. **74**(17), 7331–7354 (2015)
12. Brancati, N., Caggianese, G., Frucci, M., Gallo, L., Neroni, P.: Touchless target selection techniques for wearable augmented reality systems. In: Intelligent Interactive Multimedia Systems and Services, pp. 1–9. Springer (2015)
13. Moscato, F.: Model driven engineering and verification of composite cloud services in MetaMORP(h)OSY. In: Proceedings of 6th, International Conference on Intelligent Networking and Collaborative Systems INCoS 2014. IEEE (2014)
14. Moscato, F., Amato, F.: Thermal-aware verification and monitoring of service providers in MetaMORP(h)OSY. In: Proceedings of 6th, International Conference on Intelligent Networking and Collaborative Systems INCoS 2014. IEEE (2014)
15. Ranjan, R., Benatallah, B., Dustdar, S., Papazoglou, M.P.: Cloud resource orchestration programming: overview, issues, and directions. IEEE Internet Comput. **19**(5), 46–56 (2015)
16. Liu, C., Loo, B.T., Mao, Y.: Declarative automated cloud resource orchestration. In: Proceedings of the 2nd ACM Symposium on Cloud Computing, p. 26. ACM (2011)
17. Kumar, R., Gupta, N., Charu, S., Jain, K., Jangir, S.K.: Open source solution for cloud computing platform using openstack. Int. J. Comput. Sci. Mob. Comput. **3**(5), 89–98 (2014)
18. Plotkin, G.D.: A structural approach to operational semantics (1981)

19. Mens, T., Gorp, P.V.: A taxonomy of model transformation. Electron. Notes Theor. Comput. Sci. **152**, 125–142 (2006). Proceedings of the International Workshop on Graph and Model Transformation (GraMoT 2005), Graph and Model Transformation 2005. http://www.sciencedirect.com/science/article/pii/S1571066106001435
20. Moscato, F., Aversa, R., Amato, A.: Describing cloud use case in MetaMORP(h)OSY. In: IEEE Proceedings of the CISIS 2012 Conference, pp. 793–798 (2012)
21. Barolli, L., Chen, X., Xhafa, F.: Advances on cloud services and cloud computing. Concurr. Comput. **27**(8), 1985–1987 (2015)
22. Pop, F., Dobre, C., Cristea, V., Bessis, N., Xhafa, F., Barolli, L.: Reputation-guided evolutionary scheduling algorithm for independent tasks in inter-clouds environments. Int. J. Web Grid Serv. **11**(1), 4–20 (2015)

A GDPR-Compliant Approach to Real-Time Processing of Sensitive Data

Luigi Sgaglione(✉) and Giovanni Mazzeo

University of Naples "Parthenope", Naples, Italy
{luigi.sgaglione,giovanni.mazzeo}@uniparthenope.it

Abstract. Cyber-attacks represent a serious threat to public authorities and their agencies are an attractive target for hackers. The public sector as a whole collects lots of data on its citizens, but that data is often kept on vulnerable systems. Especially for Local Public Administrations (LPAs), protection against cyber-attacks is an extremely relevant issue due to outdated technologies and budget constraints. Furthermore, the General Data Protection Regulation (GDPR) poses many constraints/limitations on the data usage when "special type of data" is processed. In this paper the approach of the EU project COMPACT (H2020) is presented and the solutions used to guarantee the data privacy during the real time monitoring performed by the COMPACT security tools are highlighted.

Keywords: Real time processing · SIEM · SOC · Data privacy
Homomorphic encryption

1 Introduction

The advent of the Internet has been opening new opportunities for Public Administrations (PAs) to improve their efficiency while providing better services to citizens via an ever larger set of specialized network applications, including e-government, e-health, and more. This is at the heart of a European wide eGovernment action plan, whose latest update covers the years 2016 to 2020 and which also mentions the importance of trustworthiness and security as a key guiding value. Indeed, as a potential channel of accessing personal information, these specialized applications also expose the public sector to new risks.

The cybersecurity landscape is changing, and Local Public Administrations (LPAs) and Critical Infrastructures (CIs) are rapidly becoming an attractive target for cyber-criminals [1–5], who might access some sets of personal data or gain control over smartly operated city resources through LPAs/CIs infrastructures. The consequences of cyber-threats have the potential to be considerable causing business interruptions, data losses, and thefts of intellectual property, significantly impacting both individuals and organizations.

It is claimed that cyber-threats are the most significant and rising risk that public sector organizations are facing. Reports demonstrates that nearly 40% of malware attacks and in general cyber threats to which public bodies have been subject [6] are against public sector organizations [1], i.e. more than sectors (e.g. finance) which have

G. De Pietro et al. (Eds.): KES-IIMSS-18 2018, SIST 98, pp. 43–52, 2019.
https://doi.org/10.1007/978-3-319-92231-7_5

traditionally been thought of as top targets. The interconnection of operational environment systems, used by the public bodies in ever growing scale, exacerbates the problem, especially as malware distribution periods (both fixed and mobile) are becoming increasingly short [7]. The increase in severity of cyber-attacks coincides with a boom in the different types of connected devices, as well as with a huge expansion in virtualization and public clouds.

In particular, Information Commissioner's Office (ICO) reports that: "In a change to the previous quarter, the second most prevalent sector in Q4 (January to March 2016)1 was local government. The number of data security incidents in this sector increased by 34% compared to the previous quarter (from 32% in Q3 to 43% in Q4).

Coupled with the overall decrease in data security incidents during Q4, this means that the percentage of total incidents suffered by the local government sector has also increased, from 6% in Q3 to 10% in Q4." [8].

Therefore, LPAs need to understand the cyber risks to which they are exposed and take proper actions to protect their infrastructures from cyber disruptions, to safeguard citizen's and enterprises' information they manage. The DBIR 2016 report [6] provides the number of security incidents by victim industry and organization size (2015 dataset). The category "Public Industry" – which refers to PA organizations – is by far the most targeted, with 47000+ attacks out of a total of about 64000. The report also shows the distribution of incidents by patterns: the vast majority of incidents in the public sector can be rooted to: (1) miscellaneous errors (24%), (2) privilege misuse (22%), (3) stolen assets (20%), and (4) crimeware (16%).

The issues that have been identified and that hamper the ability of PA organizations of improving their cyber security level, most notably are:

1. **Lack of standardized data classification** – 45% of public sector respondents do not use standardized data classification techniques/procedures. As a consequence, LPAs run a higher risk of accidentally exposing private data in their rush to comply with emerging regulations – both at the national and at the EU level – promoting transparency of the Public Sector. Also, only 12% stated that they used standardized policies and that they proactively verify and enforce those policies.
2. **Lack of effective Non-Disclosure Agreements** (NDAs) – 40% of public sector organizations still rely on paper-based NDAs, and use them inconsistently. This amplifies risks related to the human factor, which is already one of the biggest, since malicious or disgruntled personnel with access to important information assets can be a significant threat to the security of those assets.
3. **Lack of plans for responding to security breaches and for disaster recovery** – 36% of public sector organizations do not have a plan for responding to security breaches, and only 10% of public sector organizations test for the worst-case scenario. 34% of public sector organizations do not have budgeted disaster recovery plans. These are major impairments to contain the damage, since when a security incident or a disaster occur, proper and timely action is key.
4. **Lack of uniformly enforced security policies** – 33% of public sector organizations do not have uniformly enforced security policies (this means limited application - if not complete lack - of a consistent security policy throughout the whole

organization.). This condition hinders their ability to comply with regulations, such as the European Union Data Protection Directive (EUDPD).
5. **Lack of adequate policies and practices for data disposal** – 76% of public sector organizations do not have adequate policies and practices for secure and reliable data disposal. In particular, only 16% of public sector organizations have written policies that require destruction records to be actually collected, practiced, and audited. The enforcement of strong policies to govern the proper disposal of electronic and paper records - based on sound technical and organizational guidelines and best practices - is the prerequisite for protecting private data from unauthorized disclosure.
6. **Lack of effective access control mechanisms** – 20% of public sector organizations do not use roles to manage access, and more than 26% of public sector organizations have no official procedure for terminated or reassigned employees. This create vulnerabilities, since it allows inappropriate access to resources.
7. **Large set of legacy unmaintained and undocumented systems** representing an attack surface of unknown dimension.
8. **Inappropriate management of security updates** (patches), as well as usage of out of date software in computers, mobile devices and central servers.
9. **Limited capacity, and motivation, of LPAs personnel** in detecting and reporting cyber-attacks. This is due to a number of interconnected factors including (i) the aging of the LPAs workforce, (ii) its limited technological skills and (iii) the lack of acknowledgment of employees' achievements. This makes the PA workforces less responsive to the traditional educational measures (like classroom training).

It is clear that innovative cyber security tools are needed in order to guarantee the protection of LPAs. In addition, these tools must to deal with:

(1) Limited resources in terms of both economic and structural
(2) Strong privacy requirements coming from the recent adoption of the General Data Protection Regulation (GDPR) on the protection of natural persons with regard to the processing of personal data and on the free movement of such data.

2 Backgroud – Homomorphic Encryption

2.1 Homomorphic Encryption

Homomorphic Encryption is a recent cryptographic method which allows to perform computation on encrypted data without decrypting it. This way, the confidential data can be protected not only during the storage and exchange/transfer but also during the processing. Avoiding intrusions from semi-honest or malicious cloud providers when outsourcing data processing to the Cloud is crucial for the case of sensitive data that are about to be processed in frames of the COMPACT solution.

The first HE algorithms, i.e., Partially Homomorphic Encryption (PHE) [14, 15], had the ability to carry out just one type of operations (e.g., addition, or multiplication). Clearly, the limitation in the type of executable computations hampered the usage of HE in practical contexts. Gentry et al. [16] provided the first implementation of a Fully

Homomorphic Encryption (FHE) scheme. Gentry's algorithm allows the execution of an arbitrary number of additions and multiplications over encrypted data. The security of the system is based on the noise introduced into the ciphered text. When the noise reaches some maximum amount, the ciphertext becomes undecryptable. This solution was very costly in terms of performance. It highly affects CPU and memory resources.

An attempt to simplify the method has been provided by Van Dijk et al. [17] who proposed a FHE i.e., Somewhat Homomorphic Encryption (SHE) over the integers. The price to pay with SHE is given by the limited number of mathematical operations that can be performed. However, in many real-world applications (e.g., medical, financial) this seems reasonable since – as Naehrig et al. [18] analysis reports – most of the evaluations required, i.e., one-time statistical functions, fits well with SHE constraints.

Among the aims of COMPACT are to adopt Fully Homomorphic Encryption (FHE) Schemes capable of performing any arbitrary function in an homomorphic way and to mitigate performance overheads introduced by Homomorphic computation, using recent dedicated compilation and parallelism techniques and mechanisms.

3 The COMPACT Project

COMPACT's overarching objective is to enable LPAs to become the main actors of their own cyber-resilience improvement process, by providing them with effective tools and services for removing security bottlenecks. This can be broken down into five finer-grain objectives:

- Objective #1 - Making the PA personnel aware of the basic cyber security threats they are exposed to.
- Objective #2 - Improving the skills – both technical and behavioral – of the PA personnel via innovative training techniques that are well received by the (non IT-expert) workforce.
- Objective #3 - Providing protection tools against basic cyber security threats, i.e. those with a higher impact on LPAs. These include [10–12]: phishing, ransomware, Bring Your Own Device (BYOD), jailbreaking the cloud, cross-site scripting, code (particularly SQL) injection, and more.
- Objective #4 - Creating a LPAs level information hub, for favouring reliable and timely exchange of information among LPAs on cyber security guidelines and best practices, as well as on Indicators of Compromise (IoC).
- Objective #5 - Creating a link between COMPACT LPAs level information hub and major EU level initiatives, for supporting LPAs to improve cyber-resilience in a complex European context.

To achieve its objectives, COMPACT will develop four types of tools/services (Fig. 1), which include:

1. Risk assessment tools - Tailored to the LPAs context that will allow LPAs to evaluate and monitor their exposure to the most relevant (i.e. with the highest impact) cyber treats. They will enable LPAs to prioritize the adoption of preventive

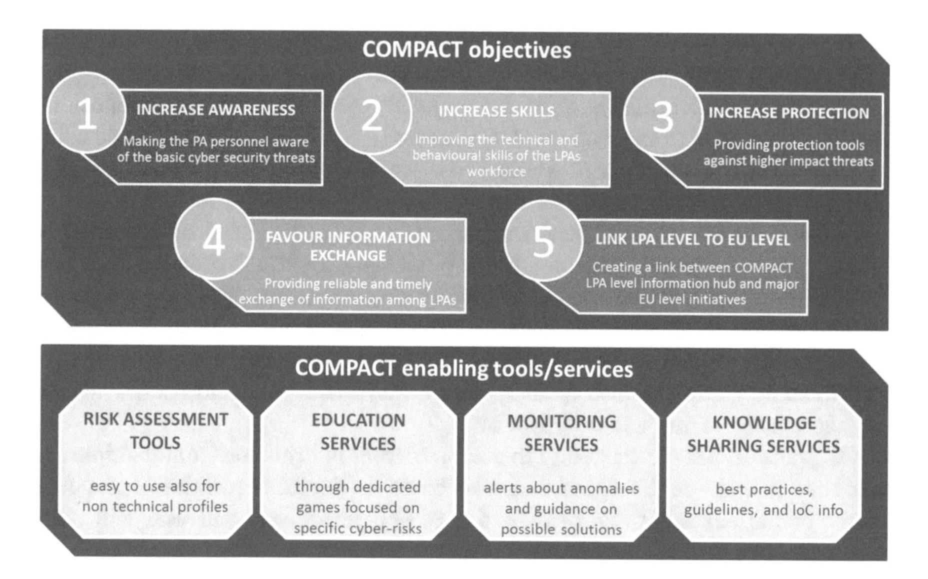

Fig. 1. COMPACT objectives

and reactive countermeasures, for maximum efficiency of resource usage for cyber protection purposes.

2. Education services - Through dedicated game-based training, focused not only on specific cyber-threats but also on psychological and behavioral factors, to maximize the effectiveness of the learning experience, while also containing the training time.
3. Monitoring services (SOC) - That continuously process events related to the status of the infrastructure and correlate them with information from threat intelligence feeds to timely spot anomalies and also suggest recovery actions that can be implemented.
4. Knowledge Sharing services - These will include best practices and guidelines, focused on the specific needs of LPAs, that can be easily adopted to quickly increase the cyber security level of the organization. Just as importantly, they are also used (i) at the Member States level as an input for the activity of national cybersecurity stakeholders (like national CERTs5) and (ii) at the EU level as an input for European boards, agencies, and initiatives (like ENISA and the CSIRT [13] network foreseen in the NIS directive)

4 COMPACT Monitoring Service

The Security Operations Centre (SOC) provides, throughout advanced Security Information and Event Management (SIEM), the real-time monitoring capability of the organization. SOC platform is an integrated technology platform that allows for accurate, timely and trustworthy detection and diagnosis of security attacks, combining information from physical and logical event sources. The platform has been implemented in a distributed loosely interoperating architecture, where components depend on each other to the least extent practicable.

The SOC is implemented as a distributed architecture that enables: (i) collection of security-relevant data from a variety of data feeds; (ii) correlation of events and context information, via combined use of stream and batch processing; and (iii) production and secure storage of incident-related evidence.

The event sources for SOC platform can be physical or logical alike. Physical event sources include physical systems that are existing in the buildings, like video surveillance system, physical access control system, fire alarm system, other physical security systems, or automation and building management systems, for example. Logical security systems can be defined to consist of software safeguards for an organization's systems, including user identification and password access, authenticating, access rights and authority levels.

SOC platform has the capability to combine event information from multiple event sources and to make sophisticated diagnosis based on the received information. As the outcome of the analysis performed by the SOC platform, the end user will receive ranked alerts and forensic evidences.

An architecture of the current solution is reported in Fig. 2. SOC platform consists of the following main components:

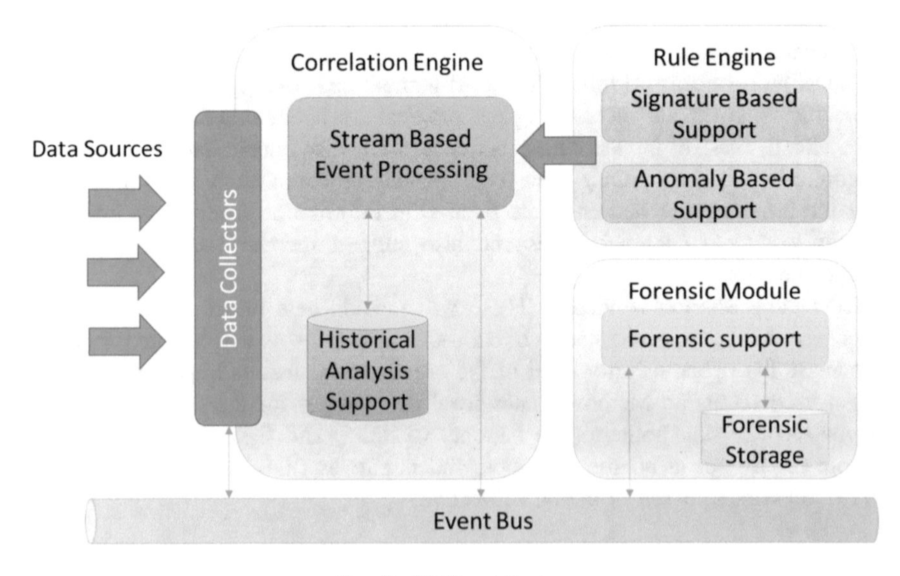

Fig. 2. SOC architecture

- Correlation Engine:

The Correlation Engine is the component in charge of the event diagnosis process. It operates by correlating a huge amount of security relevant events/information from the physical and the electronic domain in real-time, through Complex Event Processing (CEP) techniques and stream processing computing technologies.

The attack diagnosis process is driven by correlation rules that aggregate the parameters of attack symptoms, such as the attack type, the target component and the temporal proximity. Alerts are generated only when the correlation among such symptoms indicates a potential attack, thus exhibiting low false positive rates and improved detection capability w.r.t. single probes.

- Rule Engine:

The Rule Engine provides the logical rules to be followed for the Correlation Engine. The Rule Engine includes two main components, Signature Based Support and Anomaly Based Support.

- Forensic Module:

The Forensic Module provides a set of services that enables the end user (SOC operator) to trace from an event to the log data from which it was identified. The module will ensure that the events and their associated logs are stored in a forensically sound manner. It will support processes that ensure, to the greatest extent possible, that the event data will be acceptable as evidence.

In terms of data collection, the prototype is equipped with a number of adapters, for receiving events from a wide variety of Commercial Off The Shelf (COTS) products for logical and physical security monitoring. In terms of data processing, the prototype enables: (1) pre-processing of data at the edge of the system and (2) stream and batch processing in the core of the system. The business logic that drives the correlation process can be easily customized by means of a user-friendly graphical interface.

The SIEM is the main component of the SOC systems and includes:

- A runtime engine to allow the distributed streaming dataflow
- Two data processing APIs, one for the Stream Processing and one for the Batch Processing
- Three class of libraries:

  1. Complex Event Processing (CEP) to detect event patterns in an endless stream of events. Event processing combines data from multiple sources to infer events or patterns in order to highlight specific situations. The goal of complex event processing is to identify meaningful events (such as threats) and respond to them as quickly as possible. This real time elaboration can be based on a time window or event-driven approach.
  2. Machine Learning that gives SIEM the ability to learn without being explicitly programmed. It requires the use of algorithms that can learn from and make predictions on data – such algorithms overcome following strictly static program instructions by making data-driven predictions or decisions, through building a model from sample inputs.
  3. Homomorphic Data Processing to allow the processing of homomorphic encrypted data without decrypting them

The communication between the SOC component is provided by a Publish Subscribe communication channel: it is in charge of delivering the data and messages between data sources, SIEM GUI and SIEM Core (Fig. 3).

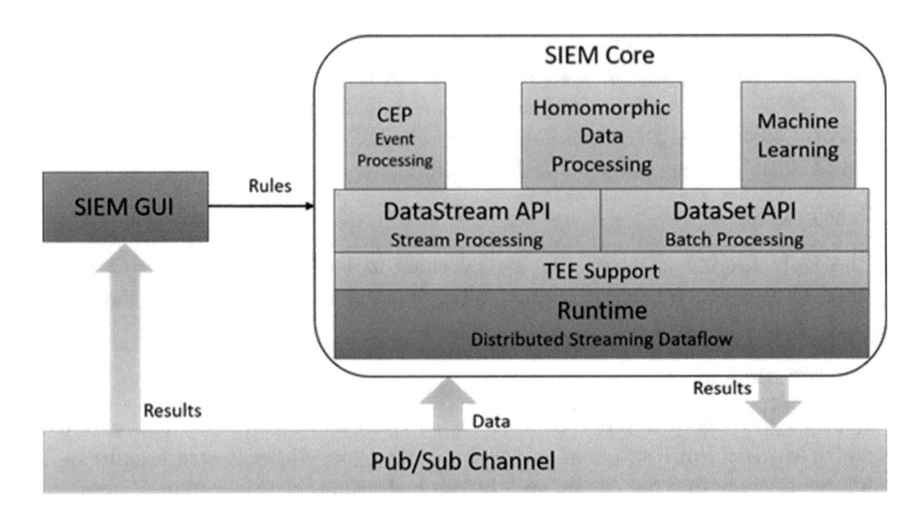

Fig. 3. SIEM components

Even a SOC prototype is already available; it will be evolved to meet the COMPACT requirements along several dimensions.

The first development will regard the improvement and adaptation of the SOC data collection to the data that must be acquired during the LPA monitoring. Many data collection features are already available in the current SOC prototype and these will be adapted to be compliant with the LPA environments; others will be developed to meet specific requirements likes the acquisition of information from the Windows Management Instrumentation tool and from other security tools (Nagios, Sophos, etc.).

The second improvement will be related to the implementation of the Data Management and Policy Enforcement component (DMPE). This component will be integrated in each data collection tool in order to fulfill the privacy requirements imposed by the LPA (to be compliant with the GDPR). In particular, the DMPE will be in charge of applying the most appropriate techniques needed to meet the privacy requirements, such as anonymization and pseudo anonymization to remove special categories of data or homomorphic encryption to hide data and process it in an encrypted form.

The third improvement is related to the technology update of the current correlation and processing features of the SOC, by exploiting a best of breed selection of Open Source technologies for CEP, machine learning, and data mining.

The fourth improvement will be related to the implementation of specific correlation operators (CEP operators) able to process the homomorphically encrypted data without decrypting it.

Finally, the SOC graphical user interface will be developed/adapted in order to meet the guidelines defined by the COMPACT consortium and to be integrated with the COMPACT unified dashboard.

5 Conclusions

In this paper, a brief overview of the COMPACT approach used for the implementation of a Security Monitoring Center tailored to LPAs has been presented. The paper also highlights how this component will guarantee the privacy of sensitive data during the processing phase.

Acknowledgments. This project has received funding from the European Union's Horizon 2020 Framework Programme for Research and Innovation under grant agreements No 74071 (COMPACT)

References

1. Time to face up to cyber risk. http://www.publicfinance.co.uk/opinion/2016/03/time-face-cyber-risk. Accessed 09 Apr 2018
2. Coppolino, L., D'Antonio, S., Romano, L.: Exposing vulnerabilities in electric power grids: an experimental approach. Int. J. Crit. Infrastr. Prot. **7** (2014). https://doi.org/10.1016/j.ijcip.2014.01.003
3. Coppolino, L., D'Antonio, S., Formicola, V., Romano, L.: Enhancing SIEM technology to protect critical infrastructures, pp. 10–21 (2013). https://doi.org/10.1007/978-3-642-41485-5_2
4. Coppolino, L., D'Antonio, S., Formicola, V., Romano, L.: Integration of a system for critical infrastructure protection with the OSSIM SIEM platform: a dam case study, vol. 6894, pp. 199–212 (2011). https://doi.org/10.1007/978-3-642-24270-0_15
5. D'Antonio, S., Coppolino, L., Elia, I., Formicola, V.: Security issues of a phasor data concentrator for smart grid infrastructure (2011). https://doi.org/10.1145/1978582.1978584
6. Data Breach Investigations Report (DBIR). http://www.verizonenterprise.com/verizon-insights-lab/dbir/2017/. Accessed 09 Apr 2018
7. CCN-CERT, Threats and Risk Analysis in Industrial Control Systems (ICS), Report IA-04/16, Centro Criptologico Nacional, Madrid, 28 January 2016. (in Spanish). https://www.ccn-cert.cni.es/informes/informes-ccncert-publicos/1381-ccn-cert-ia-04-16-amenazas-y-analisis-de-riesgos-en-sistemas-de-control-industrial-ics/file.html. Accessed 09 Apr 2018
8. Data security incident trends. https://ico.org.uk/action-weve-taken/data-security-incident-trends/. Accessed 09 Apr 2018
9. Gajli, A.: Time to face up to cyber risk, Public Finance, 31 March 2016. http://www.publicfinance.co.uk/opinion/2016/03/time-face-cyber-risk. Accessed 09 Apr 2018
10. 3 Basic cyber security threats to be aware of that people still get wrong. http://blog.scstechsolutions.co.uk/3-basic-cyber-security-threats/. Accessed 09 Apr 2018
11. Biggest cybersecurity threats in 2016. http://www.cnbc.com/2015/12/28/biggest-cybersecurity-threats-in-2016.html. Accessed 09 Apr 2018
12. Top 7 Cyberthreats to Watch Out for in 2015-2016. Kaspersky Lab
13. Computer security and incident response teams network. https://www.enisa.europa.eu/topics/national-csirt-network
14. El Gamal, T.: A public key cryptosystem and a signature scheme based on discrete logarithms. In: Proceedings of CRYPTO 84 on Advances in Cryptology. Springer, New York, pp. 10–18 (1985). http://dl.acm.org/citation.cfm?id=19478.19480
15. Paillier, P.: Public-key cryptosystems based on composite degree residuosity classes, pp. 223–238. Springer, Heidelberg (1999). https://doi.org/10.1007/3-540-48910-x_16

16. Gentry, C.: Fully homomorphic encryption using ideal lattices. In: Proceedings of the Forty-First Annual ACM Symposium on Theory of Computing, STOC 2009, pp. 169–178. ACM, New York (2009). https://doi.org/10.1145/1536414.1536440, http://doi.acm.org/10.1145/1536414.1536440
17. van Dijk, M., Gentry, C., Halevi, S., Vaikuntanathan, V.: Fully homomorphic encryption over the integers, Cryptology ePrint Archive, Report 2009/616 (2009). http://eprint.iacr.org/2009/616
18. Naehrig, M., Lauter, K., Vaikuntanathan, V.: Can homomorphic encryption be practical? In: Proceedings of the 3rd ACM Workshop on Cloud Computing Security Workshop, CCSW 2011, pp. 113–124. ACM, New York (2011). https://doi.org/10.1145/2046660.2046682, http://doi.acm.org/10.1145/2046660.2046682

Data Mining in Social Network

Flora Amato[1(✉)], Giovanni Cozzolino[1], Francesco Moscato[2], Vincenzo Moscato[1], Antonio Picariello[1], and Giancarlo Sperlì[1]

[1] Dipartimento di Ingegneria Elettrica e delle Tecnologie dell'Informazione DIETI, University of Naples "Federico II", Naples, Italy
{flora.amato,giovanni.cozzolino,vmoscato,picus,giancarlo.sperli}@unina.it

[2] Dipartimento di Scienze Politiche. DiSciPol, University of Campania "Luigi Vanvitelli", Caserta, Italy
francesco.moscato@unicampania.it

Abstract. In this paper, we propose a novel data model for Multimedia Social Networks, i.e. particular social media networks that combine information on users belonging to one or more social communities together with the content that is generated and used within the related environments. The proposed model relies on the hypergraph data structure to capture and represent in a simple way all the different kinds of relationships that are typical of social media networks, and in particular among users and multimedia content. We also introduce some user and multimedia ranking functions to enable different applications. Finally, some experiments concerning effectiveness of the approach for supporting relevant information retrieval activities are reported and discussed.

1 Introduction

Social media networks provide users an interactive platform to create and share multimedia content such as text, image, video, audio, and so on. Just as an example, each minute thousands of tweets are sent on Twitter, several hundreds of hours of videos are uploaded to YouTube, and a huge quantity of photos are shared on Instagram or uploaded to Flickr.

Within these "interest-based" networks, each user interacts with the others through a multimedia content and such interactions create "social links" that well characterize the behaviors of involved users in the networks. Here, multimedia data seems to play a "key-role" especially if we consider the *Social Network Analysis* (SNA) perspective: representing and understanding user-multimedia interaction mechanisms can be useful to predict user behavior, to model the evolution of multimedia content and social graphs, to design human-centric multimedia applications and services and so on. In particular, several research questions have to be addressed:

- It possible to exploit multimedia features and the notion of *similarity* to discover more useful links?

G. De Pietro et al. (Eds.): KES-IIMSS-18 2018, SIST 98, pp. 53–63, 2019.
https://doi.org/10.1007/978-3-319-92231-7_6

- Can all the different types of user annotations (e.g. tag, comment, review, etc.) and interactions with multimedia objects provide a further support for an advanced network analysis?
- Is it possible to integrate and efficiently manage in a unique network the information coming from different social media networks (for example, a Twitter user has usually an account also on Instagram or Flickr)?
- How can we deal with a very large volume of data?
- In this context, how is possible to model all the various relationships among users and multimedia objects [1]? Are the "graph-based" strategies still the most suitable solutions?

To capture the described issues, we adopt the term *Multimedia Social Networks* (MSNs) to indicate "*integrated social media networks that combine the information on users, belonging to one or more social communities, together with all the multimedia contents that can be generated and used within the related environments*".

Actually, the term MSN have been used over the last years in the literature together with *Social Multimedia Network* or *Social Media Network* to indicate information networks that leverage multimedia data in a social environment for several purposes: distributed resource allocation for multimedia content sharing in cloud-based systems [2], generation of personalized multimedia information recommendations in response to specific targets of interests [3], evaluation of the trust relationship among users [4], high dimensional video data distribution in social multimedia applications [5], characterization of user behavior and information propagation on the base of multimedia sharing activities [6], representation of a social collaboration network of archeologists for cultural heritage applications [7], just to cite some of the most recent proposals. In this paper, inspired by hypergraph based approaches, we propose a novel data model [1,8,9] for Multimedia Social Networks. Our model provides a solution for representing MSNs sufficiently general with respect to: (i) a particular social information network, (ii) the different kinds of entities, (iii) the different types of relationships, (iv) the different applications [10–12]. Exploiting hypergraphs, the model allows us to represent in a simple way all the different kinds of relationships that are typical of a MSN (among multimedia contents, among users and multimedia content and among users themselves [13,14]) and to enable several kinds of analytics and applications by means [15,16] of the introduction of some user [17] and multimedia (global and "topic sensitive" [18]) *ranking* functions.

We exploit functionalities of a well know framework for NLP processing, GATE [19] in order to extract relevant information from the famous online social network Yelp.

The paper is organized as in the following. Section 2 describes in details and using different examples our model with its properties and foundations. Section 3 shows the obtained experimental results using a standard Yelp dataset, while Sect. 4 reports conclusions and the future work.

2 The Ranking Model

Ranking functions can be profitably used to "rank" users and multimedia objects in a MSN in an absolute way or with respect to a given topic of interest. Let us first introduce some preliminary definitions.

Definition 1 (Distances). *We define* minimum distance *($d_{min}(v_i, v_j)$),* maximum distance *($d_{max}(v_i, v_j)$) and* average distance *($d_{avg}(v_i, v_j)$) between two vertices of a MSN the length of the shortest hyperpath, the length of the longest hyperpath and the average length of the hyperpaths between v_i and v_j, respectively. In a similar manner, we define the* minimum distance *($d_{min}(v_i, v_j \,|v_k)$),* maximum distance *($d_{max}(v_i, v_j \,|v_k)$) and* average distance *($d_{avg}(v_i, v_j \,|v_k)$) between two vertices v_i and v_j, for which there exists a hyperpath containing v_k.*

In the computation of distances, we apply a *penalty* if the considered hyperpaths contain some users: all the distances can be computed as $\tilde{d}(v_i, v_j) = d(v_i, v_j) + \log(\beta \cdot N)$, N being the number of user vertices in the hyperpath between v_i and v_j and β a scaling factor[1].

Definition 2 (λ-Nearest Neighbors Set). *Given a vertex $v_i \in V$ of a MSN, we define the* λ-Nearest Neighbors Set *of v_i the subset of vertices NN_i^λ such that $\forall v_j \in NN_i^\lambda$ we have $\tilde{d}_{min}(v_i, v_j) \leq \lambda$ with $v_j \in U$. Considering only the constrained hyperpaths containing a vertex v_k, we denote with NN_{ik}^λ the set of nearest neighbors of v_i such that $\forall v_j \in NN_{ik}^\lambda$ we have $\tilde{d}_{min}(v_i, v_j \,|v_k) \leq \lambda$.*

If we consider as neighbors only vertices belonging to user type, the NN^λ set is called *λ-Nearest Users Set* and denoted as NNU^λ, similarly in case of multimedia objects we define the *λ-Nearest Objects Set* as NNO^λ. On the top of such definitions, we are able to introduce the *ranking functions.*

Definition 3 (User Ranking function). *Given a user $u_i \in U$ and a subset of users $\widehat{U} \subseteq U (u_i \notin \widehat{U})$ of a MSN, a* user ranking function *is a particular function $\rho : U \rightarrow [0, 1]$ able to associate a specific* rank *to the user u_i with respect to the community $\widehat{U}$ that is computed as in the following:*

$$\rho_{u_i}\left(\widehat{U}\right) = \frac{\left|NNU_{u_i}^\lambda \cap \widehat{U}\right|}{\left|\widehat{U}\right|} \tag{1}$$

NNU_i^λ being the λ-Nearest Users Set of u_i.

Definition 4 (Multimedia Ranking function). *Given a multimedia object $m_i \in M$ and a subset of users $\widehat{U} \subseteq U$ of a MSN, a* multimedia ranking function

[1] Such strategy is necessary in the ranking to penalize *lurkers*, i.e. users of a MSN that are quite inactive and not directly interact with multimedia content but through user to user relationships.

is a particular function $\rho : M \rightarrow [0,1]$ *able to associate a specific* rank *to the object* m_i *with respect to the community* $\widehat{U}$ *that is computed as in the following:*

$$\rho_{m_i}\left(\widehat{U}\right) = \frac{\left|NNU^{\lambda}_{m_i} \cap \widehat{U}\right|}{\left|\widehat{U}\right|} \tag{2}$$

$NNU^{\lambda}_{m_i}$ *being the* λ*-Nearest Users Set of* m_i*.*

In a similar manner, considering only hyperpaths containing a given topic a_j we can define the *topic sensitive* user ($\rho^{a_j}_{u_i}\left(\widehat{U}\right)$) and multimedia ($\rho^{a_j}_{m_i}\left(\widehat{U}\right)$) ranking functions.

Definition 5 (Topic User Ranking function). *Given a user* $u_i \in U$ *and a subset of users* $\widehat{U} \subseteq U (u_i \notin \widehat{U})$ *of a MSN, a* topic user ranking function *is a particular function* $\rho^a_u : U \times A \rightarrow [0,1]$ *able to associate a specific* rank *to the user* u_i *with respect to the community* $\widehat{U}$ *given the topic* a_j *that is computed as in the following:*

$$\rho^{a_j}_{u_i}\left(\widehat{U}\right) = \frac{\left|NN^{\lambda}_{ij} \cap \widehat{U}\right|}{\left|\widehat{U}\right|} \tag{3}$$

$NN^{\lambda}_{u_ij}$ *being the* λ*-Nearest Users Set of* u_i *given* a_j*.*

Definition 6 (Topic Multimedia Ranking function). *Given a multimedia object* $m_i \in M$ *and a subset of users* $\widehat{U} \subseteq U$ *of a MSN, a* multimedia ranking function *is a particular function* $\rho^a_m : M \times A \rightarrow [0,1]$ *able to associate a specific* rank *to the object* m_i *with respect to the community* $\widehat{U}$ *given the topic* a_j *that is computed as in the following:*

$$\rho^{a_j}_{m_i}\left(\widehat{U}\right) = \frac{\left|NN^{\lambda}_{kj} \cap \widehat{U}\right|}{\left|\widehat{U}\right|} \tag{4}$$

$NN^{\lambda}_{m_ij}$ *being the* λ*-Nearest Users Set of* m_i *given* a_j*.*

In our model the concept of *rank* of a given node is related to the concept of *influence*, and in our vision it can be measured by the number of user nodes that are "reachable" within a certain number of steps using social paths (Fig. 1).

By similarity relationships paths can be "implicitly" instantiated: two users (that are not friend, do not belong to any group and do not share any multimedia object) have annotated two images that are very similar, or they have commented two different posts which concern similar topics.

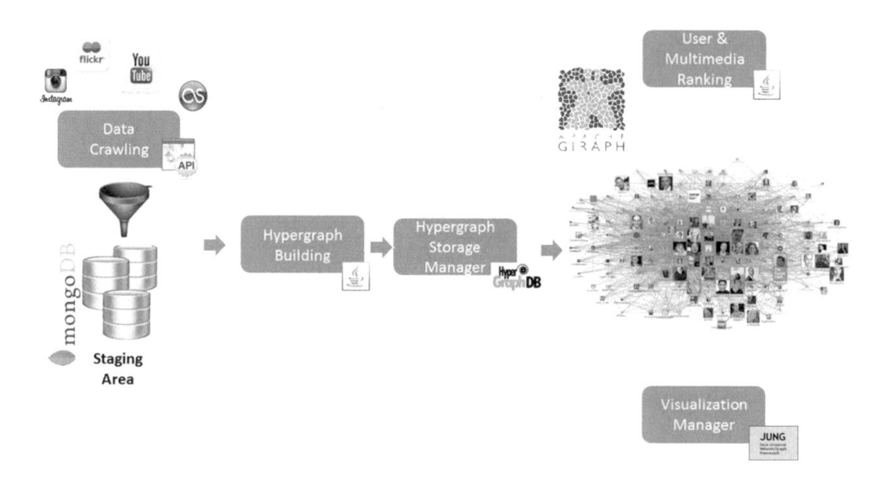

Fig. 1. Proposed prototype

3 Methodology to Extract Information by Social Network

In order to apply our ranking evaluation we have to extract information about the posts of the customers of an Online Social Network. In this section, we describe our analyses performed on Yelp social network.

The Dataset used for the experimentation is given by Yelp website, it is composed by: 4.1 millions of reviews, 947 thousands of tips posted by 1 million users for 144 thousands of businesses.

In order to obtain information about each review, we used Gate NLP Tool developed by University of Sheffield (https://gate.ac.uk).

Gate is an open source software able to solve many text-processing problems. This tool is plugin-based so is possible to customize the processing steps adding or removing modules, in order to obtain different results.

Gate components are specialized types of Java Bean and are of three type:

- Language resources (LRs): entities such as lexicons, corpora or ontologies.
- Processing resources (PRs): entities such as parsers, generators or ngram modellers.
- Visual resources (VRs): visualization and editing components

Gate's CORE, for its structure, is named CREOLE: Collection of Reusable Objects for Language Engineering.

Because Yelp reviews are encoded as a set of JSON tuples, a semi-structured data type, we needed to store this dataset into a NoSQL Database. We chose CouchDB, a Document-Oriented Database by Apache Foundation.

CouchDB is a schemaless database with an intuitive HTTP/JSON API. It speaks JSON natively so it is what we needed. To perform analysis on each review, we used an Official plugin of the GATE framework, developed for Twitter.

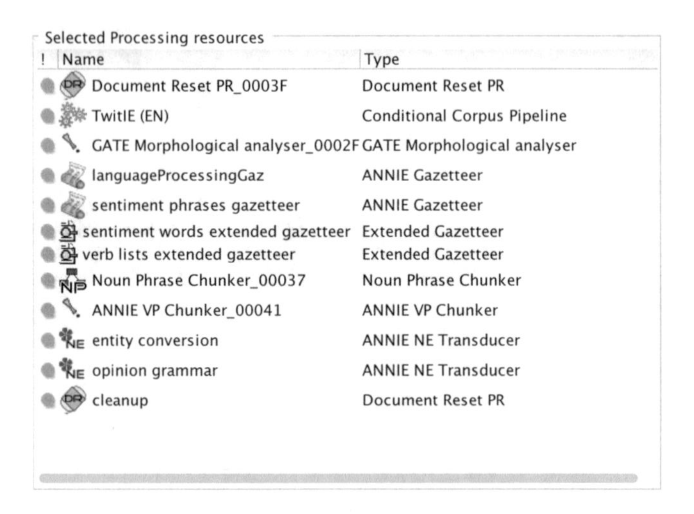

Fig. 2. Review analysis pipeline

The pipeline used in this plug-in is composed by:

- Document reset: for resetting the default annotation set;
- TwitIE: a pipeline specialized to analyze tweets.
- Gate Morphological Analyzer: taking as input a tokenized GATE document. Considering one token and its part of speech tag, one at a time, it identifies its lemma. LanguageProcessingGaz: an ANNIE Gazetteer. The role of the gazetteer is to identify entity names in the text based on lists.
- Verb Lists Extended Gazetteer: an extended version of the Gate Default List Gazetteer.
- Noun Phrase Chunker: The NP Chunker application is a Java implementation of the Ramshaw and Marcus BaseNP chunker which attempts to insert brackets marking noun phrases in text which have been marked with POS tags
- ANNIE VP Chunker: The rule-based verb chunker, based on a number of English grammars.
- Entity Conversion, ANNIE NE Transducer: a semantic tagger. It contains rules, which act on annotations assigned in earlier phases, in order to produce outputs of annotated entities.
- Opinion Grammar, ANNIE NE Transducer.
- TwitIE is a specialized pipeline that is composed by many components (Fig. 2).

The core of this application is TextCat Language Identification and a huge set of gazetteers customized to recognize hashtags and emojis.

TextCat Language Identification is necessary because our dataset is composed by reviews written in English, French and Deutsch natural language. TwitIE is the lexical and semantic analyzer in our pipeline and its results allow to perform deeper text analysis (Fig. 3).

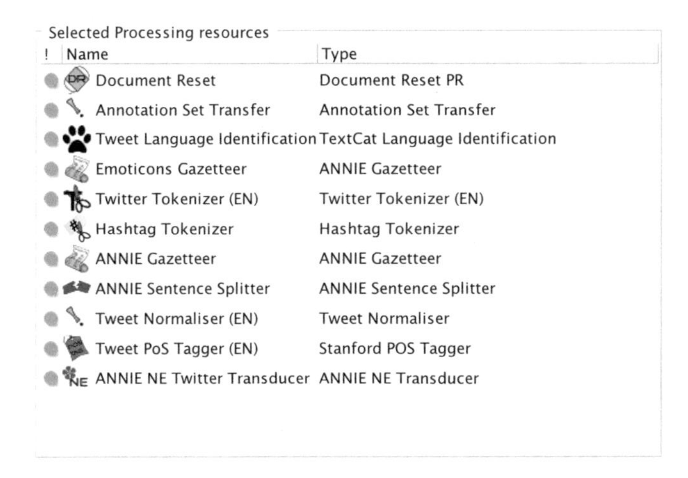

Fig. 3. TwitIE pipeline

We use the described GATE functionalities in order to analyze the set of reviews. We have to set the documents parameters.

We create a Corpus from the input Documents (Fig. 4).

We launch the system functionalities by selecting the Application English-OM and set the corpus to analyze.

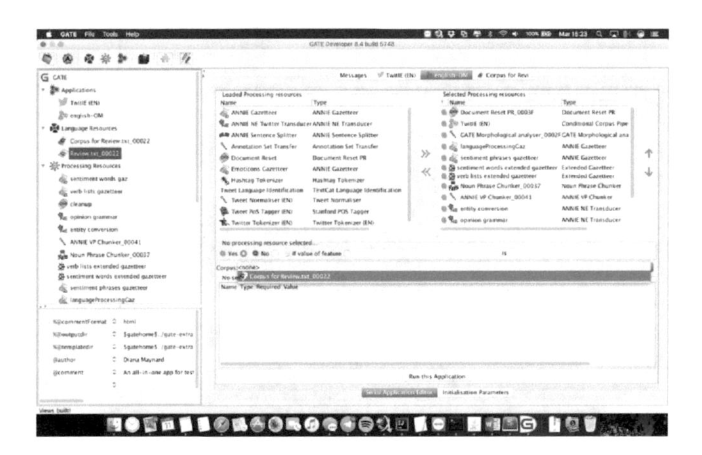

Fig. 4. Corpus selection

After the computation ended, we to check the results Double-click on Document, click on Annotation Sets and Annotation List to view tags (Fig. 5).

Each sentence that have a sentiment, will be tagged as SentenceSentiment with a set of Features, that are customizable using a JAPE Grammar: a set of phases, each of which consists of a set of pattern/action rules. The phases run

Fig. 5. Document annotations view

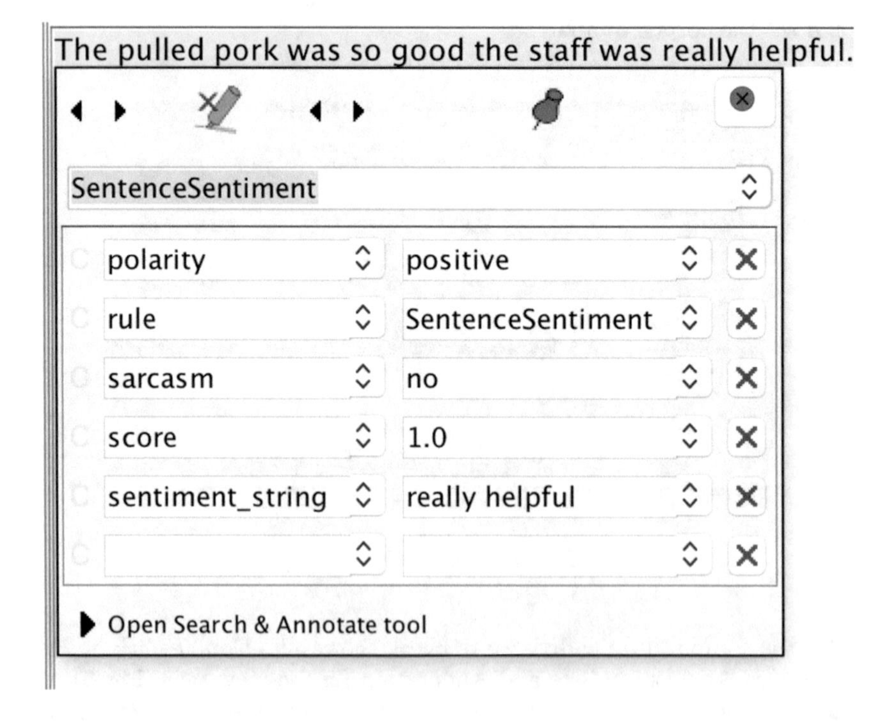

Fig. 6. Information about reviews

sequentially and constitute a cascade of finite state transducers over annotations. The left-hand-side (LHS) of the rules consist of an annotation pattern description. The right-hand-side (RHS) consists of annotation manipulation statements.

In order to manage the 4.1 millions of reviews composing the dataset, we created a batch java program that uploaded the reviews as Document on CouchDB. After that, our program perform a HTTP GET request to database to obtain, for each single file, the text of the review, executing Gate on it, load Sentiment parameters and perform the Sentiment Analysis. The last step is to update the Document on CouchDB, performing an HTTP PUT request. The obtained information are structured in two fields:

Score, representing the sentiment of the entire review. It is the mean value of the single sentences score. Sentences, which is an array of sentences that generated Score (Figs. 6 and 7).

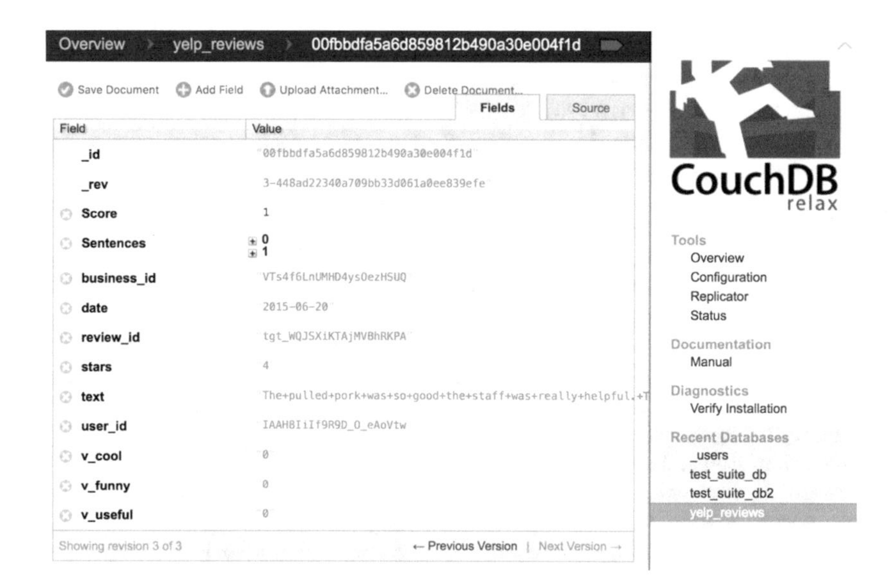

Fig. 7. Obtained output

4 Conclusions and Future Work

In this paper we described a data model for *Multimedia Social Networks*, extracting and modelling information about users. Inspired by hypergraph based approaches, our model provides a solution for representing MSNs sufficiently general with respect to: (i) a particular social information network, (ii) the different kinds of entities, (iii) the different types of relationships, (iv) the different applications.

We developed a methodology using a combination of modules applications provided by GATE NLP toolkit, that allows the extraction of relevant information from post related to the online social network Yelp.

As future work, we are planning to exploit the introduced ranking functions to support multimedia recommendation and influence analysis applications, in order to perform an experimental evaluation of the proposed model.

References

1. Moscato, F.: Exploiting model profiles in requirements verification of cloud systems. Int. J. High Perform. Comput. Networking **8**(3), 259–274 (2015)
2. Nan, G., Zang, C., Dou, R., Li, M.: Pricing and resource allocation for multimedia social network in cloud environments. Knowl.-Based Syst. **88**, 1–11 (2015)
3. Liu, D., Ye, G., Chen, C.-T., Yan, S., Chang, S.-F.: Hybrid social media network. In: Proceedings of the 20th ACM International Conference on Multimedia. ACM, pp. 659–668 (2012)
4. Zhang, Z., Wang, K.: A trust model for multimedia social networks. Soc. Netw. Anal. Min. **3**(4), 969–979 (2013)
5. Ji, X., Wang, Q., Chen, B.-W., Rho, S., Kuo, C.J., Dai, Q.: Online distribution and interaction of video data in social multimedia network. Multimedia Tools Appl. 1–14 (2014)
6. O'Donovan, F.T., Fournelle, C., Gaffigan, S., Brdiczka, O., Shen, J., Liu, J., Moore, K.E.: Characterizing user behavior and information propagation on a social multimedia network. In: 2013 IEEE International Conference on Multimedia and Expo Workshops (ICMEW), pp. 1–6. IEEE (2013)
7. Moscato, V., Picariello, A., Subrahmanian, V.: Multimedia social networks for cultural heritage applications: the givas project. In: Data Management in Pervasive Systems, pp. 169–182. Springer (2015)
8. Amato, F., Moscato, F.: A model driven approach to data privacy verification in e-health systems. Trans. Data Priv. **8**(3), 273–296 (2015)
9. Amato, F., Moscato, F.: Pattern-based orchestration and automatic verification of composite cloud services. Comput. Electr. Eng. **56**, 842–853 (2016)
10. Chianese, A., Benedusi, P., Marulli, F., Piccialli, F.: An associative engines based approach supporting collaborative analytics in the internet of cultural things, pp. 533–538 (2015)
11. Hong, M., Jung, J., Piccialli, F., Chianese, A.: Social recommendation service for cultural heritage. Pers. Ubiquit. Comput. **21**(2), 191–201 (2017)
12. Chianese, A., Piccialli, F.: SmaCH: A framework for smart cultural heritage spaces, pp. 477–484 (2015)
13. Della Vecchia, G., Gallo, L., Esposito, M., Coronato, A.: An infrastructure for smart hospitals. Multimedia Tools Appl. **59**(1), 341–362 (2012)
14. Essmaeel, K., Gallo, L., Damiani, E., De Pietro, G., Dipanda, A.: Comparative evaluation of methods for filtering kinect depth data. Multimedia Tools Appl. **74**(17), 7331–7354 (2015)
15. Albanese, M., d'Acierno, A., Moscato, V., Persia, F., Picariello, A.: A multimedia recommender system. ACM Trans. Internet Technol. (TOIT) **13**(1), 3 (2013)
16. Moscato, V., Picariello, A., Rinaldi, A.M.: Towards a user based recommendation strategy for digital ecosystems. Knowl.-Based Syst. **37**, 165–175 (2013)

17. Placitelli, A.P., Gallo, L.: Low-cost augmented reality systems via 3D point cloud sensors. In: 2011 Seventh International Conference on Signal-Image Technology and Internet-Based Systems (SITIS), pp. 188–192. IEEE (2011)
18. Colace, F., Santo, M.D., Greco, L., Amato, F., Moscato, V., Picariello, A.: Terminological ontology learning and population using latent dirichlet allocation. J. Vis. Lang. Comput. **25**(6), 818–826 (2014)
19. Cunningham, H., Maynard, D., Bontcheva, K., Tablan, V.: A framework and graphical development environment for robust NLP tools and applications. In: ACL, pp. 168–175 (2002)

Proposal of Continuous Remote Control Architecture for Drone Operations

Naoki Yamamoto[1(✉)] and Katsuhiro Naito[2]

[1] Graduate School of Business Administration and Computer Science, Aichi Institute of Technology, Nagoya, Aichi 464-0807, Japan
naoki-yamamoto@pluslab.org

[2] Department of Information Science, Aichi Institute of Technology, Toyota, Aichi 470-0392, Japan
naito@pluslab.org

Abstract. Drones have been considered for use in various fields according to the performance improvement and the price down of devices. They are expected for some applications: disaster relief, farm field, security field, transportation field, etc. Some companies will employ the autopilot system for their business. However, they have to switch to manual operation in case of emergency due to the autopilot safety is not guaranteed. Therefore, a pilot must connect with the drone continuously by the network for remote monitoring. Cellular network systems are the candidate networks for remote monitoring. However, typical design of cellular networks does not assume user equipment devices in the air because antennas of cellular networks are usually aimed downward to reduce inter-cell interference. This means that drones may fly out a communication area of cellular networks. Therefore, business drones must communicate with some cellular networks to keep continuous communication. However, IP-based application will disconnect due to change of cellular networks. As a result, practical business drones' operations require a continuous communication mechanism. This paper proposes a continuous remote control architecture for drone operations to improve safety of the autopilot function. The proposed architecture employs NTMobile technology as a seamless mobility protocol supporting continuous communication. Additionally, it also employs IP-based remote control application to control drones remotely. The evaluation system can acquire sensor information and exchange control information continuously when drones switch access networks. The proposed architecture can be a fundamental framework to realize a wide area drone operation service.

Keywords: Remote drone control · Cellular networks
Seamless mobility

1 Introduction

Drones have been used for various purposes due to price down and performance improvement. Drones are expected to be active in disaster relief, agriculture field,

G. De Pietro et al. (Eds.): KES-IIMSS-18 2018, SIST 98, pp. 64–73, 2019.
https://doi.org/10.1007/978-3-319-92231-7_7

security field and transportation field in business utilization [1]. In the agriculture field, condition of agricultural crops can be monitored with high precision cameras and various sensors. In addition, spraying pesticides by drones are under consideration [2–4]. In the security field, detecting and tracking system of suspicious individuals can be realized by multiple drones [5,6]. Furthermore, some companies try to demonstrate experiments on transportation using drones to convey some items in the transportation field [7,8]. In the above case, automatic control based on an autopilot function is required rather than manual operation.

The autopilot function can control a drone automatically by predefined way points indicating a flight path. A pilot need not operate the drone during the autopilot operation. However, the autopilot function may not work in unexpected behavior of a drone and environmental issues. Therefore, it is necessary to monitor the autopilot function remotely by drone sensor information and camera images. In addition, it is important to be able to switch control of drone into a pilot operation in case of emergency. Furthermore, a drone and a controller application need to be connected constantly in order to monitor the drone sensor information continuously [9].

Since wireless communication technology for drones has a limited communication range, so reliability and stability of communication cannot be guaranteed when it is used in urban areas. Therefore, a communication method using a cellular system is required to realize a wider operation [10]. In recent years, some demonstration service uses a cellular system to realize transportation to remote areas using drone [11,12].

The communication quality of each carrier varies depending on service design of cellular systems. In addition, communication quality may be degraded in the air because cellular systems typically assume that user equipment devices exist on the ground. Therefore, continuous communication scheme based on multiple carrier services is required. However, recent carrier service may employ a carrier grade NAT to reduce the required number of global IP addresses. In this situation, user equipment device should solve the NAT traversal problem. Additionally, each carrier service connects to a different core network. Therefore, an assigned IP address will change due to a change to a network address when the access network is switched. Typical application disconnects a session when an IP address changes due to switching of the access network. Since an application and a drone should be connected continuously to monitor and control, a continuous remote control architecture is required.

This paper proposes a remote control architecture of drones operating with a seamless mobility protocol to improve the safety of the autopilot function. As the seamless mobility protocol, we employed NTMobile that is developed by the authors [13]. The proposed architecture employs DJI N3 flight controller as a flight controller for a drone. The N3 flight controller has a serial communication interface that is connected to an on-board device. We employ Raspberry Pi 3 as the on-board device because our seamless mobility protocol is implemented on Linux OS. As the seamless mobility protocol, we install NTMobile on Raspbian OS for the drone and Linux OS for a remote access computer.

The evaluation system shows that we can access to the N3 flight controller from the remote access computer though NTMobile technology. NTMobile supports NAT traversal and seamless mobility due to change of an IP address. The proposed architecture does not require a modification of drone control application. Therefore, any applications can be used in the architecture. As a result, the prototype implementation is a candidate system model to solve the issues for the commercial drone use.

2 Drone's Control Mechanism

2.1 System Model of Drone Controller

Figure 1 shows the system model of the drone system. The drone system consists of an inertial measurement unit (IMU) for measuring an angle and an acceleration of the drone, an atmospheric pressure sensor for measuring the altitude, a GPS to identify the position, an electronic speed controller (ESC). The flight controller can control the attitude of the drone by using the ESC and the IMU. Moreover, drones are controlled according to a position based on the information from the Global Positioning System (GPS). Similarly, the drone system uses an atmospheric pressure sensor to maintain altitude.

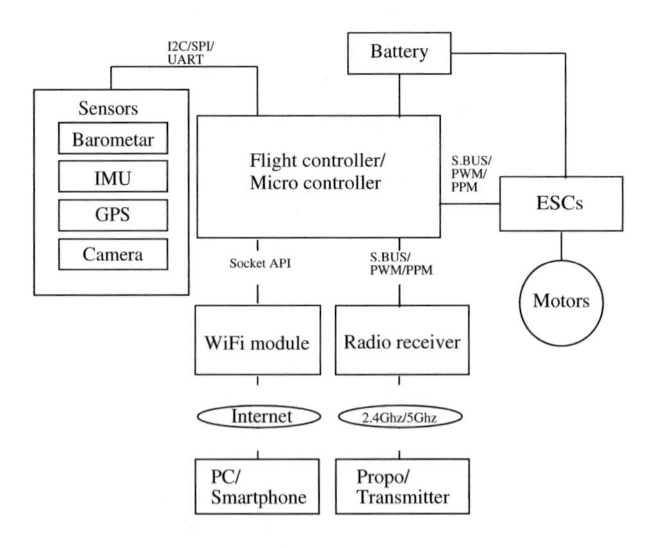

Fig. 1. System model of the drone controller

2.2 Remote Control Method

Remote Control over Wireless Communication. The controller uses the frequency band of 2.4 GHz band and 5 GHz band for remote communication between a drone and a remote controller. The 5 GHz band needs a license of amateur radio devices in Japan. On the contrary, the 2.4 GHz band is allocated to Industrial, Scientific and Medical (ISM) band. Therefore, typical drone systems employ 2.4 GHz bands even if communication stability is not maintained due to radio interference. Recently, some systems use spread spectrum technology such as FASST, S-FHSS and DMSS for communication.

A drone control system requires throttle (for rising and falling), ladder (for left and right turn), elevator (for forward and backward) and aileron (for moving left and right) for control. Therefore, the remote controller must be able to be allocated to more than 4 ch to control the drone. The receiver installed in the drone communicates with the motor system by wired communication. The wired communication control method is as follows.

1. PWM: The Pulse Width Modulation (PWM) is an analog signal that performs on/off switching and controls output power. It requires one signal line for one channel and has been used for radio control for a long time.
2. PPM: The Pulse-Position Modulation (PPM) can control eight motors with one signal line.
3. Serial: The serial is a digital signal communication. There are some standards such as S.BUS, MSP and XBUS. S.BUS can use 18 channels of signal in one cable.
4. S.BUS2: S.BUS 2 is an extensive standard of S.BUS. It supports a telemetry system that sends drone sensor information and remaining battery power from the receiver to the controller.

Remote Control over IP Networks. The method of using the IP network is commonly used like the remote controller. A remote controller that uses an IP network is usually an application on a smartphone. The pilot of a drone connects a drone and a smartphone with WiFi. The pilot generally uses the virtual button on smartphone for control of drones. This method has the advantage that it is not necessary to have a large special remote controller. However, communication distance is typically short because the transmission power of WiFi module is limited due to regulations.

The remote control application on the smartphone can display a sensor information and the remaining battery level of the drone. Furthermore, it is possible to watch the video in real time from a camera mounted on the drone. Therefore, the operator can maneuver the drone while watching the video. Currently, some drone uses WiFi module for local network connection with the controller.

2.3 Operation Management

Manual Control. The pilots generally operate the drone manually by using the remote controller. The remote controller generally has two joysticks. One of the joystick controls throttle and ladder, the other controls elevator and aileron. The remote controller also has several other buttons and can assign operations other than basic operations of the drone such as gimbal operation, camera shoot and takeoff. The flight controller operates the servomotor of the drone, according to a command from the remote controller. It uses GPS and sensor information to keep the current position, when there is no instruction from the remote controller.

Currently, the flight controller's performance is improving. Therefore, the pilot can control for drone without considering subtle attitude manipulation and wind influence.

Autopilot. An autopilot function is a flight method to operate a drone automatically. It needs to set way points before flying a drone. The way points indicate a flight path of a drone. It contains some information such as a latitude, a longitude and an altitude. Moreover, it is also possible to take a specific action, such as a takeoff and a landing, a camera shooting etc. The drone can send sensory information to a pilot using a telemetry system during an autopilot flight operation to communicable distance. The autopilot can be realized by the flight control device controlling a drone based on GPS information.

3 Proposed Architecture

Figure 2 shows the proposed architecture. The proposed architecture consists of the drone remote control application and the seamless mobility network. The drone remote control application exchanges flight data and control data over a special protocol. The seamless mobility network can convey packets of the drone remote control application. Therefore, the proposed architecture can support any drone remote control applications. The proposed architecture employs NTMobile to realize the seamless mobility network.

3.1 NTMobile

NTMobile is a seamless mobility architecture that can satisfy the mobility on various networks and the connectivity of terminals existing under NAT simultaneously. Figure 3 shows the system model of NTMobile. NTMobile consists of four elements: NTMobile nodes (NTM nodes) equipped with NTMobile, an Account Server (AS) to authenticate NTM nodes, a Direction Server (DC) to manage NTM nodes, and a Relay Server (RS) that relays communication under specific conditions. Description of the elements is shown below.

NTMobile realizes mobility by assigning a virtual IP address and Fully Qualified Domain Name (FQDN) to a NTM node. It conceals the change of the real IP address by notifying the virtual IP address for the application. The NTM node constructs a User Datagram Protocol (UDP) tunnel to the opposite node according to the instruction of the DC.

1. NTM node: The NTM node is a communication terminal equipped with the NTMobile function. It requests an authentification to AS and obtains a FQDN and a virtual IP address from DC. Moreover, it obtains the virtual IP address

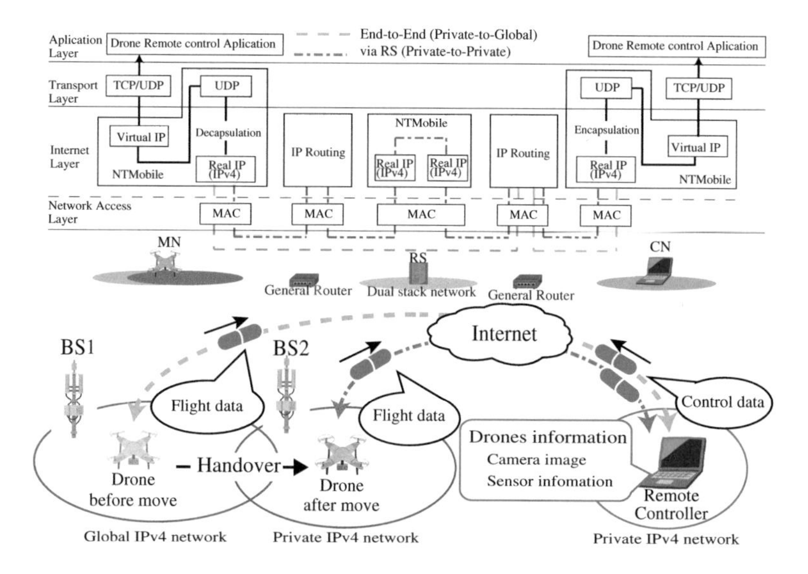

Fig. 2. Proposed architecture

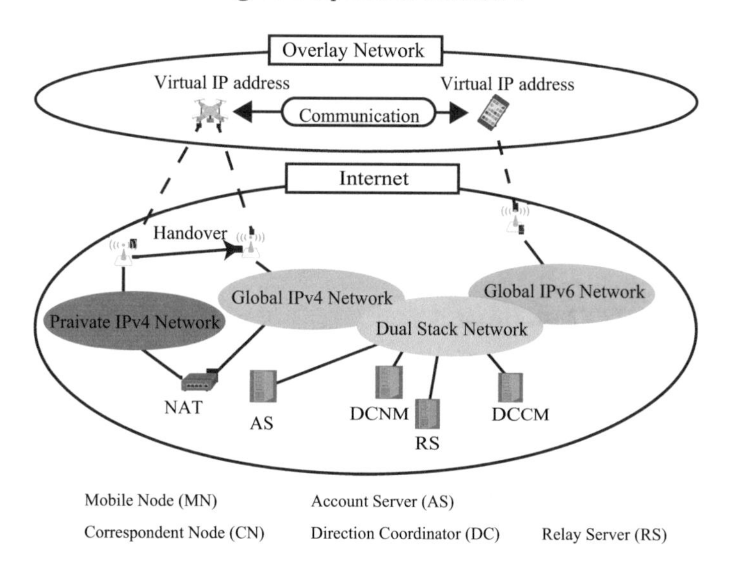

Fig. 3. NTMobile network model

of the partner by solving the FQDN of the communication partner. The application communicates with the obtained virtual IP addresses. The NTM node uses a dedicated module internally to communicate by associating the virtual IP address with the actual IP address.

2. AS: The AS distributes necessary information for information management and authentication processing of users. It also distributes necessary

information for encryption key information. The NTM node can obtain initial information for NTMobile by communicating with the AS.
3. DC: The DC assigns a virtual IPv6 address and FQDN to the NTM node. It instructs the route construction process to construct a UDP tunnel to the correspondent node when the NTM node starts communication. It is located on the dual stack networks. Therefore, any NTM nodes can connect to the DC from any network.
4. RS: RS has the role of relaying communication between NTM nodes. For example, NTM nodes cannot communicate directly when each node uses a different IP version network such as IPv4/IPv6 networks. The other use is relayed between NTM nodes behind a NAT router.

3.2 N3 Flight Controller

An N3 flight controller (N3) is Drone's flight controller released by DJI. The N3 is sold only for flight controllers, therefore users can freely create drone frames. Furthermore, it has major feature that it includes a physical API/CAN2 port. Therefore, it is possible to connect to an on-board device via the serial interface.

3.3 Onboard SDK

A DJI Onboard SDK is an open source software library. It enables direct communication between the computer and the flight controller of the DJI product via the serial interface. Moreover, it enables access to sensors, flight control, and other functions of a drone. Further, it is possible to control the camera and gimbal mounted on a drone. Developers can use the Onboard SDK to control the flight by installing the on-board devices in the drone.

3.4 Combine NTMobile and Onboard SDK

Figure 4 shows a N3 Onboard SDK system model diagram combining NTMobile. In the proposed architecture, the N3 only performs drone control. Moreover, the on-board devices operate drone and communication with a drone remote control application. The N3 connects to the on-board device via a serial interface. The on-board device sends an operation command to N3 by using the Onboard SDK. Onboard SDK and NTMobile are installed into the on-board device. In addition, the on-board device communicates with a controller that performs operation through NTMobile.

3.5 Implementation

This paper uses DJI F550 for the Drone's framework. In addition, we use Ubuntu 16.04 for a remote access computer that controls Raspberry pi 3. Moreover, we adopted Raspberry Pi 3 as an on-board device mounted on the drone. Raspberry Pi 3 is possible to install the onboard SDK and NTMobile in the same device. It connected with the N3 via the serial interface. Raspberry Pi 3 and the remote access computer install NTMobile.

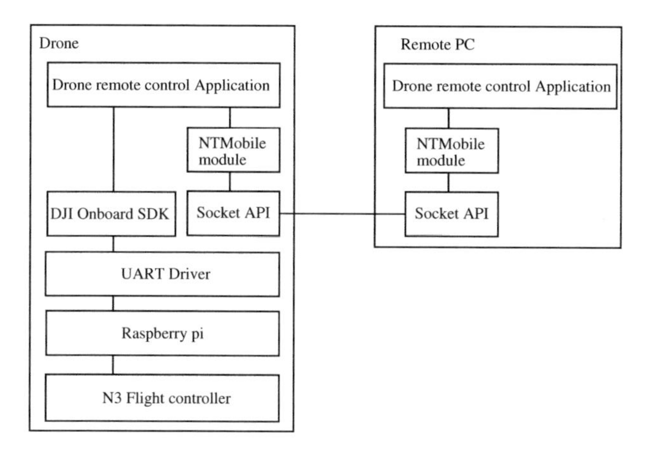

Fig. 4. Configuration diagram of NTMobile and Onboard SDK

4 Evaluation

The remote access computer accesses Raspberry Pi 3 via the NTMobile communication, and sends an operation command to the N3. Finally, we have confirmed that the N3 can receive a control message. In the evaluation, we used the simulator on the DJI Assistant 2 for flight experiment. The remote access computer and Raspberry Pi 3 logged into NTMobile network and conducted communication. It used Secure Shell to Raspberry pi 3 to launch the drone remote control application on Raspberry Pi 3, and operated the drone on the simulator.

Figure 5 shows the experimental environment and Table 1 shows the experimental data. The servers of NTMobile and remote access computer were implemented as virtual machines.

Table 2 shows the experimental results. We have evaluated the communication delay of NTMobile between the remote access computer and Raspberry Pi

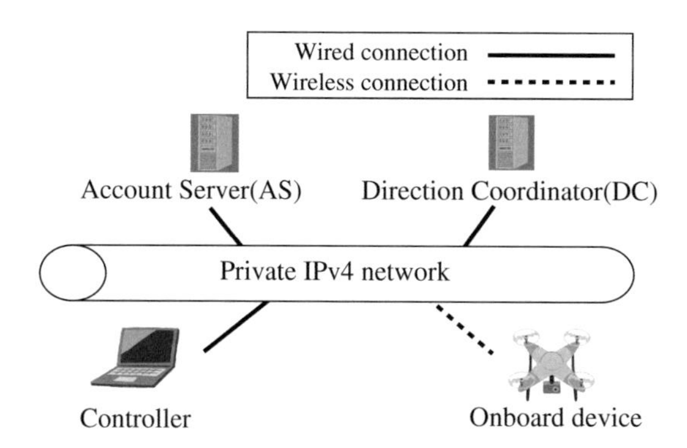

Fig. 5. Experiment environment

Table 1. Experimental data

Host Machine	
OS	OS X 10.9.5
CPU	Intel Core-i7 2.8Ghz
Memory	16 GBytes
Software	VirtualBox 5.0

Virtual Machine	
OS	Ubuntu 16.04
Memory	1 GBytes

Raspberry Pi 3	
OS	raspbian-2016-05-31
CPU	Broadcom BCM2837 1.2GHz 64-bit quad-core ARMv8 Cortex-A53
Memory	1 GBytes

Table 2. Result of UDP tunnel construction time and network throughput

UDP Tunnel building time	467.3 ms
Bandwidth(iperf)	1.517 Mbps
Rund Trip Time	73.9 ms

3. Therefore, we measured the processing time of a UDP tunnel construction in NTMobile. Moreover, we measured the throughput of NTMobile network. We assume a TCP application and evaluate the performance by Iperf. The evaluation results show the average value of ten time evaluations.

5 Conclusions

This paper has proposed the remote control architecture of drones' operating with a seamless mobility protocol to improve the safety of an autopilot function. Since the seamless mobility protocol can provide continuous communication to the drone remote control application, drone operators can monitor and control drones continuously. The evaluation system employed DJI N3 flight controller as a flight controller for a drone and Raspberry Pi 3 as the on-board device. The evaluation results demonstrated that we can access to the N3 flight controller from the remote access computer through NTMobile technology. The proposed architecture supports any drone remote control applications. Therefore, the prototype implementation is a candidate system model to solve the issues for the commercial drone use.

Acknowledgment. This work is supported in part by the collaborative research project with KDDI Research, Inc., Japan, Grant-in-Aid for Scientific Research (B)(15H02697) and (C)(17K00142), Japan Society for the Promotion of Science (JSPS), the Cooperative Research Project Program of the Research Institute of Electrical Communication, Tohoku University and the Hibi science foundation.

References

1. Camara, D.: Topology control of a network of autonomous aerial drones. URSI Radio Sci. Bull. **88**(4), 26–34 (2015)
2. Reinecke, M., Prinsloo, T.: The influence of drone monitoring on crop health and harvest size. In: 2017 1st International Conference on Next Generation Computing Applications (NextComp), pp. 5–10 (2017)
3. Murugan, D., Garg, A., Singh, D.: Development of an adaptive approach for precision agriculture monitoring with drone and satellite data. IEEE J. Sel. Top. Appl. Earth Obs. Remote Sens. **10**(12), 5322–5328 (2017)
4. Yallappa, D., Veerangouda, M., Maski, D., Palled, V., Bheemanna, M.: Development and evaluation of drone mounted sprayer for pesticide applications to crops. In: 2017 IEEE Global Humanitarian Technology Conference (GHTC), pp. 1–7 (2017)
5. Shrit, O., Martin, S., Alagha, K., Pujolle, G.: A new approach to realize drone swarm using ad-hoc network. In: 2017 16th Annual Mediterranean Ad Hoc Networking Workshop (Med-Hoc-Net), pp. 1–5 (2017)
6. Pobkrut, T., Eamsa-ard, T., Kerdcharoen, T.: Sensor drone for aerial odor mapping for agriculture and security services. In: 2016 13th International Conference on Electrical Engineering/Electronics, Computer, Telecommunications and Information Technology (ECTI-CON), pp. 1–5 (2016)
7. Ackerman, E., Strickland, E.: Medical delivery drones take flight in east Africa. IEEE Spectr. **55**(1), 34–35 (2018)
8. McFariand, M.: Google drones will deliver chipotle burritos at Virginia Tech. CNN Money (2016)
9. Katila, C.J., Di Gianni, A., Buratti, C., Verdone, R.: Routing protocols for video surveillance drones in IEEE 802.11s Wireless Mesh Networks. In: 2017 European Conference on Networks and Communications (EuCNC), pp. 1–5 (2017)
10. Koubâa, A., Qureshi, B., Sriti, M.F., Javed, Y., Tovar, E.: A service-oriented cloud-based management system for the Internet-of-Drones. In: 2017 IEEE International Conference on Autonomous Robot Systems and Competitions (ICARSC), pp. 329–335 (2017)
11. Qualcomm, Inc.: Evolving cellular technologies for safer drone operation (2016)
12. Amorim, R., Nguyen, H., Mogensen, P., Kovács, I.Z., Wigard, J., Sørensen, T.B.: Radio channel modeling for UAV communication over cellular networks. IEEE Wirel. Commun. Lett. **6**(4), 514–517 (2017)
13. Naito, K., Nishio, T., Mori, K., Kobayashi, H., Kamienoo, K., Suzuki, H., Watanabe, A.: Proposal of seamless IP mobility schemes: network traversal with mobility (NTMobile). In: 2012 IEEE Global Communications Conference (GLOBECOM) (2012)

Development of Field Sensor Network System with Infrared Radiation Sensors

Masatoshi Tamura[1](✉), Takahiro Nimura[2], and Katsuhiro Naito[2]

[1] Graduate School of Business Administration and Computer Science, Aichi Institute of Technology, Nagoya, Aichi 464-0807, Japan
masatoshi-tamura@pluslab.org

[2] Department of Information Science, Aichi Institute of Technology, Toyota, Aichi 470-0392, Japan
{taka,naito}@pluslab.org

Abstract. Information technology has been focused to estimate growth degree of plants in agriculture. This paper focuses on leaf temperature that changes according to the activity of photosynthesis. Infrared cameras are a major method to measure leaf temperature in conventional methods. However, the expensive device price causes difficulty to install many sensors in practical fields. Infrared radiation sensors are new candidate device to estimate growth state by measuring leaf temperature. Since the price of infrared radiation sensors is inexpensive, we can install a lot of sensors into fields. Additionally, the consumed power of infrared radiation sensors is relatively small comparing to Infrared cameras. These features of infrared radiation sensors are appropriate for sensor networks working with a battery. This paper proposes a field sensor network to measure growth state of plants by infrared radiation sensors. Our goal is to realize a practical and inexpensive sensor network system with typical system on chip (SoC). Therefore, we employ a reasonable price SoC supporting IEEE 802.15.4 standard to design a unique device with various sensors. In order to realize multi-hop communication with low-power consumption, we propose a routing and media access control mechanisms for the developed system. The media access control technology realizes periodic sleep operation of all devices to enable long-term operation of the system. The routing control technology can construct a multi-hop network with the minimum number of hops. The experimental results demonstrated that the development system works in the practical fields.

Keywords: Wireless sensor networks · Field sensing
Multi-hop communication · Smart agriculture · Wireless module SoC

1 Introduction

Low agricultural productivity in Japanese agriculture has been focused as a big issue to realize sustainable agriculture. Producing high-value crops are one idea to increase the agricultural productivity. On the contrary, farmers should

G. De Pietro et al. (Eds.): KES-IIMSS-18 2018, SIST 98, pp. 74–83, 2019.
https://doi.org/10.1007/978-3-319-92231-7_8

store various knowledge to improve their farming methods. Traditional way is empirical knowledge in their farming. In recent years, smart agriculture attracts attention in order to solve such issues [1,2].

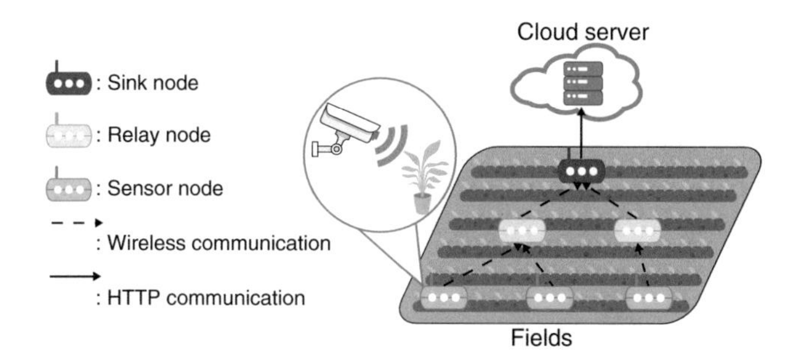

Fig. 1. System model of the field sensor network system

Smart agriculture can reduce the workloads of the farmers and high quality production by using advanced technologies such as robotic technology and information communication technology (ICT). Wireless sensor networks (WSN) are a candidate method to collect environmental information and to measure the growing state of crops [3–5]. Since there is no power supply in the typical fields, WSN should realize a battery-based operation by reducing consumed power [6–8]. In the reduction of consumed power, a hardware selection is the most important part because an operation period depends on consumed power during a deep sleep state. As the software, routing protocols and media access control mechanisms are dominant factors because they have a great influence on the deep sleep period [9,10]. In recent years, mobile sinks have been actively researching. Mobile sinks moves through a network to collect data of devices within a communication range [11,12]. On the contrary, typical agricultural fields are not suitable for mobile devices due to dirt trails.

As the measurement of growing state of crops, leaf temperature has been focused recently. Traditionally, infrared cameras are most methods to measure the leaf temperature. On the contrary, the expensive price of the infrared cameras is a big issue to install them into practical fields. Recently, infrared radiation sensors have been released. The benefits of infrared radiation sensors are inexpensive price and low-power consumption of the device even if the infrared radiation sensors have lower resolution than the infrared camera.

This paper proposes a field sensor network to measure growth state of plants by infrared radiation sensors. Our goal is to realize a practical and inexpensive sensor network system with typical system on chip (SoC). Therefore, we employ a reasonable price SoC supporting IEEE 802.15.4 standard to design a unique device with various sensors: infrared radiation sensors, temperature sensors, humidity sensors, air pressure sensors etc. In order to realize multi-hop communication with low-power consumption, we propose a routing and media access control mechanisms for the developed system. The media access control technology realizes periodic sleep operation of all devices to enable long-term operation of the system. The routing control technology can construct a multi-hop network with the minimum number of hops. The proposed system consists of three device types: sensor nodes, relay nodes, and a sink node. The sensor nodes observe leaf temperature and environment and communicate with other nodes to collect observation data. The relay nodes implement a communication function for forwarding the observation data from the sensor nodes to the sink node. The sink node has the internet connection to upload the observation data to the cloud server. The experimental results demonstrate that the development system works in the practical fields.

2 Proposed System

2.1 Field Sensor Network System

Figure 1 shows the proposed system model of the field sensor network system. Our goal is to realize a practical and inexpensive sensor network system with typical system on chip (SoC) because SoC works with a small amount of electricity. Our sensor devices support various sensors: infrared radiation sensors, temperature sensors, humidity sensors, air pressure sensors etc. The sensor network system consists of sensor nodes, relay nodes, and a sink node. We assume the sensor network system uploads sensor information to a cloud service. The sensor nodes have two functions: measurement with sensors and uploading sensor information toward the sink node. The relay nodes just forward sensor information from the sensor nodes or another relay node to the sink node. The sink node also has two functions: receiving of sensor information and uploading it to the cloud service.

2.2 Frame Structure

In the proposed communication protocol, all nodes operate according to a time frame format shown in Fig. 2. The purpose of the frame format is to avoid collisions among neighbor nodes and manage sleep operations. By synchronizing the time of all nodes, all nodes are operated according to the same time frame

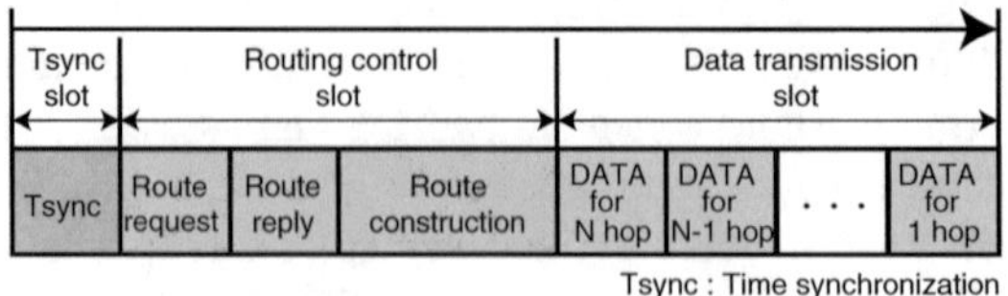

Fig. 2. Time frame format

format. This control realizes the sleep operations of each node. The time synchronization slot for time synchronization in the network is defined at the beginning of the time frame format. The sink device broadcasts special information for time synchronization. Therefore, each node can synchronize its local time according to its upstream node. When a node synchronizes the time, it also operates according to the time frame format. The routing control slot is used to construct a tree-structured network. The tree-structured network is suitable to reduce routing control packets and required routing information. In a tree-structured network, each node has only parent node and child node information, so each node can manage a route with a small amount of routing information. When a node completes the construction of a route to its upstream node, it also starts data transmission in the data slot. Hereinafter, these three slots will be described below.

- Tsync (Time synchronization) slot
  Figure 3 shows the example of an operation in the Tsync slot. In the Tsync slot, the sink node transmits a Tsync packet storing elapsed time information of the network to its own child node. The child node receiving the Tsync packet performs the time synchronization with the parent node based on the elapsed time information included in the packet and the propagation delay time. Then, the child node that received the Tsync packet further relays the Tsync packet to its own child node. This operation is repeated until the leaf node completes the time synchronization.
- Routing control slot
  Figure 4 shows an example of an operation in the routing control slot. The routing slot consists of three slots: RREQ (Route REQuest) slot, RREP (Route REPly) slot, and RCON (Route CONstruction) slot. In the routing control slot, a node without routing information constructs a route to a node with routing information. The route construction process prevents a redundant route by considering the number of hop count to the sink node. The routing control realizes construction of the tree-structured network with a minimized number of hops.
- Data transmission slot
  Figure 5 shows the example operation in the data transmission slot. In the data transmission slot, the maximum number of hops from the leaf nodes to the sink node is defined as N and it is divided into N sub-slots to control the timing of data transmission. Since N depends on the size of the field, it is definitely taking into consideration the size of the field. In the first sub-slot, the N hop nodes start transmitting a packet. In addition, the data packets are transmitted from nodes with a large number of hops in order to avoid packet collisions among neighbor nodes. When there are multiple nodes of the same hop number, the data transmission timing is controlled by random delay of each node. Then, each node sleeps until the next frame starts when its own data transmission slot is over.

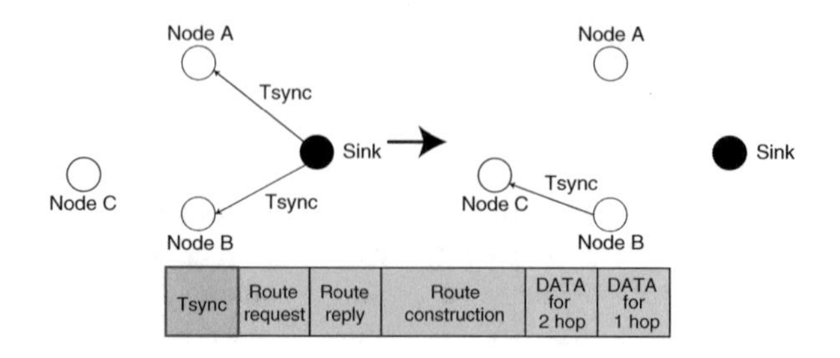

Fig. 3. Time synchronization slot

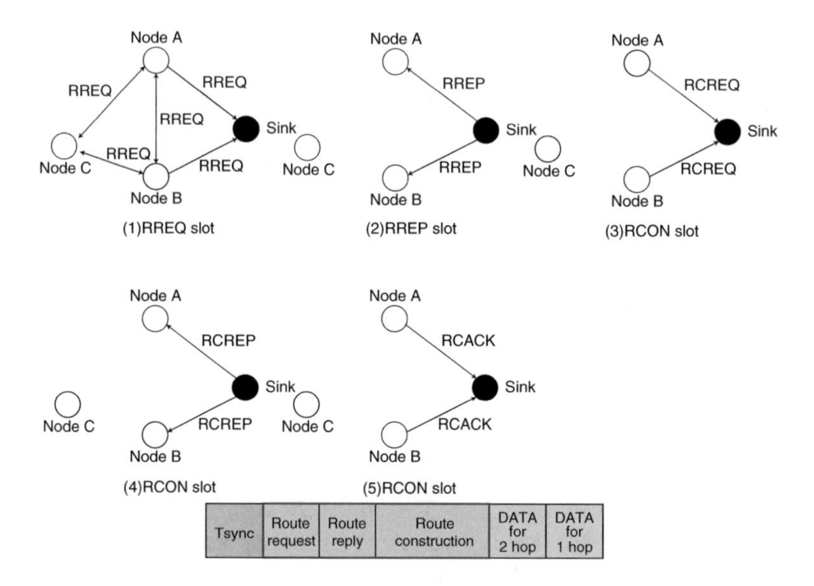

Fig. 4. Routing control slot

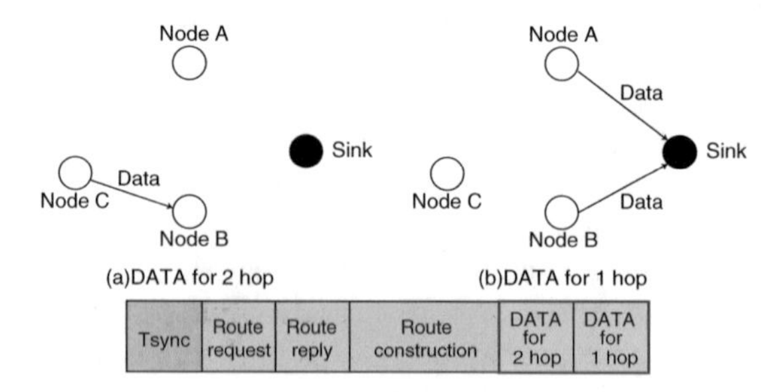

Fig. 5. Data transmission slot

2.3 Routing Control

Figure 4 shows the example of the routing control. The sink node has an only node with routing information when the system starts up. all nodes: Node A, B, and C broadcast a RREQ message in the RREQ slot because they do not have any routing information. At first step, nodes A and B construct a route to the sink node because they can communication with the sink node directly.

In the RREP slot, the sink node returns a RREP message to Node A and Node B. The proposed routing mechanism employs hop-based routing. Therefore, the minimum hop count node is selected as an upstream node. Since the sink node is 0 hop node, Node A and Node B should construct a route to the sink node even if some neighbor nodes have routing information.

Then, Node A and Node B send an RCREQ (Route Construction REQuest) message in the RCON slot to start the route construction process with the sink node. When the sink node receives the RCREQ message, it sends an RCREP (Route Construction REPly) message to Node A and Node B respectively. Finally, Node A and Node B confirm that the route has been constructed by sending an RCACK (Route Construction ACKnowledgement) message to the sink node. The above process completes the route construction. Node C constructs a route in the same process in the next time frame. As a result, our routing protocol can construct a tree-structured network from a sink node.

3 Implementation

3.1 Device Design

Figure 6 shows the developed the sink device, the relay device and the sensor device. In this paper, TWELITE DIP manufactured by Mono Wireless Inc. is used for wireless communication module and MLX 90621 manufactured by Melexis Inc. is used for infrared radiation sensors.

Figure 7 shows TWELITE DIP. TWELITE DIP is a wireless microcomputer module conforming to the wireless communication standard of IEEE 802.15.4. TWELITE DIP can operate at current consumption of about 7.2 mA in normal mode, but it can reduce current consumption up to 0.1 μA in deep sleep mode. TWELITE DIP is designed with system on chip (SoC), which has a role as a microcomputer and a wireless communication function on one chip. Therefore, miniaturization of the device and the omission of wasteful communication procedures with other modules can realize.

Figure 8 shows MLX90621. MLX 90621 is infrared radiation sensors that recognize the temperature of the object by 16 $\times$ 4 array pixels. The power supply voltage is about 2.6 V, and it can operate with current consumption less than 9 mA.

Each developed device has a role, but the relay device only relays the data. Therefore, the relay device consists only of TWELITE DIP and battery. Conversely, the sensor device and the sink device have their own implementation. Hereinafter, the design and the role of the sensor device and the sink device will be described below.

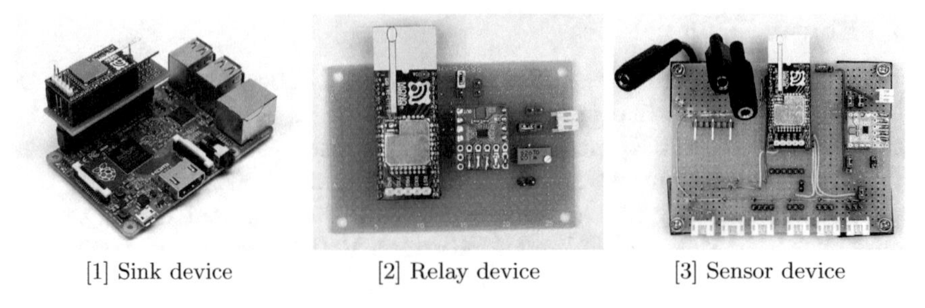

[1] Sink device [2] Relay device [3] Sensor device

Fig. 6. Three types of devices

Fig. 7. TWELITE DIP

Fig. 8. MLX90621

3.2 Sensor Device

The sensor device developed a unique board that can connect not only TWELITE DIP and MLX 90621 but also several kinds of sensors. Connectable sensors are various sensors such as temperature sensors, humidity sensors, atmospheric pressure sensors and soil moisture sensors. Sensor devices that have completed sensing realize power saving of devices by cutting off power supply to those sensors. Since various sensors are connected to the sensor device, the observation data acquired by the sensor device exceed the packet size of TWELITE DIP conforming to IEEE 802.15.4. Therefore, the sensor device implements fragmentation of data where observation data is divided into some packets.

3.3 Sink Device

The sink device is composed of TWELITE DIP for wireless communication and Raspberry Pi for sending the collected data to the cloud server. The sink device periodically acquires the time information from the NTP server through Raspberry Pi and transmits the time information to the TWELTE DIP by serial communication. When the sink device receives time information, it broadcasts the received time information to other nodes in the network. As a result, the sink device realizes sharing of time information in the network. In addition, the sink device transmits the observation data collected from the nodes in the network to the raspberry Pi by serial communication. The observation data sent to Raspberry Pi are converted to JSON format and registered in the cloud server. Therefore, the user can confirm the collected observation data by accessing the cloud server at any time.

4 Experiment and Results

In this experiment, we conducted a communication experiment in the actual fields using the communication protocols and devices developed in this paper. Figure 9 shows the arranged devices. In addition, Table 1 shows the specifications of the experiment.

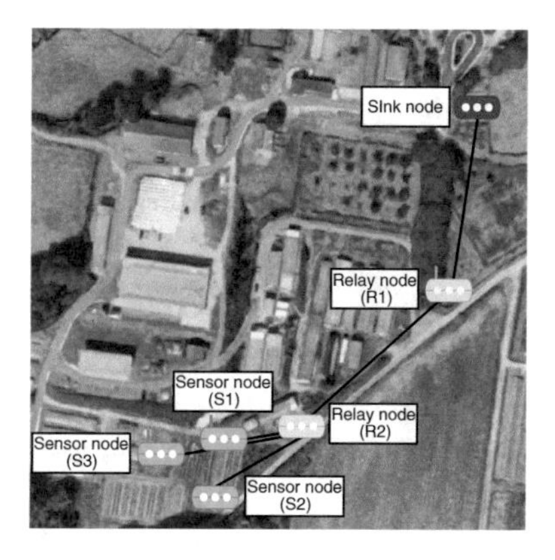

Fig. 9. Experimental field

Table 1. Specification

Experiment schedule	Jan. 16, 2018–Jan. 23, 2018
Frequency band	2.4 GHz
Transmission power	2.5 dbm
Number of retransmissions	3
Communication range between nodes	50–100 m

Table 2. Packet arrival rate

Relay node (R1)	Relay node (R2)	Sensor node (S1)	Sensor node (S2)	Sensor node (S3)
88.80%	88.73%	86.90%	88.47%	85.48%

In this experiment, observation data of three sensor nodes surrounding a field are collected to a sink node via two relay nodes. At that time, the remaining battery level of the relay node is also collected as observation data. Then, we evaluate the packet arrival rate from the total of five nodes: two relay nodes and three sensor nodes.

Table 2 shows the packet arrival rate of each node. As a result, the arrival rate from all nodes was in the higher 85% range. The experimental results demonstrate that the development system works in the practical fields. However, more higher packet arrival rate can reduce consumed power. The log data in the cloud server show that observation data from all nodes may not arrive in some cases. The reason of this problem is considered to be that the packet from R1 did not reach the sink node.

The distance between the sink node and the R1 was only 100 m. However, we confirmed that there was a slope and the difference in altitude was about 7 m. Therefore, we thought that the reason of a low packet arrival rate by all nodes is due to the difference in altitude between nodes.

In addition, the packet arrival rate of S1 and S3 was 3% lower than that of other nodes. The reason for this may be that there were shields such as weeds between R2 and S1 or S2. In addition, the line of sight communication was possible between R2 and S2. As a result, we confirmed that obstacles obviously affect communication between nodes.

Throughout this experiment, we confirmed that the difference in altitude and the obstructions have a great influence on the proposed system. However, improving the arrangement method of the device and the route construction method can solve this problem. In order to solve this problem, we need to conduct further field experiments.

5 Conclusion

This paper has proposed the field sensor network to measure growth state of plants by infrared radiation sensors. The proposed mechanisms can realize sleep operation and packet collision avoidance by using an original time frame format. The experimental results showed that the installation environment of the device influences communication between devices. However, we can solve these issues by arranging a location of each device.

Acknowledgment. This work is supported in part by the collaborative research project with KDDI Research, Inc., Japan, Grant-in-Aid for Scientific Research (B)(15H02697) and (C)(17K00142), Japan Society for the Promotion of Science (JSPS), the Cooperative Research Project Program of the Research Institute of Electrical Communication, Tohoku University and the Hibi science foundation.

References

1. Channe, H.: Multidisciplinary Model for Smart Agriculture using Internet of Things (IoT), Sensors, Cloud- Computing, Mobile-Computing & Big-Data Analysis. Sukhesh Kothari, Dipali Kadam Assistant Professors, Department of CE, PICT, Pune, India. Int. J. Computer Technology & Applications, vol. 6 (2015)
2. Dlodlo, N., Kalezhi, J.: The Internet of Things in agriculture for sustainable rural development. In: 2015 International Conference on Emerging Trends in Networks and Computer Communications, Windhoek, Namibia, pp. 13–18 (2015)
3. Langendoen, K., Baggio, A., Visser, O.: Murphy loves potatoes: experiences from a pilot sensor network deployment in precision agriculture. In: 14th International Workshop on Parallel and Distributed Real-Time Systems (WPDRTS), pp. 1–8 (2006)
4. Boselin Prabhu, S.R., et al.: Environmental monitoring and greenhouse control by distributed sensor network. Int. J. Adv. Netw. Appl. **5**, 2060–2065 (2014)
5. Anisi, M.H., Abdul-Salaam, G., Abdullah, A.H.: A survey of wireless sensor network approaches and their energy consumption for monitoring farm fields in precision agriculture. Precis. Agric. **16**(2), 216–238 (2015)
6. Bushnag, A., Alessa, A., Li, M., Elleithy, K.: Directed diffusion based on weighted grover's quantum algorithm (DWGQ). In: Systems, Applications and Technology Conference (LISAT), pp. 1–5 (2015)
7. Zaman, N., Low, T.J., Alghamdi, T.: Energy efficient routing protocol for wireless sensor network. In: 2014 16th International Conference on Advanced Communication Technology (ICACT), pp. 808–814 (2014)
8. Liu, T.H., Yi, S.C., Wang, X.W.: A fault management protocol for low-energy and efficient wireless sensor networks. J. Inf. Hiding Multimedia Sig. Process. **4**, 34–45 (2013)
9. Hoque, A., Amin, S.O., Alyyan, A., Zhang, B., Zhang, L., Wang, L.: NLSR: named-data link state routing protocol. In: Proceedings of the 3rd ACM SIGCOMM Workshop on Information-centric Networking, pp. 15–20 (2013)
10. Kiani, F., Amiri, E., Zamani, M., Khodadadi, T., Abdul Manaf, A.: Efficient intelligent energy routing protocol in wireless sensor networks. Int. J. Distrib. Sensor Netw. **2015**, 1–3 (2015)
11. Tunca, C., Isik, S., Donmez, M., Ersoy, C.: Distributed mobile sink routing for wireless sensor networks: a survey. IEEE Commun. Surveys Tuts. **16**, 877–897 (2014)
12. Khan, A.W., Abdullah, A.H., Razzaque, M.A., Bangash, J.I.: VGDRA: a virtual grid-based dynamic routes adjustment scheme for mobile sink-based wireless sensor networks. IEEE Sens. J. **15**, 526–534 (2015)

Detection of Mistaken Foldings Based on Region Change of Origami Paper

Hiroshi Shimanuki[1](✉), Toyohide Watanabe[2], Koichi Asakura[3], and Hideki Sato[3]

[1] Tokyo, Japan
simanuki@gmail.com
[2] Nagoya Industrial Science Research Institute, Nagoya, Japan
watanabe@nagoya-u.jp
[3] Daido University, Nagoya, Japan
{asakura,hsato}@daido-it.ac.jp

Abstract. This paper proposes an approach to detect mistaken foldings in computer-aided origami by using a single top-view camera. The position of the origami paper is continually tracked by matching the shape of the paper with the state model of the folding process. Further, the change of the shape is also grasped, and the folder's mistake is pointed out online. In this paper, first, we define the change ratio of the shape of the paper to measure the progress of foldings. Next, a criterion expressing the degree of difference from the correct shape is introduced to detect the mistake. Finally, threshold measurements for those criteria are investigated. Several experimental results showed the validity of our approach.

Keywords: Computer-aided origami · Mistake detection
Single camera · Top-view image · Augmented reality (AR)

1 Introduction

Origami is the art of paper folding in Japan. Japanese children do origami at home with family members and learn it at preschool and elementary school. It is particularly important for origami beginners to learn how to fold paper precisely at each step.

People often do origami while looking at illustrated instruction diagrams in origami drill books, which explain the process of folding sheets of square paper into completed origami models. However, it is not always easy for beginners to understand the diagrams and fold correctly by themselves. The beginners sometimes continuing folding despite making some mistakes, so the folded models predictably end up having the wrong shapes.

The folding mistakes are roughly divided into three categories as Fig. 1. As can be seen by these examples, outer shapes of origami paper folded correctly

G. De Pietro et al. (Eds.): KES-IIMSS-18 2018, SIST 98, pp. 84–92, 2019.
https://doi.org/10.1007/978-3-319-92231-7_9

are often different from outer shapes folded incorrectly. Therefore, this paper examines the possibility to detect the folding mistakes by using only outer shapes (silhouette). So, folding mistakes that hardly change outer shapes are exempt from this paper.

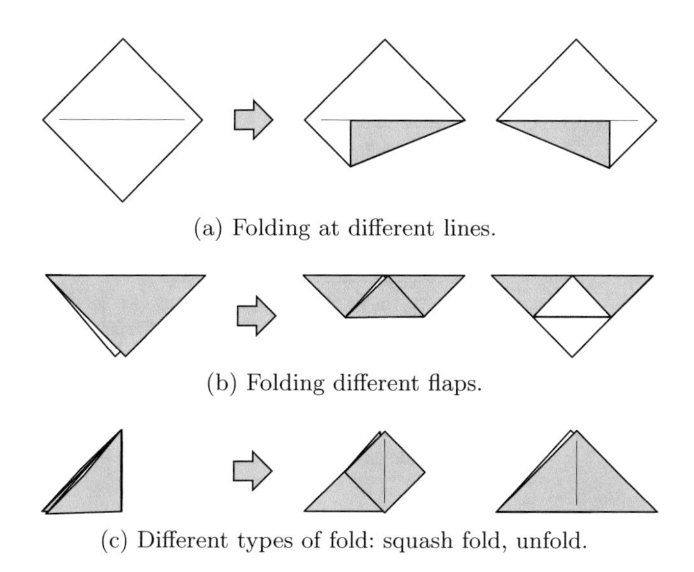

(a) Folding at different lines.

(b) Folding different flaps.

(c) Different types of fold: squash fold, unfold.

Fig. 1. Example of difference ways to fold from a origami model.

We have proposed an origami support system that recognizes shapes of origami paper on single top-view camera images and displays how to fold the paper onto the images interactively [1,4]. This system maintains an instruction model that represents folding processes constituted from only correct foldings, not incorrect ones. In this paper, we propose a method to detect mistakes in the user's folding on the basis of two criteria. One is the change ratio of the origami paper, which expresses progress of foldings. The other is a folding mistake ratio defined on the basis of the degree of difference from the correct shape. The system judges that a user has made a mistake if the mistake ratio is larger than a threshold. Therefore, the threshold is adjusted by analyzing correct and incorrect foldings in a preliminary experimentation.

2 Origami Support System

Figure 2 gives the framework of the origami support system. An instruction model that records folding processes necessary to complete origami works is manually constructed on the basis of the instruction diagrams in general drill books in advance. First, corners of the user's origami paper are extracted. Second, the location of the origami paper is estimated by using a data structure called a silhouette model that represents outer shapes of origami paper. Third, the state of the user's paper is estimated, and the validity of the performed folding is analyzed. Finally, instruction contents are overlaid onto the user's paper, so the user is able to fold accurately by looking at the given contents.

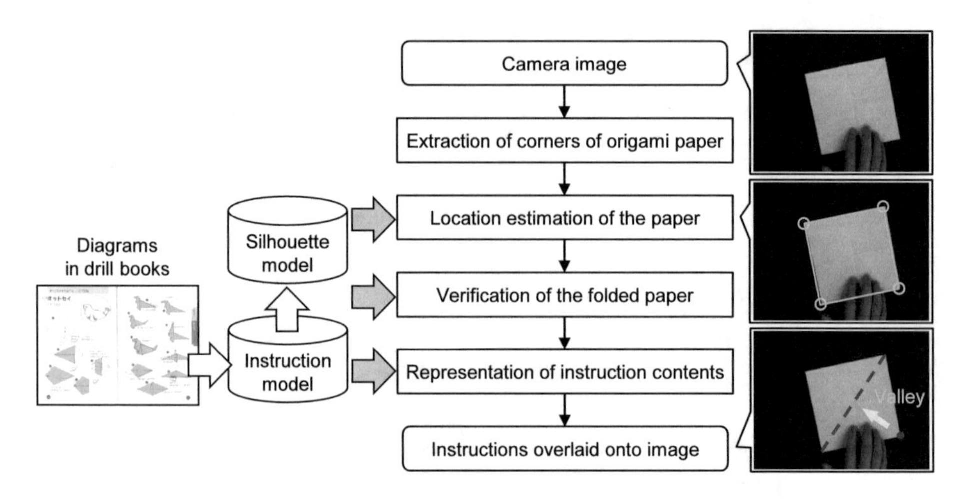

Fig. 2. Framework of origami support.

2.1 Instruction Model

In order to constitute a sequence of origami folds, a data structure defined by Miyazaki et al. [2] is used. The structure describes update history information of origami models as geometric objects (such as vertices, edges and faces) and records the overlap order of coplanar faces. On the basis of the data structure, all instruction contents to complete origami models are constituted as an instruction model. Figure 3 shows an example folding process. The node represents each state of the origami paper, and the arrow means state transition by a folding operation (OP).

2.2 Silhouette Model

In order to estimate the state of the user's origami paper fast enough, the shape of the paper and the instruction model are matched by using a silhouette of the model.

A silhouette is constituted by a projection transform of one state in the instruction model and consists of vertices and edges. A silhouette at state m is represented by $s_m = (V_m, E_m)$, where V_m and E_m denote sets of vertices and edges, respectively. One vertex has 2-D coordinates and is connected with only two edges. There are two types of vertices: one is based on the vertex in the

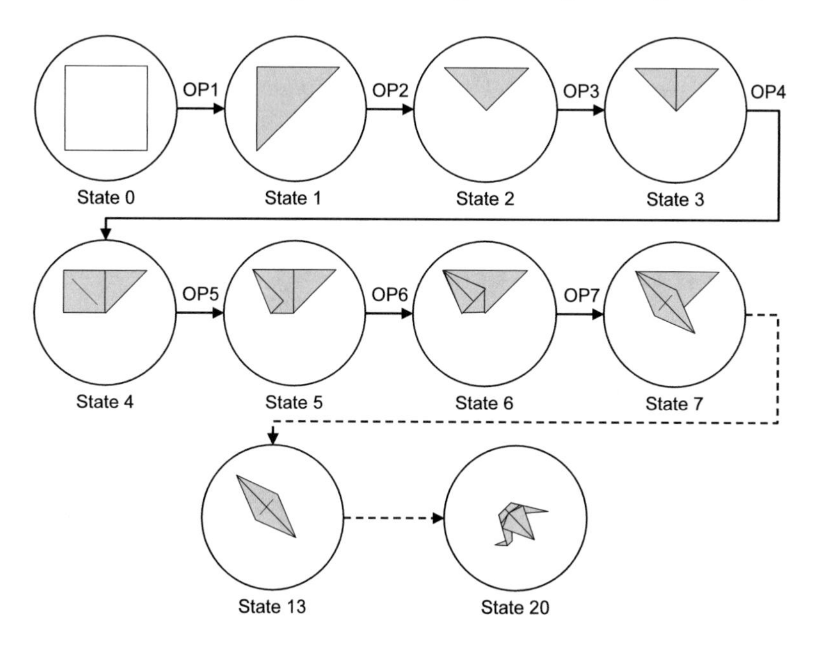

Fig. 3. Example folding process for a crane.

instruction model, and the other is an intersection point of edges in the instruction model. Edges of a silhouette do not intersect, and the graph is expressed as a simple cycle.

3 Tracking of Origami Paper

As preprocessing for location estimation of the user's origami paper, both segmentation and feature point extraction are applied to camera images [1,4]. Camera images are automatically segmented into three regions: background, user's hands, and a sheet of origami paper. First, the background region is segmented. In our system, the background is defined as the darkest region, of which pixel values are darker than a threshold. Figure 4(b) represents an example of detecting the background region (white pixels) in the image in Fig. 4(a). Next, the region of user's hands is extracted by using general detection techniques for human skin color (Fig. 4(c)). Since there are some increases in noise around a hand region, the region is made larger than the detected region on the basis of the morphological dilation. Finally, the remaining region is extracted as the region of origami paper as shown in Fig. 4(d). Let B_t be the background region at time t, and H_t be the region of user's hands. The region of the paper P_t at time t is developed as follows:

$$P_t = \overline{B_t \cup H_t}.$$

That is, origami paper of any color except for dark colors and skin colors can be used in our system.

In order to recognize the position and the direction of user's origami paper, corners of the paper are extracted as feature points from camera images by using the method for detecting intersection points among edges. Furthermore, only the points that are located onto contour of the origami paper are chosen among the detected feature points. Many of the unneeded points are intersection points between user's hands and others. Therefore, the points contained within the user's hands region H_t are removed, so only correct feature points contained within the paper region P_t are extracted.

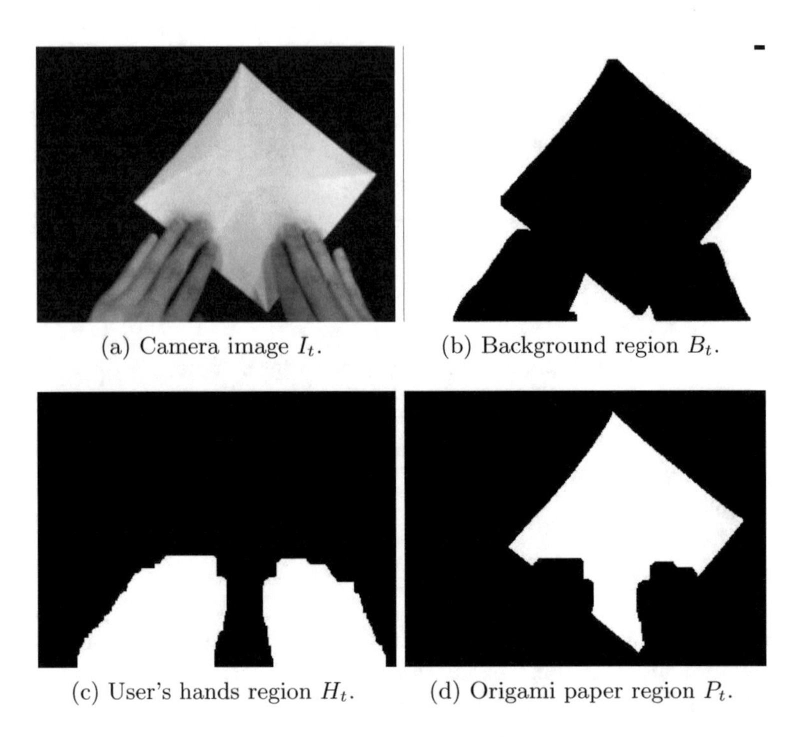

(a) Camera image I_t. (b) Background region B_t.

(c) User's hands region H_t. (d) Origami paper region P_t.

Fig. 4. Segmentation.

3.1 Location Estimation of Origami Paper

By matching among the feature points extracted from a camera image, a silhouette s_m at the current step, and another silhouette s_{m+1} at the next step, the position and the direction of the user's origami paper are estimated. The more vertices in a silhouette, the longer the time required to match. Therefore, Kinoshita et al. proposed a simple and effective matching algorithm based on the degree of similarity of shapes (overlap degree) [1,4]. The overlap degree is defined as follows:

$$R(s_m, I_t) = \frac{|S_m \cap P_t| + \alpha\ |S_m \cap H_t|}{|S_m \cup P_t|}$$

where S_m is the region enclosed by edges of a silhouette $s_m = (V_m, E_m)$ and I_t is a camera image at time t. Moreover, $|X|$ is the area of region X, and α is the weight value of the noise caused by hands.

The position and direction of the silhouette s_m for which $R(s_m, I_t)$ attains the maximum value are discovered for each camera image. Figure 5 shows an example of estimating the location of a user's paper while a square is folded into a triangle. The graph represents variations of maximum values of the overlap degrees for silhouettes s_m and s_{m+1}. Thus, if the user folds correctly, the origami paper can be tracked accurately.

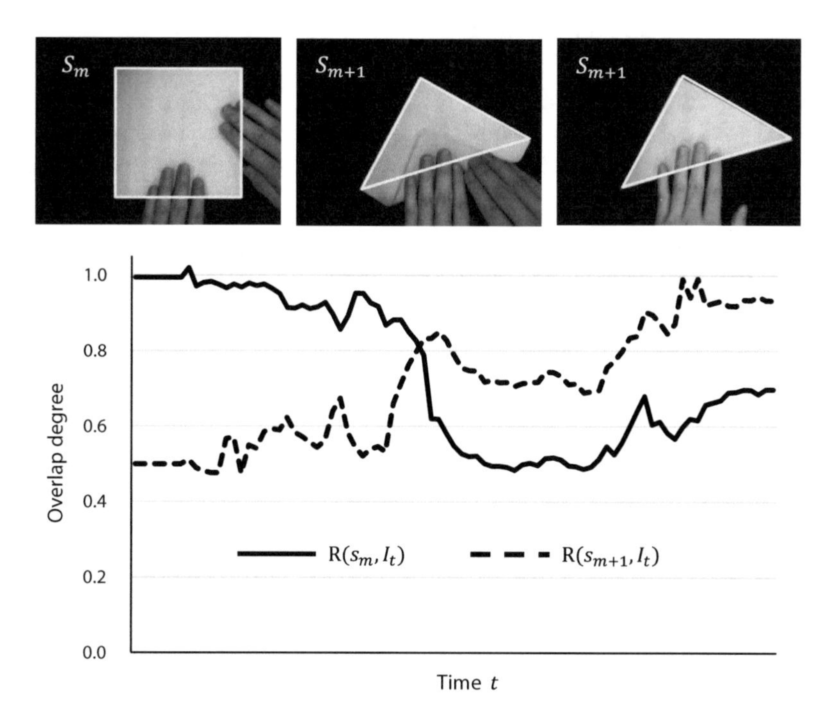

Fig. 5. Example of tracking sheet of origami paper.

4 Mistake Detection

In order to detect a user's incorrect folding, we use two criteria based on region change: the change ratio of origami paper, and the mistake ratio. The difference regions changed by a folding are pre-defined as follows:

$$\Delta P_t = (P_t \cup S_m) - (P_t \cap S_m),$$
$$\Delta S_m = (S_m \cup S_{m+1}) - (S_m \cap S_{m+1}),$$

where ΔP_t is the amount of change of the paper region from the beginning of the folding, and ΔS_m is the difference between the previous and next silhouettes.

Therefore, the change ratio of origami paper expresses progress of folding and is defined as follows:

$$\text{Change ratio } c_t = \frac{|\Delta P_t \cap \overline{H_t}| - \epsilon}{|\Delta S_m|},$$

where ϵ denotes a constant number to reduce influence by noise, because ΔP_t contains the quantization error and the error caused by distortion of paper. Moreover, the folding mistake ratio is defined as follows:

$$\text{Mistake ratio } m_t = \frac{|\Delta P_t \cap \overline{\Delta S_m} \cap \overline{H_t}|}{|\Delta P_t|}.$$

If values of c_t and m_t are larger than thresholds T_c and T_m, respectively, the performed folding is judged to be incorrect. In our work, the error caused by image processing is taken into consideration, and the judgment conditions are defined as follows:

When the change ratio $c_t > T_c$ and the mistake ratio $m_t > T_m$ on successive three images are satisfied, the performed folding is judged to be incorrect.

5 Experiments

This section shows the experimental results and evaluates the proposed methods. In order to prepare experimental data, we took videos that five Japanese college students folded sheets of origami paper into three kinds of origami works three times each work. Further, those videos were manually divided into scenes of each folding. About 300 video files were produced in the following file format:

Frame size: 320×240
Frame rate: 10 fps

In order to decide appropriate thresholds, we conducted a preliminary experiment with 100 arbitrarily chosen videos. Further, the change ratio and the mistake ratio were observed by tracking the sheet of origami paper in each video with correct silhouettes and willfully incorrect ones. Specifically, when a silhouette s_m and the next silhouette s_{m+1} are used in the correct samples, the other silhouette transited from s_m is used as the next silhouette s'_{m+1} in the incorrect samples. Figure 6 shows the distribution among the change ratio and the mistake ratio in each frame.

In the change ratio, correct samples were almost all distributed from 0.2 to 1.0, whereas incorrect samples were widely distributed. On the other hand, the mistake ratio of correct samples was quite small, and the mean and the standard deviation (SD) were 0.08 and 0.05, respectively. From the above results, the thresholds of the judgment conditions are determined on the basis of a basic outlier detection method. Specifically, let μ be the mean, and σ be the SD, and

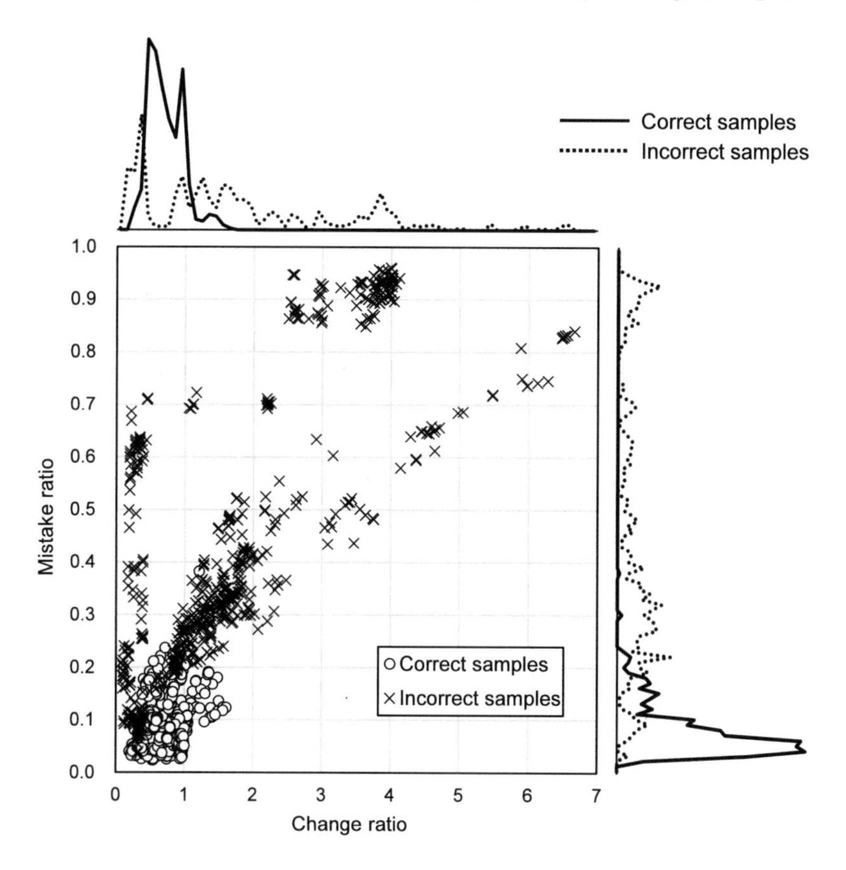

Fig. 6. Distribution among change ratio and mistake ratio.

if data x is 3σ or more away from μ, x is regarded as the outlier. Therefore, the threshold for the mistake ratio is calculated as follows:

$$T_m = 0.08 + 3 \times 0.05 = 0.23.$$

The thresholds described in Sect. 4 are set as follows:

$$T_c = 0.20,\ T_m = 0.23.$$

In order to evaluate the method of mistake detection for origami support, another 50 videos were arbitrarily chosen and tested by using the obtained thresholds. Table 1 gives the experimental results.

Table 2 shows the variation of precision, recall and F-measure by changing thresholds. Precision and recall are calculated as follows:

$$\text{precision} = \frac{\text{number of correctly detected mistakes}}{\text{number of folds judged as incorrect}}.$$

$$\text{recall} = \frac{\text{number of correctly detected mistakes}}{\text{number of trackings with incorrect silhouettes}}.$$

Table 1. Contingency table of results for mistake detection.

	Judged as correct folds	Judged as incorrect folds
Tracked using correct silhouettes	40	10
Tracked using incorrect silhouettes	14	36

Table 2. Precision, recall, and F-measure at several thresholds.

Thresholds	Precision	Recall	F-measure
$T_c = 0.20$, $T_m = 0.23$	0.78	0.72	0.75
$T_c = 0.50$, $T_m = 0.23$	0.88	0.58	0.70
$T_c = 0.20$, $T_m = 0.40$	0.88	0.46	0.61
$T_c = 0.50$, $T_m = 0.40$	1.00	0.36	0.53

In our origami support system, if a folding is judged as incorrect despite being correct, the user will be confused. Therefore, the precision value should be kept high, but this will drastically decrease the recall value.

6 Conclusion

The present work demonstrated that it is possible to detect mistakes in folding origami paper by only using region change of the paper. As future work, it is necessary to detect mistakes robustly over noise such as distortion of the origami paper and to consider a method for automatically recognizing and separating the beginning and the end of each folding operation. Moreover, we consider an effective way to point out the user's mistakes.

References

1. Kinoshita, Y., Watanabe, T.: Estimation of folding operation using silhouette of origami. IAENG Int. J. Comput. Sci. **37**(2), 177–184 (2010)
2. Miyazaki, S., Yasuda, T., Yokoi, S., Toriwaki, J.: An origami playing simulator in the virtual space. J. Visual. Comput. Animation **7**(1), 25–42 (1996). https://doi.org/10.1002/(SICI)1099-1778(199601)7:1⟨25::AID-VIS134⟩3.0.CO;2-V
3. Shimanuki, H., Watanabe, T., Asakura, K., Sato, H.: Operational support for origami beginners by correcting mistakes. In: Proceedings of the 10th International Conference on Ubiquitous Information Management and Communication, IMCOM 2016, pp. 77:1–77:8. ACM (2016). DOI https://doi.org/10.1145/2857546.2857625
4. Watanabe, T., Kinoshita, Y.: Folding support for beginners based on state estimation of origami. In: Proceedings of TENCON 2012 - 2012 IEEE Region 10 Conference (2012). https://doi.org/10.1109/TENCON.2012.6412167

Sink Nodes Deployment Algorithm for Wireless Sensor Networks Based on Geometrical Features

Koichi Asakura[1(✉)], Kengo Osuka[1], and Toyohide Watanabe[2]

[1] Daido University, Nagoya 457-8530, Japan
asakura@daid-it.ac.jp
[2] Nagoya Industrial Science Research Institute, Nagoya 464-0819, Japan
watanabe@nagoya-u.jp

Abstract. This paper proposes an algorithm for deploying sink nodes in outdoor wireless sensor networks focusing on smart meters for electricity or gas in a residential area. In this situation, the location of nodes is pre-determined and the nodes cannot be deployed freely. This algorithm calculates the number of required sink nodes and selects the appropriate nodes in order to decrease operational costs of the wireless sensor networks. Positional information on meters in a real residential area was used for experiments. Our algorithm calculated an optimal number of sink nodes.

Keywords: Node deployment · Wireless sensor networks
Sink nodes · IoT

1 Introduction

Recently, the term "Internet of Things (IoT)" has attracted attention [3]. In IoT, several technical developments are included: sensor devices, networks, big data and so on. Our focus is on technologies of networks, especially wireless sensor networks.

Wireless sensor networks (WSNs) consists of a large number of sensor nodes, which are appropriately deployed in an area [1,2,11]. Each sensor node monitors and stores physical states of the area by using sensors that are equipped in the nodes. Stored data is transferred to central servers regularly through wireless networks. For effective WSNs, configuration of wireless networks is very important. In ordinary WSNs, two types of nodes are used: sensor and sink. For effective successful communication to the central servers, optimal deployment of sink nodes is essential.

In this paper, we propose a sink node deployment algorithm for WSNs. Our algorithm is based on the geometrical features of nodes in the WSNs. Since nodes in WSNs are placed statically in a dedicated area, sink nodes are selected by their

G. De Pietro et al. (Eds.): KES-IIMSS-18 2018, SIST 98, pp. 93–102, 2019.
https://doi.org/10.1007/978-3-319-92231-7_10

positional information. In order to decrease operational cost in WSNs, our algorithm calculates the optimal number of sink nodes and selects the appropriate ones.

The rest of the paper is organized as follows. Section 2 states a sink nodes deployment problem in details. Section 3 presents our proposed algorithm. Section 4 describes experimental results. Section 5 concludes our paper and states future work.

2 Preliminaries

This section describes a sink node deployment problem in details. Before problem definition, we describe configuration of WSNs which are targeted by our problem.

2.1 Configuration of WSNs

We focused on WSNs for outdoor fields. An example of such WSNs is a network system for smart meters. Smart meters are usually used for gas and electricity [5]. They measure the quantity of gas/electricity usage digitally and send the measured data to their suppliers via wireless communication. For wireless communication between a smart meter and a supplier, a wide-area radio access network such as LTE and WiMAX is required. With these smart meters, both customers and suppliers can gather information on when and how much energy is used. Another example is an agricultural sensing system [4,10]. In this system, many sensor devices are deployed on an agricultural farm. The sensor devices periodically measure field status such as temperature, humidity and so on. The measured data is collected in the central servers and used for the effective growth of crops.

For outdoor WSNs, wireless communication between nodes and central servers is necessary. However, usage of a wide-area radio access network includes constant operational costs. Thus, multi-hop communication methods among nodes are adopted in ordinary WSNs. In multi-hop communication methods, nodes play the role of relay routers. Namely, intermediate nodes receive a packet from a node and send the packet to another node. Communication schemes such as IEEE 802.11 (WiFi) and IEEE 802.15.4 (ZigBee) are commonly used for multi-hop communication in WSNs. These schemes require no infrastructure networks. With multi-hop communication, the measured data in WSNs can be gathered into one node or several nodes, which can reduce the number of nodes required to equip the wide-area radio access network, and thus, can reduce the operational costs of WSNs.

As described above, there are two types of nodes in WSNs: nodes with and nodes without a wide-area radio access network facility. The former nodes are referred to as sink nodes, and the latter are called sensor nodes.

Sink Node. A sink node consists of four devices with the following functions: sensor, memory for data storage, multi-hop communication and wide-area radio access network. The measured data of other nodes is sent to the sink node via a multi-hop communication scheme. In order to gather the measured data of other nodes, a sufficient quantity of memory is provided in this node. Stored data is transmitted to the central servers periodically via the wide-area radio access network device.

Sensor Node. A sensor node consists of three devices with the following functions: sensor, memory and multi-hop communication. Contrastively, it has no functions for communication with central servers. The measured data taken by the sensor device is stored in the memory device and is periodically transferred to a sink node using the multi-hop communication device.

Figure 1 shows an example of the WSNs. In this figure, the sensor nodes are denoted as single black circles and the sink nodes are represented by double circles. The lines between the nodes refer to multi-hop communication. The sensing data on the sensor nodes are stored in the connected sink node. The wide-area radio access network is denoted as an antenna symbol attached to the sink nodes.

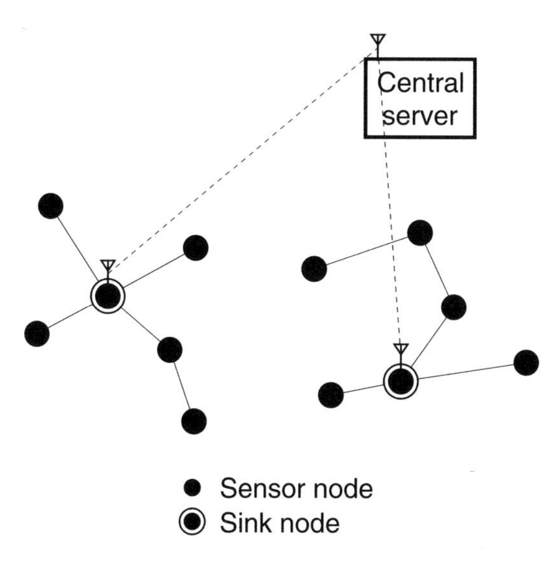

Fig. 1. Configuration of WSNs

Due to limitations in computing resources, multi-hop communication methods generally have several restrictions. For example, the maximum number of sensor nodes for one sink node is determined in accordance with the capacity of the memory device. Furthermore, to suppress communication latency, the maximum number of hops between a sink node and a sensor node is also provided.

2.2 Problem Definition

Our goal was to find an optimal deployment of sink nodes for effective WSNs. In order to achieve this, we had to solve the following two problems: operational costs and the reliability of WSNs.

Operational Cost. As described above, monthly charges are normally required to use wide-area radio access networks, while a multi-hop communication does not have such costs. Thus, in order to reduce operational costs, the number of sink nodes should be minimized.

Reliability. In order to ensure reliable WSNs, multi-hop communication methods play an important role. Here, we focused on the number of hops and sensor nodes for one sink node. Generally, fewer hops provide more reliable communication. This is because communication errors such as packet collisions occur in inter-node communications. Furthermore, the number of sensor nodes for each sink node should be balanced. If the number of sensor nodes was not balanced, communication congestion occurred and multi-hop communications became unreliable.

In the following section, we propose an algorithm for deploying sink nodes to solve the above problems. The algorithm resolves which nodes should be sink nodes when positional information on all nodes was given by setting the maximum number hop counts and sensor nodes for one sink node.

3 Algorithm

In this section, we propose an algorithm for deploying sink nodes for achieving effective WSNs. Section 3.1 expresses our approach and Sect. 3.2 describes our proposed algorithm.

3.1 Approach

In our algorithm, appropriate positioning for sink nodes is calculated geometrically. Our algorithm selects the appropriate positions based on the positional information on all nodes.

In general, a sink node should be placed near the median point of all nodes since the average number of hops decreases. Figure 2 shows a typical example. In this figure, sensor nodes are the full black circles and sink nodes are denoted as the double circles. The communication range is represented as a dotted circle. We assumed that five nodes were located linearly, and that the distance between two neighboring nodes was equal to the communication range. Thus, each node could communicate with only the neighboring nodes. When node A was selected as a sink node (shown in Fig. 2(a)), the hop counts from nodes B $\sim$ E were 1, 2, 3 and 4, respectively. In contrast, when node C was selected as a sink node (shown in Fig. 2(b)), the hop counts were 2, 1, 1 and 2. Thus, it was clear that node C was the best option for the sink node.

Additionally, we had to take the maximum number of hops into account. Especially, hop counts for the outermost nodes have to be considered. Figure 3 shows another example. Here, we assumed that the maximum number of hops was two. In Fig. 3(a), node D was selected as a sink node since the position of node D was the nearest to the median point of all nodes. However, node A could

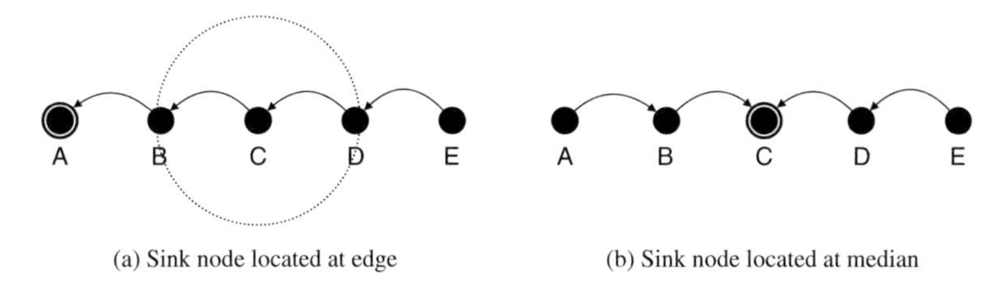

(a) Sink node located at edge (b) Sink node located at median

Fig. 2. Hop counts depending on sink node location

not send sensing data to node D because the hop count exceeded the maximum number of hops. Thus, in this example, both nodes A and G must become sink nodes, making the total number of sink nodes three. In Fig. 3(b), the required hop count for node A was considered, so node C was selected as the sink node since it was the farthest node from node A, and still met the maximum number of hops. Thus, the number of required sink nodes decreased to two in this example.

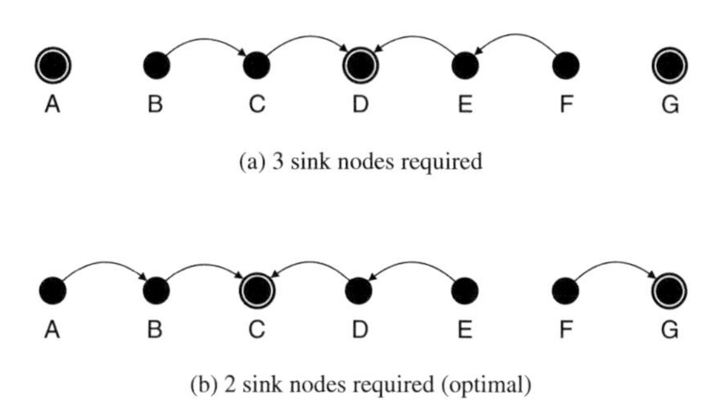

(a) 3 sink nodes required

(b) 2 sink nodes required (optimal)

Fig. 3. Maximum hop counts taken into account

Based on the above considerations, our proposed algorithm selects candidate nodes for sink nodes as follows:

1. The outermost node is selected as a starting point for calculation.
2. The farthest node from the starting point within the maximum number of hops is selected as a possible sink node.
3. The nodes which can communicate with the candidate node are selected as a group.
4. A sink node is determined based on the median point of the group.

3.2 Sink Node Deployment Algorithm

Inputs for the algorithm are as follows. The set $N = \{n_1, \cdots, n_m\}$ is the node set. The maximum number of hops is denoted as *hop*.

```
ALGORITHM SinkNodeDeployment
Input:
  N = {n1, ..., nm}: set of nodes
  hop: the maximum number of hops
Output:
  Ns: set of sink nodes
begin
 1:  Ns = { }.
 2:  while N is not empty.
 3:  begin
 4:    // start node selection
 5:    mbr = minimum bounding rectangle for N.
 6:    n = the nearest node from vertices of mbr.
 7:    remove n from N.
 8:    Nc = {n}.
 9:
10:    // candidate node set generation
11:    repeat hop times
12:    begin
13:      Ncomm = nodes in N within communication range of n.
14:      n = the farthest node from n in Ncomm.
15:      remove nodes Ncomm from N.
16:      add Ncomm to Nc.
17:    end
18:    Ncomm = nodes which can communicate with n.
19:    remove Ncomm from N.
20:    add Ncomm to Nc.
21:
22:    // sink node selection
23:    repeat
24:      median = median point of Nc.
25:      sink = nodes in Nc which are the nearest to median.
26:      Nadd = nodes which can communicate with sink.
27:      add Nadd to Nc.
28:      remove Nadd from N.
29:    until Nc is not updated.
30:    add sink to Ns.
31:  end
end
```

Fig. 4. Sink node deployment algorithm

Figure 4 describes the algorithm for deploying sink nodes. This algorithm consists of three parts: start node selection, candidate node set generation and sink node selection. First, in order to select the starting node, the minimum bounding rectangle (MBR) of nodes is calculated. The node, which is the nearest to the vertices of the MBR, is selected as the starting point. Next, a candidate node set (one sink node and sensor nodes) is generated. The farthest node from the starting point with respect to the maximum number of hops "*hop*" is selected as a candidate node for a sink node. Then, the nodes which can communicate with the candidate node within *hop* are selected as a candidate set N_c. The number of nodes in N_c must not exceed the maximum number of sensor nodes for one sink node. If there are many nodes which can connect to the candidate sink node, the nodes are selected in accordance with the distance from the vertices of the MBR. Finally, an appropriate sink node is selected from the candidate node set N_c. The median point of nodes in N_c is calculated, and the node which is the nearest from the median point is selected as a sink node. If there are new nodes which can connect to the new sink node, these nodes are selected for N_c and the potential sink node is recalculated. The algorithm stops when all nodes are grouped.

4 Experiments

In order to evaluate our proposed algorithm, we conducted simulation experiments. For the experiments, we developed a simulation system in Java, and used positional data on gas meters in a real residential area. The size of the simulation area was 300 m wide and 350 m high. There were 192 gas meters in this area. Figure 5 shows positional data for the experiments. The maximum number of sensor nodes for one sink node was set to 50. In order to determine if two nodes could communicate with each other by multi-hop communication, analysis of radio propagation in the residential area was very important. We conducted field experiments in the residential area. Based on the results of the field experiments and ITU-R recommendations [6], the communication range was set to 39.8 m. The maximum number of hops was set to four.

Fig. 5. Node distribution

4.1 Experimental Results

Figure 6 shows our experimental results. In this figure, the selected sink nodes are denoted as black squares and the sensor nodes are the colored squares. In this experiment, seven nodes were selected as the sink nodes. Table 1 shows the number of sink nodes and the average number of hops.

In this experiment, nodes were not distributed uniformly since the positional data of nodes depended on the real positions of the meters in the residential area. So, isolate nodes existed in this experiment. Thus, there are several node groups which cannot communicate with each other. This data has five individual

Fig. 6. Experimental results

Table 1. Numerical results of experiment

Characteristics	Result
# of sink nodes	7
Average hop counts	1.83

node groups. Table 2 shows the properties for each node group. The largest group had 138 nodes and other four groups had less than 50. For the largest group, three sink nodes were required since $\lceil\frac{138}{50}\rceil = 3$. Therefore, it is proved that our algorithm can calculate an appropriate number of sink nodes.

4.2 Related Work

Although there has been research done for deploying sensor nodes in WSNs [7–9], they have focused on WSNs in which the location of nodes can be determined

Table 2. Properties of node groups.

No.	Node counts	Sink node counts
1	16	1
2	138	3
3	1	1
4	34	1
5	3	1

freely. Thus, they did not have to pay attention to the problem of isolated nodes. In our case, the position of the nodes was fixed beforehand and the nodes could not be redeployed.

Furthermore, regarding problems with selecting appropriate nodes as sink nodes, there has been some research on user support tools. However, the selection of the sink nodes was performed by hand.

5 Conclusion

In this paper, we proposed an algorithm for deploying sink nodes in outdoor WSNs. We dealt with the problem in which the number of nodes and positions were fixed beforehand. Our algorithm could select appropriate nodes as sink nodes. Also, our algorithm could select the optimal number of sink nodes.

For our future work, we will look at how to handle the problem using other approaches. For example, a node clustering approach may be more suitable to our problem since one sink node can handle a limited number of sensor nodes [12]. Also, approaches for optimizing problems such as swarm intelligence should be considered.

References

1. Akyildiz, I., Su, W., Sankarasubramaniam, Y., Cayirci, E.: Wireless sensor networks: a survey. Comput. Netw. **38**(4), 393–422 (2002)
2. Akyildiz, I.F., Su, W., Sankarasubramaniam, Y., Cayirci, E.: A survey on sensor networks. IEEE Commun. Mag. **40**(8), 102–114 (2002)
3. Al-Fuqaha, A., Guizani, M., Mohammadi, M., Aledhari, M., Ayyash, M.: Internet of Things: a survey on enabling technologies, protocols, and applications. IEEE Commun. Surv. Tutorials **17**(4), 2347–2376 (2015)
4. Baggio, A.: Wireless sensor networks in precision agriculture. In: ACM Workshop on Real-World Wireless Sensor Networks (REALWSN 2005), pp. 1567–1576 (2005)
5. Depuru, S.S.S.R., Wang, L., Devabhaktuni, V.: Smart meters for power grid: challenges, issues, advantages and status. Renew. Sustain. Energy Rev. **15**(6), 2736–2742 (2011)
6. International Telecommunication Union: Propagation data and prediction methods for the planning of short-range outdoor radio-communication systems and radio local area networks in the frequency range 300 MHz to 100 GHz, ITU-R Recommendation P.1411-9 (2017)

7. Kim, H., Seok, Y., Choi, N., Choi, Y., Kwon, T.: Optimal multi-sink positioning and energy-efficient routing in wireless sensor networks. In: International Conference on Information Networking, pp. 264–274 (2005)
8. Oyman, E.I., Ersoy, C.: Multiple sink network design problem in large scale wireless sensor networks. In: IEEE International Conference on Communication, vol. 6, pp. 3663–3667 (2004)
9. Poe, W.Y., Schmitt, J.B.: Node deployment in large wireless sensor networks: coverage, energy consumption, and worst-case delay. In: Asian Internet Engineering Conference, pp. 77–84 (2009)
10. Wark, T., Corke, P., Sikka, P., Klingbeil, L., Guo, Y., Crossman, C., Valencia, P., Swain, D., Bishop-Hurley, G.: Transforming agriculture through pervasive wireless sensor networks. IEEE Pervasive Comput. **6**(2), 50–57 (2007)
11. Yick, J., Mukherjee, B., Ghosal, D.: Wireless sensor network survey. Comput. Netw. **52**(12), 2292–2330 (2008)
12. Younis, O., Krunz, M., Ramasubramanian, S.: Node clustering in wireless sensor networks: recent developments and deployment challenges. IEEE Netw. **20**(3), 20–25 (2006)

Speaker Recognition in Orthogonal Complement of Time Session Variability Subspace

Satoru Tsuge[1(✉)] and Shingo Kuroiwa[2]

[1] Daido University, Nagoya, Aichi 457-8530, Japan
tsuge@daido-it.ac.jp
[2] Chiba University, Chiba, Chiba 263-8522, Japan

Abstract. A time session variability between the enrollment data and the recognized data degrades speaker recognition performance. Hence, the time session variability is one of the most important issues in the speaker recognition technology. In this paper, we propose a robust speaker recognition method for the time session variability. The proposed method estimates a time session variability subspace. Then, the proposed method carries out the speaker recognition in the orthogonal complement of the time session variability subspace. In addition, we incorporate a linear discriminant analysis method into the proposed method. In order to evaluate the proposed method, we conducted a speaker identification experiment. Experimental results show that the proposed method improves speaker identification performance of baseline.

Keywords: Speaker recognition · i-vector · Time session variability Deflation · Orthogonal complement

1 Introduction

Even if the speaker intends to utter in the same way in every sessions, the speaker's conditions are changed under the influences of the physical conditions (short time variability), the aging (long-time variability), and so on. The change of these conditions varies the recorded speech data in every sessions. This speech variability, which is caused by the time session difference, causes the difference between the enrollment data and the recognized data. It is widely known that the speech variability caused by the time session variability degrades speaker recognition performance [1,2]. Therefore, the time session variability is one of issues for improving speaker recognition performance. Hence, some methods have been proposed to suppress the time session variability [3–6].

In recent years, it is widely known that the speaker recognition systems based on an i-vector [7] show high speaker recognition performance. The i-vector approach estimates a total variability matrix which decomposes a speaker and a channel elements and others. Then, this approach projects the GMM (Gaussian Mixture Model)-supervector to a low-dimensional space using the estimated

G. De Pietro et al. (Eds.): KES-IIMSS-18 2018, SIST 98, pp. 103–109, 2019.
https://doi.org/10.1007/978-3-319-92231-7_11

matrix. In this paper, we call this low-dimensional space constructed by the i-vector approach "i-vector space". However, the influence of the time session variability cannot be completely removed in the i-vector space. Therefore, in this paper, we propose the novel speaker recognition method which suppresses the influence of the time session variability in the i-vector space. The proposed method estimates the time session variability subspace, then removes the estimated subspace for the speaker recognition. In this paper, we assume that the time session variability subspace is common in all speakers. Hence, we estimate the time session variability subspace using a lot of a specific speaker's speech data. In order to evaluate the proposed method, we conducted the speaker recognition experiment using a part of a large speech corpus constructed by the National Research Institute of Police Science, Japan (NRIPS) [8] and Corpus of Spontaneous Japanese (CSJ) [9].

The rest of this paper is organized as follows. In Sect. 2, we describe about the proposed method. Section 3 describes the speaker identification experiment and shows the experimental results. Finally, we summarize this paper in Sect. 4.

2 Speaker Recognition Method Excluding Time Session Variability

In this section, we propose a robust speaker recognition method against the time session variability. The proposed method, first, estimates a time session variability subspace using speech data which are uttered by a specific speaker over the long-time sessions. Then, the proposed method reduces the influence of the time session variability using the deflation technique [10], i.e., the proposed method constructs the orthogonal complement of the time session variability subspace. In this paper, we use the speaker recognition method based on an i-vector [7] approach, which is a state-of-the-art speaker recognition method. Hence, the proposed method estimates the space excluding the time session variability in the i-vector space. The flow of the proposed method is described as follows:

1. Estimation of the time session variability subspace
   First, the i-vector of each recorded session is extracted using speech data that are uttered by a specific speaker over long-time sessions. Then, in order to construct the subspace with large influence of the time session variability, the proposed method carries out a principal component analysis (PCA) using the extracted i-vectors and computes principal components. We define that the i-vectors, $\boldsymbol{x}_p$ $(1 \leq p \leq P)$, are used for the PCA in this paper, and P is the number of i-vectors used for the calculation of PCA, i.e., P is the number of the recorded sessions. The number of the principal components is M. We assume that the space comprised of these principal components is the time session variability subspace.
2. Construction of the orthogonal complement of the time session variability subspace (Deflation)

The proposed method excludes the time session variability by constructing the orthogonal complement of the time session variability subspace estimated by the previous procedure. The construction of the orthogonal complement is realized by deflation [10]. We examine the following two techniques as a deflation.

- Method 1

$$\hat{\boldsymbol{x}}_s = \boldsymbol{x}_s - \sum_{m=1}^{M} w_m (\boldsymbol{p}_m^T \boldsymbol{x}_s) \boldsymbol{p}_m, \tag{1}$$

  where $\boldsymbol{p}_m$ and $\boldsymbol{x}_s$ are m-th principal component and the i-vector of speaker s, respectively. w_m indicates a contribution rate of m-th principal component. $\hat{\boldsymbol{x}}_s$ is the i-vector of speaker s after the deflation. This is used as the speaker model in the speaker recognition.
- Method 2
  The deflation is carried out for the PCA data, $\boldsymbol{x}_p$, in the following formulas.

$$\hat{\boldsymbol{x}}_p = \boldsymbol{x}_p - w_1 (\boldsymbol{p}_1^T \boldsymbol{x}_p) \boldsymbol{p}_1, \tag{2}$$

$$\boldsymbol{x}_p \leftarrow \hat{\boldsymbol{x}}_p. \tag{3}$$

  Using the i-vectors, $\boldsymbol{x}_p$, after deflation, the PCA is carried out again. The proposed method iterates the formulas (2) and (3) and PCA in M times.
  Then, using first principal components of each iteration, $\boldsymbol{p}_1^{(m)}$, and the contribution rates of these components, $w_1^{(m)}$, the proposed method removes the influence of the time session variability from the i-vector of speaker s in the following formulas.

$$\boldsymbol{x}_s^{(m)} = \boldsymbol{x}_s^{(m-1)} - w_1^{(m)} ((\boldsymbol{p}_1^{(m)})^T \boldsymbol{x}_s^{(m-1)}) \boldsymbol{p}_1^{(m)} \quad (1 \leq m \leq M), \tag{4}$$

$$\hat{\boldsymbol{x}}_s \leftarrow \boldsymbol{x}_s^{(M)}, \tag{5}$$

  where $\boldsymbol{x}_s^{(m)}$ and $\hat{\boldsymbol{x}}_s$ are the i-vector of speaker s of the m-th iterations and the i-vector of speaker s used for speaker recognition, respectively.

The proposed method employs the weighted deflations, which are formulas (1) and (4), for restraining the reduction of the speaker information in the i-vectors. Hence, the speaker recognition of the proposed method is not performed on the complete orthogonal complement of the principal components. If the speaker recognition on the complete orthogonal complement of the principal components is performed, the weight values, w_m and $w_1^{(m)}$, are set to 1.0. The principal components calculated in Method 1, $\boldsymbol{p}_m$, are the orthogonal vectors each other. However, the principal components used in Method 2, $\boldsymbol{p}_1^{(m)}$, might not be the orthogonal vectors each other because the proposed method employs the weighted deflation in the formulas (2) and (3).

3 Speaker Identification Experiments

In order to evaluate the proposed method, we conducted the speaker identification experiments. In this experiment, we used National Research Institute of Police Science, Japan (NRIPS) Japanese large speech corpus [8] and Corpus of Spontaneous Japanese (CSJ) [9] for the speaker identification experiments and AWA-LTR first version [11] for constructing the time session variability subspace.

3.1 Experimental Conditions

The speech data collected in NRIPS speech corpus, which are used in the speaker identification experiments, were down-sampled from 44.1 kHz to 16 kHz. We used 164,680 utterances uttered by 1,005 speakers, which were collected in CSJ, for UBM training and calculating the total variance matrix for extracting the i-vectors. For the enrollment data, i.e., the speaker model training data, we used 78 male speakers' speech data collected in NRIPS speech corpus. The number of the enrollment utterances of each speaker was five. These utterances were a part of ATR phonetically balanced Japanese sentences and were recorded on the second time in the first session. We also used 780 utterances (78 enrollment speakers $\times$ 10 sentences) as the evaluation data. The texts of the evaluation data were not included in the texts of the enrollment data. The recording session of these utterances was different from the enrollment data. There were about three-month intervals between the enrollment data and the evaluation data. Hence, there was the time session variability between the enrollment data and the evaluation data.

All speech data, which were sampled at 16 kHz, were segmented into overlapping frames of 20 ms, producing a frame every 10 ms. A Hamming window was applied to each frame. Mel-filtering was performed to extract 12-dimensional static MFCCs, as well as a logarithmic energy (log-energy) measure. The 12-dimensional delta MFCCs, delta-delta MFCCs, delta log-energy, and delta-delta log-energy were extracted from the static MFCCs and the log-energy, respectively. After that, we constructed a 39-dimensional feature vector (12 static MFCCs + 12 delta MFCC + 12 delta-delta MFCC + log-energy + delta log-energy + delta-delta log-energy). Cepstral Mean Subtraction (CMS) was applied on the static MFCCs, and then we selected the speech section from the feature vectors using power information. The number of mixtures of UBM was set to 2,048. The number of dimensions of i-vector of each speaker was set to 400. In addition, we applied the length normalization [12] to i-vectors. In this experiment, we used a cosine similarity as a measurement between i-vectors. In this experiment, we used Kaldi [13] for the i-vector extraction, the GMM training, and the calculation of similarity.

For constructing the orthogonal complement of the time session variability subspace, we used the 6,240 speech data of a male speaker (52 recording days $\times$ 3 times of each day $\times$ 40 utterances of each session) in AWA-LTR first version. The texts of these speech data were the same as the texts of the UBM training data, i.e., these were not included in the texts of the enrollment data and the evaluation data. The number of i-vectors used for the PCA was 156 because we

Table 1. Experimental results (IER in %)

Method	Number of principal components									
	0	1	2	3	4	5	6	7	8	9
1	9.62	**9.23**	9.36	9.36	9.36	9.36	9.49	9.36	9.49	9.49
2	9.62	**9.23**	9.62	9.62	10.13	10.38	10.38	10.38	10.38	10.38

extracted the i-vector of each recording time. The nine principal components were extracted by the PCA. The scikit-learn [14] was used as a tool of the PCA.

3.2 Experimental Results

Table 1 shows the IER (Identification Error Rate). The IERs under the condition that the number of principal components is 0 indicate the baseline results which the proposed method was not applied.

From this table, we can see that the proposed methods, Method 1 and 2, decrease the IER of baseline under the condition that the number of principal components is 1. In fact, Method 1 is the same as Method 2 under the condition that the number of principal components is 1. In this case, the percent contribution of first principal component is 17.9%. Although Method 2 can improve the baseline performance only in case that the number of principal components is 1, Method 1 can improve the baseline performance in other cases. From these results, we conclude that the proposed method can improve speaker identification performance under the condition that the time session variability between the enrollment data and the recognized data exists.

3.3 Application of the Proposed Method After LDA

The proposed method can be applied not only the i-vector space but also other spaces such as a feature vector space, a space of dimension reduction. Therefore, we apply the proposed method in the space where a Linear Discriminant Analysis (LDA) method reduces the dimension. It is known that the dimension reduction by LDA improves speaker recognition performance. In this paper, we call the space where a dimension was reduced by LDA "LDA space".

The proposed method is applied as follows in the space where LDA reduces the dimension:

1. Projection of i-vector to LDA space
   Using the enrollment data, we calculate the projection matrix by LDA. Using the calculated matrix, we project the i-vectors, which are used for estimating the time session variability subspace, into LDA space.
2. Application of the proposed method
   The proposed method computes the principal components, which compose the time session variability subspace, by PCA in LDA space. Using these

Table 2. Experimental results (Application of the proposed method in LDA space) (IER in %)

Method	Number of deflation									
	0	1	2	3	4	5	6	7	8	9
1	7.05	7.05	6.67	6.67	**6.54**	6.54	6.54	6.54	6.54	6.67

principal components, we apply the proposed method to the i-vectors of the enrollment data and the recognized data in LDA space, then the speaker identification is performed using these i-vectors.

In order to evaluate the proposed method in LDA space, we conducted the speaker identification experiment. The experimental conditions are the same as Sect. 3.1. In this experiment, we incorporate LDA into the proposed method (Method 1) and reduce the dimension of i-vector from 400 to 150 by LDA.

Table 2 shows the experimental results of the proposed method on LDA space. Comparing Tables 1 and 2 under the condition that the number of principal components is 0, we can confirm that the dimension reduction using LDA improves the speaker identification performance. We can also see from Table 2 that the proposed method improves the speaker identification performance of the baseline. Actually, the proposed method decreases the IER of baseline from 7.05% to 6.54%. The proposed method is useful for the speaker identification in LDA space.

4 Summary

In this paper, we proposed a speaker recognition method that suppresses a time session variability between the enrollment data and the recognized data. This method estimates a time session variability subspace using a principal component analysis in the i-vector space, then recognized the speaker in the space which is the orthogonal complement of the estimated space. We proposed two techniques which estimate the time session variability subspace. In addition, we incorporated a linear discriminant analysis into the proposed method.

To evaluate the proposed method, we conducted speaker identification experiments using male speech data in the National Research Institute of Police Science, Japan speech corpus and Corpus of Spontaneous Japanese. Experimental results showed that the proposed method improved the speaker recognition performance of baseline which is not applied the proposed method. Actually, the proposed method decreased the identification error rate from 9.62% to 9.23% in the i-vector space and from 7.05% to 6.54% in the LDA space. From these results, we concluded that the proposed method is useful for the speaker identification. We have not investigated the details of the space of the time session variability in this paper. We plan to investigate the details of this space, especially we investigate the differences between phonemes.

Acknowledgments. This work was supported by JSPS KAKENHI Grant Number JP16K00229.

References

1. Kinnunen, T., Li, H.: An overview of text-independent speaker recognition: from features to supervectors. Speech Commun. **52**(1), 12–40 (2010)
2. Matsui, T., Nishitani, T., Furui, S.: A study of model and a priori threshold updating in speaker verification. IEICE Trans. **J81-DII**(2), 268–276 (1998). (in Japanese)
3. Kenny, P., Boulianne, G., Ouellet, P., Dumouchel, P.: Joint factor analysis versus eigenchannels in speaker recognition. IEEE Trans. Audio Speech Lang. Process. **15**(4), 1435–1447 (2007)
4. Kenny, P., Boulianne, G., Ouellet, P., Dumouchel, P.: Speaker and session variability in GMM-based speaker verification. IEEE Trans. Audio Speech Lang. Process. **15**(4), 1448–1460 (2007)
5. Kenny, P., Ouellet, P., Dehak, N., Gupta, V., Dumouchel, P.: A study of interspeaker variability in speaker verification. IEEE Trans. Audio Speech Lang. Process. **16**(5), 980–988 (2008)
6. Kenny, P.: Bayesian speaker verification with heavy-tailed priors. In: Proceedings of Odyssey (2010)
7. Dehak, N., Kenny, P., Dehak, R., Dumouchel, P., Ouellet, P.: Front-end factor analysis for speaker verification. IEEE Trans. Audio Speech Lang. Process. **19**(4), 788–798 (2011)
8. Makinae, H., Osanai, T., Kamada, T., Tanimoto, M.: Construction and preliminary analysis of a large-scale bone-conducted speech database. IEICE Techn. Rep. Speech **107**(165), 97–102 (2007). (in Japanese)
9. Furui, S., Maekawa, K., Isahara, H.: A Japanese national project on spontaneous speech corpus and processing technology. In: Proceedings of ASR 2000, pp. 244–248 (2000)
10. Partridge, M., Calvo, R.A.: Fast dimensionality reduction and simple PCA. Intell. Data Anal. **2**, 203–214 (1998)
11. Tsuge, S., Kuroiwa, S.: AWA long-term recording speech corpus (AWA-LTR). In: Proceedings of 2013 International Workshop on Nonlinear Circuits, Communication and Signal Processing (NCSP 2013), pp. 17–20 (2013)
12. Garcia-Romero, D., Espy-Wilson, C.Y.: Analysis of i-vector length normalization in speaker recognition systems. In: Proceedings of Interspeech, pp. 249–252 (2011)
13. Povey, D., Ghoshal, A., Boulianne, G., Burget, L., Glembek, O., Goel, N., Hannemann, M., Motlicek, P., Qian, Y., Schwarz, P., Silovsky, J., Stemmer, G., Vesely, K.: The Kaldi speech recognition toolkit. In: IEEE 2011 Workshop on Automatic Speech Recognition and Understanding (2011)
14. scikit-learn, machine learning in Python. http://scikit-learn.org/stable/

Verification of Identification Accuracy of Eye-Gaze Data on Driving Video

Naoto Mukai[1(✉)], Kazuhiro Fujikake[2], Takahiro Tanaka[2], and Hitoshi Kanamori[2]

[1] Department of Culture-Information Studies, Sugiyama Jogakuen University, 17-3, Hoshigaoka-motomachi, Chikusa-ku, Nagoya, Aichi 464-8662, Japan
nmukai@sugiyama-u.ac.jp

[2] Institutes of Innovation for Future Society, Nagoya University, Furocho, Chikusa-ku, Nagoya, Aichi 464-8601, Japan
{fujikake,tanaka,hitoshi_kanamori}@coi.nagoya-u.ac.jp

Abstract. It is said that the most cause of traffic accidents is the lack of confirming the safety. Visual information from both eyes is one of the important factors for safe driving. In this paper, we collect eye-gaze data of drivers who watch a driving video, and try to develop a model of their eye movements to identify factors to enhance their safety. For the purpose of modeling, we adopted a recurrent neural network and Long Short-Term Memory (LSTM) to the collected eye-gaze data because the LSTM is able to deal with a time-series data such as the eye-gaze data. Moreover, we performed an experiment to evaluate the identification accuracy of drivers. The results indicated that the driver's intention and habit can be approximated partially by the trained network, but it was insufficient to identify a personal driver for practical use.

1 Introduction

According to the statistical report in 2017 of Traffic Bureau, Japan[1], the number of car accidents are on a downward trend, but the ratio of accidents by the elderly people accounts for a substantial fraction of them. It seems that the cognitive decline in the elderly is one of the reasons of the traffic accidents. Thus, it is important to preserve the cognitive function of elderly people to improve their safety. Moreover, we think that the most important cognitive function for safe driving is the sense of vision. We assumed that the eye-gaze data of them include the intents and habits of driving, e.g., the order of looking mirrors and obstacles in the road. The recognition and analysis of such intents and habits are the key factors to avoid traffic accidents.

In this paper, we attempt to extract the intents and habits from the eye-gaze data by using a neural network approach. We adopted **Recurrent Neural Network** and **Long Short-Term Memory (LSTM)** to model driver's eye-movement because the LSTM is suitable for a time-series data. The LSTM [1]

[1] The report is publicly available at https://www.e-stat.go.jp/.

G. De Pietro et al. (Eds.): KES-IIMSS-18 2018, SIST 98, pp. 110–119, 2019.
https://doi.org/10.1007/978-3-319-92231-7_12

is proposed by Hochreiter and Schmidhuber in 1997, but it becomes the focus of attention within recent years with the popularization of deep learning techniques. The LSTM has already achieved some powerful results in various research fields, e.g., sound recognition [2], machine translation [3], and so on. Here, we collect eye-gaze data of twenty test drivers when they watch the video of driving simulation which includes some different situations, e.g., intersection, street parking, stop sign, and so on. We used these collected eye-gaze data as a training data to the LSTM. The trained network represents the model of eye movement which includes intents and habits of driving. Moreover, the trained network has an ability to predict future eye-movement, and its ability can be used for identification of each driver. Thus, we performed an experiment to evaluate the identification accuracy, i.e., the mean squared errors between actual data and predicted data, by using the trained network.

The structure of this paper is following. Section 2 summaries the related works of this paper. Section 3 explains that the detail of eye-gaze data we collected from twenty test drivers. Section 4 shows the structure of the recurrent neural network and the example of prediction result by the trained network. Section 5 discusses the comparison of mean squared errors between actual data and predicted data. Section 6 concludes this paper and describe our future works.

2 Related Works

This section shows the related works. First, we pick up two works which target sensor data of drivers. Mima et al. proposed an estimation method of driver's state by using sensor data of brake pedal [4]. Their method consists of two techniques: clustering of brake signals by **Gaussian Mixture Model (GMM)** and estimation of driver's state by **Hidden Markov Model (HMM)**. Their discussion concludes that the state of drivers can be estimated by the sensor data of brake pedal in real-time at some level, even though it needs to consider other driving situations. Okada et al. also extracts driver's pattern from sensor data of accelerator and brake pedals [5]. They applied **Dynamic Time Warping (DTM)** to define similarity of driver's pattern. Their results indicated that the patterns of drivers are different depending on individual features and road conditions. It is important that these works pointed out that the sensor data of drivers are depending on the characteristics or drivers.

Next, we pick up three works which target on eye-gaze data. Horiguchi et al. proposed a graph structure of eye-gaze data for a train's driver, and compare with eye-gaze data of expert drivers and beginner drivers [6]. The graph structure of them is based on **Markov Cluster Algorithm (MCA)**. Their results indicated that there is a difference between experts and beginners in night-time driving. Tanaka et al. proposed a driver assist agent, and analyzed eye-gaze data when the driver is assisted by the agent [7]. They compared three conditions of agents: voice agent, virtual agent, and human-like robot. They concluded that the human-like robot is the best agent to guide elderly people. Kamisaka et al. proposed a prediction method for driver's state on the basis of eye-gaze data [8].

They applied **Support Vector Machine (SVM)** to the eye-gaze data to classify the six states (right turn, left turn, right lane change, left lane change, go straight, and stop). Their results indicated that the eye-gaze data can predict driver's state effectively. Kamisaka's approach is similar to our work, but the expression of eye-gaze data is different. They classify eye-gaze data into seven areas (left, left-front, center, right-front, right, room mirror and side mirror) according to their direction. This classification is the good way to simplify eye-gaze data, but the time relation between eye-gaze data is missing. On the other hand, in this paper, we use the original eye-gaze data, i.e., time-series data, without the missing of time relation to represent driver's model.

3 Detail of Eye-Gaze Data

This section explains the detail of eye-gaze data we used. In a general way, the test driver manipulates a driving simulator such as [7]. At the same time, their eye-gaze data are recorded with those time-stamps. However, the eye-gaze data in different situations have the same time-stamp by the above way. This is not suitable for our propose. Therefore, we made a video of a driving simulator, and recorded the eye-gaze data of test drivers during they are watching the video. The model of a driving simulator is provided by the Research Project of Agent Mediated Driving Support of Nagoya University[2]. The length of the video is 137 s (30 fps), and the video consists of various typical traffic situations. For example, Fig. 1 shows the situation of **street parking vehicle**. A driver must circumvent the street paring vehicle. Figure 2 shows the situation of a **stop sign**. A driver must stop obeying the stop sign.

Fig. 1. Street parking vehicle

Fig. 2. Stop sign

We introduced an eye tracking system **Tobii Eye Tracker 4C** developed by Tobii[3] to record eye-gaze data of test drivers. Each eye-gaze data consists of

[2] http://www.mirai.nagoya-u.ac.jp/.
[3] https://www.tobiipro.com/.

a frame number and X-Y coordinates on a display. The resolution of the display is 1920×1080. A sample of the original eye-gaze data is shown in Table 1. The shape of these data is unfitted for the training data of a neural network. Thus, we standardized the original data as shown in Table 2. The average and standard deviation of the standardized data is 0 and 1, respectively. Figures 3 and 4 show the eye-gaze data of an example test driver in 2D and 3D views, respectively. The eye-gaze data are scattered in the display, but we noticed that there is a certain trend in the data. It seems that these trends have driver's intentions or habits.

Table 1. An example of original eye-gaze data

Frame number	Gaze_X	Gaze_Y
0	340	451
1	335	457
2	340	468
3	338	463
4	341	455

Table 2. An example of standardized eye-gaze data

Frame Number	Standardized_X	Standardized_Y
0	−2.085	−0.695
1	−2.098	−0.631
2	−2.085	−0.514
3	−2.090	−0.567
4	−2.082	−0.652

4 Training of Recurrent Neural Network

This section shows the structure of the recurrent neural network we developed. We developed a recurrent neural network with LSTM by **PyBrain**[4] as shown in Fig. 5. The network consists of three layers: input layer, LSTM (hidden) layer, and output layer. The input layer receives the sequence of standardized X-Y coordinates. The LSTM layer consists of three gates (input gate, output gate, and forget gate) and memory, and receives the previous output data in addition to the next input data. This structure of LSTM layer is able to learn the long-term dependencies of the sequence data. The output layer outputs the predicted

[4] https://github.com/pybrain.

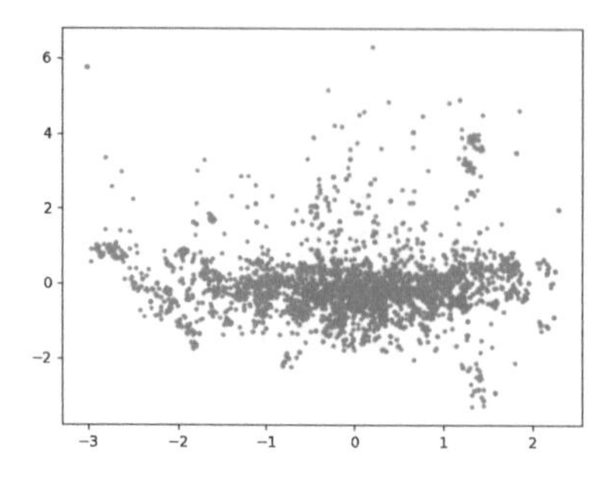

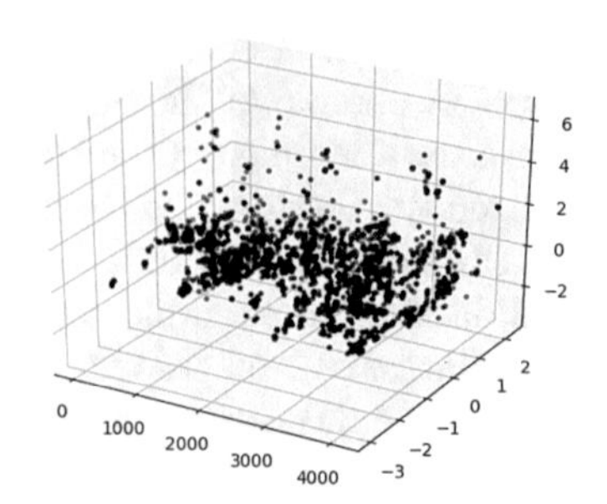

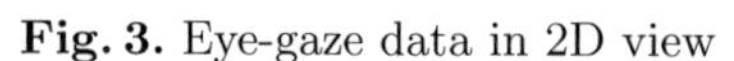

Fig. 3. Eye-gaze data in 2D view

Fig. 4. Eye-gaze data in 3D view

X-Y coordinates because the teacher signals of training data are eye-gaze data of 3 frames later. We set the parameter of the epoch, i.e., the number of learning times, to 200 times. We confirmed that the setting of the epoch is enough to converge a mean square error between input and output data for eye-gaze data as shown in Fig. 6. Figures 7 and 8 show an example of predicted results in X and Y coordinates, respectively. You can find that the trained network can almost predict the future eye-gaze coordinates with little error. We consider that the error of prediction represents the similarity of drivers to identify the personal driver. Thus, we report the comparison of identification accuracy in the next section.

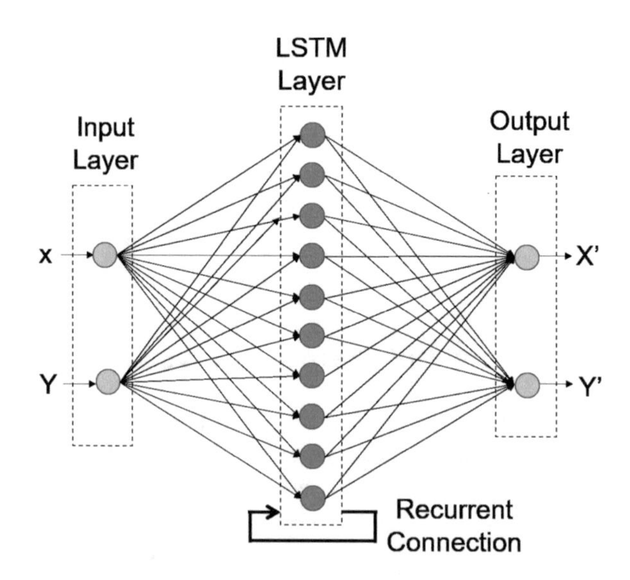

Fig. 5. Recurrent neural network and LSTM

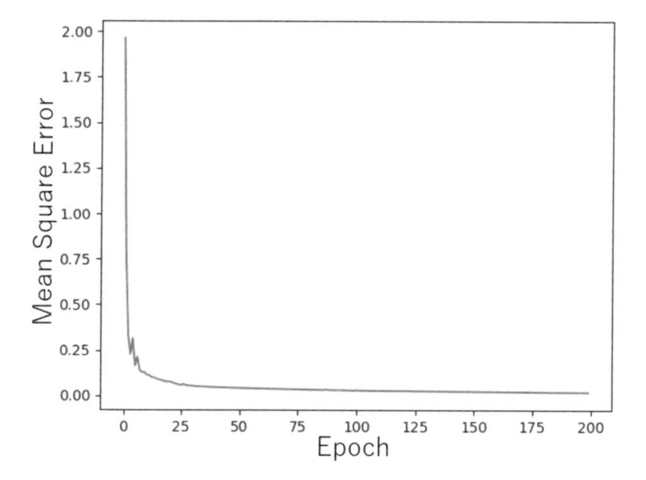

Fig. 6. Convergence of mean square error

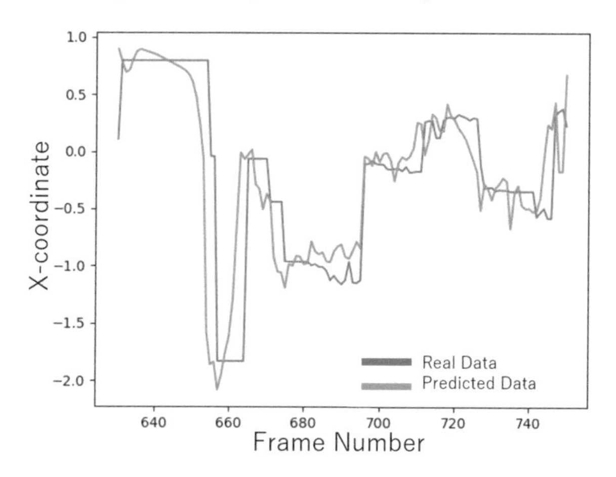

Fig. 7. Prediction of eye-gaze moving on X-coordinate

5 Comparison Results

This section shows the comparison of predicted results by the trained network, and discuss the ability of identification for drivers. We collected the eye-gaze data of twenty test drivers, who are students at Sugiyama Jogakuen University. All of the test drivers have driver's licenses. Each test driver watches the same video of driving simulator twice. The data of first time is used for training data, and the data of second time is used for evaluation. We built twenty different networks for above twenty test drivers. Each network represents the intent or habit of each driver. Here, we pick up three situations in the video: intersection (480–600 frames), street parking (630–750 frames), and stop sign (1080–1200). The video length of all the above situation is 120 frames (4s).

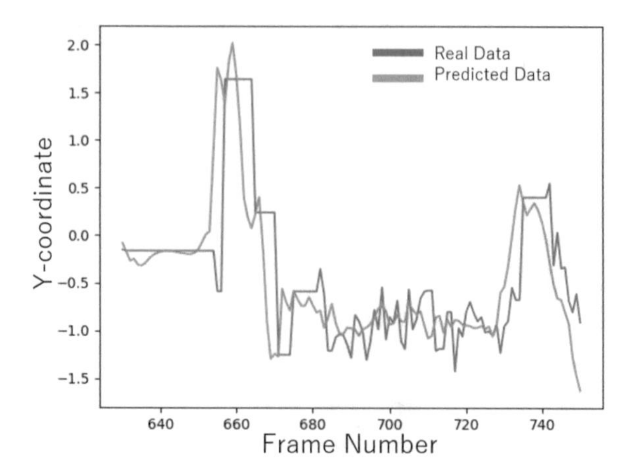

Fig. 8. Prediction of eye-gaze moving on Y-coordinate

Table 3 shows the comparison of mean squared errors in the case of an intersection. The result indicated that the errors of 17 test drivers are less than the average of errors, but they are not equal to the minimum error. It means that no one cannot be identified completely, but the trained network can partly represent the unique intention and habit of eye movement. This fact is not perfectly good, but it implies that the trained network has the potential to emulate the capability of the individual human driver. Table 4 shows the comparison of mean squared errors in the case of a street-parking vehicle. The result indicated that the errors of 14 test drivers are less than the average of errors, and only one driver can be identified by the trained network. The overall trend of this result is similar to the previous result, but the accuracy is inferior to previous one. We think that the eye movement at the situation is almost the same. In fact, drivers

Table 3. Comparison of errors at intersection (frame: 480–600)

ID	Error of target	Avg. of errors	Min. of errors	Max. of errors
1	**0.646**	1.018	0.198	2.556
2	**2.864**	3.719	0.287	16.631
3	**0.602**	0.679	0.180	1.701
4	**1.161**	1.423	0.905	2.706
5	**0.431**	1.207	0.100	4.660
6	**6.322**	10.685	0.123	43.538
7	**0.587**	2.894	0.347	11.720
8	**0.863**	2.041	0.180	6.845
9	**0.251**	0.727	0.169	2.343
10	**0.769**	0.829	0.459	2.040

(*continued*)

Table 3. (*continued*)

ID	Error of target	Avg. of errors	Min. of errors	Max. of errors
11	3.919	1.493	0.150	4.786
12	**1.749**	3.737	0.598	7.777
13	**1.008**	1.244	0.088	4.622
14	**0.451**	1.402	0.404	4.598
15	**0.690**	0.858	0.091	3.218
16	2.771	2.560	0.582	8.971
17	5.521	1.208	0.084	5.521
18	**1.095**	1.226	0.076	4.380
19	**2.940**	3.423	1.350	6.452
20	**0.174**	0.854	0.074	4.258

should always cast careful at stopping vehicle. Table 5 shows the comparison of mean squared errors in the case of a stop sign. The result indicated that the errors of 13 test drivers are less than the average of errors. The accuracy is the worst in three situations. We think that this case has the same cause of the previous one. Consequently, the trained network enables to learn the unique intention and habit of drivers, but the individual identification cannot be realized on this condition.

Table 4. Comparison of errors at street parking (frame: 630–750)

ID	Error of target	Avg. of errors	Min. of errors	Max. of errors
1	**0.539**	0.733	0.110	2.448
2	1.257	0.734	0.057	2.733
3	1.084	1.070	0.268	5.683
4	**0.439**	0.713	0.134	2.957
5	**0.559**	0.848	0.136	2.101
6	2.335	0.806	0.031	4.132
7	**0.248**	0.945	0.238	3.515
8	**0.150**	2.328	**0.150**	8.789
9	**0.580**	0.791	0.071	2.998
10	**2.096**	5.848	0.182	11.609
11	3.925	0.773	0.069	3.925
12	**0.443**	0.867	0.072	4.095
13	**0.195**	1.703	0.064	9.398
14	**0.700**	0.790	0.032	3.237
15	**0.296**	0.715	0.142	2.432
16	**0.245**	2.030	0.045	15.563
17	8.261	2.972	0.201	8.261
18	**0.398**	0.867	0.093	5.116
19	2.005	1.416	0.050	4.353
20	**0.250**	1.335	0.082	4.015

Table 5. Comparison of errors at stop sign (frame: 1080–1200)

ID	Error of target	Avg. of errors	Min. of errors	Max. of errors
1	1.550	0.987	0.170	2.662
2	1.714	1.384	0.518	3.670
3	**1.350**	1.586	0.048	5.043
4	**0.375**	0.983	0.257	2.608
5	**0.484**	0.950	0.086	2.208
6	1.084	1.054	0.227	2.441
7	**0.613**	1.200	0.114	3.733
8	**1.573**	1.709	0.299	4.855
9	**1.405**	1.769	0.386	4.671
10	**0.378**	1.994	0.378	6.076
11	3.385	1.779	0.835	3.385
12	**1.546**	4.595	0.896	16.976
13	1.674	0.942	0.101	3.516
14	1.644	1.280	0.279	2.678
15	**0.873**	1.504	0.322	3.780
16	5.335	1.095	0.128	5.335
17	**0.955**	1.787	0.342	5.546
18	**2.168**	9.182	0.292	29.782
19	**0.727**	1.341	0.156	4.101
20	**0.865**	3.118	0.075	5.937

6 Conclusions

This paper focused on eye-gaze data of drivers and developed a recurrent neural network with Long Short-Term Memory block which models driver's eye-movement. We assumed that the sequence of eye-gaze data includes the unique intent and habits of drivers, and tried to identify individual eye-gaze data by the trained network. Our results indicated that trained network represents the intentions and habits of drivers, but the ability of identification is inadequate. In order to improve the ability, we need to consider other factors of drivers such as sensors of pedals in addition to eye-gaze data.

Acknowledgment. This work is supported by the Research Project of Agent Mediated Driving Support of Nagoya University. We are truly thankful for the members of the group.

References

1. Hochreiter, S., Schmidhuber, J.: Long short-term memory. Neural Comput. **9**(8), 1735–1780 (1997)
2. Graves, A., Fernández, S., Schmidhuber, J.: Bidirectional LSTM networks for improved phoneme classification and recognition. In: Duch, W., Kacprzyk, J., Oja, E., Zadrożny, S. (eds.) Artificial Neural Networks: Formal Models and Their Applications - ICANN 2005, pp. 799–804. Springer, Heidelberg (2005)

3. Sutskever, I., Vinyals, O., Le, Q.V.: Sequence to sequence learning with neural networks. In: Ghahramani, Z., Welling, M., Cortes, C., Lawrence, N.D., Weinberger, K.Q. (eds.) Advances in Neural Information Processing Systems, vol. 27, pp. 3104–3112. Curran Associates, Inc. (2014)
4. Mima, H., Ikeda, K., Shibata, T., Fukaya, N., Hitomi, K., Bando, T.: Estimation of driving state by modeling brake pressure signals. IEICE Tech. Rep. NLP **109**(124), 49–53 (2009). (in Japanese)
5. Okada, S., Hitomi, K., Chandrasiri, N.P., Rho, Y., Nitta, K.: Analysis of driving behavior based on time-series data mining of vehicle sensor data. Proc. Forum Inf. Technol. **11**(4), 387–390 (2012). (in Japansese)
6. Horiguchi, Y., Suzuki, T., Suzuki, T., Sawaragi, T., Nakanishi, H., Takimoto, T.: Analysis of train driver's visual perceptual skills using Markov cluster algorithm. J. Jpn. Soc. Fuzzy Theor. Intell. Inform. **28**(3), 598–607 (2016). (in Japanese)
7. Tanaka, T., Fuzikake, K., Yonekawa, T., Yamagishi, M., Inagami, M., Kinoshita, F., Aoki, H., Kanamori, H.: Analysis of relationship between forms of driving support agent and gaze behavior-study on driver agent for encouraging safety driving behavior of elderly drivers. In: Proceedings of Human-Agent Interaction Symposium (HAI), p. 2 (2017). (in Japanese)
8. Kamisaka, T., Noda, M., Mekada, Y., Deguchi, D., Ide, I., Murase, H.: Prediction of driving behavior using driver's gaze information. IEICE Tech. Rep. Med. Imaging **111**(49), 105–110 (2011). (in Japanese)

Research Issues to Be Useful in Educational/Learning Field

Toyohide Watanabe(✉)

Nagoya Industrial Science Research Institute,
#203, 2-1-16 Seimeiyama, Chikusa-ku, Nagoya 464-0087, Japan
watanabe@nagoya-u.jp

Abstract. The educational/learning research field has counted up over 60 years since "Educational Technology" had been established as one of new research frontiers. For 60 years, the research viewpoints have been shifted from teaching-specific functions to learning-based functions, owing to the enormous evolution of information technology. This shift has enforced to evolve research issues with smart teaching/learning support systems. This research field is generally characterized as one of field-fusion types of existing related research fields, and may be often said that the original discussion points are not sharply recognized. In this paper, we survey our current educational/learning research viewpoints and re-consider innovative research topics over the accumulated results with a view to attaining the advanced educational/learning paradigm or framework as one step-up subject in the next age.

Keywords: Usefulness · Knowledge transfer scheme · Cognitive load
Goal achievement · Learning support · Reasoning

1 Introduction

The researches in the educational/learning field have been focused on from an engineering point of view, and the effects which encourage to organize teaching actions and learning actions constructively have been continuously tried. In particular, the information technology developed the support functions/tools for teaching/learning actions effectively, and contributed to promote the foundation and evolution of educational/learning researches. Although the initial phase of educational technology about 60 years ago was estimated to be very dependent on industrial technology, information science, human science, behavior science, etc. [1, 2] and focused on making teaching actions effective and effectual in practice, the information technology changed the educational/learning environment tremendously and enforced to look upon the learning actions as the main research view. It is so desirable or hopeful that the educational/learning field is likely looked upon as a field-fusion or an interdisciplinary, but its hope is not always successfully attained. This is dependent on various reasons: (1) the activity is dependent on the intelligence and skill, related to various characteristics of learners; (2) the activity is composed of various evaluation criteria complicatedly; (3) the activity is not transient, but is repeatedly successive.

G. De Pietro et al. (Eds.): KES-IIMSS-18 2018, SIST 98, pp. 120–129, 2019.
https://doi.org/10.1007/978-3-319-92231-7_13

In this paper, we address the research issues without being sufficiently until today as the important subjects in the educational/learning field.

2 Research View

It may be difficult to answer the questions: "What is the research in the educational/learning field?", "What are the clue discussion points in the educational/learning research?", and so on. The research viewpoints may be changeable, corresponding to the age requirement, or the science/technology evolution, the social policy. Of course, a newly discovered method, concept, approach and so on must be always clearly shown as the research result. Additionally, they must be useful or applicable to the other related researches, or the successive researches. The research result/effect is to be useful. Of course, this usefulness is not practical, but is to influence to others.

The researches in the educational/learning filed are largely categorized into the information domain, the cognitive domain, and the education domain in accordance with their own original-domain interests, and focus on the research objectives derived from their own knowledge bases [3]. Table 1 arranges the features about research points, research results, research approaches to be observed for learners from these three domains [4]. Also, Fig. 1 shows their own positions for learners: activity support process, thinking process, and working process.

Table 1. Research views in information, cognitive and education domains

	Existing field	Methodology	Viewpoint	Learner	Research interest
Cognitive domain	Humanities (cognitive science, psychological science, etc.)	Observation, interview, etc.	Modelling, conceptualization, etc.	Internal behaviors of learners	Thinking process
Education domain	Social science (education, etc.)	Investigation, observation, experiment, etc.	Data analysis, experiment survey, etc.	External results of learners	Working process
Information domain	Engineering (information engineering, etc.)	Development, improvement, design, etc.	Information technology, etc.	Support of learners'activities	Activity support process

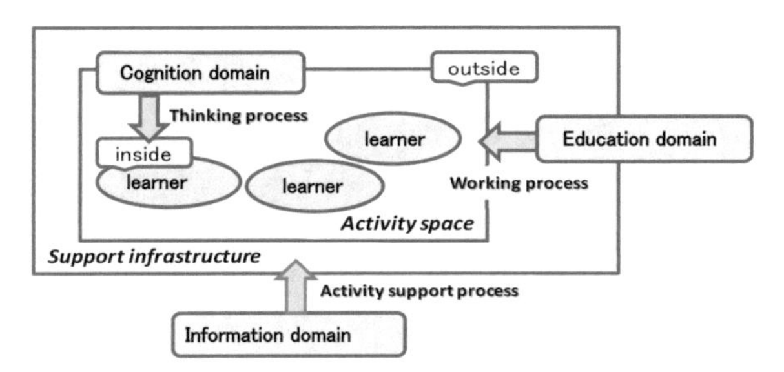

Fig. 1. Research positions in information, cognitive and education domains

The research article is different from the report. The report is to analyze the data about performed contents or investigated contents, and arrange the features, observably extracted from the analyzed results. Also, it can give comments by using only the information, extracted from analyzed data. While, the research is meaningfully required to make the performance conditions, environments, targets, etc. clear as the pre-conditions, next discuss and validate the proposal, method, approach, etc. concretely, and then evaluate the performed results for the estimated effects logically.

In case of experiment and investigation, it is important whether learners who participate into the experiments or investigations are chosen with random assignment. If not so, the results must be concluded after having made individual features of participated learners clear. It is not generally assured that the results derived from some special cases could be applicable to other cases, as they are. The results derived in some cases should be not directly applied to the others without any discussions; whether the common measurements between different cases are meaningful or not may be judged. Currently, all situations or all performances are not evaluated or taken in consideration. In order to propose the meaningful conclusions or fruitful remarks under the constructed conditions, the logical discussion or reasonable verification is necessary.

3 Research Issues

We refer to the important research issues to be directly or successively achieved in the educational/learning field. These issues are not newly found out, but have been discussed continuously until today. Our discussion points for these traditional research issues are to make the inherent research viewpoints clear as the main subject. Before our discussion we must understand one relationship between education, teaching, understanding-oriented learning and thinking-oriented learning from a viewpoint of educational/learning scope. Figure 2 shows the relationship, which was drawn on our knowledge transfer scheme [5]. The knowledge transfer scheme is basically illustrated by the learning process for learners, depending on the learning environment. Here, the

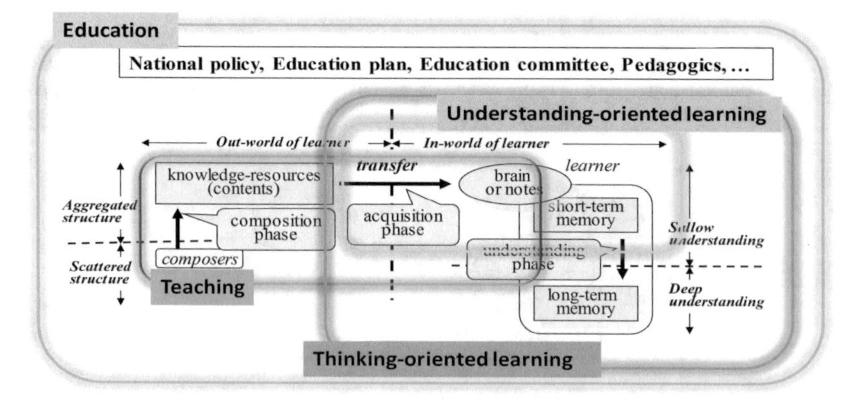

Fig. 2. Relationship among education, teaching, understanding-oriented learning, and thinking-oriented learning

generally used word "education" is corresponded to "teaching" in Fig. 2. Also, "understanding-oriented learning" is a traditional learning style, and "thinking-oriented learning" is able to cover the currently focused topic "active learning".

(1) **What are characteristics of field-specific originality?**

The research viewpoints are, more or less, dependent on the research carriers. Although this field is called as the style of field-fusion or interdisciplinary, it is not always easy to generate the original results from a viewpoint of field-fusion without the features of individual domains. When we are asked what is the originality in the educational/learning research, it is not easy for us to answer this question explicitly.

In the learning activity, learners must acquire the knowledge and understand it: the learner is the main actor in the learning activity. Thus, we must recognize again that in the research of educational/learning field the most originally focusing point is a learner. All research viewpoints must focus on learners' behaviors, and also the effectual results must evaluate and validate the actions/re-actions of learners even if any approaches as shown in Table 1 were applied to individual cases.

(2) **What is teaching action or learning action?**

Initially, the research in this educational/learning field was more or less to support the teaching actions effectually and effectively: development of authoring tools, composition of authoring resources, design of course-ware, development of lecture management system, etc. However, with the rapid expansion of information technology, the teaching actions have been replaced by the learning actions. In Fig. 2, the teaching actions can be interpreted as one means of learning actions. The teaching procedure is included into the understanding-oriented learning, except for the composition functions of knowledge resources. Although the learning action is instructed in the centralized learning environments, by the teaching action in the traditional classroom-based lecture style, the learning action in the currently distributed learning environment is performed by the learner-driven style, and is supported complementarily with help of the teaching action, if necessary.

It is important that these learning processes should be successive and stepwise, but are not transient. Namely, the teaching action or learning action on the course-ware should be implemented successively so as to make the learning process of knowledge acquisition, knowledge understanding, and knowledge utilization systematically. Of course, the process must be designed so that learners with various characteristics or different abilities can satisfy owing to their wills, their interests, and their situations.

(3) **What is learning support?**

The objective of learning support is to develop the means, methods, and mechanisms to support the learning actions effectually and effectively. The effectiveness and effectuality are different in point of achievement. The effectiveness is related to the learning effects, while the effectuality is so to the learning process. The currently reported research results do not focus on the effectiveness, but contribute to the effectuality. In particular, to develop the learning environment and learning support system with the information technology is typical to make the learning process effectual.

Concerning to this viewpoint, the goal achievement theory is proposed and the goal is composed of two views: performance goal and understanding goal (we use the term "understanding goal" in place of "learning goal" though the term "learning goal" was used in the original definition) [7, 8]. The performance goal is related to the effectuality, and the understanding goal is so to the effectiveness. In according with the goal achievement theory, the learning support must look upon these two goals as the target. The performance situation can be managed by setting the indicator on the course-ware and monitoring its progress timely. While, the understanding situation cannot easily be handled: it is difficult to deal with understanding degree and judge it. The mechanism with the learner (or student) model and its related functions, like intelligent CAI (Computer Assisted Instruction) or ITS (Intelligent Tutoring System), is a candidate for the management of understanding degree.

(4) **What is support of learner?**

This question is very similar to the previous one: here, we focus on the learning action which a learner must perform in the learning process. Namely, we must look upon the support as the mechanism which influences any interaction to learning actions. Of course, the degrees of interactions are different for individual learners.

In case of using information tools or systems with high-functionality it is questionable that learners who use personal computers or not, or learners who major in the information domain or not, will be able respectively to get the same performance. Some may do pre-exercises with a view to using the high-functionality personal computer. It is important to establish the support for means, methods, functions, mechanisms, processes, and environments, suitable to the characteristics, the interests, and the abilities of individual learners. The word "support" is often used in our human activity: it aids human activity with the cooperation between human and computer or with help of computer. "Support" is important even between human and human.

For example, we consider the driver (or driving) support or the medical-doctor (or diagnosis) support. Are these the same as "learner support"? In the driver support it is allowable that some works which the driver operates in this driving by him/her-self can be replaced by other persons or computers; and it is the best to replace all works by computer as the final goal. Also, in the medical-doctor (or diagnosis) support, the diagnosis work is composed of many sub-tasks, and it is allowable even if some of them would be performed by assistants and/or with measurement tools. To do with help of other persons or tools may be better than to do it by the medical-doctor in hem/her-self. This is because the working time can be reduced when any supports were applied. So, in both the driver support and the medical-doctor support the effects of support lead to make the working-performance such as the processing quality higher or the working-cost such as the processing time, the labor amount, etc. lower. However, is this effect or this phenomenon is similar to the learner support?

In the learner support, the work or sub-works in the learning process cannot be replaced by other persons or some tools, but all must be done by the learner in him/her-self. Of course, it is permitted that one procedure, and/or some parts of work or sub-work could be performed cooperatively with help of assistants or effectually by using tools. This situation with help of assistants or utilization of tools is "supporting" in the learner support. The requirement that all parts of work must be performed by learner may generate another problem. The main view in the driver support is to go to the destination place speedily and safely with good cost-performance, and also that in the medical-doctor support is to diagnose the patients in ill conditions effectively and correctly with good practice. The objectives of these two supports are similar to achieving the performance goal in the goal achievement theory. However, the main view in the learner support is to promote that the learner could understand the learning contents sufficiently and use them applicably. The objective is rather to achieve the understanding goal than to attain the performance goal [8].

The cognitive load theory points out as for discussing how to use the working memory in a person's activity: the person feels various loads when he/she tries to challenge some assigned tasks [9, 10]. The loads are categorized into the intrinsic load, the extraneous load and the germane load. The intrinsic load is said that the more difficult the issue is, the lower the proficiency is and also the more largely the person feels the loads of the issue. As the extraneous load is not directly related to the contents of issue, this load is un-necessary if the learner could recognize the meanings of issue smartly. However, as in many cases the learner repeats the processing steps with "try and error", he/she feels the extra load by his/her own consideration. The load which was generated un-necessarily by this behavior is called "extraneous load". On the other hand, the germane load which is a cognitive resource is similarly equal to the consciousness for the advanced or related issue. Therefore, when we consider the learner support, it is necessary and important that we should not only design and develop the tools, functions, and systems to be useful in the learning process, but also take in consideration the cognitive load of learner.

(5) **What is research evaluation?**

The research must be positioned for the objective, and verified for the approach. The evaluation is discussed together with the experiment/investigation and also estimated logically with the consideration. Even if the evaluation could not propose the necessary results it is worth as long as the logical verification can. Therefore, the research results must be derived from the objective after making the performance conditions, the characteristics of targets, and the features of application explicit.

If the members in one research laboratory developed the prototype application system and evaluated the usability and operability (but not the performance and precision), can this system be evaluated sufficiently? The answer is "No". Even if the evaluators were different from the members in the system development group, its answer is "No". If all members in the laboratory have already known the operation procedures, the usage methods, and the interaction process through the laboratory meetings, the laboratory demonstrations, etc. and moreover some may have used this system before the evaluation, it is not clear what the evaluation results can represent.

Additionally, we consider another example: in order to look for usable effects of information tools in one classroom lecture, the pupil answered the questionnaires after their exercises. In this case what can this data analysis indicate as the general result? Unfortunately, this result can report only the evaluated situation of class members, but cannot indicate the future direction or the applicability. This is because the respondents of the questionnaires are not chosen with random assignment. As one possibility about the usage for this data analysis, the group with the same properties as the questionnaire group is suitable, but it is not easy to find out such a group or organize such a group. Even if such a group were found out, is the trial evaluation meaningful?

This fact suggests that we must deal with missing data under the reasoning. The missing data is defined as a data set which cannot be measured: in the phenomenon with a front-back relationship it one was measured, another cannot be measured. Therefore, under its relationship the concept was graphically represented in Fig. 3 [11]. Moreover, concerning to the data processing based on statistics, to make the purpose of analysis clear is very important: the paradox which Yule and Simpson found out is famous. Figure 4 shows the paradox. This proves that the categorized groups derived by one attribute do not always show the same feature as a set of members, gathered from their groups. Therefore, when we analyze statistically experiment/investigation data, the features of analyzed data may be different if according to the classification objectives the attributes were not selected so as to be consistent to the objective of statistical analysis.

	Treatment group	Control group
Result with intervention	Data in treatment group	missing
Result without intervention	missing	Data in control group
Covariate term	Commonly used variables in all	

Fig. 3. Missing data

	men		women	
	passing	failure	passing	failure
No-lecture	4	3	2	3
Lecture	8	5	12	15

Merging

	men & women	
	passing	failure
No-lecture	6	6
Lecture	20	20

Fig. 4. Yule and Simpson's paradox

The research evaluation in the educational/learning field is in many cases applied to learners' questionnaires because the research view focuses on the learners' activities. The evaluation means is dependent on the quality-based values. This means that it is difficult to make the evaluation result objectively, and that the evaluation result cannot verify generally-interpreted or possibly-estimated cases by the logical reasoning. In the research which develops things, to measure the functionality and the performance of the things is commonly used, and in the engineering scope the experiments to measure or evaluate the performance and the functionality (or precision) is confident because the measured values are repeatedly reproduced. Thus, the evaluated effects are useful. On the other hand, in the person-behavior-based scope in order to measure the ability of persons or activity of persons using some criteria, the persons' features must be the same. It is not easy to satisfy this requirement.

(6) **What is research result?**

What is that the research result is useful? Is it that the result was gotten as the research objective and the estimated effect as it was? Even if the estimated effects could not be gotten or the desirable situation could not be observed sufficiently, the research will be able to give the effective influences to other related researches and successive researches when the results or the discussions related to the results generate the new directions/research-views or make the causes of the failures perspective. This indicates truly "useful research". It is never that the application-specific or practical-oriented research is useful, but fundamental or theoretical research is not so.

For example, one says that an apple drops down; another also says that an orange drops down, and so on. Each sentence represents only one instance as it was. Of course, these instances may be individually "discovery"; however, if the other person had a question "Does a pear drop down?", who can answer? When another person founded out the instance that the pear drops down, the pear is recognized like "apple", "orange", etc. so that the pear drops down. If an orange and an apple were recognized as a kind of fruits and the pear is a fruit, we can answer timely "the pear drops down". This problem is easily resolved if the facts such as "All fruits drop down.", "Things always drop down.", etc. are founded.

If the concepts of class, class hierarchy and deductive reasoning schema are established as common knowledge, it is easy to manipulate the above example. In the educational/learning field, the above results are being often reported. Though the reporting may be useful as long as the result is the first instance, it cannot become useful if not so. In the knowledge structure, the operations such as generalization and specialization, aggregation, grouping, etc. are important to refine the acquired/memorized knowledge organically [12], and make them re-useable in the high-level knowledge management/maintenance activity. The result is useful. On the other hand, it is hesitated to say that the research effects which can be derived or inherited imperatively from the already-known instances are useful.

4 Adventure to Future

Now, as one advanced topic in the school education the phrase "from memorization learning to thinking learning" is very wonderfully popular. The typical evolution issues are: (1) flipped teaching/learning; and (2) active learning. However, can these issues promote a new viewpoint in the learner's activeness? The core procedure in the flipped learning is similar to the preparation that learners study the contents, preset in the school, by themselves before the lecture in the classroom. Also, the core procedure in the active learning is a combined process with pre-learning and collaborative learning. Both issues are the same in the viewpoint that learners should perform actively by themselves. Generally, this viewpoint is a difficult assumption: if learners could do so by themselves without any controlling/monitoring means, it is no problem. It is necessary to propose a framework that learners could perform actively by themselves with any support means such as system, environment, and interface, anywhere and anytime, if necessary.

On the other hand, the topic about the diversity of person's ability is another problem. Persons grow up according to their ages, and can live with various individualities and different values in their own life styles. Also, the current/future society will request various life style to all persons. What must we consider and design for persons' individualities? So, the society enforces to shift person's ability criteria from the vertical diversity to the horizontal diversity. The vertical diversity is a criterion based on the ability of knowledge learning. While, the horizontal diversity is criterion on the basis of the corresponding axes of various intelligences and various technologies. This may lead that the next learning must be centered on the personality-oriented ability, but not to give the same process and the same contents to all.

What basic ability in the educational/learning field is needed as the critical evaluation in the horizontal diversity? The educational/learning methodology based on knowledge has been constructed as a common framework for all who participate to the social activity until today cumulatively. The learners in our social activity need to acquire the knowledge, understand it with their own existing knowledge, and use it on their own activities, and the education support system encourages the learners to do it under the support organization and the support policy. However, what is the critical evaluation ability which the person chooses when the horizontal diversity becomes real in place of current vertical diversity? The first problem is to determine what are the good abilities for individuals. However, is it possible to determine the learners' abilities before learners perform in practice? There are yet late developers.

What situation in the educational/learning activity can be estimated, 50 years after or 100 years after? How is the educational/learning field positioned for person's activity? In the age that the applied artificial intelligence with the deep learning mechanism was evolved more and more, or intelligent skilled computers with more smart abilities were developed innovatively, which contents do this field deal with? The year 2045 is called "technical singularity" and it is said that the computer ability will be over the human ability.

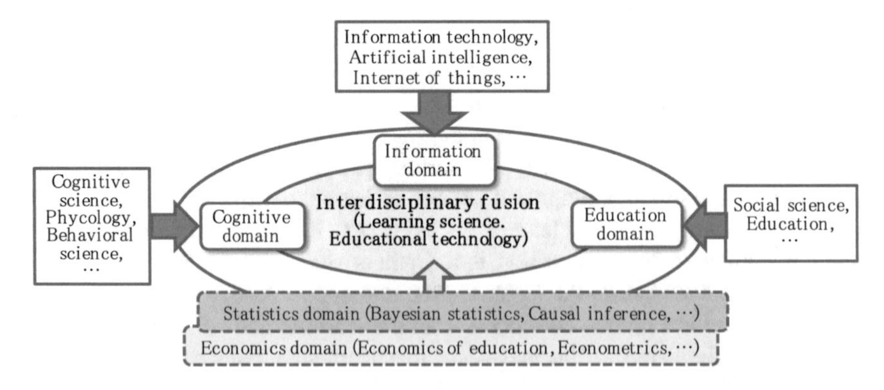

Fig. 5. A new framework of educational/learning field

5 Conclusion

The research viewpoint in the educational/learning field is first to focus on learners' behaviors and then discuss the process of knowledge acquisition, knowledge understanding (or refinement), knowledge utilization (or presentation), and knowledge thinking (or reproduction) [12] as a framework to make learners' abilities growing up.

In addition to the feature in this educational/learning field that the learner is a key discussion point, another feature is that the learning process is neither independent nor transient. The learning process is designed under the systematic and successive structure. Thus, the research view must focus on the systematic and successive structures and discuss the research evaluation and research effects on the structure with respect to the usefulness. Also, in order to be systematic and successive the learning process is not controlled as a simple procedure, but must be managed with mechanism such as the execution timing of individual tasks, the variation of task combination, the dynamic scheduling for time-variant changes and situations, etc. The research stance is not "research for research", but must be "research for usefulness".

Figure 5 shows that the educational/learning field should be organized with new two domains "statistics domain" and "economics domain", in addition to the current three domains such as information domain, cognitive domain, and education domain. This proposal view is always derived from the discussions in Sect. 3(5) [11].

References

1. AECT Task Force on Definition and Terminology: The Definition of Educational Technology, p. 16. AECT (1977)
2. Japan Society for Educational Technology (ed.): Educational Technology Dictionary. Jikkyo Shuppan Co. Ltd. (2000)
3. Watanabe, T.: Trends in teaching/learning research through analysis of conference presentation articles. In: Proceedings of KES/IIMSS 2016, pp. 309–322 (2016)
4. Watanabe, T.: Architectural framework in next learning support environment. In: Proceedings of ACM/IMCOM 2017, #S1-6 (2017)
5. Watanabe, T.: Learning support specification, based on viewpoint of knowledge management. In: Proceedings of E-Learn 2012, pp. 1596–1605. AACE (2012)
6. Watanabe, T.: Research view shift for supporting learning action from teaching action. In: Proceedings of KES/IIMSS 2017, pp. 534–543 (2017)
7. Dweck, C.S.: Motivational processes affecting learning. Am. Psychol. **41**, 1–9 (1986)
8. Watanabe, T.: A framework of information technology supported intelligent learning environment. In: Proceedings of ACM/IMCOM 2016, #11-4 (2016)
9. Sweller, J., van Merrienboer, J.J.G., Paas, F.G.W.C.: Cognitive architecture and instructional design. Educ. Psychol. Rev. **10**, 251–296 (1998)
10. Watanabe, T.: Analysis of learning-support scheme, based on cognitive load. In: Proceedings of E-Learn 2015, #45716. AACE (2015)
11. Pearl, J.: Causality: Models, Reasoning, and Inference, 2nd edn. Cambridge University Press, Cambridge (2009)
12. Watanabe, T.: Knowledge handling model to design learning activity. In: Proceedings of E-Learn 2005, pp. 1566–1571. AACE (2005)

Current State of the Transition to Electrical Vehicles

Milan Todorovic(✉) and Milan Simic

RMIT University, Melbourne, Australia
s3632981@student.rmit.edu.au

Abstract. In this research report we present the current state of the transition from traditional, internal combustion engines vehicles, to electrical vehicles. The main characteristic of this transition is that new generation cars are matching the cost and performance of traditional petrol cars. Transition to electric vehicles is driven by the environmental sustainability, in the first place, economy, government policies, inherent automotive industry dynamics and consumer preferences. Transition is presented from global perspective in addition to specificities in Australian context. The conclusion is that such transition is a major disruption affecting the whole economy. It is characterized by the convergence of mobility and energy what can bring significant benefits to the entire society.

Keywords: Electrical vehicles · Automotive industry · Transition
Environment · Policy

1 Introduction

Automotive sector is on the edge of a major transition. One of the shifts is from internal combustion engines (ICE) to electrical mobility. The other shift is a transition towards automation of driving, what ultimately will lead to the full autonomy. These changes will unsettle the market, create uncertainty and redistribute power within the industry by requiring new strategic orientations and forward looking policies.

Electrical vehicles (EV) are approaching a turning point as new generation cars are matching the cost and performance of traditional petrol cars. These cars are offering drivers everything they want but without pollution. Growth of electric vehicles' production is driven by the convergence of falling prices and improved performance of batteries, consumer demand and especially regulatory actions. One-third of the electric vehicle's cost is the cost of batteries. They are rapidly coming down in price. A 2016 study by Bloomberg New Energy Finance found that battery prices fell 65% since 2010, and 35% just in 2015, reaching US$350 per kWh. This study predicts that electric car battery costs will be below US$120 per kWh by 2030. By 2022, electric vehicles will cost the same as their petrol counterparts [1]. While previous electric cars had either the range, or the price, the new cars have both. That was the expected breakthrough which prompted series of new policies and new product development announcements, from major car companies, and new projections of future sales.

G. De Pietro et al. (Eds.): KES-IIMSS-18 2018, SIST 98, pp. 130–139, 2019.
https://doi.org/10.1007/978-3-319-92231-7_14

Total costs of an EV ownership include purchase cost and running costs, such as maintenance and energy, i.e. electricity cost. In addition to that, we have to consider offsets like governments' subsidies, tax credits, and other incentives in various markets, or countries. Cost structure is not discussed comprehensively in this paper as it will be a subject of our further research and data collection. Finding will be disseminated in the future publications. Another important issue is the cost of transition to new EV technology which includes infrastructure requirements, as well as, broad industry support to EV manufacturers. Nearly all industry will be affected, starting of course with energy sector.

Early adopters of new driving technology were primarily inspired by environmental reasons but, in the same time, they were attracted by financial savings propositions [2]. Very strong motivations for accepting electric vehicles are: new technology appeal, free supercharging across US and Europe [3], smooth and quite driving, convenience in charging at home and wireless software updates. Government policies, which provide incentives and actively support consumers' preferences, linked with good marketing campaigns, should influence faster adoption and rapid uptake of electric cars.

The transition to EVs is a part of the more broad process characterized by the convergence of mobility and energy. While the future of mobility is expected to be electric, autonomous and shared, the future of energy will be electric and decentralized. Apart from microgrids, EV, itself, is becoming a decentralized energy resource, which will provide a new manageable electricity demand, storage capacity and electricity supply when fully integrated with smart electricity grids.

Global cumulative sales of highway-legal light-duty plug-in vehicles reached 2 million units at the end of 2016 and the 3 million milestone was achieved in November 2017. Sales of light-duty plug-ins achieved a 1.3% market share of new car sales in 2017, up from 0.86% in 2016 [4]. The global ratio between battery EVs and hybrids was 66:34 in 2017, up from 61:39 in 2016 [4]. Figure 1, presents electric car stock (battery and plug-in electric vehicles) by country in 2016.

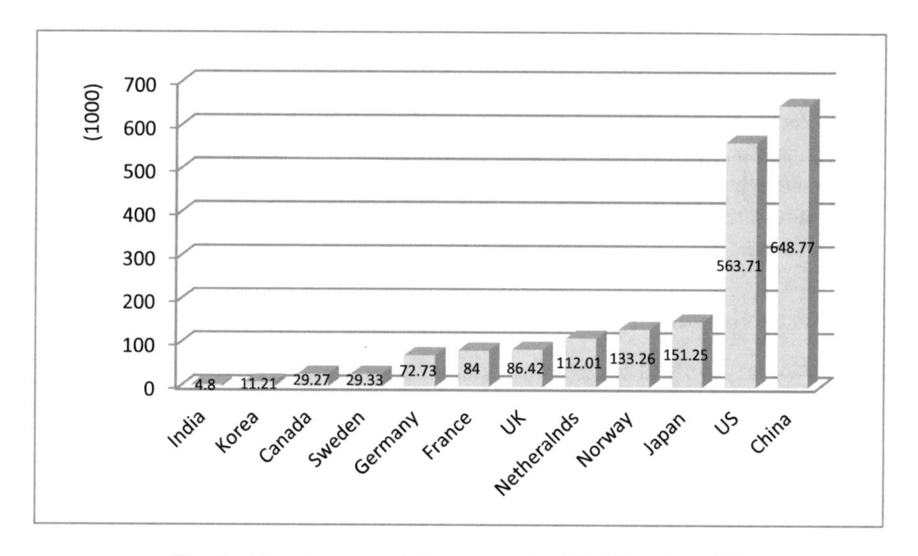

Fig. 1. Electric car stock by country in 2016 (based on [5])

2 Key Drivers of the Transition to EV

The health of the global and several national economies (China, US, EU), where the most of production facilities and EV trade markets are located, will have strong effects on the speed of the transition to EV. Government policies, global governing bodies and intrinsic automotive dynamics will foster this process based on environmental demands, public wellbeing and safety requirements. A range of new business incentives should enable economy of scale and stimulate further technology development.

2.1 Economy

There are over 1.2 billion ICE powered vehicles on the road today. If the average vehicle is worth $US20.000, this represents a potential fleet for electrifying valued at $US20 trillion [6]. This transitioning process will not be determined by technology alone, neither influenced by aggressive adoption, nor limited only on auto industry. It will simultaneously change the second largest industry in the world, the multi-trillion oil industry. Daily global oil consumption just reached 100 million barrels in June 2017 [6]. One of key benefits from the shift to electrical mobility will be lower dependence on oil considering that two-thirds of petroleum products are bound to transport. In 2015, 215 billion euros, equivalent to 88% of European market consumption, was spent for crude oil [7]. Transition to EV does not have to close down oil industry completely. We can still use diesel to generate electricity. In addition to that, some of the hybrid vehicles' solutions employ internal combustion engine to generate electricity, for the electric motor that runs the vehicle. Eventually, the oil will definitely run out in the world in some future. When will that happen depends on this EV transition. We definitely need transition to EV, with the electricity produced from the renewable sources.

There are many changes expected in the automotive job market as well. Current projections show that, in the following 10 years, there will be no reductions in employment. EU economy, for example, will benefit with 500–850,000 new job openings [7]. By 2050, forecasts predict 1% increase in GDP equivalent to the creation of 2 million additional jobs and reduction of emissions by 83% [7]. In sum, the net effect of transition to electrical vehicles will be the result of new automotive technology development which would create high quality R&D jobs, additional investments in infrastructure and higher consumer spending, based on the savings from lower EV operating costs.

2.2 Automotive Industry

The transition to EVs will cause substantial changes in automotive industry itself with emphases on value chain and required skills. There will be some job loses but overall, it is expected more job openings economy-wide. One reason for the lowering number of jobs is significantly different manufacturing process. While ICE models require about 1,400 components, EVs need only 200 building components [7]. Drive train of an EV is extremely simple, with a small number of subsystems, as it can be seen from the TESLA car, shown in Fig. 2. We can see the battery system, underneath the whole car,

Direct Current (DC) to Alternating Current (AC) converter, singe speed transmission and classical Tesla asynchronous AC motor, patented by Nikola Tesla dated February 25, 1896 [8]. Following this simplicity in construction, maintenance is extremely easier and costs much less than for an ICE vehicle.

Fig. 2. TESLA car major subsystems: battery pack, DC/AC Converter, singe speed transmission and Tesla AC motor (Photo by Milan Simic)

Transitional changes will affect primarily automotive supply chains. Traditional suppliers of ICE vehicles would need to replace gearboxes, exhaust pipes and injectors in their offerings with battery system, electric motors, braking systems. New supply chain players will emerge in forms of battery manufacturers and mining companies. It will come to the closer integrations of energy, telecommunication and transport sectors. EV manufacturer Tesla, for example, already offers lifetime free energy supercharging built in the price of their cars [9].

2.3 Policies

Many governments are considering stronger public policies to direct the car market towards electric. The breakthrough on price and performance has inspired policy makers around the world to become more visionary about electric vehicles. Norway, England, France, Scotland, India, and the Netherlands have all announced plans to phase out internal combustion engines. The combination of market and regulatory pressures are pushing car makers to move into EVs. Volvo was the first major car company to announce plans to phase out conventional engines, committing to sell only hybrid and electric cars by 2019. GM recently announced that it will sell at least 20 new models of all electric and hydrogen fuel cell vehicles globally by 2023 [10]. Ford will add 13 electric models over next several years with a five year investment of $US4.5 billion.

Altogether, Bloomberg New Energy Finance [1] expects over 120 different models of EV to be on the market by 2020 including SUVs and small vans. The International Energy Agency (IEA) [10], predicts between 40 million and 70 million electric cars on the roads by 2025. Ten member countries of IEA's EV Initiative recently launched the EV@30 campaign setting a collective goal of a 30% market share for electric vehicles by 2030. The EVs will account for 54% of new car sales by 2040 [10].

The strongest potential impact is coming from China, the largest single market for new car sales. In 2017, total Chinese vehicle sales have risen 4.3% to 17.5 million units, while US sales have fallen 2.7% to 11.35 million vehicles [11]. Last September, the Chinese government announced that they were researching the phase-out of internal combustion cars. China currently has a goal of 20% of all car sales to be from EVs and plug-in hybrids (Fig. 2) [10].

Norway is the world leader in EV adoption per capita thanks to substantial government incentives, like exemptions from the 25% tax applied to vehicle sales. Almost 42% of new cars are EVs including 27% from battery only cars [10].

In US, first regulatory policies came from California where was imposed a zero-emission/low emission vehicle (ZEV-LEV) mandate in 1990. The rule required that 10% of vehicles sold in California satisfy ZEV-LEV guidelines [10]. In spite of industry resistance, this sparked many new innovations though companies opted for hybrid variants, instead of pure electric cars. Some states followed California's lead making a target that 15% of new vehicles by 2025 would be ZEV complaint.

There is much evidence that policy actions drive the car market. The most of new developments is happening in US, China, and Europe, totaling about 90% of car sales [12]. Having strict regulations in vehicle emissions, they are massively deploying new infrastructure, creating strong incentives and conducting customer awareness campaigns to shape the market in line with their mobility strategies. Only 45,000 electric cars were sold in 2011, two million electric cars were sold globally in 2016, in 2017 was already three million with targets to reach 40 million [12]. That would be the point when electrical vehicles become dominant in car industry. Millions of orders require new investments in large battery factories to satisfy such demand and necessitate close cooperation between cars manufactures with battery suppliers, charging providers and electric power utilities.

Government policies are also dependent on the regions where electric cars are made, not only where they are sold. Just few regions in the world, namely China, US, Europe, Japan, and South Korea, produce 97% of EVs [12]. Considering that developments can take many years means that car production will remain in a limited number of regions.

Anticipated multilateral agreements will promote trade policies which also would drive the market by removing barriers for importation of electric cars and making seamless their global supply chains. Many countries already use domestic taxation policies to stimulate prevalence of electric cars. Others, India for example, unilaterally are removing duties on electric spare parts [12].

Electric vehicles and transportation are also important parts of government policies devoted to address health and environmental issues. Governments' actions in this direction should also stimulate the transition toward electric cars by supporting their sales, production and import.

3 Australian Context

Australia seems to be a late follower in the transition to low emission transport, lagging markets like China, UK, France and others in embracing electrical vehicles. While these countries have announced bans on the sale of petrol and diesel vehicles, only 0.1% (948 in 2014 or 1,369 in 2016) of new vehicle sales in Australia was electric [13]. In addition, Australia's electricity infrastructure was built without renewable energy in mind and primarily to maintain reliability for demand peaks. The recent high uptake of solar panels reduced carbon emission and grid demand by 5–10% [14]. Electric vehicles in the near future could ease the pressure on spikes in electricity prices by adding storage capacity because they are, de facto, distributed storage systems which can feed electricity back into the grid when prices are high.

According to the Office of Chief Economist, Australia produced 6 billion kilowatt hours of solar PV in 2015 [14]. This could be enough to run almost 2 million cars what is equivalent to 10% of Australia's total current passenger vehicle fleet of 20 million cars. These cars travel 280 billion kilometers annually, burning 20 billion liters of fuel at cost of $30 billion per year. The fact that crude oil import to Australia grew from 60% in 2000 to 90% in 2017 [14] contributes to this complexity. If, hypothetically, all ICE vehicles are replaced by EVs, the cost would be $15 billion.

Australia at the moment does not have a comprehensive national policy framework. However, Australian Electrical Vehicles Council [15] proposed some strategic policies in order to facilitate the transition to electric vehicles. Although considered limited, the aim was to empower consumers to learn about new technology and make informed choices, to increase use of EVs by government, commercial fleets and community, to foster growth in EV technology innovation and support rollout of infrastructure. Some additional policies should promote recycling of EV batteries, encourage EV adoption with reduced registration fees and stamp duties, allow driving in priority lines and extra taxing diesel cars because of their high pollution.

Government's purchasing of EVs in 2016 was only 2% of the total sales comparing to businesses of 64% and private buyers of 34% [16]. In developed OECD countries, considered as Australia's peers, in existence are well-developed comprehensive policy frameworks related to uptake and transitions to EVs. State governments in Australia are currently introducing some incentives, but there are still many barriers preventing rapid EV adoption including various technological, financial and market issues. The consequences of such challenges are insufficient R&D investments and deficiency in public-private collaboration, lack of innovative policy frameworks and absence of more propulsive business solutions.

4 Environmental Issues

Environmental issues are often cited as one of the primary reasons for abandoning ICE vehicles and shifting to EV. There is a long history of environmental policies in developed countries with the largest numbers of vehicles on their roads.

In Europe, emission standards were applied separately on petrol and diesel-powered cars. Euro5 standard was introduced in 2011 for all new cars in order to reduce emission of oxides and particulates which are about 20% of the total greenhouse emissions. Euro6, introduced in 2016, postulated further reduction in levels of such emissions [13].

In United States, road transportation represents nearly 18% of the energy utilization and 22% of the greenhouse emissions [17]. Regulators introduced standards more strict than Euro5 (Tier2) with the same emission levels for petrol and diesel. In 2017, they introduced even more stringent standards (Tier3) [13].

Australia had vehicle standards intended to reduce air pollution since early 1970s. Transport in Australia is currently the third biggest and second fastest developing pollution source, responsible for 8% of national greenhouse emissions [18]. Emission standards are defined by Australian Design Rules which has regulatory power given by Motor Vehicle Standards Act 1989 [13]. They introduced more strict standards in 2011 and Euro5, taken from European Union, were fully implemented in 2016.

In spite the fact that upgrade to Euro6 standard is a costly exercise, it was planned with a rationale that the costs would be reduced and value of statistical life would be improved. In this context, the prevailing opinion in EU and Australia is that upfront purchasing cost of EV and lifetime benefits are still more attractive proposition considering that EV provide a zero emission of harmful particulates and oxides.

In order to have environmental benefits from the introduction of the electrical vehicles, the electrical energy used for feeding into EVs has to be green as well. This means using hydro, solar power, wind power and fuel cells sources [19–22]. A wind farm along Hume Highway, Breadalbane in New South Wales, Australia is shown in Fig. 3.

Fig. 3. A wind farm in Australia (Photo by Milan Simic).

Increased use of renewable energy in the future will require a greater flexibility of the electricity system. There will be a necessity to retain a constant frequency of the system, to control demand, supply and storage, to enable power output regulation and to be switched on/off when needed. Based on smart charging, batteries of electric vehicles could be used as energy storages and as prospective sources of power (vehicle to everything). It means that charging will be done in a way to maximize the use of renewable energy, to avoid peak periods of demand and to be cost beneficial for users. The surplus of electricity could be resourced back to the grid or to smart buildings. Their recycled batteries might be used as decentralized storages of energy at charging hubs to reduce impacts on local electricity networks. Charging stations may well be integrated via smart buildings' microgrid with solar panels or other renewable sources in order to increase their energy efficiency. All mentioned, would require a synchronized action from economy planners, automotive industry and environmental regulatory bodies.

5 Forecasts

Some pessimistic forecasts by LMC Automotive, citing the fact that electric cars today comprise only 1% of auto sales worldwide and even less in the US, estimate that they will account for just 2.4% of US demand and less than 10% globally by 2025. The largest auto supplier In North America, Magna International, does not expect much demand for EVs over the next eight years and predicts that EVs will grow only to 3–6% of global auto sales by 2025. LMC forecasts that petrol powered engines will still make up about 85% of US new car sales in 2025 [23].

Nevertheless, that shift could be accelerated as electrified vehicles reach price parity with petrol powered cars. In total, 127 battery-electric models will be introduced worldwide in the next five years, with LMC predicting pure electric offerings will increase by more than five-fold to 75 models in the US alone [23]. GM plan to sell profitably more than one million electric vehicles per year by 2026. Tesla intends to build half of million electric cars in 2018. Ford engineers work hard on developing cost-efficient batteries, better and cheaper than today's lithium-ion versions. Toyota is working on energy-dense slid state batteries, considered as the next frontier in electric power [24].

6 Conclusion

We are witnessing how the concept of mobility is changing as electric vehicles become more affordable. Some predictions are that they will constitute almost a third of new car sales by the end of the next decade. Ride sharing continues to grow strongly, with estimates that, by 2030, it will account for more than 25% of globally driving, up from 4% today [25]. Soon, autonomous vehicles and commercial fleets of EVs will be integrated as parts of everyday life. EVs will cost significantly less per mile than ICE vehicles for personal use and could also reduce congestion and traffic incidents.

At the same time, energy is also changing in terms of worldwide evolution of energy systems, which are becoming cleaner and increasingly decentralized with energy generated, stored and distributed closer to the final customers, with more intensive use of renewables and storage technologies. In addition, digitalization allows controlling where, when and how electricity is being used. Future energy uses, including mobility, will be increasingly of electrical origin.

Transformations in the fields of energy and mobility are influenced by market factors and global trends. Their convergence creates opportunities, since EVs can be used as distributed energy resources by providing new, controllable storage capacity and electricity supply, suitable for the stability of the energy system. The mobility sector will have the prospect to develop new business models based on services, sharing models and new uses associated with EVs as decentralized energy resources. Policymakers have the legal power to promote innovation and new ways of government thinking that will make this possible.

Electric mobility was generally considered as a way to improve air quality and meet climate goals, but seldom is integrated in a broad vision for smarter cities. As urbanization increases, expectations are that additional two and a half of billions of people by 2050, i.e. 70% of the total population, will live in urban areas. Cities will experience substantial changes to create sustainable living conditions for their dwellers. Energy and mobility are key drivers of such transformation which would require radical adjustments to achieve the demographic and economic growth without increasing pollution and congestion.

EVs are still associated with traditional ownership and considered only as cars. New use models and services related to batteries, or to integrations with smart buildings, are yet insufficiently explored. Charging stations are still developed with no consideration for energy issues and without taking advantage of a full range of digital technologies. There is still a dilemma if policymakers and business leaders can utilize them in a way of making the most of their benefits for cost efficiency, economic growth, public safety, compatibility between jurisdictions and environmental sustainability. The transition to electric vehicles is a part of so-called Fourth Industrial Revolution which offers exceptional opportunities for the future.

In order to better understand contributions of all key factors involved, next step in our investigation will be to try using fuzzy logic and artificial intelligence in this global and comprehensive transition management.

Acknowledgments. This research is supported by Australian Government Research Training Program Scholarship (Project Reference Number: 30484451).

References

1. https://data.bloombeglp.com/bnef/sites/14/2017/07/BNEF_EVO_2017_Executive_Summary.pdf. Accessed 02 Mar 2018
2. https://c1cleantechnicacom-wpengine.netdna-ssl.com/files/2017/05/Electric-Car-Drivers-Report-Surveys-CleanTechnica-Free-Report.pdf
3. http://evobsession.com/ev-revolution-disruption-not-simply-incremental-transition/
4. http://evvolumes.com/Global-Plug-in-Sales-for-2017-Q4. Accessed 17 Feb 2018
5. OECD/IEA Report: Global EV Outlook 2017 - Two Million and Counting (2017)
6. https://cleantechnica.com/2017/10/07/electric-vehicles-will-drive-5-trillion-transition/
7. https://www.transportenvironment.org/sites/te/files/publications/Briefing%20-%20How%20will%20electric%20vehicle%20transition%20impact%20EU%20jobs.pdf. Accessed 02 Mar 2018
8. https://teslauniverse.com/nikola-tesla/patents/us-patent-555190-alternating-motor

9. https://evobsession.com/tesla-competitive-advantage-5-big-ones/
10. https://energytransition.org/2017/11/get-ready-for-the-next-generation-of-electricvehicles/
11. https://marketrealist.com/2017/08/to-decode-qualcomms-future-look-to-electric-vehicles
12. https://www.ictsd.org/opinion/the-transition-to-electric-cars-how-can-policy-pave-the-way
13. ClimateWorks: Stakeholder Recommendations – The Path Forward for Electric Vehicles in Australia, Melbourne, April 2016
14. http://theconversation.com/how-electric-cars-can-help-save-the-grid-73914
15. https://www.energymatters.com.au/renewable-news/queenslands-electric-vehicle-strategy/
16. https://climateworksaustralia.org/sites/files/documents/publictions/state_of_evs_final.pdf
17. Alhindawi, R., Nahleb, Y., Kumar, A., Shiwakoti, N.: Projection of the greenhouse gas emissions for road sector based on a multivariate regression model. In: 27th ARRB Conference – Linking People, Places and Opportunities, Melbourne, Victoria (2016)
18. Riesz, J., Sotiriadis, C., Ambach, D., Donovan, S.: Quantifying the costs of a rapid transition to electric vehicles. Appl. Energy **180**(15), 287–300 (2016)
19. Dou, X.X., Simic, M., Andrews, J., Mo, J.: Power splitting strategy for solar hydrogen generation. Int. J. Agile Syst. Manag. **8**(1) (2015). https://doi.org/10.1504/ijasm.2015.068609
20. Elbanhawai, M., Simic, M.: Robotics application in remote data acquisition and control for solar ponds. Appl. Mech. Mater. **11**, 252–255 (2013). https://doi.org/10.4028/www.scientific.net/AMM.253-255.705
21. Lambert, N., Simic, M., Kennedy, B.: Solar vehicle for south pole exploration. In: Subic, A., Wellnitz, J., Leary, M. (eds.) Sustainable Automotive Technologies. Springer, Heidelberg (2013)
22. Simic, M.N., Singh, R., Doukas, L., Akbarzadeh, A.: Remote monitoring of thermal performance of salinity gradient solar ponds. Paper presented at 12th Euromicro Conference on the Digital System Design, Architectures, Methods and Tools, DSD 2009, 27–29 August 2009
23. https://www.bloomberg.com/news/features/2017-12-19/the-near-future-of-electric-cars-many-models-few-buyers
24. https://www.pocket-lint.com/cars/news/140845-future-cars-upcoming-electronic-cars-of-the-future-coming-soon
25. World Economic Forum Report: Electric Vehicles for Smarter Cities - The Future of Energy and Mobility, Geneva, Switzerland, January 2018

Frequency Island and Nonlinear Vibrating Systems

Ching Nok To(✉), Hormoz Marzbani, Đại Võ Quốc, Milan Simic, M. Fard, and Reza N. Jazar

School of Engineering, RMIT University, Melbourne, Australia
catter@vtc.edu.hk, {hormoz.marzbani,milan, mohammad.fard,Reza.Jazar}@rmit.edu.au, dai.voquoc@lqdtu.edu.vn

Abstract. Piecewise linear vibration isolator system is one of the development to introduce dual rate stiffness and damping. There are several vibrating behavior if the system is designed properly. Analytical treatment of the system determines some difficulties such as jump. This investigation indicates that such strong nonlinear systems have a new phenomenon called Frequency Island in their frequency response plot. Frequency Island is a possible isolated frequency response that the vibrating system may jump into the island and stays there until the excitation frequency moves out of the range of the island.

In this student existence, appearance, growing and disappearing of frequency island will be studied and examined. Frequency Island corresponds to large amplitude vibration for certain range of system parameters and considered as a dangerous phenomena in real system. As a result, understanding its appearance will help designers and engineers to design the system to avoid Frequency Island.

Keywords: Piecewise linear system · Vibration isolator · Frequency response

1 Introduction

Wheel travel is the maximum displacement that a wheel can move towards body until a metal-to-metal contact occurred (Jazar 2018). It is also a fact that to optimize the suspension of a base excited vibrating system, the suspension characteristics must be as soft as possible (Jazar 2013; Jazar et al. 2003). The optimal characteristics of such a suspension might be calculated based on different objective functions (Narimani et al. 2004b; Jazar and Golnaraghi 2002a, b). The RMS optimization has shown its advantages in frequency and time domain very well (Narimani et al. 2004a). Its application have been validated by experiments as well as other optimization methods su as Genetic Algorithm (Christopherson and Jazar 2005a, b; Marzbani et al. 2013a, b; Alkhatib et al. 2004; Christopherson et al. 2004). However, a soft suspension has its own limits and constraints. Firstly, the suspension must carry the weight of the system, the vehicle body weight, for instance (Marzbani et al. 2012; Esmailzadeh et al. 1997; Alam et al. 2010). Secondly, soft suspension causes the wheel travel to saturate easily and a metal-to-metal contact occurs frequently (Jazar et al. 2006a, b; Agaah et al. 2005). To compensate these problems, a progressive nonlinear suspension is a solution

G. De Pietro et al. (Eds.): KES-IIMSS-18 2018, SIST 98, pp. 140–150, 2019.
https://doi.org/10.1007/978-3-319-92231-7_15

(Natsiavas 1991; Den Hartog 1932; Pogorilyi et al. 2015a). From engineering point of view, a piecewise linear suspension with dual or multiple advanced rates might also introduce a practical solution (Pogorilyi 2015b; Deshpande et al. 2006; Jazar 2008; Stahl and Jazar 2005b). The idea behind piecewise linear system is to have a soft suspension for a portion of the wheel travel and a harder suspension for the rest of the wheel travel. This way, we may have a soft and good design when the amplitude of excitation is high which is associated to high frequency excitation, and have a hard suspension to protect the system from metal-to-metal contact at high excitation amplitudes which is usually associated to low frequency excitations (Jazar 2018; Deshpande et al. 2005a; Jazar et al. 2005; Narimani et al. 2003; Christopherson et al. 2003; Jazar et al. 2002a, b).

Nonlinear progressive rate suspension as well piecewise linear suspensions fall in the category of nonlinear systems which bring the difficulties of not having a closed form solution of the differential equations (Jazar 2011; Pogorilyi et al. 2014; Jazar et al. 2006; Van del Pol 1927; Virgin 2000; Golnaraghi and Jazar 2000). Furthermore, nonlinear systems show several unpleasant phenomena such as jump, history dependency, non-smooth behavior (Nayfeh and Mook 1979; Den Hartog 1936; Jazar and Golnaraghi 2001; Deshpande et al. 2005b; Jazar et al. 2007).

There are two practical piecewise linear systems in application: Hydraulic Engine Mounts, and Piecewise Linear Suspensions (Christopherson et al. 2012; Golnaraghi and Jazar 2001; Esmailzadeh et al. 1997; Marzbani and Jazar 2013; Jazar et al. 2011; Christopherson and Jazar 2005a, b; Stahl and Jazar 2005a; Christopherson and Jazar 2006a, b, c, d; Marzbani et al. 2016; Narimani et al. 2002; Natsiavas and Gonzalez 1992; Schulman 1983). There are very good review of these suspensions and their advantages, optimizations, and problems which are out of the scope of this paper (Marzbani and Jazar 2013; Jazar and Golnaraghi 2002a; Pogorilyi et al. 2015a). In 2003, Narimani, Jazar, and Golnaraghi in University of Waterloo were working on optimization and frequency analysis of piecewise linear suspensions when they observed an unknown and unexpected behavior in their experiments. Their mathematical investigation finally discovered that unexpected behavior and they reported the new phenomenon called Frequency Island (Narimani et al. 2004b; Jazar 2008; Jazar et al. 2002a, b).

In this study we revisit this phenomenon and show its appearance and disappearance. Its problems and the way to avoid a Frequency Island occurrence in piecewise linear suspensions. Today, we know that Frequency Island may occur in other types of nonlinear systems as well.

2 Piecewise Linear Suspension and Frequency Response

Figure 1 depicts a sketch of a piecewise linear suspension system. A mass m is supported by a primary suspension with linear stiffness k_1 and damping c_1. The primary suspension carries the weight of m and supports the systems well enough in low relative amplitudes.

There is a secondary suspension with linear stiffness k_2 and damping c_2. The secondary system comes into action only when the relative displacement of m exceeds a designed limit value. This limit is called gap.

The equations of motion of the system could be written as, y is the base excitation assuming to be a harmonic function with amplitude Y, and frequency ω.

$$m\ddot{x} + g(x, \dot{x}) = f(y, \dot{y}) \tag{1}$$

where $g(x, \dot{x})$ and $f(y, \dot{y})$ are two piecewise linear functions and

$$g(x, \dot{x}) = \begin{cases} (c_1 + c_2)\dot{x} + (k_1 + k_2)x - k_2\Delta & x - y > \Delta \\ c_1\dot{x} + k_1 x & |x - y| < \Delta \\ (c_1 + c_2)\dot{x} + (k_1 + k_2)x + k_2\Delta & x - y < -\Delta \end{cases} \tag{2}$$

$$f(y, \dot{y}) = \begin{cases} (c_1 + c_2)\dot{y} + (k_1 + k_2)y & x - y > \Delta \\ c_1\dot{y} + k_1 y & |x - y| < \Delta \\ (c_1 + c_2)\dot{y} + (k_1 + k_2)y & x - y < -\Delta \end{cases} \tag{3}$$

$$y = y(t) = Y\sin(\omega t) \tag{4}$$

Employing the following parameters

$$\begin{gathered} u = x - y \quad z = \tfrac{u}{Y} \quad \delta = \tfrac{\Delta}{Y} \quad \omega_1^2 = \tfrac{k_1}{m} \quad \omega_2^2 = \tfrac{k_1 + k_2}{m} \quad \omega_3^2 = \tfrac{k_2}{m} \\ \xi_1 = \frac{c_1}{2\sqrt{k_1 m}} \quad \xi_2 = \frac{c_1 + c_2}{2\sqrt{(k_1 + k_2)m}} \quad \xi_3 = \frac{c_2}{2\sqrt{k_2 m}} \\ \xi_2\omega_2 = \xi_1\omega_1 + \xi_3\omega_3 \quad \omega_2^2 = \omega_1^2 + \omega_2^2 \end{gathered} \tag{5}$$

we may rearrange the equation to,

$$\ddot{z} + \omega_1^2 z = \omega^2\sin(\omega t) + \varepsilon g_3(z, \dot{z}) \tag{6}$$

$$g_3(z, \dot{z}) = \begin{cases} -2\xi_2\omega_2\dot{z} - \omega_3^2 z + \omega_3^2\delta & z > \delta \\ -2\xi_1\omega_1\dot{z} & |z| < \delta \\ -2\xi_2\omega_2\dot{z} - \omega_3^2 z - \omega_3^2\delta & z < -\delta \end{cases} \tag{7}$$

We consider a solution in the following form as the steady state response of the system,

$$z = A(t)\sin(\omega t + \beta(t)) \quad \varphi = \omega t + \beta(t) \quad \omega^2 = \omega_1^2 + \varepsilon\sigma \tag{8}$$

Using method of averaging, we are able to determine the frequency response of the system as (Nayfeh and Mook 1979):

$$\begin{aligned} &4A^2\left(\omega_1^2 + \sigma\right)\left\{(\xi_1\omega_1 + \xi_3\omega_3)\left(\pi - \sin\left(2\sin^{-1}\left(\tfrac{\delta}{A}\right)\right) - 2\sin^{-1}\left(\tfrac{\delta}{A}\right)\right)\right. \\ &\left. + \xi_1\omega_1\left(\sin\left(2\sin^{-1}\left(\tfrac{\delta}{A}\right)\right) + 2\sin^{-1}\left(\tfrac{\delta}{A}\right)\right)\right\}^2 \\ &+ \left\{\omega_3^2\left(2A\sin^{-1}\left(\tfrac{\delta}{A}\right) - A\sin\left(2\sin^{-1}\left(\tfrac{\delta}{A}\right)\right) + 4\delta\cos\left(\sin^{-1}\left(\tfrac{\delta}{A}\right)\right)\right) + \pi A\left(\sigma - \omega_3^2\right)\right\}^2 \\ &- \left(\omega_1^2 + \sigma\right)^2\pi^2 = 0 \end{aligned} \tag{9}$$

and rewrite this as a second-degree equation in terms of detuning frequency σ,

$$Z_1\sigma^2 + 2Z_2\sigma + Z_3 = 0 \tag{10}$$

where Z_1, Z_2, and Z_3 are presented in Appendix. Now we can find frequency σ as a function of the amplitude in nonlinear domain, A with parametric dependent to δ , ω_1, ω_3, ξ_1, ξ_3.

3 Kinematic and Dynamic Vehicle Rotation Centers

Let us rearrange Eq. (10) to find the frequency σ as a function of amplitude A

$$\sigma = \frac{-Z_2 \pm \sqrt{Z_2^2 - Z_1 Z_3}}{Z_1} \tag{11}$$

Depending on the values of the parameters of the system embedded in Z_1, Z_2, and Z_3 we may find two values for σ, saying σ_1 and σ_2, as two functions of A. As long as δ is in the less than the maximum amplitude of the linear frequency response P of the primary without existence of the secondary system, the second stage interacts and a nonlinear resonance zone appears. On the frequency response plot, the nonlinear response amplitude, A starts at $A = \delta$, and extends to a maximum value, A_{max} where Eq. (10) provides only one σ. Linear and nonlinear resonance zones, along with P and A_{max} are illustrated in Fig. 2. If there are more than one A at which we have equal roots for Eq. (10), then it is possible to have another steady state frequency response separated from the main resonant zone. If the new frequency response is a closed path, we call the new zone a Frequency Island. Existence of more than one equal roots for Eq. (11) indicates appearance of a frequency Island.

To have an Island, the Point equation Γ must have more than one roots for A.

$$\Gamma = Z_{10}A^4 + Z_{11}A^3 + Z_{12}A^2 + Z_{13}A + Z_{14} \tag{12}$$

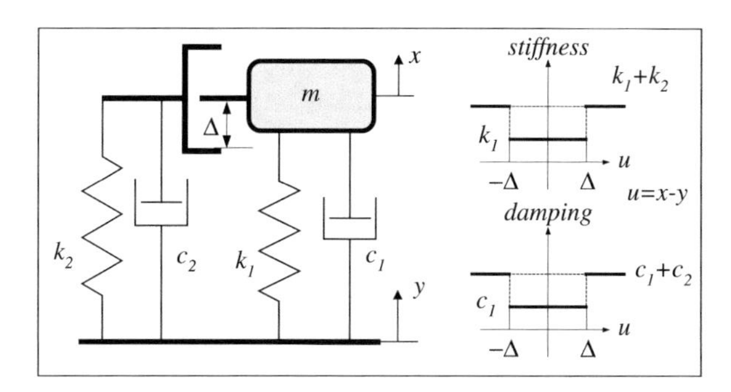

Fig. 1. A sketch of a mechanical model of a piecewise linear suspension system.

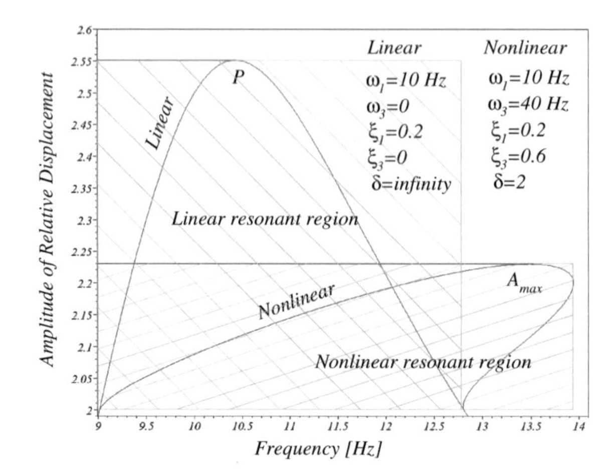

Fig. 2. Comparison of the relative displacement frequency response of the nonlinear system (perturbation result) with linear system, for $A > \delta$.

Let us assume δ as a control parameter of the system then, Γ may be assumed as a two-dimensional surface of variables A and δ. The point values of A would be at the intersection of the Γ surface and $\Gamma = 0$.

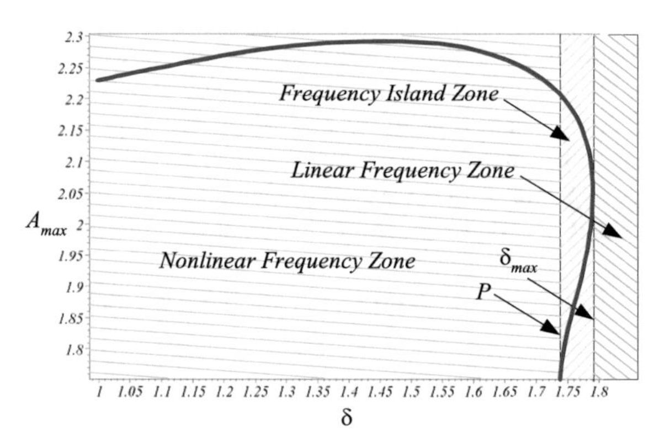

Fig. 3. A sample of the intersection curve indicating the peak amplitude, A_{max} as a function of the gap size, δ.

As long as $\Gamma > 0$, there are two real values for A indicating the secondary suspension interacts and therefore, there exists a nonlinear resonant zone. There is no real solution for Eq. (11) in when $\Gamma < 0$ indicating the secondary suspension will not interact and the system remains in on primary suspension and linear response. If we imagine a three dimensional surface of Γ as functions of A and δ, we may determine the cross-section of the surface with plane $\Gamma = 0$. The intersection curve shows a relationship between the point values of A and the gap size δ. Figure 3 illustrates a sample of the intersection curve. The curve shows A_{max} for $\delta > 1$ when the secondary

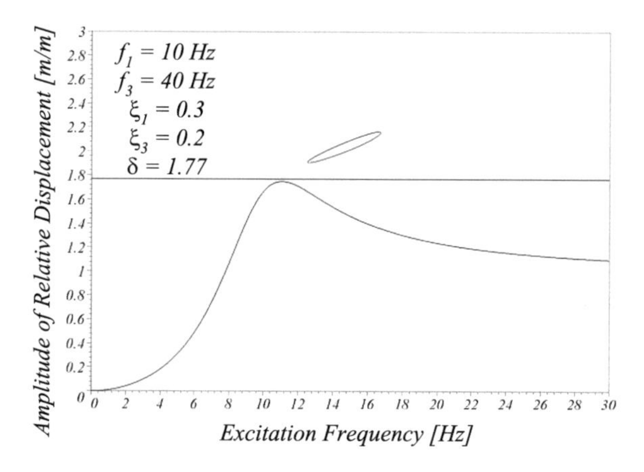

Fig. 4. A sample of frequency response a Frequency Island exists.

suspension interacts. There is a region on horizontal axis that there exist two values for A. When the system is in this region then it is possible that the system to vibrate at two different stable amplitudes. The upper one is on a frequency island and the lower value is on the main frequency response. Figure 4 depicts the frequency response of the system when a Frequency Island exists. The horizontal solid line indicates the gap.

4 Frequency Island Appearance and Disappearance

Consider vibrating system that only has a primary system. The frequency response of such a system is the same as a linear system in elementary vibration with a peak value depending on the damping. Let's us assume all parameters as well as excitation characteristics of such a system is constant. As a result, the shape of frequency response is also constant. Now we may assume that there exists a secondary system the gap size δ is just equal to the maximum amplitude of the system at resonance. Such a system, technically should not interact with the secondary suspension and work only on the primary. However, we may assume that because of a disturbance the amplitude jumps over the peak value and secondary system comes into action. A frequency response of linear system is stable. That means if the amplitude jumps over or be diminished from the steady state value, the system will lose or gains energy until the amplitude goes back to its steady state frequency response value at the associated excitation frequency. However, in this case that a secondary system exists, the system will go to the nolinear regime and it may or may not get back to its original linear response depending on cross-section Fig. 3. In case the system parameters are such that the system is in the range of two A values then system will remain in the upper value of A and vibrates at the amplitude associated to the Frequency Island.

The maximum size of the Frequency Island is when the gap is equal to the peak value of linear amplitude. Now assume that we increase the gap size of such a vibrating system. The size of the Frequency Island will shrink by increasing gap size, until at a certain value the Frequency Island disappears. These situations are illustrated in Fig. 5.

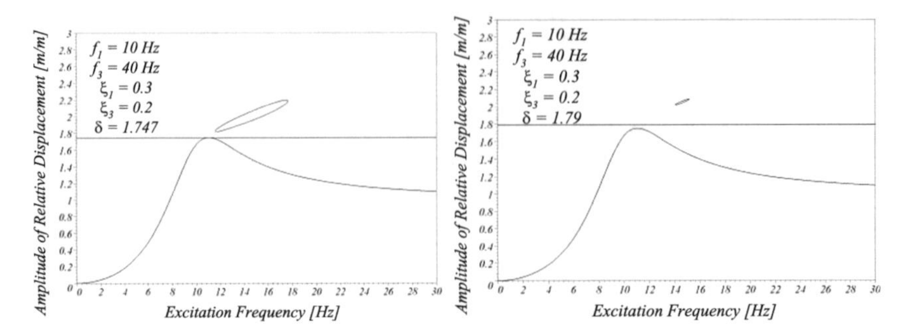

Fig. 5. Appearance and disappearance of Frequency Island by increasing gap size.

5 Conclusion

Piecewise linear suspension falls in the category of nonlinear systems which has been in under investigations since the science of two centuries ago (Minorski 1949). It showed its application in vibration isolation system where we wished to spend the wheel travel with dual function. A soft suspension for low amplitude and a hard suspension for high amplitude vibrations. Although the mechanical model of the system is promising, its experiment shows that there are some hidden behaviors that might be unpleasant. Firstly by making the stiffness and damper to be dual linear, the nonlinear phenomenon of jump may appears. The other phenomenon which we reviewed in this paper, is Frequency Island. It is a separate possible steady state amplitude separated from the main frequency response which starts from zero excitation frequency to infinity. We have shown that Frequency Island appears for a range of excitation frequency above the resonance frequency and also it appears when the gap size is slightly above the linear peak amplitude.

It indicates that to design such piecewise linear suspensions, we must pick the values of the parameters of the primary and secondary suspensions in a way that the system avoids jump and Frequency Island. Both are important in application as these phenomena are usually harmful to the system and cause sudden change amplitude of vibrations.

Appendix

Parameters of the analytical equation to predict the frequency response of a piecewise linear suspension system

$$Z_1 = \pi^2 (A^2 - 1)$$

$$Z_2 = \pi A\left[\omega_3^2\left(4\delta\,\cos\left(\sin^{-1}(\frac{\delta}{A})\right) - A\,\sin\left(2\sin^{-1}(\frac{\delta}{A})\right) - \pi A + 2A\,\sin^{-1}(\frac{\delta}{A})\right)\right]$$
$$+\,2A^2\left[(\xi_1\omega_1+\xi_3\omega_3)\left(\pi - \sin\left(2\sin^{-1}(\frac{\delta}{A})\right) - 2\sin^{-1}(\frac{\delta}{A})\right)\right.$$
$$\left.+\,\xi_1\omega_1\left(\sin\left(2\sin^{-1}(\frac{\delta}{A})\right) + 2\sin^{-1}(\frac{\delta}{A})\right)\right]^2 - \omega_1^2\pi^2$$

$$Z_3 = 4\omega_1^2A^2\left[\xi_1\omega_1\left(\sin(2\sin^{-1}\left(\frac{\delta}{A}\right)) + 2\sin^{-1}\left(\frac{\delta}{A}\right)\right)\right.$$
$$\left.+\,(\xi_1\omega_1+\xi_3\omega_3)\left(\pi - \sin(2\sin^{-1}\left(\frac{\delta}{A}\right)) - 2\sin^{-1}\left(\frac{\delta}{A}\right)\right)\right]^2 + \omega_1^4\pi^2$$
$$+\,\left(-\omega_3^2A\,\sin(2\sin^{-1}\left(\frac{\delta}{A}\right)) - \omega_3^2\pi A + 2\omega_3^2\sin^{-1}\left(\frac{\delta}{A}\right)A + 4\omega_3^2\delta\,\cos(\sin^{-1}\left(\frac{\delta}{A}\right))\right)^2$$

$$Z_6 = 2A\left(\omega_1^2+\sigma\right)\left[(\xi_1\omega_1+\xi_3\omega_3)\left(2\sin^{-1}\left(\frac{\delta}{A}\right) - \pi - \sin(\sin^{-1}\left(\frac{\delta}{A}\right))\right)\right]$$

$$Z_7 = 2A\left(\omega_1^2+\sigma\right)\xi_1\omega_1\left(2\,\sin^{-1}\left(\frac{\delta}{A}\right) + \sin(\sin^{-1}\left(\frac{\delta}{A}\right))\right)$$

$$Z_8 = \pi A\sigma + \omega_3^2A\left(2\sin^{-1}\left(\frac{\delta}{A}\right) - \pi - \sin(2\sin^{-1}\left(\frac{\delta}{A}\right))\right) + 4\delta\omega_3^2\cos(\sin^{-1}\left(\frac{\delta}{A}\right)$$

$$Z_{10} = 4f_3^4\xi_3^4Z_{15}^4 + 32\pi^2f_1\xi_1f_3^3\xi_3^3Z_{15}^3 + 4\pi\xi_3^2f_3^2Z_{15}^2\left(f_3^2Z_{16} + 4\pi^3f_1^2\left(6\pi^4\xi_1^2 - 1\right)\right)$$
$$+\,16\pi^3\xi_1\xi_3f_1f_3Z_{15}^2\left(f_3^2 - 4\pi^3f_1^2\left(1 - 2\xi_1^2\right)\right) + 16\pi^5\xi_1^2f_1^2f_3^2Z_{16}$$

$$Z_{11} = 64\pi^3\delta\sqrt{1-\frac{\delta^2}{A^2}}\left(\xi_3^2f_3^4Z_{15}^2 + 4\pi^2\xi_1\xi_3f_1f_3^3Z_{15} + 4\pi^4\xi_1^2f_1^2f_3^2\right)$$

$$Z_{12} = \pi^2f_3^4Z_{15}^2 - 8\pi^5f_1^2f_3^2Z_{15} + 16\pi^8f_1^4 \qquad Z_{13} = 32\pi^4\delta\sqrt{1-\frac{\delta^2}{A^2}}f_3^2\left(f_3^2Z_{15}^2 - 8\pi^3f_1^2\right)$$

$$Z_{14} = 256\pi^6f_3^4\delta^4\left(1-\frac{1}{A^2}\right) \qquad Z_{15} = -2\pi\,\sin\left(2\sin^{-1}(\frac{\delta}{A})\right) + 2\pi^2 - 4\pi\left(\sin^{-1}(\frac{\delta}{A})\right)^2$$

$$Z_{16} = -4\sin\left(2\sin^{-1}(\frac{\delta}{A})\right) - 8\pi^3 + 8\pi^2\left(\sin^{-1}(\frac{\delta}{A})\right)^2$$

References

Aagaah, M.R., Mahmoudian, N., Jazar, R.N., Mahinfalah, M.: Frequency response of vibration isolators with saturation spring element. In: 20th ASME Biennial Conference on Mechanical Vibration and Noise, Long Beach, CA, 24–28 September 2005 (2005)

Alam, F., Jazar, R.N., Faheem, A., Smith, L.V.: A study of vehicle ride performance using a quarter car model and half car model. In: ASME International Mechanical Engineering Congress and Exposition, Vancouver, British Columbia, Canada, 12–18 November 2010 (2010)

Alkhatib, R., Jazar, R.N., Golnaraghi, M.F.: Optimal design of passive linear mounts with genetic algorithm method. J. Sound Vibr. **275**(3–5), 665–691 (2004)

Christopherson, J., Jazar, R.N.: Development of a new design hydraulic engine mount. In: ASME International Mechanical Engineering Congress and Exposition, Orlando, Florida, 5–11 November 2005 (2005a)

Christopherson, J., Jazar, R.N.: Dynamic behavior comparison of passive hydraulic engine mounts, Part 1: mathematical analysis. J. Sound Vibr. **290**, 1040–1070 (2006a)

Christopherson, J., Jazar, R.N.: Dynamic behavior comparison of passive hydraulic engine mounts, Part 2: finite element analysis. J. Sound Vibr. **290**, 1071–1090 (2006b)

Christopherson, J., Jazar, R.N.: Optimization of classical hydraulic engine mounts based on RMS method. J. Shock Vibr. **12**(2), 119–147 (2005b)

Christopherson, J., Gustin, J., Jazar, R.N., Mahinfalah, M.: RMS optimization of a standard hydraulic engine mount. In: The 2004 ASME International Mechanical Engineering Congress and Exposition, Anaheim, CA, 14–19 November 2004 (2004)

Christopherson, J., Jazar, R.N., Mahinfalah, M.: Chaotic behavior of hydraulic engine mount. In: ASME International Mechanical Engineering Congress and Exposition, Chicago, Illinois, 5–10 November 2006 (2006c)

Christopherson, J., Jazar, R.N., Mahinfalah, M.: Dynamic analysis and nonlinear characteristics of suspended decoupler hydraulic engine mounts. In: CSME Forum 2006, Symposium on Intelligent Vehicles and Transportation Systems, Calgary, Alberta, Canada, 21–23 May 2006 (2006d)

Christopherson, J., Mahinfalah, M., Jazar, R.N.: Frequency response of a base excited system with hydraulic mount. In: ASME-IMECE Symposium on Dynamic Response of Advanced Materials and Structures, Washington D.C., 15–21 November 2003 (2003)

Christopherson, J., Mahinfalah, M., Jazar, R.N.: Suspended decoupler: a new design of hydraulic engine mount. Adv. Acoust. Vibr. **2012**, 11 (2012). Article ID 826497

Den Hartog, J.P., Heiles, R.M.: Forced vibration in nonlinear systems with various combinations of linear springs. Trans. ASME J. Appl. Mech. **58**, 127–130 (1936)

Den Hartog, J.P., Mikina, S.J.: Forced vibrations with nonlinear spring constraints. Trans. ASME **54**, 157 (1932). Paper APM-54-15

Deshpande, S., Mehta, S., Jazar, R.N.: Jump avoidance conditions for piecewise linear vibration isolator. In: 20th ASME Biennial Conference on Mechanical Vibration and Noise, Long Beach, CA, 24–28 September 2005 (2005a)

Deshpande, S., Mehta, S., Jazar, R.N.: Optimization of secondary suspension of piecewise linear vibration isolation systems. Int. J. Mech. Sci. **48**(4), 341–377 (2006)

Deshpande, S., Mehta, S., Jazar, R.N.: Sensitivity of jump avoidance condition of piecewise linear vibration isolator to dynamical parameters. In: ASME International Mechanical Engineering Congress and Exposition, Orlando, Florida, USA, 5–11 November 2005b (2005b)

Esmailzadeh, E., Jazar, R.N., Mehri, B.: Vibration of road vehicles with nonlinear suspension. Int. J. Eng. **10**(4), 209–218 (1997)

Esmailzadeh, E., Mehri, B., Jazar, R.N.: Periodic solution of a second order autonomous, nonlinear system. J. Nonlinear Dyn. **10**(4), 307–316 (1996)

Golnaraghi, M.F., Jazar, R.N.: Development and analysis of a simplified nonlinear model of a hydraulic engine mount. J. Vibr. Control **7**(4), 495–526 (2001)

Golnaraghi, M.F., Jazar, R.N.: Development and analysis of a simplified nonlinear model of a hydraulic engine mount. In: Eight Conference of Nonlinear Vibrations, Stability, and Dynamics of Structures, University of Virginia, Blacksburg, 23–27 July 2000 (2000)

Jazar, R.N.: Advanced Dynamics: Rigid Body, Multibody, and Aerospace Applications. Wiley, New York (2011)

Jazar, R.N.: Advanced Vibrations: A Modern Approach, 3rd edn. Springer, New York (2013)

Jazar, R.N., Golnaraghi, M.F.: Steady state vibrations of a piecewise linear system. In: 18th ASME Biennial Conference on Mechanical Vibration and Noise, Pittsburgh, Pennsylvania, USA, 9–12 September 2001 (2001)

Jazar, R.N., Golnaraghi, M.F.: Engine mounts for automotive applications: a survey. Shock Vibr. Digest **34**(5), 363–379 (2002a)

Jazar, R.N., Golnaraghi, M.F.: Nonlinear modeling, experimental verification, and theoretical analysis of a hydraulic engine mount. J. Vibr. Control **8**(1), 87–116 (2002b)

Jazar, R.N.: Frequency island in nonlinear vibration isolator. In: 3rd Canadian Conference on Nonlinear Solid Mechanics, Toronto, Ontario, Canada, 25–29 June 2008 (2008)

Jazar, R.N., Houim, R., Narimani, A., Golnaraghi, F.: Nonlinear passive engine mount, frequency response and jump avoidance. J. Vibr. Control **12**(11), 1205–1237 (2006a)

Jazar, R.N., Mahinfalah, M., Aagaah, M.R., Nazari, G.: Comparison of exact and approximate frequency response of a piecewise linear vibration isolator, DETC2005-84522. In: 20th ASME Biennial Conference on Mechanical Vibration and Noise, Long Beach, CA, 24–28 September 2005 (2005)

Jazar, R.N., Mahinfalah, M., Alimi, M.A., Khazaei, A.: Periodic behavior of a nonlinear third order vibrating system. In: International Congress and Exposition, ASME-IMECE 2002, New Orleans, Louisiana, 17–22 November 2002 (2002a)

Jazar, R.N., Mahinfalah, M., Deshpande, S.: Design of a piecewise linear vibration isolator for jump avoidance. IMechE Part K J. Multi-Body Dyn. **221**(K3), 441–450 (2007)

Jazar, R.N., Mahinfalah, M., Christopherson, J.: Suspended decoupler design of hydraulic engine mount. In: ASME International Mechanical Engineering Congress and Exposition (IMECE 2011), Denver, Colorado, 11–17 November 2011 (2011)

Jazar, R.N., Mahinfalah, M., Narimani, A., Golnaraghi, M.F.: Smart passive suspension design consideration for one DOF systems. In: CSME Forum 2006, Symposium on Intelligent Vehicles and Transportation Systems, Calgary, Alberta, Canada, 21–23 May 2006 (2006b)

Jazar, R.N., Narimani, A., Golnaraghi, M.F., Swanson, D.A.: Practical frequency and time optimal design of passive linear vibration isolation mounts. J. Veh. Syst. Dyn. **39**(6), 437–466 (2003)

Jazar, R.N., Narimani, A., Golnaraghi, M.F.: Sensitivity analysis of frequency response of a piecewise linear system in frequency island. In: Ninth Conference of Nonlinear Vibrations, Stability, and Dynamics of Structures, Blacksburg, Virginia, 28 July–1 August 2002 (2002b)

Jazar, R.N.: Vehicle Dynamics: Theory and Application, 3rd edn. Springer, New York (2018)

Marzbani, H., Fard, M., Jazar, R.N.: Chaotic behavior of hydraulic engine mount. In: 20th International Conference on Knowledge Based and Intelligent Information and Engineering Systems, KES 2016, New York, United Kingdom, 5–7 September 2016 (2016)

Marzbani, H., Jazar, R.N., Fard, M.: Comparison between hydraulic engine mounts and piecewise linear vibration isolators, science and motor vehicles 2013. In: JUMV International Automotive Conference 2013 Belgrade, Serbia, 23–24 April 2013 (2013a)

Marzbani, H., Jazar, R.N., Fard, M.: Hydraulic engine mounts: a survey. J. Vibr. Control **19**(16), 1439–1463 (2013b)

Marzbani, H., Jazar, R.N., Khazaei, A.: Smart passive vibration isolation: requirements and unsolved problems. J. Appl. Nonlinear Dyn. **1**(4), 341–386 (2012)

Narimani, A., Golnaraghi, M.F., Jazar, R.N.: Frequency response of a piecewise linear system. J. Vibr. Control **10**(12), 1775–1894 (2004a)

Narimani, A., Jazar, R.N., Golnaraghi, M.F.: Sensitivity analysis of frequency response of a piecewise linear system in frequency island. J. Vibr. Control **10**(2), 175–198 (2004b)

Narimani, A., Golnaraghi, M.F., Jazar, R.N.: Optimizing piecewise linear isolator for steady state vibration. In: ASME-IMECE Symposium on Dynamic Response of Advanced Materials and Structures, Washington D.C., 15–21 November 2003 (2003)

Narimani, A., Jazar, R.N., Golnaraghi, M.F.: Vibration analysis of a piecewise linear system. In: USNCTAM14 Fourteenth U.S. National Congress of Theoretical and Applied Mechanics, Blacksburg, Virginia, 23–28 June 2002 (2002)

Natsiavas, S., Gonzalez, H.: Vibration of Harmonically Excited Oscillators with Asymmetric Constraints. Trans. ASME J. Appl. Mech. **59**, S284–S290 (1992)

Natsiavas, S.: Dynamic of piecewise linear oscillators with Van der Pol type damping. Int. J. Non-Linear Mech. **26**(¾), 349–366 (1991)

Nayfeh, A.H., Mook, D.: Nonlinear Oscillations. Wiley, New York (1979)

Pogorilyi, O., Trivailo, M., Jazar, R.N.: Challenges in exact response of piecewise linear vibration isolator. In: Dai, L., Jazar, R.N. (eds.) Nonlinear Approaches in Engineering Applications, vol. 3. Springer, New York (2015a)

Pogorilyi, O., Trivailo, M., Jazar, R.N.: On the piecewise linear exact solution. Nonlinear Eng. Model. Appl. **3**(4), 189–196 (2014)

Pogorilyi, O., Trivailo, M., Jazar, R.N.: Sensitivity analysis of piecewise linear vibration isolator with dual rate spring and damper. Nonlinear Eng. Model. Appl. **4**(1), 1–13 (2015b)

Schulman, J.N.: Chaos in piecewise linear systems. Phys. Rev. A **28**(1), 477–479 (1983)

Stahl, P., Jazar, R.N.: Frequency response analysis of piecewise nonlinear vibration isolator. In: 20th ASME Biennial Conference on Mechanical Vibration and Noise, Long Beach, CA, 24–28 September 2005 (2005a)

Stahl, P., Jazar, R.N.: Stability analysis of a piecewise nonlinear vibration isolator. In: ASME International Mechanical Engineering Congress and Exposition, Orlando, Florida, 5–11 November 2005 (2005b)

Van del Pol, B.: Phil. Mag. **3**(65) (1927)

Virgin, L.N.: Introduction to Experimental Nonlinear Dynamics. Cambridge University Press, New York (2000)

Software Development for Autonomous and Social Robotics Systems

Chong Sun, Jiongyan Zhang, Cong Liu, Barry Chew Bao King, Yuwei Zhang, Matthew Galle, Maria Spichkova(✉), and Milan Simic

RMIT University, Melbourne, Australia
{s3557753,s3589957,s3556054,s3584485,s3492095,s3491364}@student.rmit.edu,
{maria.spichkova,milan.simic}@rmit.edu.au

Abstract. One of the core features of social robotics system is a physical interaction between humans and humanoid robots. This provides additional challenges, both from safety and usability prospectives. When dealing with human-robot interaction, human safety has the highest priority. While in industrial environment we have robot cells to protect humans, in social robotics, that we consider, physical contact is possible, as well as other interactions, with consequences that might be in psychological areas. For example, the conversation with children might have different requirements in comparison to the conversation with adults, the behavioural assumptions might be different, etc. This paper summarises the core results of a project on social robotics system, where an autonomous humanoid robot guides visitors through a lab tour. The results of our work were implemented on the humanoid PAL REEM robot. The implementation includes a web-application to support the management of robot-guided tours. The application also provides recommendations for the users as well as allows for a visual analysis of historical data on the tours.

1 Introduction

Social robotics, autonomous agents and autonomous robots, are emerging research areas. Over the last years there were many publications on applications of robotics for healthcare and rehabilitation, household service, healthcare and rehabilitation, companionship, etc., cf. [34]. The core function of social robots is assisting people through social interaction, in many cases involving also a physical interaction. A highly cited[1] paper of Feil-Seifer and Mataric [13] defines the concept of socially assistive robotics. Another highly cited[2] paper of Duffy [9] discusses the use of anthropomorphic paradigms to augment the functionality and behavioural characteristics of a robot use of human-like features for social interaction with people.

To understand the impact and capabilities of autonomous robots, it is crucial to identify, observe and measure human-robot interaction (HRI), as well as to

[1] 462 citations according to the Google Scholar, retrieved 20 December 2017.
[2] 592 citations according to the Google Scholar, retrieved 20 December 2017.

G. De Pietro et al. (Eds.): KES-IIMSS-18 2018, SIST 98, pp. 151–160, 2019.
https://doi.org/10.1007/978-3-319-92231-7_16

develop systems that support these observations and measurement. The focus of our project is on social robotics: analysis of interaction between humans and humanoid robots, as well as the corresponding support in the development and maintenance of humanoid robotics systems that are acting autonomously.

The work was conducted in collaboration with Commonwealth bank (CBA) under support of the Australian Technology Network (ATN). This project was a part of the ATN CBA Robotics Education and Research program, and continued our previous research on the topic of social robotics using a humanoid PAL REEM robot to conduct the experiments: The first project was dedicated to the development of a general framework for a REEM guided tour as well as its implementation for the REEM robot, cf. [7].

Contributions: The current project extends the developed framework by new features, such as (1) providing a web-based application for navigating the robot during the phase of collecting the spatial information, (2) creating and editing the tour files in a user-friendly manner, (3) providing recommendations for the users, as well as (4) allowing for a visual analysis of the data on previous tours. In this paper we present a solution that allows lab assistants to interact with the robotics system without having any technical knowledge about the system. It can be operated by any exhibition, lab staff member or social psychologist. The paper is based on a technical report [31].

The project presented in this paper is a part of the RMIT University activities on enhancing learning experience by collaborative industrial projects [6,25,28–30]. The core results of our previous project on social robotics are presented in [7]: We focused on the Lab tours use case, where the robot takes guests on tours of our Innovation Labs and answers related questions. The framework presented in [7] is based on a formal framework for modelling and analysis of autonomous systems and their compositions [27], and can be applied to any kind of guided tours, as changing the application domain would mean changing only on the content of information provided about exhibits. In the project that we present in this paper, we went further to extend the framework with web-based interface providing many useful features. While the old version with the voice commands is more human-oriented, the new web interface can be useful for noisy environments.

Outline: The rest of the paper is organised as follows. Section 2 presents related work. The architecture of the developed system as well as its core functionalities are introduced in Sect. 3. Section 4 summarises the paper and introduces directions of our future work.

2 Related Work

Duffy et al. [10] presented the concept of Social Robot Architecture, which integrates the key elements of agenthood and robotics in a coherent and systematic manner. The ethical and social implications of robotics were discussed by Lin et al. in [18]. Young et al. [33] examined social-psychology concepts to apply them to the human-robot interaction.

Eyssel et al. [12] presented a case study where they analysed the effects of robot features (human-likeness and gender) and user characteristics on the HRI acceptance and psychological anthropomorphism. Salem et al. [23] analysed the effects of gesture on the perception of psychological anthropomorphism, by conducting a case study using the Honda humanoid robot. Trovato et al. [32] conducted a cross-cultural study on generation of culture dependent facial expressions of humanoid robot. Sabanovic et al. [22] discussed the use of observational studies of human-robot social interaction in open human-inhabited environments. Klein and Cook [15] analysed and compared the findings in the UK and Germany on robot-therapy with emotional robots as a treatment approach for people with cognitive impairments.

There were also a number of surveys and literature reviews on the related topics. A survey on social robots for long-term interaction was presented in [17]. A systematic review on application of social robotics in the Autism Spectrum Disorders treatment was presented in [21]. Cabibihan et al. [5] presented a survey on the roles and benefits of social robots in the therapy of children with autism.

Alemi et al. [1] examined the effect of robot assisted language learning on the anxiety level and attitude in English vocabulary acquisition amongst Iranian EFL junior high school students. The results demonstrated that application of social robotics in this context can increase learners' engagement as well as satisfaction from the education process. Shimada et al. [24] used a social robot as a teaching assistant in a class for children's collaborative learning, and concluded that a robot can increase children's motivation, but cannot increase their learning achievement. Glas et al. [14] introduced a design framework enabling the development of social robotics applications by cross-disciplinary teams of programmers and interaction designers.

From computer science point of view, our social robots are software agents. Software agent is a running computer program which is performing tasks for a user, i.e. on behalf of him. In order to do it, agent should have knowledge about the tasks that it is expected to execute. Those are intelligent systems that could be autonomous, work with humans, or other agents. If mobile robots are involved, i.e., capabilities of performing motion, apart from those task operational duties, they have to sense and navigate environment. An adaptive roadmap approach, for autonomous robots path planning, is already investigated comprehensively, as given in [11].

3 System Architecture and Core Features

The architecture of the proposed system is demonstrated in Fig. 1. The core physical component of the system is the REEM robot controlled by the Robotics Operating System (ROS) to enable precise control from high-level programs. ROS provides services for Web-ROS communication, cf. Fig. 2: ROS side can launch a service while the web-interface can call a service. In this project, we focused on the Tour and Motion Services.
Like in our previous project [7], our work was divided into two phases:

- Phase 1 was conducted in the RMIT University VXLab (Melbourne, Australia). The introduction to the VXLab facilities can be found in [3,4,26]. The web-based interface was developed using a simulated environment provided by a ROS robot software development framework.
- Phase 2 was conducted in the CBA Labs (Sydney, Australia). A number of experiments were conducted to apply the developed web-based interface to a real REEM robot and to simulate the scenario when an operator prepares an exhibition/lab tour and executes it, both in simulated environment and on a real robot.

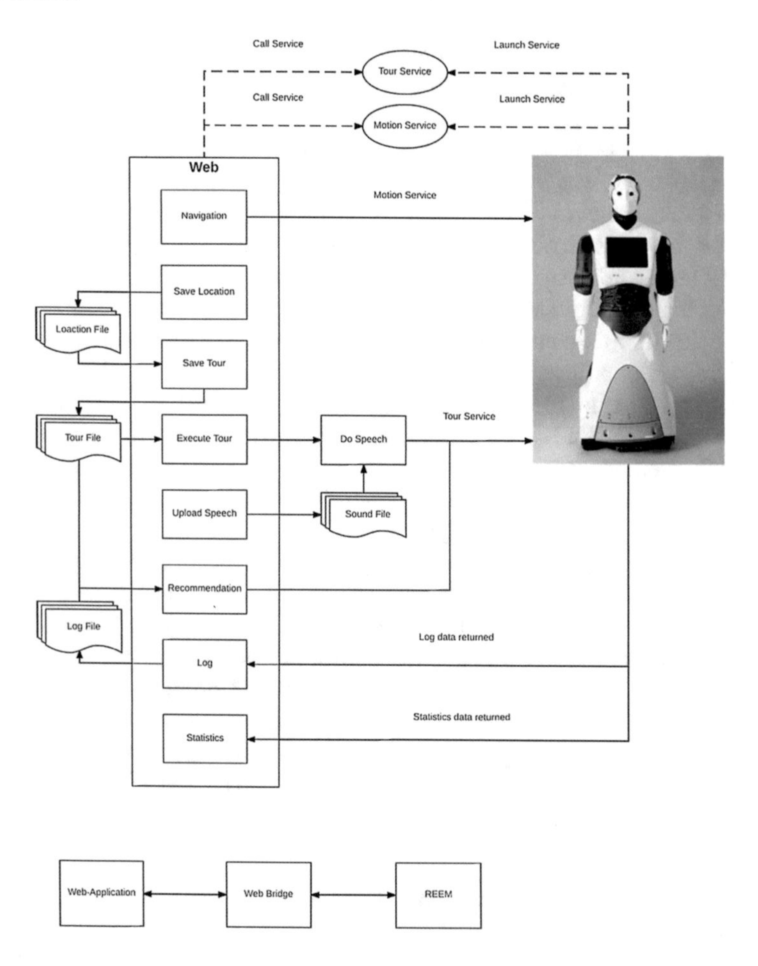

Fig. 1. System Architecture

To develop the web-based interface, we applied React.js, an open-source JavaScript library. To execute the JavaScript code server-side we applied Node.js, an open-source JavaScript runtime environment. Node.js provided the management of dependencies to certain web packages that were required for certain features to be used such as UI elements and the ROS-bridge API.

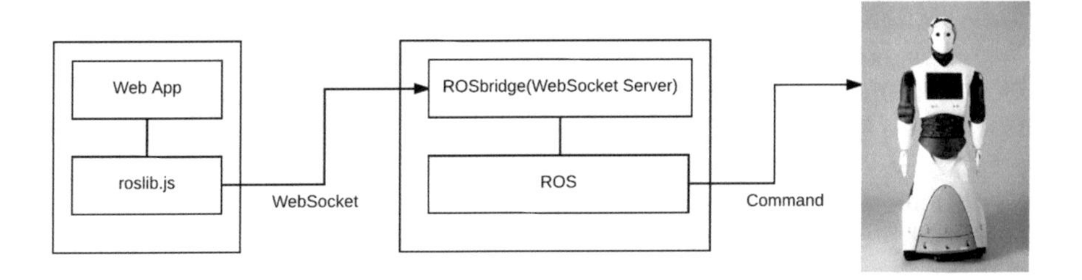

Fig. 2. Command flow within the system

ROS (Hydro Medusa) was used as the robot operating system that provides interfaces to the REEM robot's sensors, motors, actuators and speakers, by utilising Python and C++ libraries. The robot gesture, movement and navigation functionalities relied on ROS libraries. The experiments were conducted under the Ubuntu 12.04 platform. The Gazebo simulator 2.2.3 was used as the simulation environment to test the capabilities of the robot. During simulation, the movements of the robot were portrayed through control of the RViz visualisation. Converting text to speech was conducted using

- IBM Watson text-to-speech (TTS) service for the simulation, and
- on-board Acapela TTS for the experiments on the real robot.

In the simulation environment, TTS relied on Watson for wave files generation and then playing them back through sound player. When deploying on the actual robot this process is handled by Acapela, a TTS engine from Acapela Group.

Figure 3 presents the control page, where the movements of the robot can be controlled by using the corresponding menu items. This provides the functionality necessary to create the lab tours: to navigate the robot, and to store the current locations of the robot.

Fig. 3. Navigation management

The tour management page allows to display all the existing tours and all the locations within each tour, as well as to manage them. We can edit the content of an existing tour, add new tours, copy existing tours, search for particular tours or locations, etc. The web application also provides the functionality to edit the information on stored locations. The text within the description field will be a part of the speech within the guided tour: the *text-to-speech* module of the robot system will transform the text to the speech when the robot approaches the location.

Figure 4 presents an example of the visualisation features that are provided by the application: (1) statistics on the tours within the previous 6 months, (2) the distribution of he tours by their types, both in tabular and graphical format. The users can also obtain more detailed information on a particular tour, cf. Fig. 5 for an example. Figure 6 demonstrates how the developed system provides recommendations to the users based on the tour popularity.

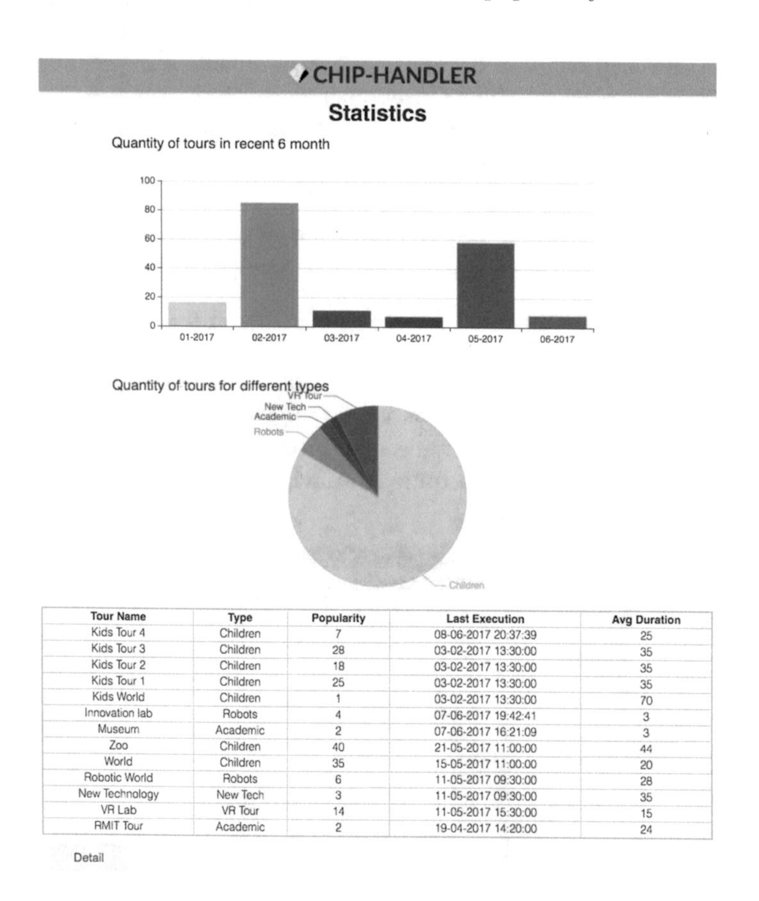

Tour Name	Type	Popularity	Last Execution	Avg Duration
Kids Tour 4	Children	7	08-06-2017 20:37:39	25
Kids Tour 3	Children	28	03-02-2017 13:30:00	35
Kids Tour 2	Children	18	03-02-2017 13:30:00	35
Kids Tour 1	Children	25	03-02-2017 13:30:00	35
Kids World	Children	1	03-02-2017 13:30:00	70
Innovation lab	Robots	4	07-06-2017 19:42:41	3
Museum	Academic	2	07-06-2017 16:21:09	3
Zoo	Children	40	21-05-2017 11:00:00	44
World	Children	35	15-05-2017 11:00:00	20
Robotic World	Robots	6	11-05-2017 09:30:00	28
New Technology	New Tech	3	11-05-2017 09:30:00	35
VR Lab	VR Tour	14	11-05-2017 15:30:00	15
RMIT Tour	Academic	2	19-04-2017 14:20:00	24

Fig. 4. Analysis of the data on the previous tours

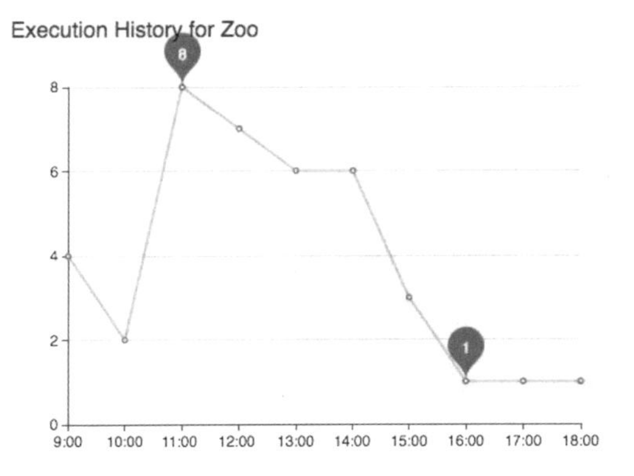

No.	Start Time	End Time	Duration
1	21-05-2017 11:00:00	15-05-2017 11:30:00	30
2	20-05-2017 13:00:00	15-05-2017 13:30:00	30
3	19-05-2017 12:00:00	15-05-2017 12:30:00	30
4	18-05-2017 13:00:00	15-05-2017 13:30:00	30
5	17-05-2017 12:00:00	15-05-2017 12:30:00	30
6	16-05-2017 12:00:00	15-05-2017 12:30:00	30
7	12-05-2017 11:30:00	12-05-2017 11:55:00	25

Fig. 5. Analysis of the data on the tour *Zoo*

4 Discussion and Conclusions

In this paper, we presented the core results of the project on software development for social robotics systems: We developed a web-based solution that supports the management of robot-guided tours, including the collection of the spatial information for the tours within noisy environments. To summarise, the developed web-based interface provides the following features:

- Navigate the robot,
- Store the current locations of the robot,
- Manage saved locations,
- Create customised tours based on saved location,
- Embed speeches for customised tours,
- Store information about a customised tour, such as the tour type, duration, etc.,
- Provide recommendations for the users, and
- Visual tool to analyse the data on previous tours.

The results of our work were implemented to the humanoid PAL REEM robot, but their core ideas can be applied for other types of humanoid robots and autonomous systems.

We plan our future work on this project in the following directions:

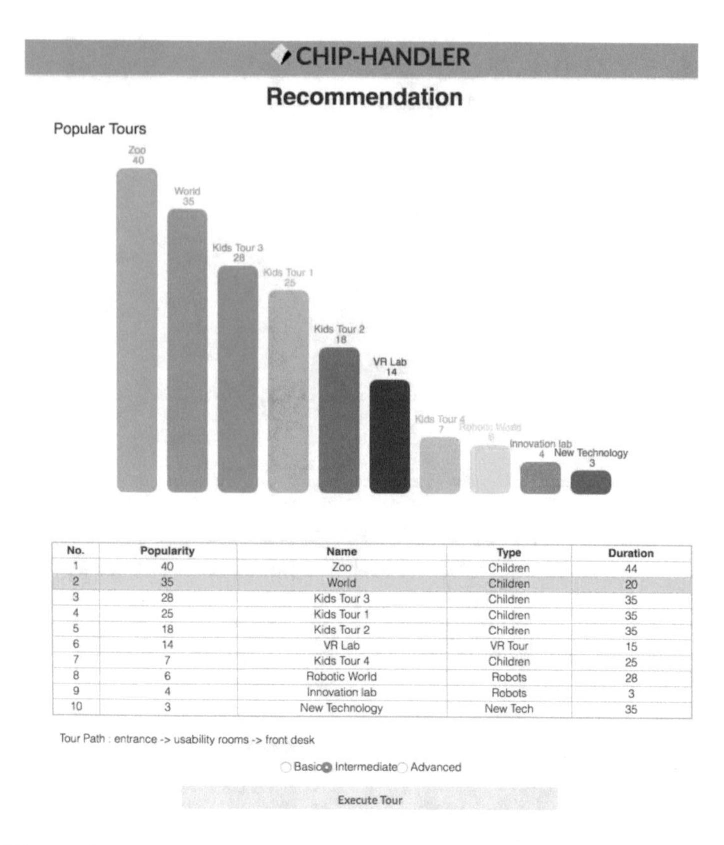

No.	Popularity	Name	Type	Duration
1	40	Zoo	Children	44
2	35	World	Children	20
3	28	Kids Tour 3	Children	35
4	25	Kids Tour 1	Children	35
5	18	Kids Tour 2	Children	35
6	14	VR Lab	VR Tour	15
7	7	Kids Tour 4	Children	25
8	6	Robotic World	Robots	28
9	4	Innovation lab	Robots	3
10	3	New Technology	New Tech	35

Fig. 6. Recommendations for the users based on the tour popularity

- To embed into the developed REEM framework the efficient testing methods, e.g. [16,19,20] as well as the model-based hazard, impact analysis and optimisation methods [2,8];
- To apply the developed framework to another type of ROS-based robot, Baxter, hosted in the RMIT University VXLab;
- To expand the developed guided tour features to involve game activities, as this would make the tours for children more entertaining and increase the children engagement;
- At this stage of our research, we have concentrated on a single intelligent robot working in human environment. Next step will be to explore scenario when our intelligent robot exists and perform tasks in the environment where, apart from humans, we have other intelligent agents as well. We will consider multi-agent systems for various applications.

Finally, but extremely important, we will comprehensively investigate safe Human-Robot interactions, current strategies and methods in practice, and possibly come up with the new solutions for the particular work environments and scenarios.

Acknowledgements. The project was sponsored by the Commonwealth Bank of Australia (CBA), Stockland Corporation Limited and the Australian Technology Network of Universities (ATN). We would like to thank William Judge (CBA) and Alec Webb (ATN) for numerous discussions and support.

References

1. Alemi, M., Meghdari, A., Ghazisaedy, M.: The impact of social robotics on L2 learners anxiety and attitude in English vocabulary acquisition. In: Social Robotics, pp. 523–535 (2015)
2. Bauer, V., Broy, M., Irlbeck, M., Leuxner, C., Spichkova, M., Dahlweid, M., Santen, T.: Survey of modeling and engineering aspects of self-adapting & self-optimizing systems. TU München, Technical report (TUM-I130307) (2013)
3. Blech, J.O., Spichkova, M., Peake, I., Schmidt, H.: Cyber-virtual systems: simulation, validation & visualization. In: ENASE (2014)
4. Blech, J.O., Spichkova, M., Peake, I., Schmidt, H.: Visualization, simulation and validation for cyber-virtual systems. In: Maciaszek, L., Filipe, J. (eds.) Evaluation of Novel Approaches to Software Engineering, pp. 140–154. Springer, Cham (2015)
5. Cabibihan, J.-J., Javed, H., Ang, M., Aljunied, S.M.: Why robots? A survey on the roles and benefits of social robots in the therapy of children with autism. Int. J. Soc. Robot. **5**(4), 593–618 (2013)
6. Christianto, A., Chen, P., Walawedura, O., Vuong, A., Feng, J., Wang, D., Spichkova, M.: Software engineering solutions to support vertical transportation. CoRR (2017)
7. Clunne-Kiely, L., Idicula, B., Payne, L., Ronggowarsito, E., Spichkova, M., Simic, M., Schmidt, H.: Modelling and implementation of humanoid robot behaviour. In: 21st International Conference on Knowledge-Based and Intelligent Information & Engineering Systems, pp. 2249–2258. Elsevier (2017)
8. Dobi, S., Gleirscher, M., Spichkova, M., Struss, P.: Model-based hazard and impact analysis. Technische Universität München, Technical report (TUM-I1333) (2013)
9. Duffy, B.R.: Anthropomorphism and the social robot. Robot. Auton. Syst. **42**(3), 177–190 (2003)
10. Duffy, B.R., Rooney, C., O'Hare, G.M., O'Donoghue, R.: What is a social robot? In: 10th Irish Conference on AI&CS (1999)
11. Elbanhawi, M., Simic, M., Jazar, R.: Autonomous robots path planning: an adaptive roadmap approach. Appl. Mech. Mater. **373**, 246–254 (2013)
12. Eyssel, F., Kuchenbrandt, D., Bobinger, S., de Ruiter, L., Hegel, F.: 'If you sound like me, you must be more human': on the interplay of robot and user features on human-robot acceptance and anthropomorphism. In: 7th ACM/IEEE International Conference on Human-Robot Interaction, pp. 125–126. ACM (2012)
13. Feil-Seifer, D., Mataric, M.J.: Defining socially assistive robotics. In: International Conference on Rehabilitation Robotics, pp. 465–468. IEEE (2005)
14. Glas, D., Satake, S., Kanda, T., Hagita, N.: An interaction design framework for social robots. In: Robotics: Science and Systems, vol. 7, p. 89 (2012)
15. Klein, B., Cook, G.: Emotional robotics in elder care-a comparison of findings in the UK and Germany. In: Social Robotics, pp. 108–117 (2012)
16. Laali, M., Liu, H., Hamilton, M., Spichkova, M., Schmidt, H.W.: Test case prioritization using online fault detection information. In: Bertogna, M., Pinho, L., Quiñones, E. (eds.) Reliable Software Technologies-Ada-Europe 2016, pp. 78–93. Springer International Publishing, Cham (2016)

17. Leite, I., Martinho, C., Paiva, A.: Social robots for long-term interaction: a survey. In: Social Robotics, pp. 291–308 (2013)
18. Lin, P., Abney, K., Bekey, G.A.: Robot Ethics: The Ethical and Social Implications of Robotics. MIT Press, Cambridge (2011)
19. Liu, H., Spichkova, M., Schmidt, H.W., Sellis, T., Duckham, M.: Spatio-temporal architecture-based framework for testing services in the cloud. In: 24th Australasian Software Engineering Conference, pp. 18–22. ACM (2015)
20. Liu, H., Spichkova, M., Schmidt, H.W., Ulrich, A., Sauer, H., Wieghardt, J.: Efficient testing based on logical architecture. In: 24th Australasian Software Engineering Conference, pp. 49–53. ACM (2015)
21. Pennisi, P., Tonacci, A., Tartarisco, G., Billeci, L., Ruta, L., Gangemi, S., Pioggia, G.: Autism and social robotics: a systematic review. Autism Res. **9**(2), 165–183 (2016)
22. Sabanovic, S., Michalowski, M., Simmons, R.: Robots in the wild: observing human-robot social interaction outside the lab. In: Advanced Motion Control, pp. 596–601. IEEE (2006)
23. Salem, M., Eyssel, F., Rohlfing, K., Kopp, S., Joublin, F.: Effects of gesture on the perception of psychological anthropomorphism: a case study with a humanoid robot. In: Social Robotics, pp. 31–41 (2011)
24. Shimada, M., Kanda, T., Koizumi, S.: How can a social robot facilitate children's collaboration? In: Social Robotics, pp. 98–107 (2012)
25. Simic, M., Spichkova, M., Schmidt, H., Peake, I.: Enhancing learning experience by collaborative industrial projects. In: ICEER 2016, pp. 1–8 (2016)
26. Spichkova, M., Schmidt, H., Peake, I.: From abstract modelling to remote cyber-physical integration/interoperability testing. In: Improving Systems and Software Engineering Conference (2013)
27. Spichkova, M., Simic, M.: Towards formal modelling of autonomous systems. In: Damiani, E., Howlett, R., Jain, L., Gallo, L., De Pietro, G. (eds.) Intelligent Interactive Multimedia Systems and Services, pp. 279–288. Springer, Cham (2015)
28. Spichkova, M., Simic, M.: Autonomous systems research embedded in teaching. In: De Pietro, G., Gallo, L., Howlett, R., Jain, L. (eds.) Intelligent Interactive Multimedia Systems and Services, pp. 268–277. Springer, Cham (2017)
29. Spichkova, M., Simic, M., Schmidt, H.: Formal model for intelligent route planning. Procedia Comput. Sci. **60**, 1299–1308 (2015)
30. Spichkova, M., Simic, M., Schmidt, H., Cheng, J., Dong, X., Gui, Y., Liang, Y., Ling, P., Yin, Z.: Formal models for intelligent speed validation and adaptation. Procedia Comput. Sci. **96**, 1609–1618 (2016)
31. Sun, C., Zhang, J., Liu, C., King, B.C.B., Zhang, Y., Galle, M., Spichkova, M.: Towards software development for social robotics systems. arXiv preprint arXiv:1712.08348 (2017)
32. Trovato, G., Kishi, T., Endo, N., Hashimoto, K., Takanishi, A.: A cross-cultural study on generation of culture dependent facial expressions of humanoid social robot. In: Social Robotics, pp. 35–44 (2012)
33. Young, J.E., Hawkins, R., Sharlin, E., Igarashi, T.: Toward acceptable domestic robots: applying insights from social psychology. In: Social Robotics, pp. 95–108 (2009)
34. Yumakulov, S., Yergens, D., Wolbring, G.: Imagery of disabled people within social robotics research. In: Social Robotics, pp. 168–177 (2012)

Local Saliency Estimation and Global Homogeneity Refinement for Video Saliency Detection

Rahma Kalboussi[1,2,3(✉)], Mehrez Abdellaoui[1,2(✉)], and Ali Douik[1,2(✉)]

[1] Networked Objects Control and Communication Systems Laboratory, Pôle technologique de Sousse, Route de Ceinture Sahloul, 4054 Sousse, Tunisia
rahma.kalboussi@gmail.com, mehrez.abdellaoui@gmail.com, ali.douik@gmail.com

[2] University of Sousse, Sousse, Tunisia

[3] Higher Institute of Computer Sciences and Communications Technology, Hammam Sousse, Sousse, Tunisia

Abstract. Saliency detection aims to segment the object of interest from the rest of the scene. While there has been a big number of saliency detection methods in still images, video saliency is in its early stages. In this paper, we propose a two stages video saliency detection method using local saliency estimation and global homogeneity refinement. Starting from a patch, the problem of saliency detection is modeled as a growing region which starts from a patch with high saliency information to the background. Local saliency is measured by combining spatial priors presented by local surrounding contrast with temporal information issued from the motion estimation feature. Temporal and spatial information are fused and then used to label each patch as foreground and background patches and produce the final saliency maps. Finally, Global homogeneity refinement is used to refine the saliency results by evaluating the foreground and background probabilities ratio propagated from the patches. Experiments have proved that the proposed method outperforms state-of-the-art methods over two benchmark datasets.

1 Introduction

In a natural image, some regions attract human gaze more than others. Scientific researches are trying to understand why some areas are more attractive than others. To answer such mysterious question several experiments were realized: different pictures are shown to different observers and their eye-movements were recorded. Such experiments have driven to different theories of eye-movements pattern. A very popular theory assumes that human gaze is directed to a specific location called salient region or salient location. At that moment, the human brain incorporates the salient region in a map called saliency map.

Generally, saliency map computation starts from selecting the salient object then segment it from the whole scene. In the saliency map, every pixel's value represents its saliency degree (its probability to belong to the salient object).

G. De Pietro et al. (Eds.): KES-IIMSS-18 2018, SIST 98, pp. 161–170, 2019.
https://doi.org/10.1007/978-3-319-92231-7_17

Since the visual world is well structured, the human visual system selects the salient region according to the different stimuli presented in the observed image. So, a very interesting question can be asked: How does our brain order our gaze? The stimuli properties play a fundamental role in salient regions selection. When observing a natural image, gaze is moved to regions with high local contrast like edges and borders, and warm colors. In a successive set of frames, observer does not have enough time to examine different location in the frame, so the gaze is always focused in a particular location which contains the dynamic object.

In the recent years saliency computation has obtained more attention thanks to its different utilization. It is used as a preprocessing step in different image processing and computer vision fields like video summarization and person identification [2], object detection [19], image classification and editing [1], [25].

Saliency detection has been widely studied and a big number of computational methods have been developed. Itti et al. proposed one of the earliest saliency model based on the Feature Integration Theory for visual attention proposed by Treisman and Gelade which assumes that the input scene can be decomposed into different feature maps [23]. They supposed that local features like color, intensity and orientation can be used by center-surround operations to produce feature maps which will be fused into one final saliency map. Itti's model inspired many researches.

In this paper we only review the main methods on video saliency, for an excellent review of saliency methods in still images we refer to [3]. As color, texture, contrast etc., cues are used for detecting salient objects in still images, moving objects attract observer's attention which makes motion as the main cue in video saliency detection. Most of the existing methods tried to combine an image saliency model with motion cue. The model proposed by Gao et al. [7] is an extension of the image saliency model [8] where they add a motion channel. Also, Mahadevan and Vasconcelos [16] used the saliency model in [8] to produce a spatiotemporal model using dynamic texture. In [21], Seo et al. used the local kernel regression from a given video frame to measure the likelihood between pixels, then extract a feature vector which includes temporal and spatial information. Rahtu et al. [20] proposed a saliency model for both natural images and videos which combines saliency measure formulated by using local features and statistical framework, with a conditional random field model. The model proposed by Fu et al. [5] which is a region saliency model, computes the saliency of a cluster using spatial, contrast and global correspondence cues. Zhong et al. [27] developed a video saliency by fusing a spatial saliency map inherited from a classical bottom-up spatial saliency model and temporal saliency map resulting from a new optical flow model. Based on Bottom-up saliency, Wang et al. [24] developed a video saliency object segmentation model using the geodesic distance with spatial edges and temporal motion boundaries which are used as foreground indicators. Later, Mauthner et al. [18] used the Gestalt principle of figure-ground segregation for appearance and motion cues to predict video saliency. Singh et al. [22] incorporated color dissimilarity, motion difference, objectness measure, and boundary score feature, of superpixels into a video saliency framework for saliency detection. Beside motion, Fang et al. [4] have used color, luminance and texture to produce saliency model in compressed domain. Lee et al. [13] combine a set of spatial saliency features including rarity, compactness, and center prior with temporal features of motion intensity and motion contrast into an SVM regressor to detect each video frame's salient object. Kim et al. [12] developed a novel approach based on the random walk with restart to detect salient regions. First, a temporal saliency distribution is found using the motion distinctiveness, Then, that temporal saliency distribution is used as a restarting distribution of the random walker. The spatial features are used to design a transition probability matrix for the walker, to estimate the final spatiotemporal saliency distribution.

In this paper we propose a new video saliency detection method by fusing motion cues with surrounding contrast cue. The first step consists on decomposing the input frame into a set of patches, then, for each path a local motion estimation using optical flow measure is computed. Local motion estimation is fused with surrounding contrast cue to produce a spatio-temporal local estimator which will be used to indicate whether a patch belongs to foreground or to background. Finally, giving foreground and background patches saliency computation is derived from a probability ratio measures according to foreground/background likelihood.

In Sect. 2 we will provide more details about saliency maps generation. In Sect. 3 we will explain our experiments and discuss our results. Section 4 is dedicated to conclusions and future works.

2 Methodology

While spatial cues like color, texture, contrast, etc., are very effective in saliency prediction in still images, they can not detect the object of interest in videos because of the complex scenes like high texture or low color distinctiveness of the foreground-background. In that case, motion information have to be included. In this paper, we propose a video saliency detection method which makes use of spatial and temporal information to produce a coherent saliency maps.

First, the input frame into a set of patches. For each patch we compute the spatial saliency information where we use the contrast measure, then the temporal information by computing our proposed motion feature. Local Spatial and temporal features are fused together then used to label each patch as foreground or background patches. Saliency scores of each patch are derived from the foreground/background probabilities. Finally, a global homogeneity refinement is performed to refine the saliency maps.

2.1 Local Motion Estimation

When an observer watches a video he does not have enough time to examine the whole scene so his gaze will be directed to a specific region which contains the dynamic target. In this section we will present how to estimate the motion between each pair of frames and estimate the saliency.

First we compute the optical flow V_f of a given frame f using [15] which provides orientation and magnitude of each flow vector. Then, we need to highlight the exact boundaries of the dynamic object which can be defined as a brutal change in the optical flow. Usually motion boundaries correspond to the boundaries of the physical salient object. We define the motion boundaries strength at patch P_i as fallows

$$M(P_i) = \frac{\sum_{x=1}^{X} (1 - exp(-\lambda||V_f(x)||))}{X} \tag{1}$$

X is the total number of pixels in the patch P_i and λ is a controlling parameter which is set to 0.5 in the experiments.

$M(P_i)$ measures the motion boundaries at every pixel (x) of the patch p and is close to 1 when the pixels of the patch change position rapidly, which can be a good saliency indicator.

But in case of low motion, optical flow can produce wrong measures. So, it will necessary to estimate the motion considering the neighboring pixels.

We introduce $O(P_i)$ a second estimator which is based on direction difference between a given pixel and its neighboring N in a giving patch p. Let be

$$O(P_i) = 1 - exp(-\beta\theta(P_i)) \tag{2}$$

where β is a controlling parameter, $\theta(P_i)$ is the maximum orientation angle between a giving pixel and all its neighbors from the patch P_i. The idea comes from the assumption that if a pixel is moving with a height or low velocity and has a different direction than the background, then, it belongs essentially to the object boundaries.

Given a the aforementioned measures, the local motion estimation can be computed as follow

$$LM(P_i) = \begin{cases} M(P_i) & if\, M(P_i) > \gamma \\ M(P_i) \times O(P_i) & if\, M(P_i) \leq \gamma \end{cases} \tag{3}$$

Local motion estimation depends on the velocity, in case of modest velocity, motion strength will be around 0.5 and wont provide good measure, so it will be necessary to add orientation based boundaries estimator and to set γ to 0.5. Figure 1 shows how important our proposed motion estimator for saliency detection.

2.2 Surrounding Contrast

Usually a salient region is distinctive from the rest of the scene. While motion is the main cue to detect saliency in videos, in case of low movement, extra cues should be considered. In our method we use the local contrast measure. However, contrast detector will measure the distinctiveness regarding the rest of the scene. Local contrast is defined as the brutal change of color independently of the spatial distance between different patches. While salient patches generally are spatially grouped, spatial distance is considered for a good local contrast representation. In this context, surrounding contrast cue is presented by [9] which assumes that not only color distinctiveness is necessary for saliency detection but also the surrounded patches characteristics. To do so, local contrast distinctiveness for each patch P_i is defined as follow

$$LC(P_i) = \sum_{P_j \forall j} \frac{D_c(P_i, P_j)}{1 + \alpha . D_p(P_i, P_j)} \tag{4}$$

where α is a variable to control the weight rate of color/spatial distance, $D_c(P_i, P_j)$ is the euclidean distance between P_i and P_j in the CIE L*a*b color space and $D_p(P_i, P_j)$ is the Euclidean distance between P_i and P_j positions.

2.3 Saliency Map Generation and Refinement

Generally, patches with higher motion attract the attention. In the last section we defined local motion estimator to measure motion feature. Now local motion estimation will be fused with surrounding contrast to produce spatio-temporal local estimator which is defined as follows

$$ST_e(P_i) = LC(P_i) \times LM(P_i) \tag{5}$$

The local spatial and temporal estimation will be used to select foreground and background patches similar to [26]. The first thing to do is to sort the ST_e values in the ascending order, then patches are ranked according to their ST_e degree, where patches with higher spatio-temporal degree is marked as foreground patches and patches with lower ST_e degree are marked as background patches. More precisely, the P_F are the first $\delta_f\%$ patches and the P_B are the last $\delta_b\%$ patches.

Given the foreground and background patches, we define for each the foreground-background probability as

$$Pr(P|P_B) = \frac{1 - ST_e(P)}{|P_B|} \sum_{X \in P_B} exp(-\frac{D_p(P, X)}{\sigma_p}) \tag{6}$$

and

$$Pr(P|P_F) = \frac{ST_e(P)}{|P_F|} \sum_{Y \in P_F} exp(-\frac{D_p(P, Y)}{\sigma_p}) \tag{7}$$

Then, we define a probability ratio R, given by

$$R = \frac{Pr(P|P_B)}{Pr(P|P_F)} \tag{8}$$

Foreground and background probabilities of a given patch P_i depend on the distance in the space domain regarding the other patches of the whole frame and on the spatio-temporal local estimator The probability ratio will serve to compute saliency scores of each patch which is defined as

$$S(P_i) = \frac{\sum\limits_{Y \in P_F, X \in P_B} exp(-\frac{D_p(P,Y) \times D_p(P,X)}{\sigma_p})}{1 + R} \tag{9}$$

Global homogeneity refinement using the foreground-background ratio is used to refine the results by finding the other salient patches of the object based on their similarities to the foreground patches in the color and temporal domain.

3 Experiments

In this section we evaluate the performance of our method on video saliency detection on two benchmark datasets and compare our results to six state-of-the-art video saliency methods in terms of Precision-Recall.

3.1 Datasets

Our results will be evaluated on two benchmark datasets which are used by most video saliency detection methods.

SegTrack v2 dataset [14] is a video segmentation and tracking dataset. It contains 14 videos with 976 frames where some videos contain one dynamic object, some others have more. Each video object has specific characteristics that can be Slow motion, Motion blur, change in Appearance, Complex deformation, Occlusion, and Interacting objects. In addition to video frames, a binarized ground truth for each frame is provided.

Fukuchi dataset [6] is a video saliency dataset which contains 10 video sequences with a total of 936 frames with a segmented ground truth.

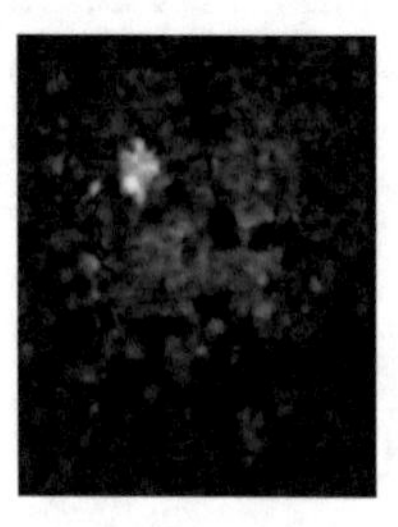

Fig. 1. Impact of our motion feature on the saliency map. From left to right: input frame, ground truth, saliency map with local motion estimation, saliency map without local motion estimation

3.2 Evaluation

We evaluate the performance using the precision-recall (PR) curve.

Pecision-Recall curve plots the Precision against the recall. To do so, each saliency map is binarized using a fixed set of thresholds varying from 0 to 255. The precision and the recall are then computed by comparing the binarized map S to the ground-truth G see Eqs. 10 and 11.

$$\text{precision} = \frac{\sum_{\mathrm{x,y}} \mathrm{S(x,y)G(x,y)}}{\sum \mathrm{S(x,y)}} \tag{10}$$

$$\text{recall} = \frac{\sum_{\mathrm{x,y}} \mathrm{S(x,y)G(x,y)}}{\sum \mathrm{G(x,y)}} \tag{11}$$

S is the binarized estimated saliency map and G is the binary ground truth. The Presion-Recall curve can be computed by varying the threshold which is used to binarize $S(x, y)$.

3.3 Results

Our results will be compared to six state-of-the-art video saliency methods in terms of **Precision-Recall**. On **SegTrack v2** and **Fukuchi** datasets we outperform all other approaches with a big gap in term of best precision rates. The nature of video sequences of the SegTrack v2 dataset are quite different. In Segtrackv2 dataset some videos contain more than one dynamic object, which means that motion cue alone wont be enough to predict saliency which obliges us to add spatial cue (color). Spatial cue fused with temporal cue helps to separate the salient object from background and to highlight it.

On Fukuchi dataset, all videos contain only one dynamic object which explains the finer precision shape. The GVS [24] computes spatial and temporal edges of each object in the video frame. This method is very efficient when the video frame has one moving object that's why when applying this method to the SegTrack v2 which includes video frames with different conditions (as we explained in the last paragraph), results decreased a little bit. On Fukuchi dataset which all videos contain one dynamic object, results are better.

GB [10] PR curve are not good because this method does not include motion as a saliency cue. Even if this method is used for video saliency detection is more suitable for image saliency.

Mancas video saliency method RR [17] exploits the optical flow to select motion in a crowd. For videos with stable camera like video surveillance applications, this method is very effective but does not produce good results with video frames of SegTrack v2 and Fukuchi datasets where camera motion causes noise.

Precision-Recall curves on Segtrackv2 and Fukuchi datasets are reported in Fig. 2 and Fig. 2 where our proposed method outperforms other methods. The recall values of **RR**[17] and **GVS** [24] are very small when we varied the threshold to 255 and even can go down to 0 for **ITTI** [11], **RT** [20] and **GB** [10].

On fukuchi dataset we have best precision rate which shows that our proposed method is very efficient and provides very precise salient objects.

On Segtrackv2 dataset, we have competitive results compared to **GVS** [24] which also indicate that our saliency maps are informative of the region on interest. Also, on Segtrackv2 dataset, the minimum value of recall does not go down to zero which means that in its worst cases, our method detects the region of interest with a good response values.

We presented in Fig. 3 a visual comparison of the saliency maps produced by our approach against state-of-the-art methods.

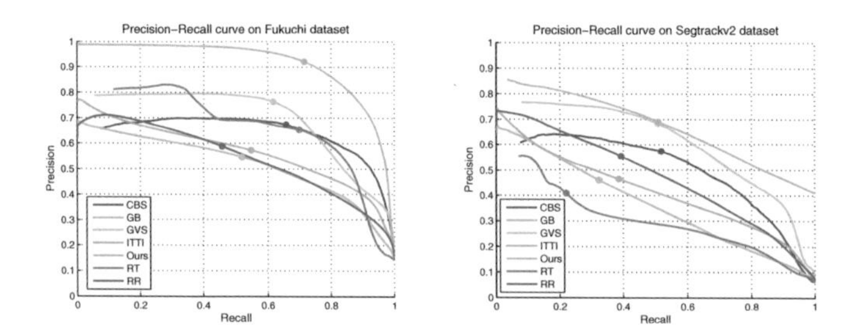

Fig. 2. Precision-Recall curve on Fukuchi and Segtrackv2 dataset

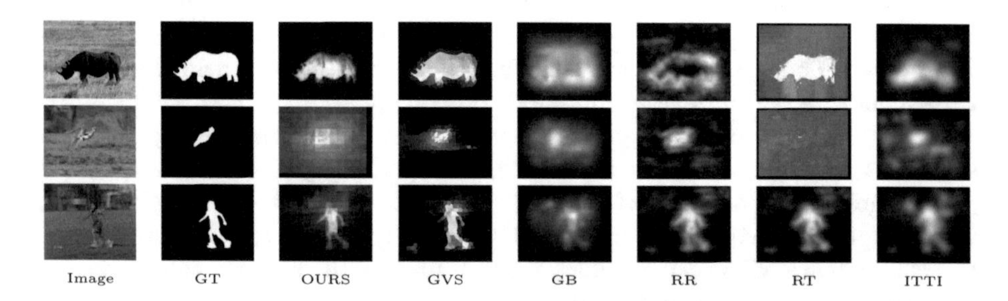

Fig. 3. Visual comparison of saliency maps generated from 6 different methods, including our method, GVS [24], GB [10], RR [17], RT [20] and ITTI [11]

4 Conclusion

In this paper we propose a video saliency detection method by fusing temporal cues with spatial information of a local patch and using global refinement to produce consistent saliency maps. Our contribution consists on measuring the local motion estimation of a patch from a video frame. The experiments have demonstrated that our method outperforms state-of-the-art methods in terms of Precision-Recall metrics on two benchmark datasets. In future work, we will try to extend our method to be a real time application which can be installed on a mobile device and to be used for various interests.

References

1. Avidan, S., Shamir, A.: Seam carving for content-aware image resizing. ACM Trans. Graph. (TOG) **26**, 10 (2007)
2. Bi, S., Li, G., Yu, Y.: Person re-identification using multiple experts with random subspaces. J. Image Graph. **2** (2014)
3. Borji, A., Cheng, M.-M., Jiang, H., Li, J.: Salient object detection: a benchmark. IEEE Trans. Image Process. **24**, 5706–5722 (2015)
4. Fang, Y., Lin, W., Chen, Z., Tsai, C.-M., Lin, C.-W.: A video saliency detection model in compressed domain. IEEE Trans. Circ. Syst. Video Technol. **24**, 27–38 (2014)
5. Fu, H., Cao, X., Tu, Z.: Cluster-based co-saliency detection. IEEE Trans. Image Process. **22**, 3766–3778 (2013)
6. Fukuchi, K., Miyazato, K., Kimura, A., Takagi, S., Yamato, J.: Saliency-based video segmentation with graph cuts and sequentially updated priors. In: 2009 IEEE International Conference on Multimedia and Expo, pp. 638–641. IEEE (2009)
7. Gao, D., Mahadevan, V., Vasconcelos, N.: The discriminant center-surround hypothesis for bottom-up saliency. In: Advances in Neural Information Processing Systems, pp. 497–504 (2008)
8. Gao, D., Vasconcelos, N.: Bottom-up saliency is a discriminant process. In: IEEE 11th International Conference on Computer Vision, ICCV 2007, pp. 1–6. IEEE (2007)
9. Goferman, S., Zelnik-Manor, L., Tal, A.: Context-aware saliency detection. IEEE Trans. Pattern Anal. Mach. Intell. **34**, 1915–1926 (2012)
10. Harel, J., Koch, C., Perona, P., et al.: Graph-based visual saliency. In: NIPS, vol. 1, p. 5 (2006)
11. Itti, L., Baldi, P.: A principled approach to detecting surprising events in video. In: IEEE Computer Society Conference on Computer Vision and Pattern Recognition, CVPR 2005, vol. 1, pp. 631–637. IEEE (2005)
12. Kim, H., Kim, Y., Sim, J.-Y., Kim, C.-S.: Spatiotemporal saliency detection for video sequences based on random walk with restart. IEEE Transa. Image Process. **24**, 2552–2564 (2015)
13. Lee, S.-H., Kim, J.-H., Choi, K.P., Sim, J.-Y., Kim, C.-S.: Video saliency detection based on spatiotemporal feature learning. In: 2014 IEEE International Conference on Image Processing (ICIP), pp. 1120–1124. IEEE (2014)
14. Li, F., Kim, T., Humayun, A., Tsai, D., Rehg, J.M.: Video segmentation by tracking many figure-ground segments. In: Proceedings of the IEEE International Conference on Computer Vision, pp. 2192–2199 (2013)
15. Lucas, B.D., Kanade, T., et al.: An iterative image registration technique with an application to stereo vision. In: IJCAI, vol. 81, pp. 674–679 (1981)
16. Mahadevan, V., Vasconcelos, N.: Spatiotemporal saliency in dynamic scenes. IEEE Trans. Pattern Anal. Mach. Intell. **32**, 171–177 (2010)
17. Mancas, M., Riche, N., Leroy, J., Gosselin, B.: Abnormal motion selection in crowds using bottom-up saliency. In: 2011 18th IEEE International Conference on Image Processing (ICIP), pp. 229–232. IEEE (2011)

18. Mauthner, T., Possegger, H., Waltner, G., Bischof, H.: Encoding based saliency detection for videos and images. In: Proceedings of the IEEE Conference on Computer Vision and Pattern Recognition, pp. 2494–2502 (2015)
19. Navalpakkam, V., Itti, L.: An integrated model of top-down and bottom-up attention for optimizing detection speed. In: 2006 IEEE Computer Society Conference on Computer Vision and Pattern Recognition (CVPR 2006), vol. 2, pp. 2049–2056. IEEE (2006)
20. Rahtu, E., Kannala, J., Salo, M., Heikkilä, J.: Segmenting salient objects from images and videos. In: Computer Vision-ECCV 2010, pp. 366–379 (2010)
21. Seo, H.J., Milanfar, P.: Static and space-time visual saliency detection by self-resemblance. J. Vis. **9**, 15–15 (2009)
22. Singh, A., Chu, C.-H.H., Pratt, M.: Learning to predict video saliency using temporal superpixels. In: 4th International Conference on Pattern Recognition Applications and Methods, pp. 201–209 (2015)
23. Treisman, A.M., Gelade, G.: A feature-integration theory of attention. Cogn. psychol. **12**, 97–136 (1980)
24. Wang, W., Shen, J., Porikli, F.: Saliency-aware geodesic video object segmentation. In: Proceedings of the IEEE Conference on Computer Vision and Pattern Recognition, pp. 3395–3402 (2015)
25. Wu, R., Yu, Y., Wang, W.: Scale: supervised and cascaded Laplacian eigenmaps for visual object recognition based on nearest neighbors. In: Proceedings of the IEEE Conference on Computer Vision and Pattern Recognition, pp. 867–874 (2013)
26. Yeh, H.-H., Liu, K.-H., Chen, C.-S.: Salient object detection via local saliency estimation and global homogeneity refinement. Pattern Recogn. **47**, 1740–1750 (2014)
27. Zhong, S., Liu, Y., Ren, F., Zhang, J., Ren, T.: Video saliency detection via dynamic consistent spatio-temporal attention modelling. In: AAAI, pp. 1063–1069 (2013)

Innovation in Medicine and Healthcare (KES-InMed-18) Introduction

Innovation in Medicine and Healthcare (KES-InMed-18) Introduction

The 6th KES International Conference on Innovation in Medicine and Healthcare (InMed-18) was held on 20–22 June 2018 in Gold Coast, Australia, organized by KES International. All submissions were carefully reviewed by at least two reviewers of the International Program Committee. Finally 11 papers were accepted to be presented as chapters in this book.

The major areas include:

(1) Digital Architecture for Internet of Things, Big data, Cloud and Mobile IT in Healthcare;
(2) Advanced ICT for Medical and Healthcare.

The chapters contained here form an informative and useful snap-shot of the state of the art in this subject.

Yen-Wei Chen
Satoshi Tanaka
Robert J. Howlett
Lakhmi C. Jain
Ljubo Vlacic

Organization

InMed-18 International Programme Committee

Arnulfo Alanis	Departamento de Sistemas y Computación Instituto Tecnológico de Tijuana, Mexico
Ahmad Taher Azar	Faculty of Computers and Information, Benha University, Egypt
Alfonso Barros-Loscertales	University Jaume I, Spain
Smaranda Belciug	University of Craiova, Romania
Vitoantonio Bevilacqua	DEI - Polytechnic University of Bari, Italy
Isabelle Bichindaritz	State University of New York at Oswego, USA
Christopher Buckingham	Aston University, UK
Yen-Wei Chen	Ritsumeikan University, Japan
Luis Enrique Sánchez Crespo	University of Castilla-la Mancha, Spain
Massimo Esposito	ICAR-CNR, Italy
Jesús M. Doña Fernández	Andalusian Health Service, Spain
Cecilia Dias Flores	Universidade Federal de Ciências da Saúde de Porto Alegre - UFCSPA, Brazil
María del Rosario Baltazar Flores	Instituto Tecnológico de León, México
Amir H. Foruzan	Biomedical Engineering Group, Faculty of Engineering, Shahed University, Tehran, Iran
Arfan Ghani	Coventry University, UK
Juan Manuel Górriz	University of Granada, Spain
Florin Gorunescu	University of Piteşti/University of Medicine and Pharmacy of Craiova, Romania
Manuel Graña	Universidad del Pais Vasco, Spain
Ioannis Hatzilygeroudis	University of Patras, Greece
Tomohiro Kuroda	Kyoto University, Japan
Lenin G. Lemus-Zúñiga	Universitat Politècnica de València, España
Giosue' Lo Bosco	Universita' di Palermo, Dipartimento di Matematica e Informatica, Italy
Yoshimasa Masuda	Keio Graduate School, Japan
Rashid Mehmood	King Abdulaziz University, Jeddah, Saudi Arabia
Aniello Minutolo	Institute for High Performance Computing and Networking, ICAR-CNR, Italy
Stefania Montani	DISIT, Computer Science Institute, Universita' del Piemonte Orientale, Alessandria, Italy
Louise Moody	Coventry University, UK
Marek Ogiela	AGH University of Science and Technology, Poland
Wiesław Paja	University of Rzeszów, Poland
Dorin Popescu	University of Craiova, Romania

Luigi Portinale	University of Piemonte Orientale, Italy
Marco Pota	High Performance Computing and Networking (ICAR) - National Research Council of Italy (CNR)
Javier Ramirez	University of Granada, Spain
Ana Respício	University of Lisbon, Portugal
John Ronczka	SCOTTYNCC Independent Research Scientist, Australia
Yves Rybarczyk	Universidad de Las Américas, Portugal
Marina Sokolova	Southwest State University, Russia
Ruxandra Stoean	University of Craiova, Romania
Catalin Stoean	University of Craiova, Romania
Kenji Suzuki	Medical Imaging Research Center, Illinois Institute of Technology, USA
Satoshi Tanaka	Ritsumeikan University, Japan
Eiji Uchino	Yamaguchi University, Japan
Shuichiro Yamamoto	Nagoya University, Japan

A Vision for Open Healthcare Platform 2030

Yoshimasa Masuda[1,3(✉)], Shuichiro Yamamoto[2], and Seiko Shirasaka[1]

[1] Graduate School of System Design and Management, Keio University, Yokohama, Japan
ymasuda@australia.cmu.edu
[2] Graduate School of Information Science, Nagoya University, Nagoya, Japan
[3] Carnegie Mellon University, Pittsburgh, USA

Abstract. Internets of Things (IoT) applications and services have spread and are rapidly being deployed in the information services of the healthcare and financial industries, etc. However, the previous paper suggested that the current IoT services were individually developed, therefore, the open platform and architecture for the above IoT services of the healthcare industries should be deemed necessary. An open healthcare platform is expected to promote and implement the digital IT applications for healthcare communities efficiently. In this paper, we suggest that the open platform for healthcare related IoT services will be proposed and verified by the research initiative named "Open Healthcare Platform 2030 – OHP2030". In addition, the vision for the OHP2030 research initiative is expressed.

Keywords: Digital healthcare · Enterprise architecture · Internet of Things
Digital platform · Digital IT

1 Introduction

Many global corporations have encountered a variety of changes, such as progress of new technologies, globalization, shifts in customer needs, and new business models. Significant changes in cutting-edge IT technology due to recent developments in Cloud computing and Mobile IT (such as progress in big data technology) have emerged as new trends in information technology. Furthermore, major advances in the abovementioned technologies and processes have created a "Digital IT economy," bringing about both business opportunities and business risks and forcing enterprises to innovate or face the consequences [2]. Enterprise Architecture (EA) should be effective because it contributes to the design of large integrated systems, which represents a major technical challenge toward the era of Cloud/Mobile IT/Big Data/Digital IT in Digital Transformation. From a comprehensive perspective, EA encompasses all enterprise artifacts, such as business, organization, applications, data, and infrastructure, to establish the current architecture visibility and future architecture/roadmap. On the other hand, EA frameworks need to embrace change in ways that consider the emerging new paradigms and requirements affecting EA, such as Mobile IT/Cloud [1, 3].

G. De Pietro et al. (Eds.): KES-IIMSS-18 2018, SIST 98, pp. 175–185, 2019.
https://doi.org/10.1007/978-3-319-92231-7_18

Furthermore, considering the above background, the previous study proposed the "Adaptive Integrated EA framework," which should align with IT strategy promoting Cloud/Mobile IT/Digital IT, and verified this in the case study [15]. The author of this paper has named the EA framework suitable for the era of Digital IT as "Adaptive Integrated Digital Architecture Framework – AIDAF" [23].

This paper is organized as follows: the next section presents the background of this study, followed by the description of the research methodology and an overview of the AIDAF application for the cross-functional healthcare community, and healthcare community case. Finally, the challenges and final thoughts for this OHP2030 are outlined.

2 The Direction of EA and IoT

2.1 Related Work and Direction of Cloud/Mobile IT/Big Data

In the past decade, EA has become an important method for modeling the correlation for overall images of corporate and individual systems. In ISO/IEC/IEEE42010:2011, architecture framework is defined as "principles, and practices for the architecture descriptions established within a specific domain of application and/or community." Furthermore, EA visualizes the current corporate IT/business landscape to promote a desirable future IT model [3]. It is not a simple support activity [1], and it offers many benefits to companies, such as coordination, communication and planning between business and IT, and reduction in the complexity of IT [14]. For the delivery of these benefits, EA frameworks need to cope with the emerging new paradigms such as Cloud computing or enterprise mobility [1].

Mobile IT computing is an emerging concept using Cloud services provided over mobile devices [11]. In addition, Mobile IT applications are composed of Web services. Many studies discuss the integration of EA with Service Oriented Architecture (SOA), except for Mobile IT. The SOA architecture pattern defines the four basic forms of business service, enterprise service, application service, and infrastructure service [13]. The OASIS, which is a public standards group [9], introduces an SOA reference model. Many organizations have invested in SOA as an approach to manage rapid change [4]. Meanwhile, attention has been focused on Microservices architecture, which allows rapid adoption of new technologies, such as Mobile IT applications and Cloud computing [12]. SOA and Microservice vary greatly from service characteristics perspective [13]. Microservice is an approach for dispersed systems that is defined by the two basic forms of functional services through an API layer and infrastructure services. Multiple Microservices cooperating to work together enable the implementation as a Mobile IT application [5].

For Cloud Computing, the NIST defined three cloud service models such as SaaS, PaaS, and IaaS [7]. PaaS is an IaaS platform that includes both system software and an integrated development environment. SaaS is a software application developed, implemented, and operated on a PaaS foundation. IaaS accommodates PaaS and SaaS by offering infrastructure resources, such as computing network storage memory through specific centers [7]. Many Mobile IT applications also operate with SaaS Cloud-based software [11]. The integration and relationship between EA and Cloud computing is discussed rarely in literature. Considering the recent dynamic moves in Cloud computing, it is necessary for companies to link the service characteristics of EA and Cloud computing [8]. The traditional approach takes months to develop an EA realizing a Cloud adoption strategy, and organizations will demand adaptive enterprise architecture to iteratively develop and manage an EA adaptive to the Cloud technology [6].

Moreover, according to previous research [10], when promoting Cloud/Mobile IT in a strategic manner, it is proposed as a good option that a company that applies TOGAF or FEAF can adopt the integrated framework with the Adaptive EA framework supporting elements of Cloud computing.

Furthermore, in terms of Big Data, new computing trends require data with far greater volume, velocity, and variety than ever before. Big data is utilized in ingenious methods to predict customer buying behaviors, detect fraud and waste, analyze product opinion, and react quickly to changes in business conditions (a driving force behind new business opportunities) [16]. The term "big data" refers to data that is so large, it is difficult to process using currently-available IT systems. There is a growing opportunity for analysis, visualization, and distributed processing software to enable users to extract useful information from such data [2]. Sources of big data include the following.

- Corporate data in SQL databases
- Data in cloud-based SQL or NoSQL databases
- Data provided by social networks
- Data provided by sensors or object identifiers in the internet-of-things (IoT)

Big data applications may include visualization functionality for effective user presentation of analytical results. Furthermore, big data applications should leverage web services that make the results of their analyses available to other applications through APIs; objects in the IoT can be data generators [2].

Existing big data reference architectures have been shepherded by NIST, which helped create the big data interoperability framework, including a reference architecture volume [17].

LinkedIn, for example, collects data from users and offers services, such as skill endorsements or newsfeed updates to users based on data analysis. Additionally, Twitter uses collected data for real-time query suggestion [22]. Therefore, most solutions exist in the Big Data Application Provider component and should be categorized as Specific Application Layers on Cloud and Mobile IT platforms. Technology vendors such as Oracle [19], IBM [20], and Microsoft [21] have also developed Big Data Reference Architectures [18]. These vendors publish practical Reference Architectures for Big Data toward EA practitioners in corporations and other groups.

2.2 Internet of Things

2.2.1 Internet of Things Architecture

The term of "Internet of Things (IoT)" is used to mean "the collection of uniquely identifiable objects embedded in or accessible by Internet hosts" [2]. A "uniquely identifiable object" can be described as follows: these objects are connected with real world interaction devices, smart homes and cars, and other SmartLife scenarios. The IoT fundamentally revolutionizes digital strategies with innovative business operating models [33], and holistic governance models for business and IT [34], under fast changing markets [32].

- A sensor, such as a temperature sensor (thermometer)
- A control; for example, to control a valve in a heating system
- A combination of sensor and control (for example, a thermostat)
- An object identifier, such as an RFID tag or a barcode

The current state of research for the Internet of Things architecture [27] lacks an integral understanding of EA and Management [28–31], and shows a number of physical standards, methods, tools and a large number of heterogeneous IoT devices [32]. A first reference architecture (RA) for the IoT is proposed by [35] as shown in Fig. 1 below, and can be mapped to a set of open source products. This RA covers aspects like "cloud server-side architecture," "monitoring/management of IoT devices, services," "specific lightweight RESTful communications," and "agent, code on small low power devices." Layers can be instantiated by suitable technologies for the IoT [32].

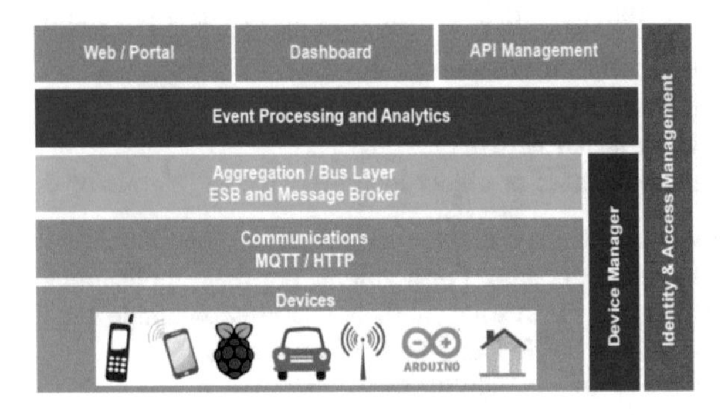

Fig. 1. Internet of Things reference architecture [WS15]

On the other hand, in the field of IoT, platforms are designed to deploy digital IT applications that monitor, manage, and control connected IoT devices. McKinsey also shows the layered model of the "Internet of Things tech stacks" as shown in Fig. 2 below.

2.2.2 IoT Middleware

Based on [36] architecture categories, comparative analysis, and key issues of IoT middleware were clarified. Ngu et al. defined "(1) lightweight and portability," "(2) composition engine," "(3) semantic service discovery" and "(4) the guarantee of security" as issues of IoT middleware [37].

First, during development, an IoT middleware, such as IoT devices and gateways, must be available in the cloud as well as on the edge to support all kinds of IoT applications. This requires the middleware to be portable and lightweight. Only Calvin [38], Node-RED [39], and Ptolemy [40] are designed to be portable and light-weight [36].

Second, to create IoT applications targeted to their context, a composition engine should be provided for consumers. The visual tools provided by Node-RED [39] and Ptolemy [40] can be used as composition tools in for this purpose [36].

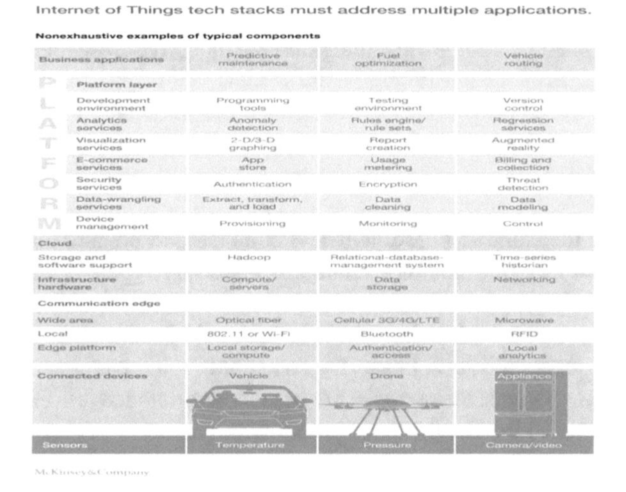

Fig. 2. McKinsey IoT stacks

Third, to provide a semantic service discovery for IoT services/devices, it is necessary to be capable of discovering or querying for compatible services at the right time and at the right place both at the time of design and at runtime [36]. Open IoT [43], GSN [42], and Hydra [41] can provide semantic service discovery as the IoT middleware [37].

Finally, to guarantee the security of IoT applications and protect the privacy of users, many digital applications need to utilize a reliable IoT infrastructure. GSN [42] and Hydra [41] can provide such security guarantee functions [37].

2.2.3 Platforms for Healthcare IoT

There are sensors that are being developed by the researchers at Monash University to monitor aged people in their homes. The product is called as a "Super Sensor" which has many sensors including Passive Infrared (PIR), light, vibration, temperature sensors [44].

Another potential solution is MySignals by Libelium. It is a development platform for medical devices and eHealth applications. It offers an out of the box open software platform that allows developers to write Software to utilize its hardware features [45].

2.3 Adaptive Integrated Digital Architecture Framework – Aligned with Digital IT Strategy

Our previous research promoted the strategic use of cloud and mobile IT, suggesting that corporate entities defining Cloud/Mobile IT/Big Data/Digital IT strategies implementing EA by applying frameworks, such as TOGAF and FEAF, could adopt a framework integrating an Adaptive EA framework to provide further support for cloud elements [10]. Accordingly, the preliminary research of this paper proposes an Adaptive Integrated EA framework depicted in Fig. 1 of the preliminary research paper, based on this suggestion, which should meet with IT strategy promoting

Cloud/Mobile IT/Big Data/Digital IT, and verified this in the case study [15]. The proposed model is an EA framework integrating an adaptive EA cycle with TOGAF or a simple EA framework 1 for different business units in the upper part of the diagram in [15]. The author of the previous paper mentioned above named this EA framework as "Adaptive Integrated Digital Architecture Framework (AIDAF) [23]."

The adaptive EA cycle makes provisions for project plan documents (including architecture designs) for new Digital IT related projects drawn up on a short-term basis. This begins with the Context Phase, which is conducted to prepare the Defining Phase (e.g., architectural design guidelines related to necessary types of Security/Digital IT aligned with IT strategy) per business needs. During the Assessment/Architecture Review Phase, the Architecture Board reviews the architecture in the initiation documents for the IT project. In the Rationalization Phase, the stakeholders and Architecture Board decide upon information systems to be replaced or decommissioned by the proposed new information system. In the Realization Phase, the project team begins to implement the new IT project agreed upon after deliberating issues and action items [15, 23].

The Adaptive EA cycle enables the corporate entity to adopt an EA framework capable of flexibly adapting to new Digital IT projects continuously. Moreover, the TOGAF and simple EA framework, based on an operational division unit in the top part of the Fig. 1 of [15, 23], can respond to differing strategies in business divisions in the mid-long-term. This part of the framework has a structure that can select an EA framework based upon the characteristics of division's operational processes and future architectures, while enabling applications. Furthermore, the framework should align EA guiding principles with those business division's principles to keep consistency among the adaptive EA cycle, the TOGAF, and simple EA framework. Furthermore, in the Defining Phase, the Architecture Board promotes the appropriate architectural design of each Digital IT-related system by sharing the architectural guidelines for Security/Digital IT, etc., to align with the IT strategy [15, 23].

3 Research Methodology

In this paper, the authors first state research questions to understand and formulate the layers and specifications for IoT digital platforms at the middleware and application layer in the healthcare industry. Then, the authors will evaluate these research questions using a case study in the cross-functional healthcare community in Asia and globally. The following research questions are evaluated in the case study.

RQ1: How can a layered architecture for the IoT be defined considering existing IoT architecture models?
RQ2: How can layers and specifications for digital IoT platforms in the middleware and application layer be clarified and formulated for healthcare industries?

The authors will investigate a case study within the cross-functional healthcare community globally and in Asia and Global, where the author should build and implement the "Adaptive Integrated Digital Architecture Framework - AIDAF" and start the Architecture Board. In the Architecture Board, all new IS/IT project architecture designs will be reviewed and action items for next steps will be raised by architects and top management and PMO members. After the Architecture Board should be held, the authors need to compare two kinds of IoT architecture models and define layers and specifications for digital IoT platforms in the middleware and application layers for healthcare industries.

4 AIDAF Application for Cross-Functional Healthcare Community

The author of this paper proposed an adaptive integrated EA framework to align with IT strategy, promoting Cloud/Mobile IT/Big Data/Digital IT, and verified by our case study [15]. Furthermore, the author of this paper has named the EA framework suitable for the era of Digital IT as an "Adaptive Integrated Digital Architecture Framework – AIDAF" [23]. Figure 2 illustrates this AIDAF proposed model in the Open Healthcare Platform 2030 (OHP2030) community. The OHP2030 community is comprised of healthcare companies such as pharmaceutical, medical development and aged healthcare companies, hospitals as well as the OHP2030 initiative and government as depicted in Fig. 3. AIDAF will be applied to the above-mentioned cross-functional healthcare community in the OHP2030. AIDAF begins with the context phase, while referencing the defining phase (i.e., architecture design guidelines related to digital IT aligned with IT strategy in the above healthcare community in the OHP2030). During the assessment and architecture review, the architecture board reviews the initiation documents and related architectures for the IT project in the above healthcare community in the OHP2030.

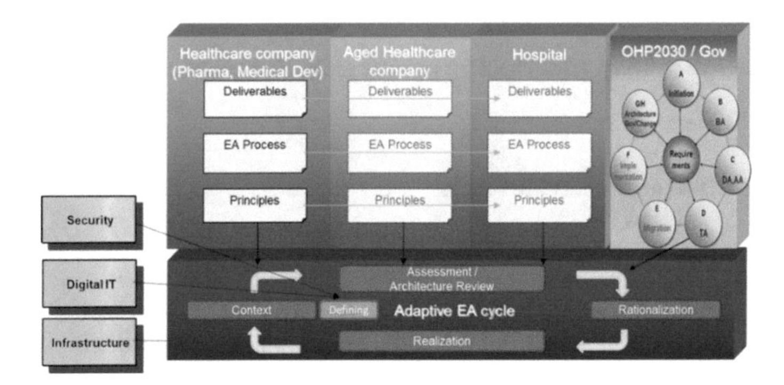

Fig. 3. AIDAF proposed model in the Healthcare community in the OHP2030 (ex: TOGAF and Adaptive EA framework)

5 Healthcare Community Case

In OHP2030, a particular Architecture Board will be formulated in the aforementioned cross-functional healthcare community. In the case study of EA rollout in the healthcare community, they will handle IoT/Big Data/Cloud strategic projects and systems well by structuring and implementing EA with the above-mentioned AIDAF to be consistent with the IT strategy focusing on IoT/Big Data/Digital IT in the above cross-functional Healthcare community.

Furthermore, in the case study of the cross-functional healthcare community, the author of this paper assumes that use cases of "IDC's Worldwide Digital Transformation Use Case Taxonomy, 2017: Healthcare," which were defined by IDC, should be applied in the context phase and assessment/architecture review phase as well as the defining phase of the AIDAF for new IoT projects in the healthcare community [24, 25]. According to [24, 25], the above use cases and digital missions in healthcare are based on the creation of value-based healthcare systems from sick care to healthcare management as shown in Fig. 4 above [26].

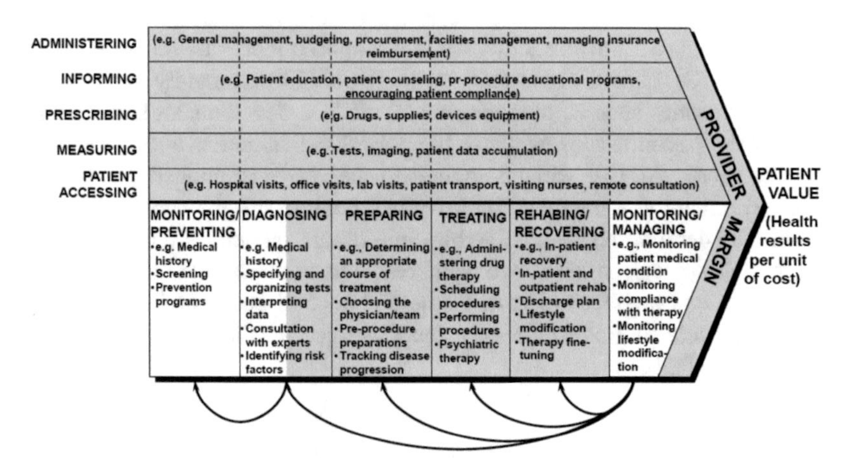

Fig. 4. Healthcare delivery value chain for practice areas (Source: Michael Porter and Elizabeth, 2005)

6 Discussions and Challenges

6.1 Future Issues

To address technological issues, we need to embody quality assurance methods such as, reliability and security for a digital healthcare IoT platform, while implementing IoT demonstration systems in the field of digital healthcare.

From the standpoint of social issues, first, the commercialization of the above platform needs to be simplified by establishing collaborative platforms for IoT services. Second, for the purpose of grasping issues in terms of realizing IoT services/systems on the above-mentioned platform, the IoT reference model capability index needs to be designed. Third, we will proceed with the international standardization of a digital healthcare IoT platform, while having discussions in conferences related to OHP2030 geared towards domestic and/or international organizations of research and development. In so doing, we will extensively understand these issues in terms of the social implementations for digital healthcare applications on the proposed platform.

7 Final Thoughts

In this paper, we have described our vision for the collaborative research initiative of OHP2030. This research initiative aims at an exploration and definition of a digital healthcare IoT platform in the middleware layer and above. Can we define and implement a digital platform, in healthcare communities, that can promote and support digital healthcare related IoT applications, while ensuring information security?

References

1. Alwadain, A., Fielt, E., Korthaus, A., Rosemann, M.: A comparative analysis of the integration of SOA elements in widely-used enterprise architecture frameworks. Int. J. Intell. Inf. Technol. **9**(2), 54–70 (2014)
2. Boardman, S., Harrington, E.: Open Group Snapshot - Open Platform 3.0™. The Open Group (2015)
3. Buckl, S., Matthes, F., Schulz, C., Schweda, C.M.: Exemplifying a framework for interrelating enterprise architecture concerns. In: Sicilia, M.A., Kop, C., Sartori, F. (eds.) Ontology, Conceptualization and Epistemology for Information Systems, Software Engineering and Service Science, vol. 62, pp. 33–46. Springer, Heidelberg (2010). https://doi.org/10.1007/978-3-642-16496-5_3
4. Chen, H., Kazman, R., Perry, O.: From software architecture analysis to service engineering: an empirical study of methodology development for enterprise SOA implementation. IEEE Trans. Serv. Comput. **3**(2), 145–160 (2014). https://doi.org/10.1109/TSC.2010.21
5. Familiar, B.: Microservices, IoT and Azure: leveraging DevOps and Microservice architecture to deliver SaaS solutions. Apress, Berkeley (2015)
6. Gill, A.Q., Smith, S., Beydoun, G., Sugumaran, V.: Agile enterprise architecture: a case of a cloud technology-enabled government enterprise transformation. In: Proceedings of the 19th Pacific Asia Conference on Information Systems (PACIS 2014), pp. 1–11 (2014)
7. Gill, A.Q.: Adaptive Cloud Enterprise Architecture, Intelligent Information Systems, vol. 4. World Scientific Publishing Co., Singapore (2015)
8. Khan, K.M., Gangavarapu, N.M.: Addressing cloud computing in enterprise architecture: issues and challenges. Cutter IT J. **22**(11), 27–33 (2009)
9. MacKenzie, C.M., Laskey, K., McCabe, F., Brown, P.F., Metz, R.: Reference model for service oriented architecture 1.0. Technical report, Advancing Open Standards for the Information Society (2006)
10. Masuda, Y., Shirasaka, S., Yamamoto, S.: Integrating mobile IT/cloud into enterprise architecture: a comparative analysis. In: Proceedings of the 21st Pacific Asia Conference on Information Systems (PACIS 2016), Taiwan, Paper 4 (2016)
11. Muhammad, K., Khan, M.N.A.: Augmenting mobile cloud computing through enterprise architecture: survey paper. Int. J. Grid Distrib. Comput. **8**(3), 323–336 (2015)
12. Newman, S.: Building Microservices. O'Reilly Media Inc., Sebastopol (2015)
13. Richards, M.: Microservices vs. Service-Oriented Architecture, 1st edn. O'Reilly Media, Inc., Sebastopol (2015)
14. Tamm, T., Seddon, P.B., Shanks, G., Reynolds, P.: How does enterprise architecture add value to organizations? Commun. Assoc. Inf. Syst. **28**, 141–168 (2011). Article 10, 2011.42

15. Masuda, Y., Shirasaka, S., Yamamoto, S., Hardjono, T.: An adaptive enterprise architecture framework and implementation: towards global enterprises in the era of cloud/mobile IT/digital IT. Int. J. Enterp. Inf. Syst. IJEIS. **13**(3), 1–22, 22 p. (2017). https://doi.org/10.4018/ijeis.2017070101
16. Chappelle, D.: Big Data & Analytics Reference Architecture. Oracle Corp., September 2013
17. US Department of Commerce, NIST Big Data Interoperability Framework: Reference Architecture Version 1 (2015)
18. Kein, J., Buglak, R., Blockow, D., Wuttke, T.: A reference architecture for big data systems in the national security domain. In: The 2nd International Workshop on Big Data Software Engineering (2016)
19. Oracle, Information Management & Big Data. White Paper (2014). http://www.oracle.com/technetwork/database/bigdataappliance/overview/bigdatarefarchitecture-2297765.pdf
20. Mysore, D., Khupat, S., Jain, S.: Big data architecture and patterns. IBM, White Paper (2013)
21. Microsoft, Microsoft Big Data Solution Brief. http://download.microsoft.com/download/F/A/1/FA126D6D-841B-4565-BB26-D2ADD4A28F24/Microsoft_Big_Data_Solution_Brief.pdf
22. Pääkkönen, P., Pakkala, D.: Reference architecture and classification of technologies, products and services for big data systems. Big Data Research. Elsevier (2015)
23. Masuda, Y., Shirasaka, S., Yamamoto, S., Hardjono, T.: Architecture board practices in adaptive enterprise architecture with digital platform: a case of global healthcare enterprise. Int. J. Enterp. Inf. Syst. **14**(1), P.1–P.20 (2018)
24. Dunbrack, L., Burghard, C., Lohse, S., Rivikin, J., Fitzgerrald, S.: IDC's Worldwide Digital Transformation Use Case Taxonomy: Healthcare. IDC (2017)
25. Ellis, S., Fitzgerald, S., Knickie, K., Santagate, J., Hojlo, J., Ashton, H., Parker, R.: IDC's worldwide digital transformation use case taxonomy, 2017: brand-oriented value chains in the manufacturing industry. IDC (2017)
26. Porter, M.E.: Redefining health care: creating value-based competition on results. Harvard Business School, Spring Leadership Meeting, Boston, MA, June 2005
27. Patel, P., Cassou, D.: Enabling high-level application development for the internet of things. J. Syst. Softw. 1–26 (2015). CoRR abs/1501.05080
28. Iacob, M.E., et al.: Delivering Business Outcome with TOGAF® and ArchiMate®, eBook BiZZdesign (2015)
29. Johnson, P., et al.: IT Management with Enterprise Architecture. KTH, Stockholm (2014)
30. The Open Group, TOGAF Version 9.1, Van Haren Publishing (2011)
31. The Open Group, Archimate 2.0 Specification, Van Haren Publishing (2012)
32. Zimmermann, A., Schmidt, R., Sandkuhl, K., Jugel, D.: Digital enterprise architecture – transformation for the internet of things. In: IEEE 19th International Enterprise Distributed Object Computing Workshop (EDOCW) (2015)
33. Ross, J.W., Weill, P., Robertson, D.C.: Enterprise Architecture as Strategy. Harvard Business School Press, Brighton (2006)
34. Weill, P., Ross, J.W.: IT Governance: How Top Performers Manage It Decision Rights for Superior Results. Harvard Business School Press, Boston (2004)
35. WSO2, White Paper: A Reference Architecture for the Internet of Things, Version 0.8.0 (2015). http://wso2.com
36. Ngu, A.H., Gutierrez, M., Metsis, V., Nepal, S., Sheng, Q.Z.: IoT middleware: a survey on issues and enabling technologies. IEEE Internet Things J. **4**(1), 1–20 (2017)
37. Yamamoto, S., Masuda, Y., Hayashi, M.: Issues on the IoT service platform reference architecture. Research Meeting on KSN, Japanese Society for Artificial Intelligence (2017)
38. Persson, P., Angelsmark, O.: Calvin – Merging Cloud and IoT. https://www.ericsson.com/assets/local/publications/conference-papers/calvin-open-access.pdf

39. Node-RED. https://developer.ibm.com/recipes/tutorials/getting-started-with-watson-iot-platform-using-node-red/
40. Ptolemy. https://ptolemy.eecs.berkeley.edu/
41. Hydra. http://www.hydramiddleware.eu/news.php
42. GSN. http://lsir.epfl.ch/research/current/gsn/
43. OpenIot. https://github.com/OpenIotOrg/openiot/wiki
44. Gutierrez, P.: Aged care receives helping hand from IoT. IoT Hub (2017). https://www.iothub.com.au/news/aged-care-receives-helping-hand-from-iot-415031
45. My-signals.com. MySignals - eHealth and Medical IoT Development Platform (2017). http://www.my-signals.com/#buy-mysignals

Vision Paper for Enabling Digital Healthcare Applications in OHP2030

Tetsuya Toma[1], Yoshimasa Masuda[1](✉), and Shuichiro Yamamoto[2]

[1] Graduate School of System Design and Management, Carnegie Mellon University, Yokohama, Japan
ymasuda@australia.cmu.edu

[2] Graduate School of Information Science, Nagoya University, Nagoya, Japan

Abstract. Internets of Things (IoT) and Big Data applications and services have spread and are rapidly being deployed in the information services of the healthcare and financial industries, etc. However, the previous paper suggested that the current IoT services were individually developed, therefore, the open platform and architecture for the above IoT services of the healthcare industries should be deemed necessary, while the Big Data applications prevail in healthcare industry gradually. An open healthcare platform is expected to promote and implement the digital IT applications for healthcare communities efficiently. In this paper, we suggest that various IoT and Big Data applications will be designed and verified while the open platform for healthcare related IoT services should be proposed and verified by the research initiative named "Open Healthcare Platform 2030 – OHP2030". In addition, the vision paper for enabling Digital Healthcare applications in the above OHP2030 research initiative is explained.

Keywords: Digital healthcare · Enterprise Architecture · Internet of things Big data · Digital platform · Digital IT

1 Introduction

Many global corporations have encountered a variety of changes, such as progress of new technologies, globalization, shifts in customer needs, and new business models. Significant changes in cutting-edge IT technology due to recent developments in Cloud computing and Mobile IT (such as progress in big data technology) have emerged as new trends in information technology. Furthermore, major advances in the abovementioned technologies and processes have created a "Digital IT economy," bringing about both business opportunities and business risks and forcing enterprises to innovate or face the consequences [2]. Enterprise Architecture (EA) should be effective because it contributes to the design of large integrated systems, which represents a major technical challenge toward the era of Cloud/Mobile IT/Big Data/Digital IT in Digital Transformation. From a comprehensive perspective, EA encompasses all enterprise artifacts, such as business, organization, applications, data, and infrastructure, to establish the current architecture visibility and future architecture/roadmap. On the other hand, EA frameworks need to

G. De Pietro et al. (Eds.): KES-IIMSS-18 2018, SIST 98, pp. 186–197, 2019.
https://doi.org/10.1007/978-3-319-92231-7_19

embrace change in ways that consider the emerging new paradigms and requirements affecting EA, such as Mobile IT/Cloud [1, 3].

Furthermore, considering the above background, the previous study proposed the "Adaptive Integrated EA framework," which should align with IT strategy promoting Cloud/Mobile IT/Digital IT, and verified this in the case study [15]. The author of this paper has named the EA framework suitable for the era of Digital IT as "Adaptive Integrated Digital Architecture Framework – AIDAF" [23].

This paper is organized as follows: the next section presents the background of this study, followed by the description of the research methodology and an overview of the AIDAF application for the cross-functional healthcare community, and healthcare community case. Finally, the challenges and final thoughts for this OHP2030 are outlined.

2 The Direction of EA and Digital IT

2.1 Related Work and Direction of Cloud/Mobile IT/Big Data/Internet of Things

In the past decade, EA has become an important method for modeling the correlation for overall images of corporate and individual systems. In ISO/IEC/IEEE42010:2011, architecture framework is defined as "principles, and practices for the architecture descriptions established within a specific domain of application and/or community." Furthermore, EA visualizes the current corporate IT/business landscape to promote a desirable future IT model [3]. It is not a simple support activity [1], and it offers many benefits to companies, such as coordination, communication and planning between business and IT, and reduction in the complexity of IT [14]. For the delivery of these benefits, EA frameworks need to cope with the emerging new paradigms such as Cloud computing or enterprise mobility [1].

Mobile IT computing is an emerging concept using Cloud services provided over mobile devices [11]. In addition, Mobile IT applications are composed of Web services. Many studies discuss the integration of EA with Service Oriented Architecture (SOA), except for Mobile IT. The SOA architecture pattern defines the four basic forms of business service, enterprise service, application service, and infrastructure service [13]. The OASIS, which is a public standards group [9], introduces an SOA reference model. Many organizations have invested in SOA as an approach to manage rapid change [4]. Meanwhile, attention has been focused on Microservices architecture, which allows rapid adoption of new technologies, such as Mobile IT applications and Cloud computing [12]. SOA and Microservice vary greatly from service characteristics perspective [13]. Microservice is an approach for dispersed systems that is defined by the two basic forms of functional services through an API layer and infrastructure services. Multiple Microservices cooperating to work together enable the implementation as a Mobile IT application [5].

For Cloud Computing, the NIST defined three cloud service models such as SaaS, PaaS, and IaaS [7]. PaaS is an IaaS platform that includes both system software and an integrated development environment. SaaS is a software application developed, implemented, and operated on a PaaS foundation. IaaS accommodates PaaS and SaaS by offering infrastructure resources, such as computing network storage memory through specific centers [7]. Many Mobile IT applications also operate with SaaS Cloud-based software [11]. The integration and relationship between EA and Cloud computing is discussed rarely in literature. Considering the recent dynamic moves in Cloud computing, it is necessary for companies to link the service characteristics of EA and Cloud computing [8]. The traditional approach takes months to develop an EA realizing a Cloud adoption strategy, and organizations will demand adaptive enterprise architecture to iteratively develop and manage an EA adaptive to the Cloud technology [6].

Moreover, according to previous research [10], when promoting Cloud/Mobile IT in a strategic manner, it is proposed as a good option that a company that applies TOGAF or FEAF can adopt the integrated framework with the Adaptive EA framework supporting elements of Cloud computing.

Furthermore, in terms of Big Data, new computing trends require data with far greater volume, velocity, and variety than ever before. Big data is utilized in ingenious methods to predict customer buying behaviors, detect fraud and waste, analyze product opinion, and react quickly to changes in business conditions (a driving force behind new business opportunities) [16]. The term "big data" refers to data that is so large, it is difficult to process using currently-available IT systems. There is a growing opportunity for analysis, visualization, and distributed processing software to enable users to extract useful information from such data [2]. Sources of big data include the following.

- Corporate data in SQL databases
- Data in cloud-based SQL or NoSQL databases
- Data provided by social networks
- Data provided by sensors or object identifiers in the internet-of-things (IoT)

Big data applications may include visualization functionality for effective user presentation of analytical results. Furthermore, big data applications should leverage web services that make the results of their analyses available to other applications through APIs; objects in the IoT can be data generators [2].

Existing big data reference architectures have been shepherded by NIST, which helped create the big data interoperability framework, including a reference architecture volume [17].

LinkedIn, for example, collects data from users and offers services, such as skill endorsements or newsfeed updates to users based on data analysis. Additionally, Twitter uses collected data for real-time query suggestion [22]. Therefore, most solutions exist in the Big Data Application Provider component and should be categorized as Specific Application Layers on Cloud and Mobile IT platforms. Technology vendors such as Oracle [19], IBM [20], and Microsoft [21] have also developed Big Data Reference Architectures [18]. These vendors publish practical Reference Architectures for Big Data toward EA practitioners in corporations and other groups.

The term of "Internet of Things (IoT)" is used to mean "the collection of uniquely identifiable objects embedded in or accessible by Internet hosts" [2]. A "uniquely identifiable object" can be described as follows: these objects are connected with real world interaction devices, smart homes and cars, and other SmartLife scenarios. The IoT fundamentally revolutionizes digital strategies with innovative business operating models [33], and holistic governance models for business and IT [34], under fast changing markets [32].

- A sensor, such as a temperature sensor (thermometer)
- A control; for example, to control a valve in a heating system
- A combination of sensor and control (for example, a thermostat)
- An object identifier, such as an RFID tag or a barcode

The current state of research for the Internet of Things architecture [27] lacks an integral understanding of EA and Management [28–31], and shows a number of physical standards, methods, tools and a large number of heterogeneous IoT devices [32]. A first reference architecture (RA) for the IoT is proposed by [35], and can be mapped to a set of open source products. This RA covers aspects like "cloud server-side architecture," "monitoring/management of IoT devices, services," "specific lightweight RESTful communications," and "agent, code on small low power devices." Layers can be instantiated by suitable technologies for the IoT [32].

2.2 Adaptive Integrated Digital Architecture Framework – Aligned with Digital IT Strategy

Our previous research promoted the strategic use of cloud and mobile IT, suggesting that corporate entities defining Cloud/Mobile IT/Big Data/Digital IT strategies implementing EA by applying frameworks, such as TOGAF and FEAF, could adopt a framework integrating an Adaptive EA framework to provide further support for cloud elements [10]. Accordingly, the preliminary research of this paper proposes an Adaptive Integrated EA framework depicted in Fig. 1 of the preliminary research paper, based on this suggestion, which should meet with IT strategy promoting Cloud/Mobile IT/Big Data/Digital IT, and verified this in the case study [15]. The proposed model is an EA framework integrating an adaptive EA cycle with TOGAF or a simple EA framework 1 for different business units in the upper part of the diagram in [15]. The author of the previous paper mentioned above named this EA framework as "Adaptive Integrated Digital Architecture Framework (AIDAF) [23]."

The adaptive EA cycle makes provisions for project plan documents (including architecture designs) for new Digital IT related projects drawn up on a short-term basis. This begins with the Context Phase, which is conducted to prepare the Defining Phase (e.g., architectural design guidelines related to necessary types of Security/Digital IT aligned with IT strategy) per business needs. During the Assessment/Architecture Review Phase, the Architecture Board reviews the architecture in the initiation documents for the IT project. In the Rationalization Phase, the stakeholders and Architecture Board decide upon information systems to be replaced or decommissioned by the proposed new information system. In the Realization Phase, the project team begins to implement the new IT project agreed upon after deliberating issues and action items [15, 23].

The Adaptive EA cycle enables the corporate entity to adopt an EA framework capable of flexibly adapting to new Digital IT projects continuously. Moreover, the TOGAF and simple EA framework, based on an operational division unit in the top part of the Fig. 1 of [15, 23], can respond to differing strategies in business divisions in the mid-long-term. This part of the framework has a structure that can select an EA framework based upon the characteristics of division's operational processes and future architectures, while enabling applications. Furthermore, the framework should align EA guiding principles with those business division's principles to keep consistency among the adaptive EA cycle, the TOGAF, and simple EA framework. Furthermore, in the Defining Phase, the Architecture Board promotes the appropriate architectural design of each Digital IT-related system by sharing the architectural guidelines for Security/Digital IT, etc., to align with the IT strategy [15, 23].

3 AIDAF Application for Cross-Functional Healthcare Community

The author of this paper proposed an adaptive integrated EA framework to align with IT strategy, promoting Cloud/Mobile IT/Big Data/Digital IT, and verified by our case study [15]. Furthermore, the author of this paper has named the EA framework suitable for the era of Digital IT as an "Adaptive Integrated Digital Architecture Framework – AIDAF" [23]. Figure 1 illustrates this AIDAF proposed model in the Open Healthcare Platform 2030 (OHP2030) community. The OHP2030 community is comprised of healthcare companies such as pharmaceutical, medical development and aged healthcare companies, hospitals as well as the OHP2030 initiative and government as depicted in Fig. 1. AIDAF will be applied to the above-mentioned cross-functional healthcare community in the OHP2030. AIDAF begins with the context phase, while referencing the defining phase (i.e., architecture design guidelines related to digital IT aligned with IT strategy in the above healthcare community in the OHP2030). During the assessment and architecture review, the architecture board reviews the initiation documents and related architectures for the IT project in the above healthcare community in the OHP2030.

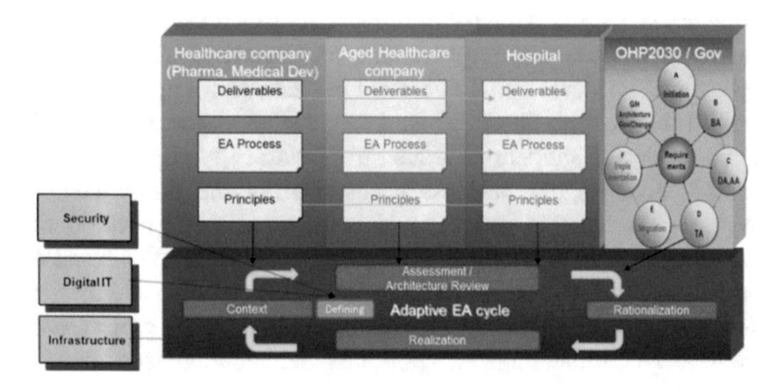

Fig. 1. AIDAF proposed model in the Healthcare community in the OHP2030 (ex: TOGAF and Adaptive EA framework)

4 Healthcare Community Case

In OHP2030, a particular Architecture Board will be formulated in the aforementioned cross-functional healthcare community. In the case study of EA rollout in the healthcare community, they will handle IoT/Big Data/Cloud strategic projects and systems well by structuring and implementing EA with the above-mentioned AIDAF to be consistent with the IT strategy focusing on IoT/Big Data/Digital IT in the above cross-functional Healthcare community.

Furthermore, in the case study of the cross-functional healthcare community, the author of this paper assumes that use cases of "IDC's Worldwide Digital Transformation Use Case Taxonomy, 2017: Healthcare," which were defined by IDC, should be applied in the context phase and assessment/architecture review phase as well as the defining phase of the AIDAF for new IoT projects in the healthcare community [24, 25]. According to [24, 25], the above use cases and digital missions in healthcare are based on the creation of value-based healthcare systems from sick care to healthcare management as shown in Fig. 2 above [26].

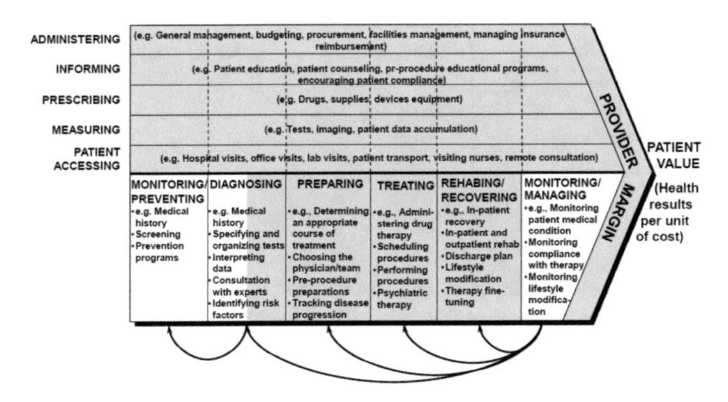

Fig. 2. Healthcare delivery value chain for practice areas (Source: Michael Porter and Elizabeth, 2005)

5 Cases of Enabling Digital Healthcare Applications

5.1 Preventive Analysis Related Case

According to [24], as the digital healthcare, a value-based health system is created, which is focused on preventive care/analysis and population health management. In this section, the scenario of preventive care and analysis is described with the related challenges.

Use Case. The use case scenario of preventive care and analysis in a value-based healthcare system is shown in the following Table 1, according to [24].

Table 1. Use case scenario for preventive care and analysis

Use case	Current situation	Goals and objectives	Technology deployment	Use case summary
Prevenive Analysis, At-risk patient identification	Traditional methods of predicting at-risk patients do not take into account clinical conditions, social determinants, or impact ability	Identify patients at risk of the onset of a chronic illness. Identify patients who would most benefit from a care/execute personalized plans	AI-driven advanced analytics for predictive modeling, machine learning, and prebuilt algorithms/models	Today's analytics are limited in the data on reporting. The use of AI-based predictive models can generate the data to create personalized care plans
Social determinants	Clinical care determines about 20% of health/disease, based on living situation, and environment	Address the full impact of health and disease to solve underlying problems	– Access to unstructured data and external sources of data	Additions of social determinants will provide the drivers of health to mitigate barriers to health

Challenge. There should be some issues of system performance in the operational aspect, and the issue regarding analytical methods will exist as the functional aspect in this kind of digital healthcare applications with IoT, Big Data.

Summary. The preventive analysis of digital healthcare applications can be performed with appropriate analytical methods, while determining with collecting external source of data such as medical authority's opinions from social tools.

5.2 Healthcare Management Related Case

According to [24], as the digital healthcare, healthcare management is also emphasized in a value-based health system. In this section, the scenario of the healthcare management is described with the related challenges.

Use Case. The use case scenario of healthcare management in a value-based healthcare system is shown in the following Table 2, according to [24].

Challenge. There should be the consideration points of patient's data structure in the viability aspect, and the issue regarding patients' data privacy should exist as the operational aspect in this kind of digital healthcare applications with mobile IT and Big Data, IoT.

Table 2. Use case scenario for the healthcare management

Use case	Current situation	Goals and objectives	Technology deployment	Use case summary
Chronic conditions management	– One-third of adults live with one or more chronic conditions – The increasing prevalence of chronic conditions is a worldwide problem	Encourage consumers to be more active participants in managing their own health and chronic conditions (e.g., blood pressure)	– Remote health monitoring devices (weight scales, and blood pressure cuffs) – Consumer-facing mobile applications and portals	Providing consumers with medical monitoring devices and clinical support when empowering consumers to better manage their chronic conditions

Summary. The preventive analysis of digital healthcare applications can be performed with appropriate analytical methods, while determining with collecting external source of data such as medical authority's opinions from social tools.

5.3 Pharma Manufacturing Related Case with Predictive Analysis

In this section, the scenario of the healthcare management is described with the related challenges. The example of solution architecture for this pharmaceutical manufacturing related system with predictive analysis is shown as the following Fig. 3, at this time.

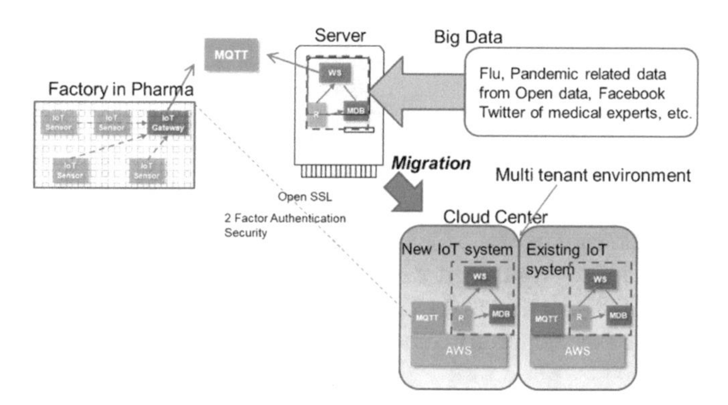

Fig. 3. The solution architecture for pharmaceutical manufacturing systems in digital healthcare

Use Case. The use case scenario of pharmaceutical manufacturing with predictive analysis in a value-based healthcare system is shown in the following Table 3, according to [25].

Challenge. There should be some issues in terms of security for the IoT through MQTT in the operational aspect, and the issue regarding compliance and validation for communications between social tools and analytical services will exist as the viability related aspect in this kind of digital healthcare applications with IoT, Big Data.

Table 3. Use case scenario for the pharma manufacturing case

Use case	Current situation	Goals and objectives	Technology deployment	Use case summary
Cognitive inventory	Inventory is inflexible. Even though companies generally have a good handle on the overall inventory number, the specific remains elusive	Reduce the overall dollar value of the inventory while improving its ability to meet demand and buffer needs	IoT, Big Data Analysis	There is real-time assessment of inventory across all locations, connecting to demand and capacity. Cognitively enabled inventory means that the system learns how to best meet business variables
Smart capacity allocation	Most Pharmaceutical companies have a general sense of available capacity at any one point in time. The ability to adjust capacity utilization in real time is a guess at best	Better utilize capacity, leading to the ability to reduce out of stock issues or lower overall capacity costs	IoT, mobile and Big Data Analysis	The instrumentation of assets includes real- time capacity monitoring at the machine level that provides consumption models that allow for optimization calculations across facilities

Summary. This kind of pharmaceutical manufacturing system can monitor the data of manufacturing site regarding temperature, humidity and quality, while collecting data related to flu, pandemic from open data, social tools such as Facebook, Twitter of medical experts, which can lead to the decision making for the urgent manufacturing of enormous amount of medicines to cope with the above flu, pandemic, etc. for several months.

6 Discussions and Challenges

6.1 Future Issues

To address architectural issues, we need to consider the functional aspects such as patients' data structure and analytical methods for a digital healthcare systems of IoT, Big Data, while considering the alignment with the Value based Healthcare Delivery model in each digital IT systems.

At the same time, for the purpose of coping with architectural issues, we need to design with coping with the operational aspects such as security and privacy of patients' data, security of IoT systems through MQTT protocol, compliance and validation regarding social analytics,system performance, etc. Furthermore, we will proceed with the international systematization of a digital healthcare application systems such as IoT, Big Data, while proceeding and analyzing each digital healthcare related systems/projects in OHP2030.

7 Final Thoughts for Enabling Digital Healthcare Applications in OHP2030

In this paper, we have described the vision for enabling digital healthcare applications in OHP2030 with several examples. This research initiative named OHP2030 aims at an exploration and definition of a digital healthcare platform such as IoT, Big Data in the middleware layer and above. Furthermore, we would like to systematize the digital healthcare application systems in healthcare industry, while ensuring information security, privacy and compliance, validation.

References

1. Alwadain, A., Fielt, E., Korthaus, A., Rosemann, M.: A comparative analysis of the integration of SOA elements in widely-used enterprise architecture frameworks. Int. J. Intell. Inf. Technol. **9**(2), 54–70 (2014)
2. Boardman, S., Harrington, E.: Open Group Snapshot - Open Platform 3.0™. The Open Group (2015)
3. Buckl, S., Matthes, F., Schulz, C., Schweda, C.M.: Exemplifying a framework for interrelating enterprise architecture concerns. In: Sicilia, M.A., Kop, C., Sartori, F. (eds.) Ontology, Conceptualization and Epistemology for Information Systems, Software Engineering and Service Science, vol. 62, pp. 33–46. Springer, Heidelberg (2010). https://doi.org/10.1007/978-3-642-16496-5_3
4. Chen, H., Kazman, R., Perry, O.: From software architecture analysis to service engineering: an empirical study of methodology development for enterprise SOA implementation. IEEE Trans. Serv. Comput. **3**(2), 145–160 (2014). https://doi.org/10.1109/TSC.2010.21
5. Familiar, B.: Microservices, IoT and Azure: Leveraging DevOps and Microservice Architecture to Deliver SaaS Solutions. Apress, Berkeley (2015)
6. Gill, A.Q., Smith, S., Beydoun, G., Sugumaran, V.: Agile enterprise architecture: a case of a cloud technology-enabled government enterprise transformation. In: Proceedings of the 19th Pacific Asia Conference on Information Systems (PACIS 2014), pp. 1–11 (2014)
7. Gill, A.Q.: Adaptive cloud Enterprise Architecture. Intelligent Information Systems, vol. 4. World Scientific Publishing Co., Singapore (2015)
8. Khan, K.M., Gangavarapu, N.M.: Addressing cloud computing in enterprise architecture: issues and challenges. Cut. IT J. **22**(11), 27–33 (2009)
9. MacKenzie, C.M., Laskey, K., McCabe, F., Brown, P.F., Metz, R.: Reference model for service oriented architecture 1.0, Technical report, Advancing Open Standards for the Information Society (2006)

10. Masuda, Y., Shirasaka, S., Yamamoto, S.: Integrating mobile IT/cloud into enterprise architecture: a comparative analysis. In: Proceedings of the 21st Pacific Asia Conference on Information Systems (PACIS 2016), Taiwan, paper 4 (2016)
11. Muhammad, K., Khan, M.N.A.: Augmenting mobile cloud computing through enterprise architecture: survey paper. Int. J. Grid Distrib. Comput. **8**(3), 323–336 (2015)
12. Newman, S.: Building Microservices. O'Reilly Media, Inc., Sebastopol (2015)
13. Richards, M.: Microservices vs. Service-Oriented Architecture, 1st edn. O'Reilly Media, Inc., Sebastopol (2015)
14. Tamm, T., Seddon, P.B., Shanks, G., Reynolds, P.: How does enterprise architecture add value to organizations? Commun. Assoc. Inf. Syst. **28**, 141–168 (2011). Article 10
15. Masuda, Y., Shirasaka, S., Yamamoto, S., Hardjono, T.: An adaptive enterprise architecture framework and implementation: Towards global enterprises in the era of cloud/mobile IT/digital IT. Int. J. Enterp. Inf. Syst. IJEIS **13**(3), 1–22 (2017). https://doi.org/10.4018/ijeis.2017070101. 22 p.
16. Chappelle, D.: Big Data & Analytics Reference Architecture. Oracle Corp, September 2013
17. US Department of Commerce, NIST Big Data Interoperability Framework: Reference Architecture Version 1 (2015)
18. Kein, J., Buglak, R., Blockow, D., Wuttke, T.: A Reference architecture for big data systems in the national security domain. In: The 2nd International Workshop on Big Data Software Engineering (2016)
19. Oracle: Information Management & Big Data. White Paper (2014). http://www.oracle.com/technetwork/database/bigdataappliance/overview/bigdatarefarchitecture-2297765.pdf
20. Mysore, D., Khupat, S., Jain, S.: Big Data Architecture and Patterns. IBM, White Paper (2013)
21. Microsoft: Microsoft Big Data Solution Brief. http://download.microsoft.com/download/F/A/1/FA126D6D-841B-4565-BB26-D2ADD4A28F24/Microsoft_Big_Data_Solution_Brief.pdf
22. Pääkkönen, P., Pakkala, D.: Reference architecture and classification of technologies, products and services for big data systems. Big Data Res. **2**, 166–186 (2015)
23. Masuda, Y., Shirasaka, S., Yamamoto, S., Hardjono, T.: Architecture board practices in adaptive enterprise architecture with digital platform: a case of global healthcare enterprise. Int. J. Enterp. Inf. Sys. **14**(1), 1–20 (2018)
24. Dunbrack, L., Burghard, C., Lohse, S., Rivikin, J., Fitzgerrald, S.: IDC's Worldwide Digital Transformation Use Case Taxonomy, 2017: Healthcare. IDC (2017)
25. Ellis, S., Fitzgerald, S., Knickie, K., Santagate, J., Hojlo, J., Ashton, H., Parker, R.: IDC's Worldwide Digital Transformation Use Case Taxonomy, 2017: Brand-Oriented Value Chains in the Manufacturing Industry. IDC (2017)
26. Porter, M.E.: Redefining Health Care: Creating Value-Based Competition on Results. Harvard Business School, Spring Leadership Meeting, Boston, MA, June 2005
27. Patel, P., Cassou, D.: Enabling High-level Application Development for the Internet of Things. CoRR abs/1501.05080, submitted to Journal of Systems and Software (2015)
28. Iacob, M.E., et al.: Delivering Business Outcome with TOGAF® and ArchiMate®, eBook BiZZdesign (2015)
29. Johnson, P., et al.: IT Management with Enterprise Architecture. KTH, Stockholm (2014)
30. The Open Group: TOGAF Version 9.1. Van Haren Publishing (2011)
31. The Open Group: Archimate 2.0 Specification. Van Haren Publishing (2012)
32. Zimmermann, A., Schmidt, R., Sandkuhl, K., Jugel, D.: Digital enterprise architecture – transformation for the internet of things. In: IEEE 19th International Enterprise Distributed Object Computing Workshop (EDOCW) (2015)

33. Ross, J.W., Weill, P., Robertson, D.C.: Enterprise Architecture as Strategy. Harvard Business School Press (2006)
34. Weill, P., Ross, J.W.: IT Governance: How Top Performers Manage It Decision Rights for Superior Results. Harvard Business School Press (2004)
35. WSO2: White Paper: A Reference Architecture for the Internet of Things, Version 0.8.0 http://wso2.com (2015)

e-Healthcare Service Design Using Model Based Jobs Theory

Shuichiro Yamamoto[1(✉)], Nada Ibrahem Olayan[1], and Junkyo Fujieda[2]

[1] Graduate School of Information Science, Nagoya University, Nagoya, Japan
yamamotosui@icts.nagoya-u.ac.jp
[2] Research Environment for Global Information Society, Inc., YAMADA Bldg., Shinjuku 1-1-14, Sinjuku-ku, Tokyo, Japan

Abstract. The demand on the innovation of enterprise services are rapidly increased. There are various modeling methods that can be applied to design digital transformation of services. However, there are few visual modeling methods to design innovations. Without consistent visual modeling methods, it is difficult to integrate service innovations and digital transformations.

In this paper, we propose a Model Based Jobs Theory (MBJT) which can visually model the Jobs Theory of Christensen. Moreover, we examine the applicability of MBJT by designing an e-Healthcare service.

Keywords: Digital healthcare · Enterprise architecture
Model Based Jobs Theory · Service design

1 Introduction

Christensen's new book of Jobs Theory [1] showed that Jobs Theory can effectively design the innovation. This paper proposes a Model Based Jobs Theory (MBJT) based on the Jobs Theory of Christensen by using Goal Oriented Requirements Engineering Approaches [2].

At first the paper explains the basic concepts of Jobs Theory. Then, the job analysis table is proposed by the consideration on the relationship between Jobs Theory and Requirements Engineering. Moreover, Model based Jobs Theory (MBJT) is proposed by mapping the jobs analysis table into goal models using ArchiMate® [3–6]. ArchiMate is a registered trademark of The Open Group (http://www.opengroup.org/). Then, an e-Healthcare service is designed by MBJT.

The rest of this paper is organized as follows. Section 2 describes related work. Section 3 proposes MBJT. An example of e-Healthcare service design is provided by using MBJT in Sect. 4. The effectiveness of MBJT is discussed in Sect. 5. Finally, Sect. 6 summarizes and provides the future work of the paper.

G. De Pietro et al. (Eds.): KES-IIMSS-18 2018, SIST 98, pp. 198–207, 2019.
https://doi.org/10.1007/978-3-319-92231-7_20

2 Related Work

2.1 Jobs Theory

Jobs Theory [1] focuses the consumer jobs to be done. There are the cause effect relationships between the jobs to be done and the solutions, because consumers accomplish jobs by using the product solutions which companies provide.

If the jobs as activities to resolve the causes of problematic situations of consumers were not clear, then the products to be developed as solutions never be consumed.

The jobs to be done are defined as the process that derives progress which consumers tried to accomplish. The specific context that jobs are created is the situation of jobs. The situation shows something that consumer takes pains.

Jobs continuously repeats when consumers need progresses. The hire is defined as that consumers use products to resolve jobs.

The purchasing and using products are the two types of hire. Consumers purchase products only once. The purchased product is using repeatedly for each jobs.

The big hire means that consumer purchases products. The little hire means that consumer uses products.

Christensen pointed the necessity of emotional and social aspects in addition to the functional aspect. Aspects are the qualities which derive better solutions. Aspects are decomposed as inevitable elements of jobs, and integrated into the total story that customers experience in the course of jobs. Solutions are the means to achieve jobs. It is necessary to discover means for jobs that only have insufficient solutions so far. Table 1 shows elements of the jobs theory.

Table 1. Elements of jobs theory

Elements	Explanations
Job	Process that derives progress which consumers tried to accomplish
Progress	What customers achieve by jobs
Hire	Use of products by customers to solve jobs
Situation	Specific context where jobs occur
Aspect	Qualities which derive better solution
Solution	Means to achieve jobs

2.2 Requirements Engineering

Christensen said that Jobs Theory (JT) is a common language for innovation. As described below, requirements engineering is also a kind of activities on innovation.

Authors used a problem analysis table (PAT) to teach requirements modeling. PAT contains concerns, problematic situations, cause analysis, target system to be developed, and solution. Table 2 shows the comparison between elements of Jobs Theory and PAT. Table 2 shows the high compatibility between JT and PAT. Therefore, PAT can be used to analyze jobs for designing innovations based on JT. JT did not provide tools like PAT.

Table 2. Comparison of JT and PAT

JT	PAT
Progress	Target system to be developed
Cause	Cause analysis
Situation	Problematic situation
Aspect	Concern
Solution	Solution

2.3 Job Analysis Example Using PAT

GM provides the car information communication service, named "on star [1]." Job Analysis Table, JAT, is a PAT for job analysis. The JAT for on star is showed in Table 3.

Table 3. Comparison of JAT and PAT

JAT	PAT
Concern	Inward peace while driving
Problematic situation	Drivers do not know what to do and may become uneasy
Cause analysis	When a driver met with a difficult situation, there is not the means that driver can talk about with people
Progress	By the support from others, drivers resolve uneasiness while driving
Solution	Solution

The on-star service is the information communication service to provide appropriate information and help for drivers to resolve uneasy while driving.

At first, on star provided a road guidance service on restaurants and a voice conversation service to ask what clothes you had dressed as value added services. To think these functions is fan for developers.

Both functions are fun to think for developers. However, drivers wanted the road guidance service for the ease of mind. The problematic situation of drivers is that they do not know what they do and they become uneasy. The problematic situation was not that they cannot find restaurants and they want to be asked questions on their dresses. The cause of the situations was that there is no means to consult around them while drivers were faced to difficult things. Therefore, what drivers want to be is to resolve uneasy things in mind while driving with the support of the on-star service. In summary, the solution is to provide the automotive information communication service to reduce uneasy things of drivers caused by the car mechanisms and natural phenomena as well as road environment.

2.4 e-Healthcare Service Examples

Schiltz and others proposed use cases for the business models of e-Health service [7]. For example, there was a use case of cloud based image archive service named PACS (Picture Archiving and Communication System) utilizing Radiology Information System (RIS).

Concept of Operation [8] for the National personally controlled electronic health record (PCEHR) system [9] was created for the purpose of delivering a high level picture of the system and how it works in a moderate level of detail in order to easily communicate with stakeholders. The version of ConOps is created following the framework for a national electronic health record system agreed by the Australian Health Ministers Conference in April 2010.

3 Model Based Jobs Theory

3.1 Representation of JT Using ArchiMate

The JAT can be described by using ArchiMate [3–6]. The Table 4 shows the mapping between elements of JAT and the corresponding elements of ArchiMate.

Table 4. JAT and ArchiMate

JAT	ArchiMate
Concern	Value
Problematic situation	Driver
Cause analysis	Assessment
Progress	Goal
Solution	Requirements

Concern, Problematic situation, Cause analysis, Progress and Solution of JAT are corresponding to value, driver, assessment, goal and requirements, of ArchiMate, respectively. By using the mapping, a generic form of JT is represented as showing in Fig. 1.

The diagram described by using JAT is called Jobs Analysis Diagram (JAD). JT itself did not provide diagrams and tables to analyze and design innovations. By using JAT and ArchiMate, JT can be visualized.

3.2 Example of MBJT

By using the mapping of Table 4, the on-star service of GM is represented in ArchiMate as shown in Fig. 2. Figure 2 is developed as follows. First, the problematic situation of car drivers is described by using the driver node of ArchiMate. The label of driver node is “Drivers do not know what to do and may become uneasy.”

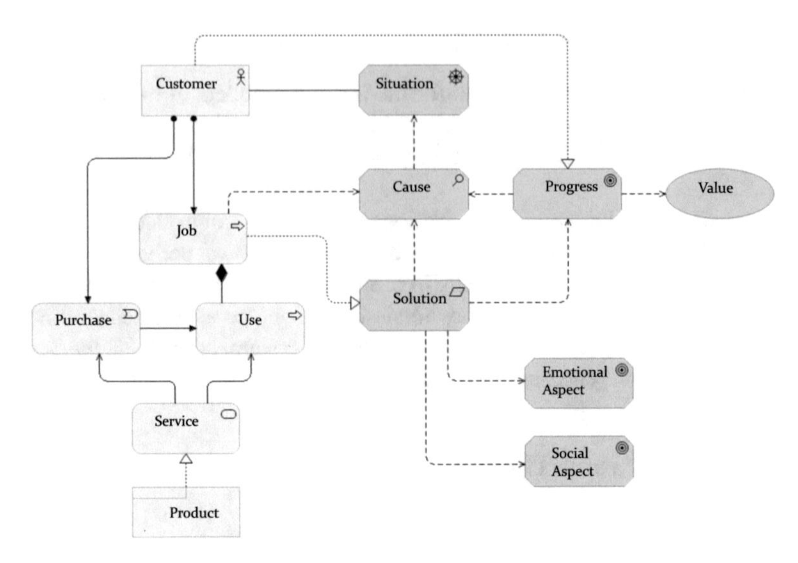

Fig. 1. A generic form of JT in ArchiMate

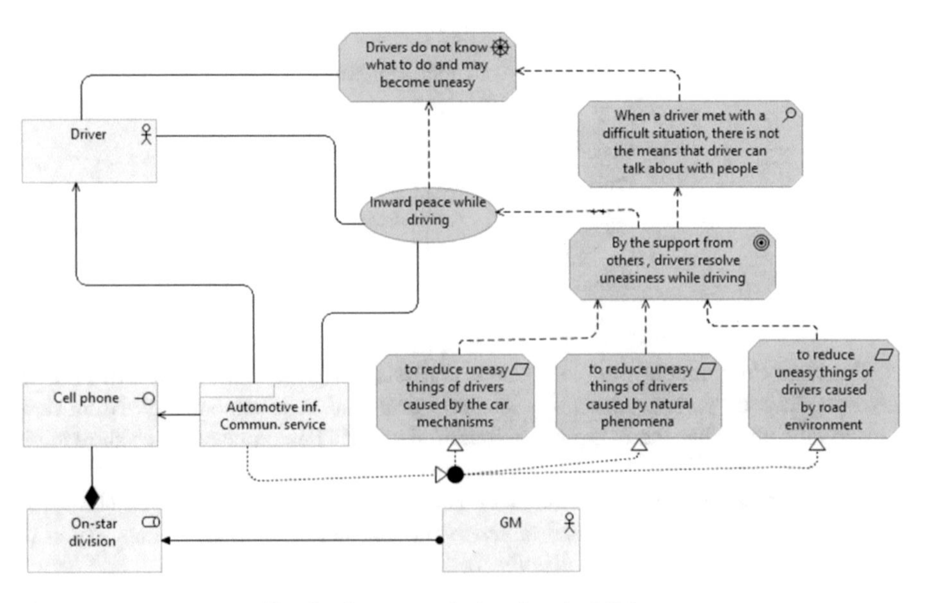

Fig. 2. On star analysis using ArchiMate

Then, the concern of the situation is described as the value node. The label of the node is "Inward peace while driving." The influence relation from the value node to the driver node is added.

The cause analysis is described by the assessment node labeled by "When a driver met with a difficult situation, there is not the means that driver can talk about." The influence relation from the assessment node to the driver node is added.

The progress is described by the goal node labeled by "Drivers resolve uneasiness while driving, by the support from others." The influence relation from the goal node to value and assessment nodes are added.

Finally, solutions is described by the requirements node labeled by "reduce uneasy things of drivers caused by the car mechanisms," "reduce uneasy things of drivers caused by natural phenomena," and "reduce uneasy things of drivers caused by road environment." The realization relations from these requirements nodes to the goal node are added.

Figure 2 also shows the automotive information communication service as the product node. The product node is connected to the requirements node by the realization relation. Figure 2 contains actor nodes for drivers and GM. The on-star service division is described as the role node. The role node provides the automotive information communication service by the interface node which represents mobile phone.

4 Application of MBJT to e-Healthcare

The business model of PACS by using MBJT can be designed as follows. Actors of PACS are patient, medical service provider, and THS (Technology for Healthcare Service provider) service provides. JAT for medical service provider is analyzed as shown in Table 5.

Table 5. JAT

JAT items	Explanation
Concern	Convenient use of medical image technology
Problematic situation	Medical image technology is hard to use
Cause analysis	There is no easy means to use medical image technology
Progress	To realize open use of medical image information
Solution	Medical image collaboration process to share and diagnose medical image remotely

The concern of medical service providers is convenient use of medical image technology. The problematic situation of medical service providers is that medical image technology is hard to use. The cause of the problem is that there is no easy means to use medical image technology. The progress to be achieved is to realize open use of medical image information. The solution is the medical image collaboration service to share and diagnose medical image remotely. If the medical image collaboration service is realized by the medical technology service provider the following processes are achieved. First, the medical service provider takes X-ray diagnose images and stores them into the medical image remote collaboration service. Then, the medical service provider retrieves X-ray images on the service. The medical service provider also diagnoses based on the X-ray images.

The MBJT model for the above scenario is shown in Fig. 3.

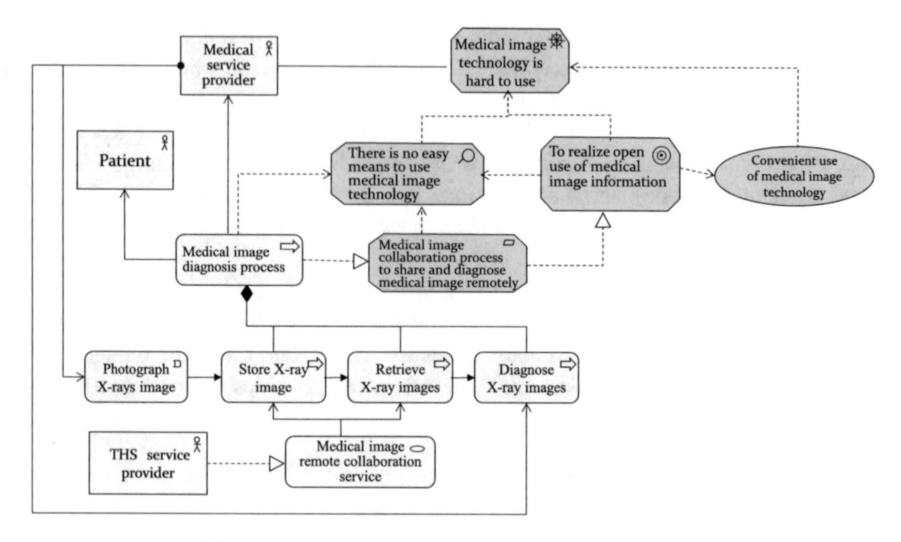

Fig. 3. Example of an e-Health service using MBJT

5 Discussions

5.1 Jobs Theory and Enterprise Architecture

In Jobs Theory, there is a "hire" relation between "Jobs to be done" and the product. In requirements engineering, there is a realization relation between requirements of customers and system. In this way, there is a relationship between Jobs Theory and Requirements Engineering.

Traditional social science researches focus on the analysis and their goals are explanations of social phenomena including innovation. In the other hands, Engineering studies focus on the synthesis and their goals are designing artificial objects. Although traditional innovation theories only explained what kinds of innovations occurred, these theories could not design innovations.

Christensen spent 20 years for writing the book of Jobs Theory to design successful innovations [1]. The Jobs Theory provides the opportunity to connect with the innovation research to software engineering which aims to synthesis. The MBJT integrates innovation and enterprise architecture by using ArchiMate.

5.2 Meta Model of MBJT

Figure 4 shows the meta-model of Model based Jobs Theory. The meta-model can be compared with those of other models, such as Business Model Canvas (BMC) [10] and ConOps. For example, the meta-model of ConOps is described as shown in Fig. 5. We can easily integrate MBJT and ConOps by comparing their meta-models.

As a Personally Controlled Electronic Health Record System (PCEHRS) was described in ConOps, e-Healthcare services defined by MBJT can easily be integrated with PCEHRS based on the integrated meta-model.

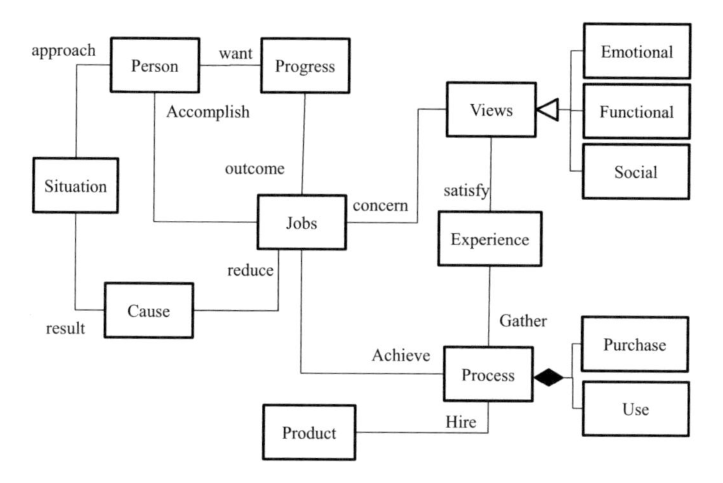

Fig. 4. Meta-model of MBJT

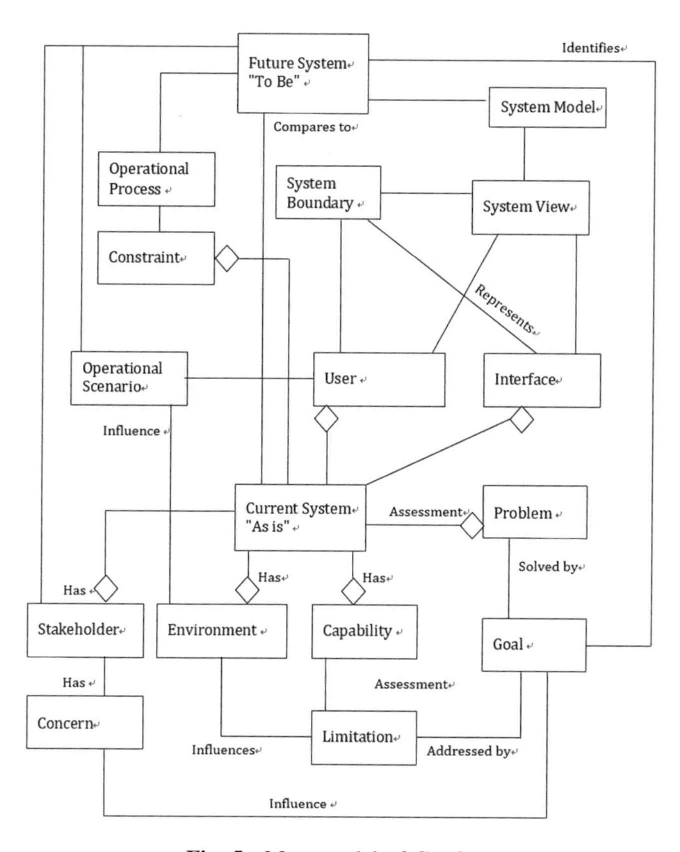

Fig. 5. Meta-model of ConOps

5.3 Applicability of MBJT

Section 4 showed MBJT was able to visualize the business model of an e-Health service described by BMC (Business Model Canvas) and business scenario. Table 6 shows the comparison between elements of BMC and MBJT. Except for the cost structure and benefit chain, MBJT can describe elements of BMC. BMC cannot describe situation, hire, cause, jobs, and product.

Table 6. Comparison of BMC and MBJT

BMC	MBJT
Cost structure	–
Revenue streams	–
Channels	Service
Key resources	Service
Key partnerships	Actor
Key activities	Process
Value propositions	Progress
Customer relationships	Experience
Customer segment	Customer
–	Situation
–	Hire
–	Cause
–	Jobs
–	Product

5.4 Limitations

This paper only showed an application of MBJT for a use case on the e-Healthcare service. More case studies are necessary to show the effectiveness of MBJT.

Schiltz and others showed 15 use cases in [7]. The remaining use cases are candidates of MBJT application.

6 Summary

This paper proposed the MBJT by integrating the Jobs Theory and the goal oriented requirements model by using ArchiMate. Moreover, a case study for applying MBJT to an e-Healthcare use case was described. The meta-model of MBJT has also proposed and compared those of ConOps and BMC. The result showed the possibility of the integration among ConOps, BMC and MBJT.

Future work includes more case studies on e-Healthcare innovation approaches based on MBJT.

References

1. Christensen, C., Hall, R., Dillson, K., Duncan, D.: Competing Against Luck. HarperCollins Publishers LLC, New York (2016)
2. Yamamoto, S., Kaiya, H., Cox, K., Bleistein, S.: Goal´oriented requirements engineering – trends and issues. IEICE Trans. Inf. Syst. **89**(11), 2701–2711 (2006)
3. The Open Group: ArchiMate 3.0 Specification, C162. Van Haren (2016)
4. Wierda, G.: Mastering ArchiMate – A Serious Introduction to the ArchiMate Enterprise Architecture Modeling Language, Edition II. The Netherlands Published by R&a (2014)
5. Lankhorst, M., et al.: Enterprise Architecture at Work – Modeling Communication and Analysis, 3rd edn. Springer, Heidelberg (2013)
6. Archi. http://archi.cetis.ac.uk/
7. Schiltz, A., Rouille, P., Zabir, S., Genestier, P., Ishigure, Y., Maeda, Y.: Business model analysis of ehealth use cases in europe and in Japan. J. Int. Soc. Telemed. eHealth **1**(1), 30–43 (2013). http://journals.ukzn.ac.za/index.php/JISfTeH/article/view/26/121
8. IEEE Std 1362™-1998 (R2007), IEEE Guide for Information Technology—System Definition—Concept of Operations (ConOps) Document (1998)
9. Australian Government, National E-Health Transition Authority, Concept of Operations: Relating to the introduction of a Personally Controlled Electronic Health Record System, NEHTA Version Number: 0.14.18 (2011)
10. Osterwalder, A., Pigneur, Y.: Business Model Generation. Wiley, Hoboken (2010)

Abdominal Organs Segmentation Based on Multi-path Fully Convolutional Network and Random Forests

Yangzi Yang[1], Huiyan Jiang[1(✉)], and Yenwei Chen[2]

[1] Northeastern University, Shenyang 110819, China
hyjiang@mail.neu.edu.cn
[2] Ritsumeikan University, Shiga 525-8577, Japan

Abstract. Fully convolutional network has predicted multiple class dense outputs in CT image labels and obtained significant improvements in segmentation tasks. In this paper, we present a joint multi-path fully convolutional network (MFCN) with random forests (RF) architecture for abdominal organs segmentation automatically. First, in coarse segmentation step, three FCNs are trained respectively with three orthogonal directions which consider contextual and spatial information of fusion layers adequately. In classification step, using features extracted from different layers of network and normalizing them to mean value as supervoxel representation to train RF. This allows the computation of supervoxel at each orientation achieve high efficiency. Finally, we aggregate the results of MFCN and RF on voxel-wise and perform conditional random fields (CRF) focuses on smoothing borders of fine segmentation regions. We exceeds the state-of-the-art methods and get achievable DSC values for our work is 90.1%, 88.4%, 88.0%, 88.6% represent liver, right and left kidney, spleen respectively.

Keywords: Abdominal organs segmentation · Fully convolutional network
Random forest · Supervoxel

1 Introduction

Organ segmentation is still a challenge for computer aided diagnosis in precision medicine. Many methods are proposed for individual organs segmentation in previous work [1–4]. To make work more efficient, multi-organ segmentation becomes common. Okada et al. built hierarchical multi-organ statistical atlases for segmentation of the liver and related organs [5]. Glocker et al. proposed a new joint classification-regression model based on decision forests for abdominal multi-organ segmentation [6]. Liu et al. considered a maximum a posterior framework for automated abdominal multi-organ localization and segmentation [7]. Shimizu et al. employed an abdominal cavity standardization process and atlas guided segmentation with parameters estimated by EM algorithm [8]. Most of the above approaches generally normalized abdominal cavity or organs to reduce the variability of different organs. But the normalization would produce error in pre-processing stage and make computation less efficient.

G. De Pietro et al. (Eds.): KES-IIMSS-18 2018, SIST 98, pp. 208–215, 2019.
https://doi.org/10.1007/978-3-319-92231-7_21

Known as deep learning architectures, convolutional neural networks (CNNs) have achieved successfully in segmentation tasks, especially for medical images. Roth et al. utilized both image patches and regions on multi-level deep ConvNets for pancreas segmentation [9]. Zhou et al. [10] proposed a novel segmentation approach which described as 3D integration from 2D proposals based only on fully convolutional network. Dou et al. presented 3D deeply supervised network with a fully convolutional architecture for automatic 3D liver segmentation [11]. On the other hand, CNNs as features extractor have generally applied to classification tasks. Razavian et al. used features extracted from the network as a generic image representation to handle the diverse range of recognition tasks of object image classification [12]. Krizhevsky et al. trained a deep convolutional neural network to classify high-resolution images in the ImageNet [13]. Socher et al. introduced a model based on a combination of convolutional and recursive neural networks (CNN and RNN) for learning features and classifying RGB-D images [14].

Particularly, building upon work on CNN, fully convolutional network (FCN) transformed fully connected layers into convolution layers and defined the skip architecture to fuse multifarious information [15]. Unlike patch-wise training methods [16, 17], the loss function in this architecture is computing the entire image and predicting correspondingly-sized output for segmentation [18]. In this paper, we consider a model for automatic abdominal organs segmentation combines two key insights: multi-path fully convolutional network (MFCN) segmentation and random forests (RF) classification. The benefit of our model is evident:

(1) Three FCNs are trained individually with different directions which not only consider contextual and location information, but also augment training data sets.
(2) The deep architecture in our study serves as segmentation and classification, providing more abundant information features without manual intervention and preprocessing operations.
(3) To make deep learning features fully utilized, high quality feature vectors from fusion layers of MFCN are picked up to train RF on supervoxels, meanwhile feature vectors express supervoxels let compute more efficient than before.

Based on the segmentation generated by MFCN and RF, we perform boundary refinement via connected conditional random field (CRF) [19, 20] to further improved segmentation results.

2 Method

2.1 Overview

In this paper, a framework is constructed for abdominal segmentation as briefly shown in Fig. 1. In the segmentation step, we divide CT volumes into sequence slices and transform them into three directions as the input of networks. Then, MFCN with multiple layers are trained end-to-end, and gain the probability of each pixel as coarse segmentation results. In classification step, CT volumes are divided into supervoxels described by features which extracted from MFCN to classify voxels. Finally, we

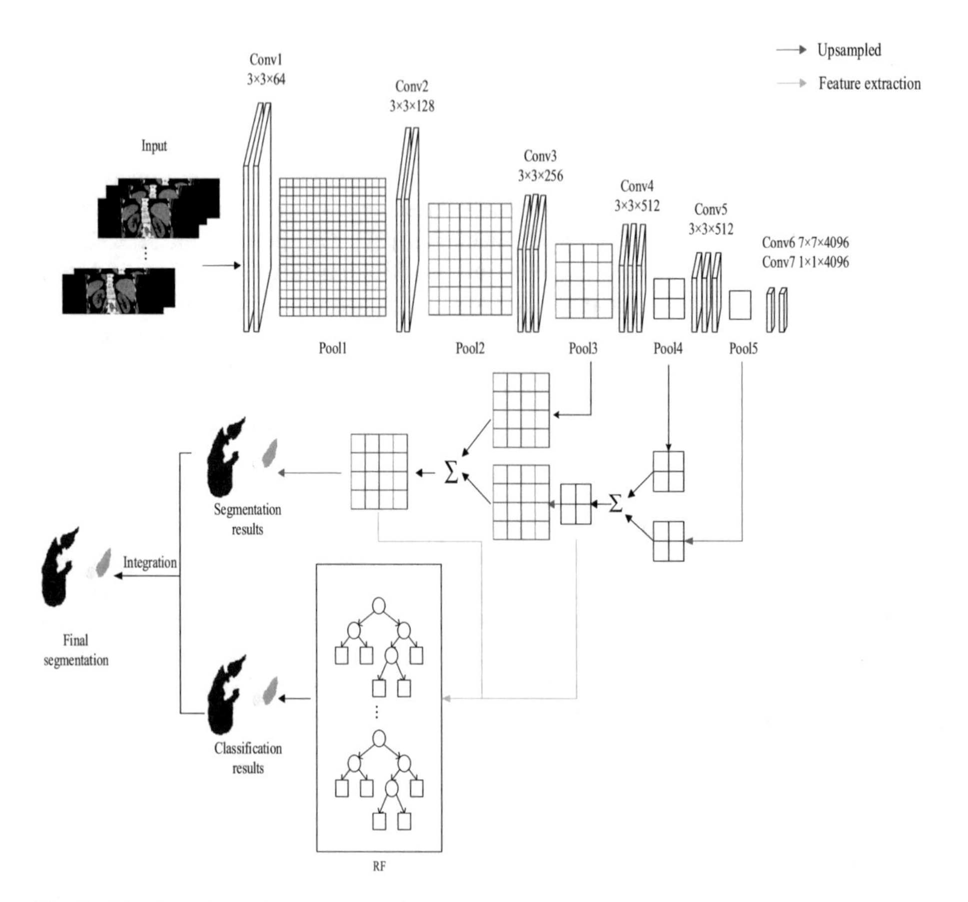

Fig. 1. The flowchart of coronal slice inputs. Axial and sagittal slices are the same as coronal slice.

integrate the results of MFCN and RF and provide border optimization by CRF to elevate the performance of our segmentation method. In this way, fine segmentation is achieved.

2.2 Fully Convolutional Networks for Segmentation

In this part, we specially learn both interior and border image-labeling of abdominal organs via networks separately with three orthogonal directions (axial, sagittal, corona). Our networks try to address two important issues: (1) Take arbitrary size images and produce correspondingly-sized output for pixel-wise prediction. (2) Acquire deeper image features via filter kernels after each fusion layer.

Our architecture as reported in Fig. 1, including convolution, pooling, upsampling and fusion layers, derived from VGG16 [21], whose fully connected layers are replaced by convolution layers. We append a convolution with channel dimension 4 to predict scores of liver, spleen, kidney and background [22]. The training process repeats feed-forward computation and back-propagation to minimize the loss function, which is defined as the sum of the pixel-wise losses between the network prediction and the label map [10]. After that, probabilistic outputs of different orientations on pixel-wise in each FCN are firstly integrated. Then the coarse segmentation results on voxel-wise are gotten by stacking pixel-wise slices prediction into the original CT volumes to forecast final class of which organs belong to. Apart from traditional layers, relying on upsampling and fusion layers, location information is gathered for outputting equal-sized map prediction. Through orthogonal orientations, spatial information of images can be learned to increase segmentation accuracy. Moreover, various directional inputs make it possible to augment data that is an integrant part in network training when only few datasets existence.

2.3 Application to Abdominal Organs Classification via Random Forest

Features Extraction

FCN regard as an effective feature extractor that describe image characteristic in classification. To make better use of features, behind fusion layers, there are 40 features (4 map in each fusion layer $\times$ 3 orientations $\times$ 3 nets + gray value + locations on three planes) of each voxel is extracted from MFCN. Besides extracted features from networks, gray and locations features are also contained within 40 features that we join manually afterwards. In this way, features can reused and get substantial information in high layers.

Random Forest Training and Testing by Supervoxel

Supervoxel classification, a classifier is trained to assign a class label to a region that comprises several voxels. Therefore, we exploit simple linear iterative clustering (SLIC) [23] to generate supervoxels.

In the training part, we calculate the features mean value of all voxels in each supervoxel as representative feature for the supervoxel to train RF classifier. Besides, the class which belongs to organ is determined by class of majority voxels in each supervoxel. In the testing part, each supervoxel is classified via well-trained RF using extracted features from well-trained MFCN. Lastly, we get the probabilities of all supervoxels. We utilize this way to gain the fine segmentation based on the aggregation of each voxel's probability values from learned predication maps MFCN and RF classification respectively. With the optimally supervoxel-wise labels and features, higher computational efficiency is achieved, thereby availably learning their combinatorial connection as well. In addition, we perform CRF in order to smooth the segmentation results and neighboring regions across CT volumes. The CRF weighting coefficient between the boundary and the unary regional term is calibrated by grid-search [9].

3 Results and Discussion

We evaluate our model using 10 cancer patients from the same CT scanner in local hospital. Our experiments segment three organs (liver, kidneys, spleen) based on the ground-truth labels, which rectified by a radiologist. CT volumes are divided into 2644 sequential slices altogether (fixed-size: 512×512) for training MFCN (FCN-8s architecture is used because of its highest performance). But we don't experiment a single FCN with only one plane, this is because the performance achieved by a single net is inferior to the triplanar net in [24]. Besides, CT volumes are taken as input to generate supervoxel which network feature vectors describe for training RF as well. Figure 2 gives the segmentation results obtained from our method. Seeing that compare with FCN-8s, our results not only fills vacancy area but also eliminate some scattered regions. Moreover, we build overall shape of an object and the segmented organs are close to ground truth through our method. Beyond that, no calibrated operation is done both in training and testing stages which make our method to be more accurate. That is also reflected in computationally efficient our model offers an adaptive way of promoting segmentation performance.

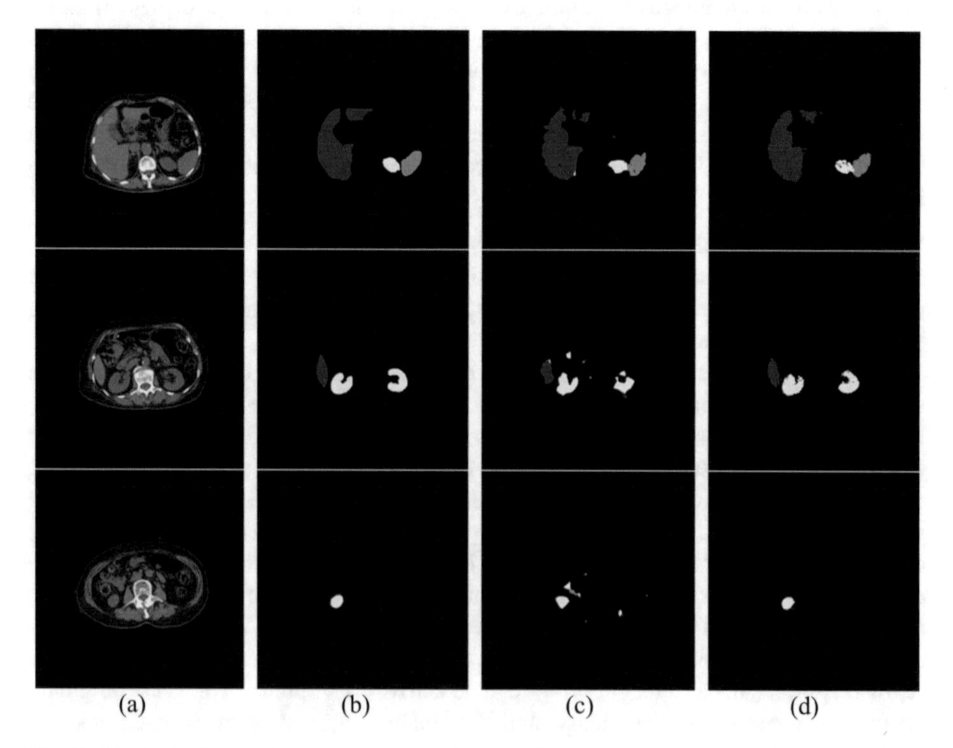

Fig. 2. Typical slices of organs segmentation results. (a) Input images (b) Ground truth (c) FCN-8s, (d) MFCN + RF.

To verify the performance of our proposed method, we make the comparison with state-of-the-art methods based on Dice similarity index [25]. The Dice index measures the volume overlap between the ground truth and segmentation results, which defined as:

$$Dice = \frac{2TP}{2TP + FP + FN} \tag{1}$$

where TP, FP and FN denote the number of voxels relevantly classified, the number of voxels irrelevantly classified and the number of voxels in ground truth that ignored. The results are reported in Table 1, our scheme performs better compared to other methods and further illustrates the potential capability of abdominal organs segmentation we have.

Table 1. Comparisons with the state-of-the-art methods based on Dice similarity index (%).

Method	Dice similarity index			
	Liver	Right kidney	Left kidney	Spleen
Shimizu [8]	89.0	85.0	78.0	84.0
Liu [7]	77.0	74.0	73.0	69.0
Okada [5]	89.1	88.2	87.4	82.5
Glocker [6]	85.0	62.0	68.0	67.0
Zhou [10]	91.0	86.0	84.0	86.0
Our method	90.1	88.4	88.0	88.6

Seeing that liver precision in our experiment is not exceed Zhou [10] whose training samples are 230, but our liver accuracy obtained is close to them using a small number of data. When cancer regions grow on the organs' border, some pustules appears which have effect on segmentation performance as well. Important to note is that our system possesses higher potential for segmentation on account of holding the shapes and locations of the organ well.

4 Conclusion

This paper proposes a structure that simultaneously minimizes segmentation error and improves the segmentation precision for abdominal organs segmentation automatically. Our system has two crucial processes, which aims to segment liver, kidneys and spleen in abdominal CT scans. One is training three ends-to-ends, pixels-to-pixels networks separately for coarse segmentation. The other is training RF on supervoxel-wise using features of networks for classification. Specifically, CRF further improves the segmentation accuracy by producing more precise contours. And the experiment comparison results demonstrate that our method improves segmentation performance in accordance with test accuracy and efficiency. In the future, we will extend our study on the segmentation of more organs and orientations in the deep learning architecture.

Acknowledgement. This work was supported by the National Natural Science Foundation of China (No. 61472073, No. 61272176).

Competing Interest
The authors declare that they have no competing interests.

References

1. Ling, H., Zhou, S.K., Zheng, Y., et al.: Hierarchical, learning-based automatic liver segmentation. In: Computer Vision and Pattern Recognition 2008. CVPR, pp. 1–8. IEEE (2008)
2. Kitrungrotsakul, T., Han, X.H., Iwamoto, Y., et al.: Robust hepatic vessel segmentation using multi deep convolution networks, In: Proceedings of SPIE, vol. 10137, Medical Imaging 2017, Florida, USA (2017)
3. Foruzan, A.H., Chen, Y.W., Zoroofi, R.A., et al.: Segmentation of liver in low-contrast images using k-means clustering and geodesic active contour algorithms. IEICE Trans. Inf. Syst. **96**(4), 798–807 (2013)
4. Xu, Z., Li, B., Panda, S., et al.: Shape-constrained multi-atlas segmentation of spleen in CT. In: Proceedings of SPIE–the International Society for Optical Engineering, vol. 9034, p. 903446. NIH Public Access (2014)
5. Okada, T., Linguraru, M.G., Yoshida, Y., et al.: Abdominal multi-organ segmentation of CT images based on hierarchical spatial modeling of organ interrelations. In: International MICCAI Workshop on Computational and Clinical Challenges in Abdominal Imaging, pp. 173–180. Springer, Heidelberg (2011)
6. Glocker, B., Pauly, O., Konukoglu, E., et al.: Joint classification-regression forests for spatially structured multi-object segmentation. In: Computer Vision–ECCV 2012, pp. 870–881 (2012)
7. Liu, X., Linguraru, M.G., Yao, J., et al.: Organ pose distribution model and an MAP framework for automated abdominal multi-organ localization. In: International Workshop on Medical Imaging and Virtual Reality, pp. 393–402. Springer, Heidelberg (2010)
8. Shimizu, A., Ohno, R., Ikegami, T., et al.: Segmentation of multiple organs in non-contrast 3D abdominal CT images. Int. J. Comput. Assist. Radiol. Surg. **2**(3–4), 135–142 (2007)
9. Roth, H.R., Lu, L., Farag, A., et al.: Deeporgan: multi-level deep convolutional networks for automated pancreas segmentation. In: International Conference on Medical Image Computing and Computer-Assisted Intervention, pp. 556–564. Springer, Heidelberg (2015)
10. Zhou, X., Ito, T., Takayama, R., et al.: First trial and evaluation of anatomical structure segmentations in 3D CT images based only on deep learning. Med. Imaging Inf. Sci. **33**(3), 69–74 (2016)
11. Dou, Q., Chen, H., Jin, Y., et al.: 3D deeply supervised network for automatic liver segmentation from CT volumes. In: International Conference on Medical Image Computing and Computer-Assisted Intervention, pp. 149–157. Springer, Heidelberg (2016)
12. Sharif Razavian, A., Azizpour, H., Sullivan, J., et al.: CNN features off-the-shelf: an astounding baseline for recognition. In: Proceedings of the IEEE Conference on Computer Vision and Pattern Recognition Workshops, pp. 806–813 (2014)
13. Krizhevsky, A., Sutskever, I., Hinton, G.E.: Imagenet classification with deep convolutional neural networks. In: Advances in Neural Information Processing Systems, pp. 1097–1105 (2012)

14. Socher, R., Huval, B., Bath, B., et al.: Convolutional-recursive deep learning for 3D object classification. In: Advances in Neural Information Processing Systems, pp. 656–664 (2012)
15. Long, J., Shelhamer, E., Darrell, T.: Fully convolutional networks for semantic segmentation. In: Proceedings of the IEEE Conference on Computer Vision and Pattern Recognition, pp. 3431–3440 (2015)
16. Ganin, Y., Lempitsky, V.: N^ 4-fields: neural network nearest neighbor fields for image transforms. In: Asian Conference on Computer Vision, pp. 536–551. Springer, Cham (2014)
17. Pinheiro, P., Collobert, R.: Recurrent convolutional neural networks for scene labeling. In: International Conference on Machine Learning, pp. 82–90 (2014)
18. Ben-Cohen, A., Diamant, I., Klang, E., et al.: Fully convolutional network for liver segmentation and lesions detection. In: International Workshop on Large-Scale Annotation of Biomedical Data and Expert Label Synthesis, pp. 77–85. Springer, Heidelberg (2016)
19. Lafferty, J.D., McCallum, A., Pereira, F.C.N.: Conditional random fields: probabilistic models for segmenting and labeling sequence data. In: Eighteenth International Conference on Machine Learning, pp. 282–289. Morgan Kaufmann Publishers Inc. (2001)
20. Krähenbühl, P., Koltun, V.: Efficient inference in fully connected CRFs with gaussian edge potentials. In: Advances in Neural Information Processing Systems, pp. 109–117 (2011)
21. Simonyan, K., Zisserman, A.: Very deep convolutional networks for large-scale image recognition. Computer Science (2014)
22. Yang, Y., Jiang, H., Sun, Q.: A multiorgan segmentation model for CT volumes via full convolution-deconvolution network. Biomed. Res. Int. **2017**, 6941306 (2017)
23. Achanta, R., Shaji, A., Smith, K., et al.: SLIC superpixels compared to state-of-the-art superpixel methods. IEEE Trans. Pattern Anal. Mach. Intell. **34**(11), 2274–2282 (2012)
24. Prasoon, A., Petersen, K., Igel, C., et al.: Deep feature learning for knee cartilage segmentation using a triplanar convolutional neural network. In: International Conference on Medical Image Computing and Computer-Assisted Intervention, pp. 246–253. Springer, Heidelberg (2013)
25. Pereira, S., Pinto, A., Alves, V., et al.: Deep convolutional neural networks for the segmentation of gliomas in multi-sequence MRI. In: International Workshop on Brainlesion: Glioma, Multiple Sclerosis, Stroke and Traumatic Brain Injuries, pp. 131–143. Springer, Cham (2015)

Interactive Liver Segmentation in CT Volumes Using Fully Convolutional Networks

Titinunt Kitrungrotsakul[1], Yutaro Iwamoto[1], Xian-Hua Han[2], Xiong Wei[3], Lanfen Lin[4], Hongjie Hu[5], Huiyan Jiang[6], and Yen-Wei Chen[1,4(✉)]

[1] Graduate School of Information Science and Engineering, Ritsumeikan University, Kyoto, Japan
chen@is.ritsumei.ac.jp
[2] Faculty of Science, Yamaguchi University, Yamaguchi, Japan
[3] Institute for Infocomm Research, Singapore, Singapore
[4] College of Computer Science and Technology, Zhejiang University, Hangzhou, China
[5] Radiology Department, Sir Run Run Shaw Hospital, Medical School, Zhejiang University, Hangzhou, China
[6] Software College, Northeastern University, Shenyang, China

Abstract. Organ segmentation is one of the most fundamental and challenging task in computer aided diagnosis (CAD) systems, and segmenting liver from 3D medical data becomes one of the hot research topics in medical analysis field. Graph cut algorithms have been successfully applied to medical image segmentation of different organs for 3D volume data which not only leads to very large-scale graph due to the same node number as voxel number. Slice by Slice liver segmentation method is one of the technique that normally used to solve the memory usage. However, the computation times are increased and reduce the accuracy. In this paper we propose an interactive organ segmentation using fully convolutional networks. The network will perform slice by slice which only 1 slice of seed points in each volume. To validate effectiveness and efficiency of our proposed method, we conduct experiments on 20 CT volumes, focus on liver organ and most of which have tumors inside of the liver, and abnormal deformed shape of liver. Our method can segment with 0.95401 dice accuracy with better than stage-of-the-art methods.

Keywords: Fully convolutional networks · Interactive · Segmentation Liver · Seed points

1 Introduction

Medical image processing has been a research field attracting researcher from various field. It has been an important part and process of clinical routine. Various methods are researched and applied on computer tomography (CT), and magnetic resonance imaging (MRI), such as, segmentation, extraction, analysis, visualization, computer-aided diagnostic (CAD), and surgical planning to support doctor [1]. There are mainly two types of medical segmentation approaches: (1) automatic and (2) semi-automatic or interactive

G. De Pietro et al. (Eds.): KES-IIMSS-18 2018, SIST 98, pp. 216–222, 2019.
https://doi.org/10.1007/978-3-319-92231-7_22

methods. Due to large variance and shape deformation produced by abnormality, it is difficult to acquire accurate segmentation via completely automatic method. To the best of our knowledge, the automatic approaches give us some acceptable results for the liver with small intensity variation and an availability of the well-defined prior knowledge, i.e., the well-defined initial contour for level-set introduced by Osher [1] and the anatomical information for model based approach [2, 3]. However, it tends to fail for the samples with large variance and abnormal diseases.

By giving initial seeds from users, some under-segmentation or over-segmentation parts can be corrected. Therefore, graph cuts and random walk based interactive segmentation methods have been successfully applied in medical image processing field [4–7]. However, the aforementioned methods have two drawbacks: the segmentation process consumes huge amount of memory due to the requirement of very large-scale graph construction, and it also takes long time for the optimization process to obtain high quality segmentation result.

Recently, the deep learning has been demonstrated the powerful ability in computer vision tasks by automatically learn hierarchies of relevant feature directly from the input data. The deep convolutional neural network has been successfully applied for image classification and object detection, especially for ImageNet classification competition, which is the most successful network for image classification since 2012 [8]. Fully Convolutional Networks (FCN) are powerful methods that use a CNN as a base network to perform segmentations [9]. The method requires huge training data to generate segmentation model which difficult to obtain for medical data.

In this paper, we proposed an interactive slice by slice medical image segmentation using FCN. Our method require user put seed points only one slice for each volume and employ the previously segmented slice as the seed points for automatic segmentation of the other slices.

The paper is organized as follows. The proposed network is introduced in Sect. 2, and experimental results of our proposed segmentation method are described in Sect. 3. Conclusions are given in Sect. 4.

2 Proposed Method

In this paper, we concentrate our efforts in the following two-fold: (1) perform segmentation using interactive Fully Convolutional Networks (iFCN); (2) integrating the result from iFCN with graph cut to refine segmentation result.

2.1 Interactive Fully Convolutional Networks (iFCN)

The interactive fully convolutional network is motivated by FCN. The goal is to make the network able to segment liver from medical data with high accuracy. The original FCN require only 2D image as input and network learn the features from input image to perform the segmentation, however the method requires huge annotated dataset to perform high accuracy. Based on this problem we include the user interaction (seed points) into the network which can reduce the number of training dataset. Figure 1 show the iFCN architecture, the base network is AlexNet with additional 2

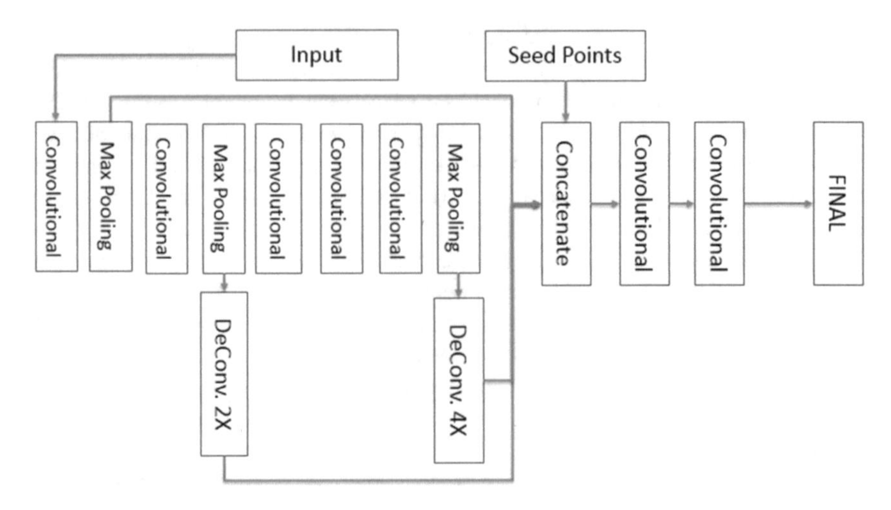

Fig. 1. Overview of our proposed iFCN

convolutional layers. Each layer consists of 32 filters and the filter size is 3 × 3 with padding added to remain the same size after adding seed points into the network.

To apply seed points into the network, we transform ground truth into seed point using skeletonize technique for training process as shown in Fig. 2. For testing process, the user is required to put seed points on only one slice. Our method will automatically generate seed points for next slice using a skeletonization method from the current slice segmentation result.

The result *iFCN(A)* of the network will be the probability of each pixel/voxel belong to liver region or background region which required in the next sub-section.

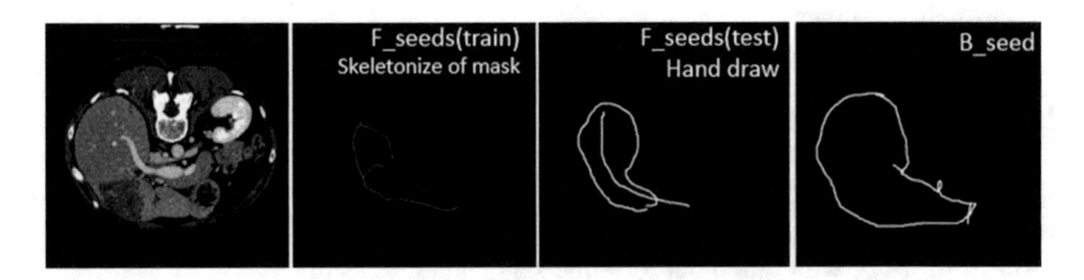

Fig. 2. Automatic seed points generating using skeletonize of mask image.

2.2 Refinement of Segmentation Result with Graph Cuts

To refine the result of network we apply a graph cut approach as a refinement method. In the graph cut approach, the region term is replaced with a network term *iFCN(A)*, which is described in previous sub-section. The method performs a minimum cut to minimize the energy with an energy function $E(A)$ and generate foreground (liver) and background regions.

Let's denote the pixel/voxel in image as $A = (A_1, \ldots, A_p, \ldots, A_{|P|})$, with which the total energy function $E(A)$ can be constructed for being minimized. Each A_p can be either "object" or "background" pixel/voxel.

$$E(A) = \mu \cdot iFCN(A) + B(A) \tag{1}$$

The coefficient μ specifies a relative importance of the network term *iFCN(A)*, versus boundary term B(A).

3 Experimental Result

To validate the effectiveness of our proposed method, we performed segmentation experiments using public IRCAD 20 datasets [13] and our original 10 datasets. We compared our method with state-of-the-art methods. The quantitative measures of Volumetric Overlap Error (VOE), and Relative Volume Difference (RVD) are used for evaluation on IRCAD dataset and the Dice coefficient is used for our original dataset. RVD is used to calculate the difference the volume of segmentation result and ground truth while VOE measure the error which based on the Jaccard's coefficient.

$$\mathrm{DICE} = \frac{2|\mathrm{A} \cap \mathrm{B}|}{|\mathrm{A}| + |\mathrm{B}|} \tag{2}$$

$$\mathrm{VOE} = \left(1 - \frac{|\mathrm{A} \cap \mathrm{B}|}{|\mathrm{A} \cup \mathrm{B}|}\right) \tag{3}$$

$$\mathrm{RVD} = \frac{|\mathrm{A} - \mathrm{B}|}{|\mathrm{B}|} \tag{4}$$

The leave-one-out method is used in our experiments. We used the Adam optimization method to train our model. The learning rate for Adam in our network is start with $1e^{-6}$ and change into $1e^{-7}$ after 20,000 training samples also transfer learning from ImageNet has been used in our experiments.

3.1 Comparison with the State-of-the-Art Methods on IRCAD Dataset

We compared our method with the state-of-the-art methods [10–12]. We also compared with the conventional FCN [9]. Comparison results on IRCAD dataset are summarized in Table 1. The calculated mean ratios of VOE, and RVD for our proposed method are 9.6%, and 0.9%, respectively. As shown in Table 1, the proposed method outperformed the conventional FCN [9] and the methods [10, 11]. Though the method [12] is a little better than our method, there is no significant difference in spite of [12] combining several models (shape model, active contour and graph cut). The computation of our method is around 60 s on CPU and it will be reduced to around 5–6 s if we use GPU.

Table 1. Quantitative comparison of our proposed method with the state-of-the-art methods on IRCAD dataset.

Method	VOE [%]	RVD [%]	Time (sec)
Chung and Delingette [10]	12.99	−5.66	NA
Erdt et al. [11]	10.34	1.55	45
Li et al. [12]	9.15	−0.07	97
FCN [9]	19.175	6.5508	51
Our method	9.691	0.9902	92

The visualization results of liver segmentation (red) overlay on ground truth (white) are shown in Fig. 3. As shown in Fig. 3, our interactive FCN can extract the liver more correctly and smoother than FCN. Our method can fill more missing part and also remove other parts which not belonging to liver region.

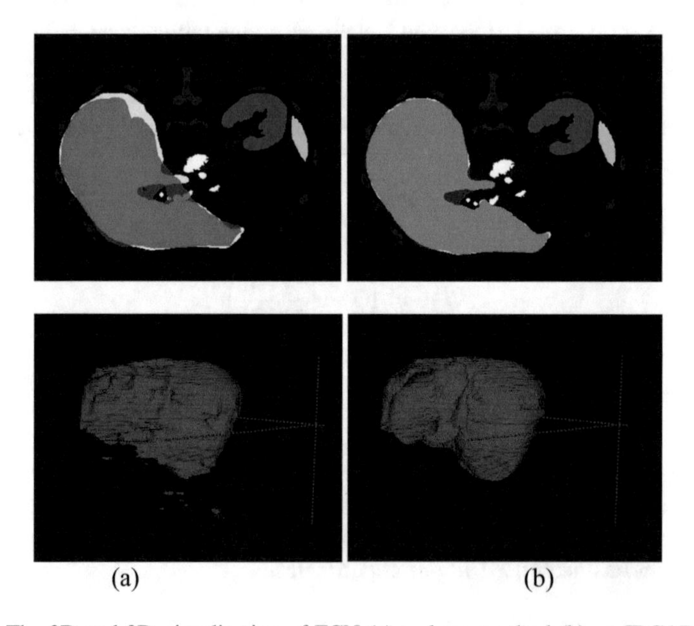

Fig. 3. The 2D and 3D visualization of FCN (a) and our method (b) on IRCAD dataset

3.2 Evaluation of Post-processing with Graph Cuts

In this section, we compared our method (iFCN+graph cuts) with only iFCN, conventional FCN, FCN with graph cuts and graph cuts on our original dataset. The results are summarized in Table 2. As shown in Table 2, we can see that both iFCN with graph cuts and FCN with graph cuts are better than iFCN and FCN, respectively. It means that using the post-processing with graph cuts can significantly improve the segmentation results. We also can see that both our proposed methods (iFCN and iFCN with graph cuts) also outperformed other methods on our original dataset.

Table 2. Quantitative comparison of our proposed postprocessing with FCN and Graph Cuts

Method	DICE
Graph Cut [4]	0.79184
FCN [9]	0.89427
FCN with Graph Cuts	0.92741
iFCN	**0.93043**
iFCN with Graph Cuts	**0.95401**

4 Conclusion

We have proposed an interactive convolution neural network using FCN for semi-automatic segmentation of medical images. The proposed network requires seed points only for the first slice and it propagated the segmentation result to generate seed points for next slice. Experimental results demonstrated that our proposed method is superior than the state-of-the-art methods such as FCN (deep learning), or other non-deep learning methods. Combining with graph cuts as a refinement method (post-processing), the result can be significantly improved around 2–3%. However, the computation time of our method is much more than original FCN because of the graph cut that we use.

Acknowledgement. This work is supported in part by Japan Society for Promotion of Science (JSPS) under Grant No. 16J09596 and KAKEN under the Grant Nos. 16H01436, 17H00754, 17K00420, 18H03267; and in part by the MEXT Support Program for the Strategic Research Foundation at Private Universities, Grand No. S1311039 (2013–2017), and also partially supported by A*STAR Research Attachment Program.

References

1. Osher, S., Sethian, J.A.: Fronts propagating with curvature dependent speed: algorithms based on Hamilton-Jacobi formulations. J. Comput. Phys. **79**, 12–49 (1988)
2. Lamecker, H., Lange, T., Seebass, M.: A statistical shape model for the liver. In: International Conference on Medical Image Computing and Computer-Assisted Intervention, pp. 412–427 (2002)
3. Dong, C., et al.: Segmentation of liver and spleen based on computational anatomy models. Comput. Biol. Med. **67**, 146–160 (2015)
4. Boykov, Y., Jolly, M.: Interactive graph cuts for optimal boundary & region segmentation of object in N-D images. In: International Conference on Computer Vision, pp. 105–112 (2001)
5. Kitrungrotsakul, T., Han, X.-H., Chen, Y.-W.: Liver segmentation using superpixel-based graph cuts and regions of shape constraints. In: Proceedings of IEEE International Conference on Image Processing (ICIP2015), pp. 3368–3371 (2015)
6. Grady, L.: Random walks for image segmentation. IEEE Trans. PAMI **28**(11), 1768–1783 (2006)
7. Dong, C., et al.: Simultaneous segmentation of multiple organs using random walks. J. Inf. Process. Soc. Jpn. **24**(2), 320–329 (2016)

8. Krizhevsky, A., Sutskever, I., Hinton, G.E.: ImageNet classification with deep convolutional neural networks In: Advances in Neural Information Processing Systems, pp. 1097–1105 (2012)
9. Long, J., Shelhamer, E.: Fully convolutional models for semantic segmentation. In: Proceedings of CVPR 2015 (2015)
10. Chung, F., Delingette, H.: Regional appearance modeling based on the clustering of intensity profiles. Comput. Vis. Image Underst. **117**(6), 705–717 (2013)
11. Erdt, M., Steger, S., Kirschner, M., Wesarg, S.: Fast automatic liver segmentation combining learned shape priors with observed shape deviation. In: Proceedings of the 26th IEEE International Symposium on Computer-Based Medical Systems, pp. 249–254 (2010)
12. Li, G., Chen, X., Shi, F., Zhu, W., Tian, J.: Automatic liver segmentation based on shape constraints and deformable graph cut in CT images. IEEE Trans. Image Process. **24**(12), 5315–5329 (2015)
13. IRCAD dataset. http://www.ircad.fr/. Accessed 30 Jan 2018

Kinect-Based Real-Time Gesture Recognition Using Deep Convolutional Neural Networks for Touchless Visualization of Hepatic Anatomical Models in Surgery

Jia-Qing Liu[1], Tomoko Tateyama[2], Yutaro Iwamoto[1], and Yen-Wei Chen[1(✉)]

[1] Information Science and Engineering, Ritsumeikan University, Kusatsu, Shiga, Japan
chen@is.ritsumei.ac.jp
[2] Department of Computer Science, Hiroshima Institute of Technology, Hiroshima, Japan

Abstract. In this paper, we present a novel touchless interaction system for visualization of hepatic anatomical models in surgery. Real-time visualization is important in surgery, particularly during the operation. However, it often faces the challenge of efficiently reviewing the patient's 3D anatomy model while maintaining a sterile field. The touchless technology is an attractive and potential solution to address the above problem. We use a Microsoft Kinect sensor as input device to produce depth images for extracting a hand without markers. Based on this representation, a deep convolutional neural network is used to recognize various hand gestures. Experimental results demonstrate that our system can significantly improve the response time while achieve almost same accuracy compared with the previous researches.

Keywords: Deep learning · Kinect · Real-time touch-less interaction Visualization of hepatic anatomical models

1 Introduction

Understanding of the patient's particular hepatic anatomical models is important and essential for successful liver surgery [1, 2]. Though the visualization of the reconstructed anatomic model on computers can provide detailed and useful anatomic information for surgery, the surgeon usually needs to use some contacting devices such as mouse, keyboard or touch panel to display the medical images during the surgical operation. After operating the visualization device, re-sterilization is necessary in order to maintain hygiene, which is an inefficient and un-effective process for surgery. So a touchless technology is a potential and useful solution to solve the above problems.

Though some touchless surgery support system using Kinect have been developed [3–5], it has to be noted that lots of publications introducing very similar approaches like using the fundamental functions of Kinect to produce gesture interface for medical image interaction in a special application [6]. Our research therefore focus on further

G. De Pietro et al. (Eds.): KES-IIMSS-18 2018, SIST 98, pp. 223–229, 2019.
https://doi.org/10.1007/978-3-319-92231-7_23

intuitive custom-built gestures specialized to increase the degree of freedom of operations and fulfill the requirements for the surgery application. In our previous work [8], we used histogram of oriented gradients (HOG) features and principal component analysis (PCA) to recognize hand gestures from the depth images. The limitation is its slow response time. In order to improve the response time, we developed our second version of the system [9]. In the second version, we combined three hand states (open, close and lasso), which are automatically detected by Kinect, with their hand movements to control the visualization instead of hand gesture recognition. The second version system can realize a real-time visualization, but the degrees of freedom and the number of gestures are limited. So it lacks flexibility.

Since 2012, deep learning-based approaches have consistently shown best-in-class performance in major computer vision tasks [7]. The present work is an improved version of the previous systems. In our new version, we use a deep learning technique to recognize hand gestures. A rapidly responding and flexible Kinect-based touchless visualization has been realized.

The paper is organized as the following. Section 2 describes our touchless visualization system including detailed description about hand gesture recognition using deep learning. Experimental results are shown in Sect. 3 and conclusion is given in Sect. 4.

2 Proposed System

The diagram in Fig. 1 summarizes our system architecture that includes two modules: interaction module and visualization module. When the Kinect sensor detects that user's gesture becomes available state (i.e., the nearest user's right hand is above waist for 45 cm), it performs a fast and flexible hand gesture recognition using trained deep learning model. The classified gesture and its movement are processed by command module and send to visualization module through a socket. Finally, the module responds to the command and performs the corresponding operation like rotation, opacity adjustment, zoom in/out, fusion and selection of vessels.

2.1 Hand Image Pre-processing

We utilize the depth information and skeleton tracking provided by the Kinect to generate the depth image of hand. First, we acquire a depth image of the user (Fig. 2(a)). Then, we do calibration between color and depth camera and using the right hand joint point as the center, chip out a 100×100 square region as a ROI of hand region (Fig. 2(b)). The depth image with a range from $d - 30$ cm to $d + 5$ cm is defined as a hand image, where d is the depth of the right hand joint point. The segmented hand image is shown in Fig. 2(c). Since the hand image has other regions' pixels with remained as noise, we apply an opening operator and a median filter to remove the noise (Fig. 2(d)). To reduce computation time and to focus on our analysis on regions of the hand shape, we cropped the ROI of hand region and resized it to 32×32, which is used as an input of the deep convolutional neural network.

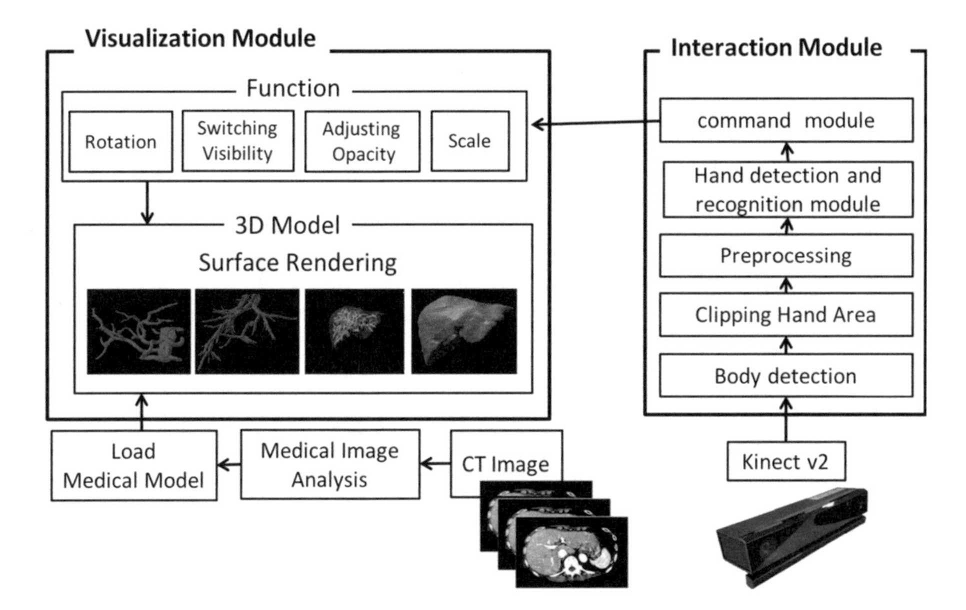

Fig. 1. Diagram of our proposed system.

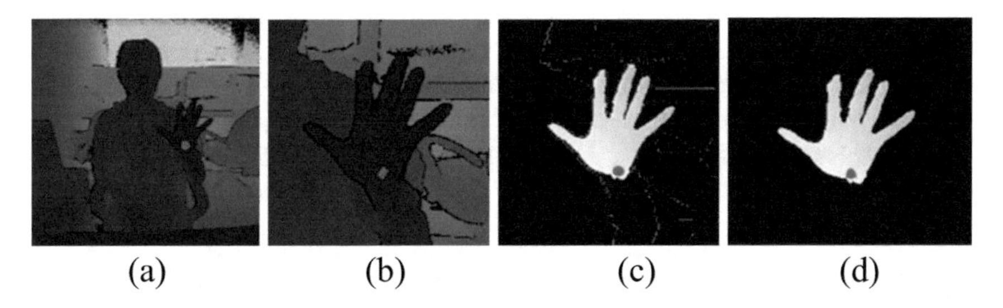

Fig. 2. The depth hand image for gesture recognition (the rad point is the right hand joint point detected by Kinect). (a) Depth image from Kinect. (b) Decision ROI of depth hand image. (c) Segmented depth hand image including noise. (d) Depth hand image excluding noise.

2.2 Recognition of Hand Gestures Using Deep Learning

In our framework, we adopt LeNet [10] as our deep network architecture. The network architecture of LeNet consists of two convolutional layers, each followed by pooling and three fully-connected layers. The first layer uses 6 kernels and the second 16, both with the same size 5×5. The output is 9 classes of gestures. The LeNet for hand gesture recognition is shown in Fig. 3.

2.3 Visualization Module

In the visualization module, surface models of hepatic anatomical models including hepatic artery, hepatic portal vein, hepatic vein and liver parenchyma (Fig. 4) are

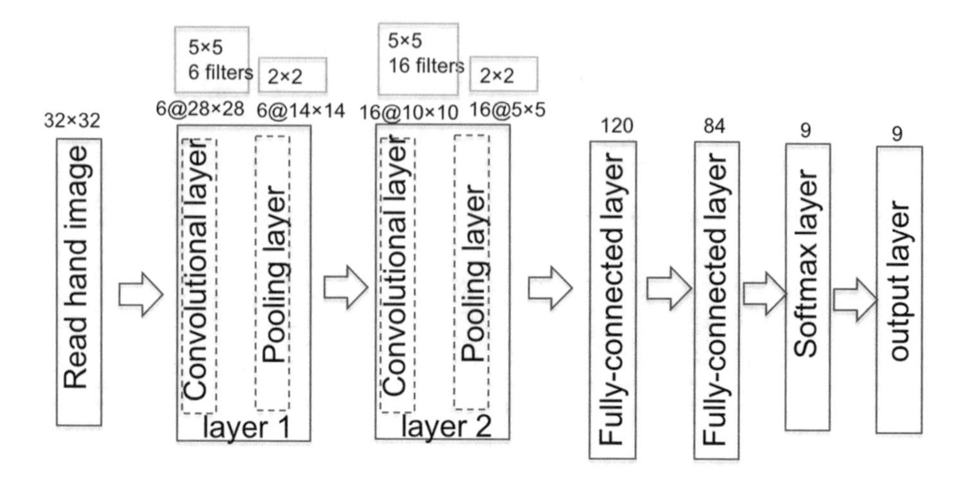

Fig. 3. LeNet for hand gesture recognition

generated by converting each corresponding volume data to a triangulated mesh surface by the use of marching cube algorithms. Each volume data is segmented semi-automatically from CT images under the guidance of a physician [1, 2]. Compared with traditional slice-by-slice visualization and review techniques, the surgeon can easily recognize the liver geometry, its vessels structures and locations during the surgery with the 3D surface rendering of hepatic anatomical models as shown in Fig. 5. Please refer [1, 2] for detailed information about CT data and segmented liver and vessel data. Our system has four visualization modes: rotation, zoom in/out, adjustment of opacity, fusion and selection of vessels.

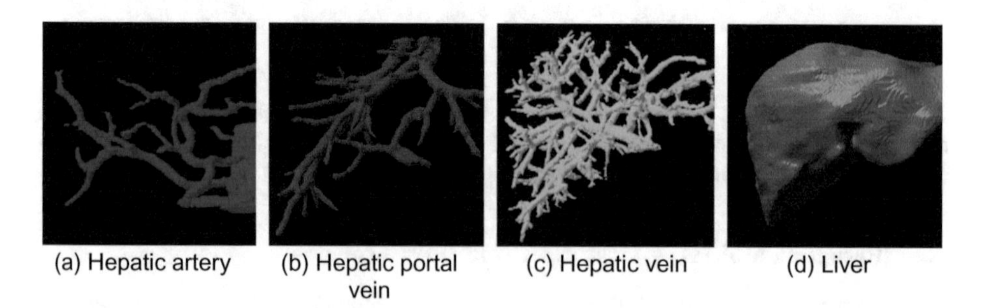

Fig. 4. Visualization of liver and its vessels.

3 Experimental Results

3.1 Dataset

The dataset was collected from 10 participants in advance, each person has 100 pieces of depth images of 9 kinds of hand shapes in Fig. 6. As a verification method, we used

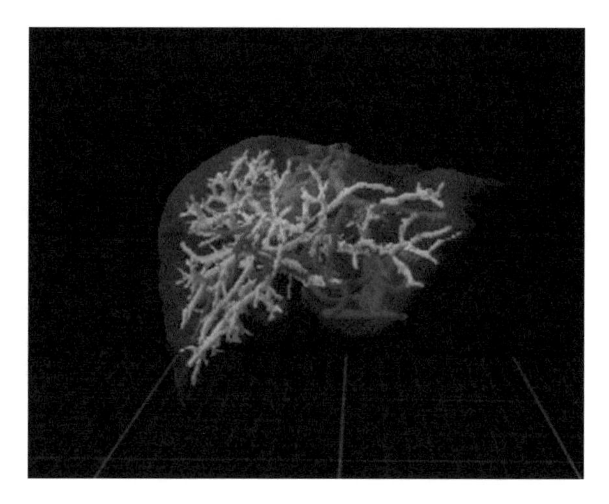

Fig. 5. Visualization of fused liver and its vessel structure.

Fig. 6. Examples of 9 hand gestures

10-fold cross-validation. 9 persons were used as training data, and 1 person was used as test data. Twenty percent of the training data are selected randomly as validation set for fine tuning the hyper-parameters. The number of training samples is 8100 and the number of test samples is 900. We repeated it 10 times in total and verify the results of all cases.

3.2 Results

Table 1 provides a comparison of the proposed method with previous work [8] in recognition accuracy. Table 2 provides a comparison of the proposed method with previous work [8] in prediction time. We can see that though the recognition accuracy of the proposed system is lower than previous system by 3.2%, the recognition response time is highly reduced and a real-time recognition is achieved.

Table 1. Comparison of the proposed method (indicated as 1) with previous method [8] (indicated as 2) in recognition accuracy (%), the number in bold means higher score.

	Hand open	Hand close	Grasp	Finger up	Finger down	Finger left	Finger right	Palm up	Palm down	Mean value
1	86.7	81	84.7	**94.1**	77.4	85	**86.2**	**84.7**	78.9	84.3
2 [8]	**91**	**91**	**87**	89	**95**	**86**	84	82	**83**	**87.5**

Table 2. Recognition time

Proposed method	Previous method [8]
0.002856 s	0.123 s

It should be noted that although the recognition time of the deep learning method is about 0.003 s, the response rate (or display rate) is limited to 30 fps since the frame rate of the Kinect is 30 fps. It should also be noted that the response rate of the previous system [8] is only 8 fps. We can improve the recognition accuracy by taking an average within a temporal window of 3 frames (temporal smoothing). In this case, the response rate is decreased to 10 fps, which is still faster than previous system (8 fps). The comparison results in both response rate and recognition accuracy are shown in Fig. 7. It can be seen that with the temporal smoothing our new system based on deep learning outperformed the previous system in both accuracy and response time.

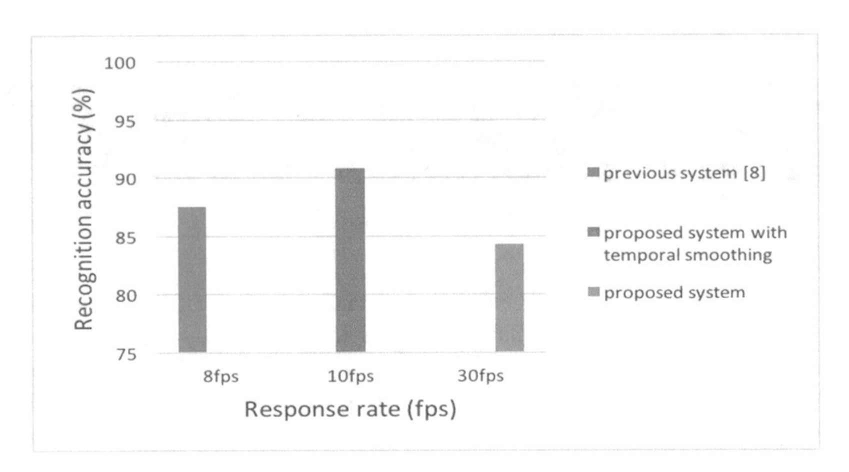

Fig. 7. The comparison results in both response rate and recognition accuracy.

4 Conclusion

In this paper, we proposed a deep learning-based gesture recognition for a touchless hepatic surgery support system. We designed a range of hand gestures, by using our system, the surgeon can operate and check the information of hepatic anatomic models without touching any devices. The experiments demonstrated that the high response time has been achieved, while maintain almost the same recognition rate. Further improvements on recognition rate by using deeper networks will be our future work. Though the current device (Kinect) can capture depth images in 30 fps is satisfy the surgeons' requirements, we are going to use more higher response device like Real-Sense Depth Camera [11], which can output depth frames in 90 fps. More temporal information will be used as pre-processing for recognition of various hand gestures to achieve more accurate and robust touchless visualization.

Acknowledgment. Authors would like to thank Dr. M. Kaibori of KANSAI Medical University for providing medical images and advice on surgical support systems. This work is supported in part by the Grant-in Aid for Scientific Research from the Japanese Ministry for Education, Science, Culture and Sports (MEXT) under the Grant Nos. 16H01436, 15K16031, 17H00754, 17K00420, 18H03267; in part by the MEXT Support Program for the Strategic Research Foundation at Private Universities, Grant (2013–2017).

References

1. Kaibori, M., Chen, Y.W., Matsui, K., Ishizaki, M., Tsuda, T., Nakatake, R., Sakaguchi, T., Matsushima, H., Miyawaki, K., Shindo, T., Tateyama, T., Kwon, A.H.: Novel liver visualization and surgical simulation system. J. Gastrointest. Surg. **17**, 1422–1428 (2013)
2. Tateyama, T., Kaibori, M., Chen, Y.W., et al.: Patient-specified 3D-visualization for liver and vascular structures and interactive surgical planning system. Med. Imaging Technol. **31**, 176–188 (2013). (in Japanese)
3. Gallo, L.: Controller-free exploration of medical image data: experiencing the Kinect. National Research Council of Italy Institute for High Performance Computing and Networking (2011)
4. Yoshimitsu, K., Muragaki, Y., Iseki, H., et al.: Development and initial clinical testing of "OPECT": an innovative device for fully intangible control of the intraoperative image-displaying monitor by the surgeon. Neurosurgery **10**(Suppl 1), 46–50 (2014)
5. Ruppert, G.C., Coares, C., Lopes, V., et al.: Touchless gesture user interface for interactive image visualization in urological surgery. World J. Urol. **30**, 687–691 (2012)
6. Mewes, A., Hensen, B., Wacker, F., et al.: Touchless interaction with software in interventional radiology and surgery: a systematic literature review. Int. J. CARS **12**, 291 (2017)
7. Krizhevsky, A., Sutskever, I., Hinton, G.: ImageNet classification with deep convolutional neural networks. In: NIPS 2012 Proceedings of the 25th International Conference on Neural Information Processing Systems (NIPS), vol. 1, pp. 1097–1105 (2012)
8. Liu, J.Q., Fujii, R., Tateyama, T., Iwamoto, Y., Chen, Y.W.: Kinect-based gesture recognition for touchless visualization of medical images. Int. J. Comput. Electr. Eng. **9**(2), 421–429 (2017)
9. Liu, J.Q., et al.: A kinect-based real-time hand gesture interaction system for touchless visualization of hepatic structure in surgery. Med. Imaging Inf. Sci. (2018, submitted)
10. LeCun, Y., Bottou, L., Bengio, Y., Haffner, P.: Gradient-based learning applied to document recognition. Proc. IEEE **86**(11), 2278–2323 (1998)
11. Intel RealSence Depth camera D400 series. https://software.intel.com/en-us/realsense/d400

Advanced Transmission Methods Applied in Remote Consultation and Diagnosis Platform

Zhuofu Deng[1,2(✉)], Yen-wei Chen[2], Zhiliang Zhu[1], Yinuo Li[1(✉)], He Li[2], and Yi Wang[2]

[1] College of Software Engineering, Northeastern University, Shenyang, China
dengzf@swc.neu.edu.cn

[2] College of Information Science and Engineering, Ritsumeikan University, Kyoto, Japan

Abstract. Remote consultation and diagnosis platform has been widespread in market for group diagnosis, education and etc. It provides convenience for people enjoying superior quality of medical service. However, most of platforms cannot meet increasing demands in practice. The most important reason is that they have troubles of navigating in complicated Internet condition. In general, remote rendering is preferred to support data transmission, by which server renders images locally, captures screenshot and delivers them to clients synchronously. As a result of neighborhood DICOM slices serving a high similarity, following frame is compressed according to the difference with last one. Even though necessary demand of bandwidth has been reduced a lot, remaining volume proves too large especially remote 3D volume rendering. In this paper, we proposed a novel method bit difference compression transmission. Compared with traditional algorithm, it redesigns a new data structure, which gives more significances to bit. Consequently, demands of network fall into a desirable amount. Specifically, we improved solution of 3D images sharing and collaboration. Instead of terrible screenshots, synchronization of cameras in clients has been introduced. With a minimal cost our platform realized the cumbersome process. Collected data from experiments demonstrated our proposed methods have an obvious superiority and run robust in different environment of Internet.

1 Introduction

Computer-aided diagnosis with digital images plays an essential role in clinic [5]. To provide convenience, picture archiving and communication system (PACS) has been proposed [4]. Clients can have access to PACS and review images and related medical records by Internet. This way helps radiologists to get rid of fixed workplace. Invoked by this idea, people wish that experienced experts would have opportunities to share their knowledge and diagnosis anytime and anywhere. Consequently, a solution called remote consultation and diagnosis

G. De Pietro et al. (Eds.): KES-IIMSS-18 2018, SIST 98, pp. 230–237, 2019.
https://doi.org/10.1007/978-3-319-92231-7_24

platform appeared [1,21,25]. In details, various form of images are shared, at the same time some easy collaborations are supported on-line, as if participants were in the same meeting room. This strategy is of great significance. For instance, in undeveloped area, it balances distribution of poor medical resources [6,26]. Besides, when extremely dangerous SARS outbroke, Chinese hospitals prevented infections in treatment depending on similar platforms [12,14].

For image sharing, many platforms adopt remote rendering that server is responsible for rendering and delivering screenshot to clients [11]. Pressure is taken over by server in order to alleviate terminals. In contrast, network has to guarantee enough space for transmission, especially when high quality of images or high concurrency exists [19,24]. Although kinds of platforms are reported, in fact, their popularity hardly becomes what we anticipated.

2 Related Work

In communication area, some platforms complied a DICOM files network protocol WADO [9,10,13,18,20,23]. WADO brings a strong compatibility but pays little attention on how to reduce streaming flow. According to condition of Internet, [2,3,8] sacrifices image resolution to win the cost of transmission. But sometimes lost quality of images is unaccepted by radiologists. Recently, on-demand methods are reported frequently. [2,7,20,22] introduced an algorithm of tile-based and on-demand, which divides a single DICOM into many blocks. In fact, it always benefits high resolution image but not very much in CT or MR image. Other papers took difference transmission strategy [15,20]. Although differences between neighborhoods in series are considerable, the method still has improved potentials in future.

For remote 3D volume rendering in collaboration, it is found that previous difference method cannot get agreement because of low similarity between adjacent RGB slices. Seriously, triple volume of DICOM file increases burden of Internet. In investigation, current platforms never afford it at all. Consequently, a new solution is urgent to give up screenshots.

In this paper, we proposed a novel compression method bit difference compression. Compared with other algorithms, it abandons two integers for record and count in run-length coding. A new defined short int is chosen as replacement. The smaller data structure denotes an significant compression and great potential applied in remote consultation and diagnosis platform. For remote 3D image collaboration, update of cameras status help us to synchronize screens and interactions. Instead of large frames, only focal, position and direction three vectors are required for broadcast in group.

3 Methods

3.1 Bit Difference Compression Transmission

We have discussed traditional difference compression method. It uses two integers to record repeated intensity sequentially. Although there are lots of long section with same value after a subtracted action, they are discrete and compressed ratio is around 70%. A reasonable explanation argues that two integers waste excessive memory, when they represent limited length of section especially an independent unit. If that, most bits are zero and should be simplified.

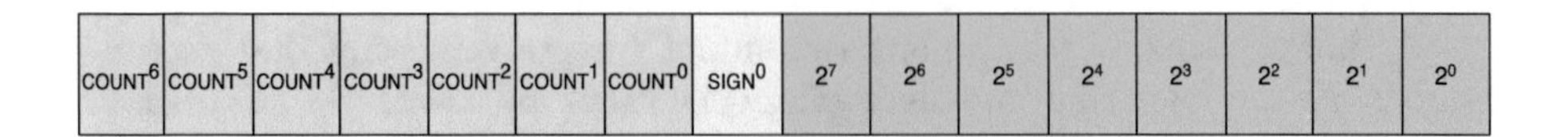

Fig. 1. Our proposed data structure applied in bit difference compression

Accordingly, we present an advanced edition of a new short int. In our scheme, we gave meanings to every bit together sixteen, seen in Fig. 1. There are three components in new protocol. The first section eight bits is used to record subtracted value. In addition, we provides intermediate bit to designate difference positive or negative. At last, 7 bits of count section tell us the length. Numerical extent is between 0 and 127. Compared with traditional algorithm, bit difference compression cut down nearly a half. Although new data structure could not manifest more than 127 units, it has a superior performance for slices transmission. Since seldom sections exist with a overlong distance.

3.2 Synchronization of Cameras for 3D Images Collaboration

With development of computer graphics, 3D reconstruction of medical image is emphasized in diagnosis, which could reflect true condition of patient. Experts wish platform is able to produce volumes that can be discussed and interacted with. Previous applications have a barrier that when image is interacted and even changes a little, neighborhood slices hold a low similarity and any compression method makes little effort. This problem becomes a disaster, because size of RGB screenshots is overlarge and infeasible in delivery.

We plan to get rid of screenshots and avoid from being trapped into traffic jam of Internet. Another optimization is presented and focuses on camera status sharing. As we known, in scenario camera like our eyes captures content in shot. If platform makes sure coincidence of all clients' camera status, our screens and interactions will be synchronized in real-time. There are two prerequisites: terminals support local rendering; part or full slices have located in clients. In our work, we chose VTK 7.0 framework as rendering engine that can support various operating systems. Besides, in habit of diagnosis, clinicians prefer to review 2D slices at first and then seek deep knowledge. Therefore, we would like to start asynchronous threads to download data in background. With bit difference compression, this process could be accelerated more quickly. Even though datasets is not complete, remaining data has possibility to give an acceptable result for reviewing. Process of this algorithm is revealed in Algorithm 1.

In details, three vectors to describe camera status are focal, position and direction. Each of them is actually three doubles. Even though a high fps there is

Algorithm 1. Advanced collaboration for 3D volume rendering

1: **if** 2D mode collaboration selected **then**
2: **Listening:** privileged controller selects a slice for reviewing;
3: **if** ID exists **then**
4: shown through slice dictionary;
5: **else**
6: download with bit difference compression transmission and insert it into slice dictionary;
7: **end if**
8: **end listening**
9: **end if**
10: **if** 3D mode collaboration selected **then**
11: **if** Direct volume rendering mode selected **then**
12: server renders volume locally and produce screenshots;
13: in three channels respectively, apply bit compression transmission;
14: every 30 s server will resend original RGB frame;
15: **end if**
16: **if** fast synchronization mode selected **then**
17: **if** data integrity==false **then**
18: **repeat**
19: download remaining slices in asynchronous thread with bit compressed difference;
20: insert into slice dictionary;
21: **until** all slices have existed in memory;
22: **end if**
23: ensure clients' cameras with same position, focalPoint and direction;
24: **server Listening:** privileged controller's interactive instructions of camera;
25: coming events in two seconds are packaged and pushed into one eventsStack.;
26: according to mobile hardware condition, selectively ignores intermediate events according to $omitInterval = eventNumbers/ECQ$;
27: sever broadcasts preserved events to update status of clients' camera;
28: **end listening**
29: **end if**
30: **end if**

little data necessary in network rather than screenshots. Therefore, this method has great predominance in transmission for 3D collaboration.

4 Results and Discussion

4.1 Data Description

We got 10 groups of adult abdominal CT images for evaluation. Each set was manufactured by SOMATOM Definition AS+, SIEMENS with 512×512 resolution. Pixel spacing of slice is 0.390625 mm and thickness 0.6 mm.

4.2 Quantitative Evaluation of Bit Difference Compression Transmission When Slices Review

In this section, our proposed bit difference compression is evaluated with quantity of necessary transmission. And we joined traditional difference compression, H.264, blocks idea [15,20] in comparison. H.264 is a popular block-oriented and motion-compensation-based video compression method [16,17]. The evaluation standard has close relations with network occupation in a mobile client. Table 1 shows different algorithms' performances in slice transmission.

Table 1. Evaluation of transmission methods in slices delivery

Performances	Traditional difference	8 × 8 blocks	64 × 64 blocks	H.264	Our proposed
No.1(kb)	258	258	258	258	258
No.2(kb)	153.2	240	233	134.2(32.7)	78.8
No.3(kb)	212.8	244	238.1	195.2(44.1)	106.5
No.4(kb)	228	244	234.2	203.2(46.6)	121.2
No.5(kb)	216.2	240	233.9	201.2(43.7)	111.7
No.6(kb)	205	240	231.3	193.6(40.1)	103.2
No.7(kb)	211.2	240	226.3	172.5(36.1)	99.9

In experiments, we select a slice's size is about 527 KB, which is too large for delivery of massive ones. This is because high-precision sampling results in two bytes between −32768 and 32767. Actually, overlarge intensity is helpless for display in mobile. We could minimize to one byte instead. At last, a slice was reduced to 258 KB. In Table 1, serial number refers a interactive action during test. Quantity of data is measured with KB. In statistics, necessary transmission data of our proposed method is nearly half of traditional difference, 0% of blocks and 55% of H.264. In investigation, adjacent difference almost locates in most blocks, which leads to not desirable results. H.264 serves a better result than traditional difference as excellent forward and backward deduction. Due to short int, less competition would be found than bit compression transmission. Bracketed value is the data by a prediction frame providing reference for clients. Unfortunately, little motion coherence between medical frames causes lower ability of prediction.

4.3 Quantitative Evaluation of Remote Collaboration in 3D Volume Rendering

To avoid massive screenshot in Internet, according to Algorithm 1 we implemented the process of camera status sharing, seen in Table 2. Collected data represents different methods' output and fps set 10 frames per second. Obviously, our method has great predominance in collaboration of 3D images. In details, only three doubles need to be delivered in per frame.

Table 2. Quantitative evaluation of remote collaboration in 3D volume rendering

Performances	Traditional difference	Bit compression	64 × 64 blocks	H.264	Our proposed
No.1(kb)	1333	720	1367	892	Less than 0.1
No.2(kb)	1458	729	1475	930	Less than 0.1
No.3(kb)	993	471	1010	762	Less than 0.1
No.4(kb)	1271	689	1411	979	Less than 0.1
No.5(kb)	1017	589	1191	891	Less than 0.1

According to Algorithm 1, there is ECQ to control quality of interactive process. If hardware has limitations, ECQ helps to drop some intermediate frames to compensate user experience. At this experiment, we ignored ECQ impacts and keeps all events invoked by hand gestures. Three doubles denote out success in complicated network.

5 Conclusion

In this paper, we proposed a novel compression method bit difference compression. Compared with common methods with run-length coding, it adopts a short int instead of two integers for record and count. Therefore, the necessary scale of transmission data has been reduced nearly a half. This method cannot be only applied in slices remote sharing, but also for some easy interactive collaboration like marks. However, remote volume rendering is not suitable for this optimization. Too small similarity limits effect of compression. In order to get rid of screenshots, we presented another way synchronization of cameras. If their status are updated to coincidence in real-time, every user would see the same screen and interactions. Our methods improve the potential of remote consultation and diagnosis platform, which has a promising serve for hospitals in future.

Acknowledgments. This research was supported in part by the Grant-in Aid for Scientific Research from the Japanese Ministry for Education, Science, Culture and Sports (MEXT) under the Grant No. 16H01436, in part by the MEXT Support Program for the Strategic Research Foundation at Private Universities (2013–2017).

References

1. Adewale, O.S.: An internet-based telemedicine system in Nigeria. Int. J. Inf. Manage. **24**(3), 221–234 (2004)
2. Arka, I.H., Chellappan, K.: Collaborative compressed i-cloud medical image storage with decompress viewer. Proc. Comput. Sci. **42**, 114–121 (2014)
3. Bernabé, G., García, J.M., González, J.: A lossy 3D wavelet transform for high-quality compression of medical video. J. Syst. Softw. **82**(3), 526–534 (2009)
4. Cooke Jr, R.E., Gaeta, M.G., Kaufman, D.M., Henrici, J.G.: Picture archiving and communication system, US Patent 6,574,629, 3 June 2003

5. Doi, K.: Diagnostic imaging over the last 50 years: research and development in medical imaging science and technology. Phys. Med. Biol. **51**(13), R5 (2006)
6. Dong, L.X.G.C.L., Yunshan, C.: Carrying out extension and popularization of agricultural technology to promote sustainable development of rural economy, based on investigation and analysis of extension and popularization of agricultural technology in Anhui province. Manage. Agric. Sci. Technol. **1**, 024 (2010)
7. Dragan, D., Ivetić, D.: Request redirection paradigm in medical image archive implementation. Comput. Methods Programs Biomed. **107**(2), 111–121 (2012)
8. Erickson, B.J.: Irreversible compression of medical images. J. Digital Imaging **15**(1), 5–14 (2002)
9. He, L., Ming, X., Liu, Q.: A medical application integrating remote 3D visualization tools to access picture archiving and communication system on mobile devices. J. Med. Syst. **38**(4), 44 (2014)
10. He, L., Xu, L., Ming, X., Liu, Q.: A web service system supporting three-dimensional post-processing of medical images based on wado protocol. J. Med. Syst. **39**(2), 6 (2015)
11. Johnson, P.T., Zimmerman, S.L., Heath, D., Eng, J., Horton, K.M., Scott, W.W., Fishman, E.K.: The ipad as a mobile device for CT display and interpretation: diagnostic accuracy for identification of pulmonary embolism. Emerg. Radiol. **19**(4), 323–327 (2012)
12. Junping, Z., Zhenjiang, Z., Huayuang, G., Yi, L., Wanguo, X., Yunqi, C.: E-health in China, our practice and exploration. In: Engineering in Medicine and Biology Society, 2009. EMBC 2009, Annual International Conference of the IEEE, pp. 4888–4893. IEEE (2009)
13. Kammerer, F.J., Hammon, M., Schlechtweg, P.M., Uder, M., Schwab, S.A.: A web based cross-platform application for teleconsultation in radiology. J. Telemed. Telecare **21**, 1357633X15575237 (2015)
14. Mei, Z., Zirong, Y.: Design of epidemic monitoring platform based on ArcGIS. In: 14th International Symposium on Distributed Computing and Applications for Business Engineering and Science (DCABES), 2015, pp. 380–383. IEEE (2015)
15. Mitchell, J.R., Sharma, P., Modi, J., Simpson, M., Thomas, M., Hill, M.D., Goyal, M.: A smartphone client-server teleradiology system for primary diagnosis of acute stroke. J. Med. Internet Res. **13**(2), e31 (2011)
16. Panayides, A., Antoniou, Z.C., Mylonas, Y., Pattichis, M.S., Pitsillides, A., Pattichis, C.S.: High-resolution, low-delay, and error-resilient medical ultrasound video communication using H. 264/AVC over mobile wimax networks. IEEE J. Biomed. Health Inform. **17**(3), 619–628 (2013)
17. Panayides, A., Pattichis, M.S., Pattichis, C.S., Loizou, C.P., Pantziaris, M., Pitsillides, A.: Atherosclerotic plaque ultrasound video encoding, wireless transmission, and quality assessment using H. 264. IEEE Trans. Inf. Technol. Biomed. **15**(3), 387–397 (2011)
18. Qiao, L., Li, Y., Chen, X., Yang, S., Gao, P., Liu, H., Feng, Z., Nian, Y., Qiu, M.: Medical high-resolution image sharing and electronic whiteboard system: a pure-web-based system for accessing and discussing lossless original images in telemedicine. Comput. Methods Programs Biomed. **121**(2), 77–91 (2015)
19. Renambot, L., Rao, A., Singh, R., Jeong, B., Krishnaprasad, N., Vishwanath, V., Chandrasekhar, V., Schwarz, N., Spale, A., Zhang, C., et al.: Sage: the scalable adaptive graphics environment. In: Proceedings of WACE, vol. 9, pp. 2004–2009. Citeseer (2004)

20. Shen, H., Ma, D., Zhao, Y., Sun, H., Sun, S., Ye, R., Huang, L., Lang, B., Sun, Y.: Miaps: a web-based system for remotely accessing and presenting medical images. Comput. Methods Programs Biomed. **113**(1), 266–283 (2014)
21. Tachakra, S., Wang, X., Istepanian, R.S., Song, Y.: Mobile e-health: the unwired evolution of telemedicine. Telemed. J. E-health **9**(3), 247–257 (2003)
22. Tian, Y., Cai, W., Sun, J., Zhang, J.: A novel strategy to access high resolution DICOM medical images based on JPEG2000 interactive protocol. In: Medical Imaging, p. 691912. International Society for Optics and Photonics (2008)
23. Verhoeven, F., van Gemert-Pijnen, L., Dijkstra, K., Nijland, N., Seydel, E., Steehouder, M.: The contribution of teleconsultation and videoconferencing to diabetes care: a systematic literature review. J Med. Internet Res. **9**(5), e37 (2007)
24. Xu, Y., Mao, S.: A survey of mobile cloud computing for rich media applications. IEEE Wireless Commun. **20**(3), 46–53 (2013)
25. Yoo, J.c.: Remote-medical-diagnosis system method, US Patent 8,630,867, 14 January 2014
26. Yu, Z.w., Zhou, G.h., Wang, B.: Analysis on the starting point and path of informatization in the underdeveloped rural areas: a survey of 52 undeveloped rural villages of ningbo. In: Advances in Computer Science and Education Applications, pp. 217–224. Springer (2011)

A Collaborative Telemedicine Platform Focusing on Paranasal Sinus Segmentation

Yinuo Li(✉), Yonghua Li, Zhuofu Deng, and Zhiliang Zhu

College of Software Engineering, Northeastern University, Shenyang, China
840436993@qq.com

Abstract. Telemedicine is an important diagnostic auxiliary tool. This field has recently begun a period of explosive growth. In this paper, we combine mobile devices and image processing algorithms to develop a real-time collaborative image processing telemedicine platform for mobile devices. This C/S mode platform is based on C++, which is mainly implemented by VTK and ITK. In addition to implementing image transmission, 3D visualization and remote rendering, we focus on paranasal sinus CT and adopt automatic medical image segmentation function using the DRLSE algorithm. Besides, collaboration function ensures that users can process images in real time using mobile devices, which benefits communication between medical experts. Through testing, the platform is proved to be able to maintain stable bandwidth demand even in crowded network. According to the current research, this is the first platform to combine paranasal sinus CT image analysis with telemedicine. Therefore, our platform outperforms conventional teleradiology platform in functional completeness. Our platform helps radiologists and medical specialists to make correct diagnoses.

Keywords: Telemedicine · Collaboration · Paranasal sinus CT
Image segmentation

1 Introduction

Telemedicine can be broadly defined as the use of telecommunications technologies to provide medical information and services [1]. The main effect is that telemedicine has the capacity to substantially transform health care in both positive and negative ways and to radically modify personal face-to-face communication [2]. Telemedicine overcomes the limitations of time and space. The combination of telemedicine and digital radiology has been widely used in image transmission, image processing, real-time meeting and other fields.

In the 1990s, teleradiology platforms focused on image transmission between remote areas and authoritative hospitals by PC [3–6]. The platform is a kind of remote meeting tool which is responsible for displaying images asynchronously. In the late 1990s, PACS [12] was popularized as an image archiving and communication system. Nevertheless, these platforms only apply to PC with functions of browsing and simple image processing. Later, advanced compression algorithms are introduced to reduce the size of medical images [7]. Thus, it is possible to use mobile devices to process medical

G. De Pietro et al. (Eds.): KES-IIMSS-18 2018, SIST 98, pp. 238–247, 2019.
https://doi.org/10.1007/978-3-319-92231-7_25

images. There are many teleradiology platforms implemented by web technology or applications in mobile phone or iPad [8–11]. Although the display devices are portable and visualization is reliable, the current platform design is insufficient to simulate the real collaborative face-to-face meeting [13, 14] introduced synchronous telemedicine platforms. They addressed issues such as real-time communication and broadcast, but these platforms only applied to PC. In conclusion, it is a valuable research to combine the convenience of mobile devices and real-time collaboration.

In this paper, we borrowed from work of [27, 28]. They respectively introduced a collaborative telemedicine and a semi-automatic paranasal sinuses CT images segmentation algorithm. We have solved with three challenges in constructing a collaborative telemedicine platform for paranasal sinus CT image analysis in mobile devices. Firstly, in this platform, each member who participates in an online real-time meeting can collaborate on the same image processing by mobile devices. Each member's modification to the image will be broadcasted to other users. Secondly, sinusitis is a common condition. It affects between about 10% and 30% of people each year in the United States and Europe [15]. While the majority of current medical images search focus on brain MR and ventricle MR, we put emphasis on paranasal sinus medical images. Thirdly, we implemented the image segmentation function. We adopted LiChunming's DRLSE algorithm [19] to realize automatic ROI segmentation. DRLSE algorithm takes the whole curve energy instead of several points' energy [18] into account, and it can achieve the desired result after fewer iterations. Besides it adapts topological changes. Therefore, we added DRLSE algorithm as a plug-in to the platform.

2 Materials and Method

2.1 Architecture and Protocol

In this paper, the platform adopts C/S mode. There are two usual modes in software architect: C/S mode and B/S mode. C/S mode means that the whole system is divided into two parts, and tasks are assigned to the client side and server side reasonably. B/S is similar to C/S, but the user interface is implemented through the WWW browser on the client side. Although B/S mode is applicable to various operating systems by browser, C/S mode has strong interactivity, safe access mode, low network traffic, and fast response. Therefore, C/S mode is suitable for processing large amounts of medical image data. The platform architecture is divided into two parts: the mobile client side and the server side.

The whole platform system is low coupled and high scalable. The functions of the platform are divided into modules. The server side is composed of 8 modules. The server side's image display and interaction functions are implemented based on the Visualization Toolkit (VTK) [16] and ITK. Collaboration module makes it possible to guarantee synchronism. To realize synchronization, the server side will broadcast the current screenshot image to all users of the platform in a fixed time interval. Auxiliary tools module contains marking and other basic image processing functions. Rendering computing module is responsible for loading images and rendering images relying on VTK.

VTK not only supports usual medical image processing functions such as visualization and interaction, but also implements diagnostic auxiliary function such as segmentation [22]. Because the platform contains remote rendering and remote image automatic segmentation, the sever side bears almost all the computation work. Visualization module can realize 3D visualization using volume rendering method. Segmentation module will be introduced in Sects. 2.2 and 2.3.

At the mobile client side, we use QT, a cross-platform C++ graphical user interface application development framework, to implement mobile client user interface containing interaction functions like displaying 2D and 3D images, rotating 3D images, marking ROI and etc. The mobile client side contains 6 modules and the function of them are similar to corresponding modules in server side. Nevertheless, these modules only send command and receive results without specific operation.

The communication between the mobile client side and the server side is implemented by the HTTP protocol and the Socket protocol transferring XML file. Communication is mainly responsible for transmitting medical images and maintaining consistency among multiple users. In this platform, image processing is done jointly by the client side and the server side. We set JPEG as the standard format for image data transmission. JPEG typically achieves 10:1 compression with little perceptible loss in image quality [17]. It has been widely used in image and video processing field. We adopt the Socket protocol to transmit images, because it not only reduces the analysis of HTTP request parsing, but also improves the speed of transmission. In addition, the platform also needs to keep the consistency of the information received by each user, which means that each user sees the same change in the mobile device. We use the HTTP protocol to transfer several parameters between client side and server side to ensure consistency (Fig. 1).

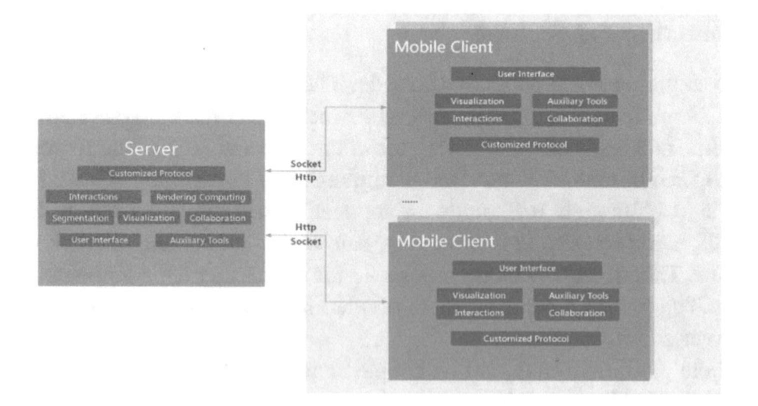

Fig. 1. Platform architect

2.2 Image Segmentation Algorithm

As mentioned above, we focus on paranasal sinus CT segmentation. Nowadays, the segmentation method of CT image has the following four methods: region-based method, threshold-based method, edge-based method and neural network-based method. In this paper, we adopt level-set algorithm, a kind of edge-based method.

Kass et al. first proposed the active contour model [21]. The basic idea is to simplify the image segmentation problem to minimize the energy functional of the closed curve problem. The shape and initial curve of the curve are respectively constrained by internal energy and external energy. The energy functional is as follow:

$$E_{snake}(v) = \alpha \int_{0}^{1} E_{cont}(v(s))ds + \beta \int_{0}^{1} E_{curv}(v(s))ds + \gamma \int_{0}^{1} E_{img}(v(s))ds \quad (1)$$

Sethian and Osher [23] proposed the level-set method on the basis of the active contour model. In this method, the closed curve is represented by implicit method, and the target curve is determined by the evolution of the level set function. Li et al. [19] proposed a new level set method called DRLSE to eliminate the initialization step. In our platform, we adopt Li Chunming's algorithm.

Dynamic contour can be described like $C(s,t)$, and level set function is $\phi(x,y,t)$. The energy formulation with distance regularization is defined as follow:

$$\varepsilon(\phi) = \mu \Re_p(\phi) + \lambda L_g(\phi) + \alpha A_g(\phi) \quad (2)$$

Where L_g is defined as $L_g(\phi) \triangleq \int_{\Omega} g\delta(\phi)|\nabla\phi|d\mathrm{x}$ which means line integral of the function along the $A_g(\phi)$ is defined as $A_g(\phi) \triangleq \int_{\Omega} gH(-\phi)d\mathrm{x}$ which means a weighted area of the region. Edge indicator function g can be presented as $g \triangleq \frac{1}{1+|\nabla G_\sigma * I|^2}$.

The energy is minimized when the zero level contour of ϕ is located at the object. And gradient flow for energy minimization is as follow:

$$\frac{\partial\phi}{\partial t} = -\frac{\partial F}{\partial\phi} \quad (3)$$

The specific DRLSE algorithm process is as followings (Fig. 2):

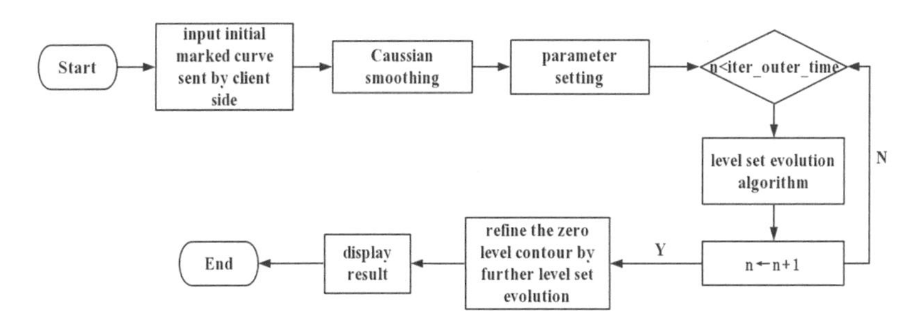

Fig. 2. DRLSE process

The core content of image segmentation is level set evolution algorithm. It can be described as follow:

Algorithm 1. level set evolution algorithm

```
Input: level set function parameters
Output: updated level set function
1 for k = 1 to iter do
2    // Make a function satisfy Neumann boundary condi
        tion, phi represents level set function
3    phi = NeumannBoundCond(phi)
4    calculate the gradient of phi in the horizontal
     and vertical coordinates: phi_x, phi_y
5    calculate divergence: curvature
6    if potentialFunction =='single-well' do
7       // compute distance regularization term in with
        the single-well potential p1.
8       distRegTerm = 4*del2(phi)-curvature
9    else
10      // compute distance regularization term in with
        the double-well potential p1.
11      distRegTerm = distReg_p2(phi)
12    end if
13    // Dirac delta function
14    diracPhi = Dirac(phi,epsilon)
15    calcalate balloon forces of the weighted length
      term and the weighted area term: edgeTerm, ar-
      eaTerm
16    phi = phi + timestep*( dis-
            tance_regularization_weight *distRegTerm +
            length_weight *edgeTerm + area_weight
            *areaTerm)
17 end while
```

2.3 Image Segmentation Interaction

In our collaboration platform, image segmentation is done by multiple client sides and server side. Since the C/S mode is adopted, the image segmentation operation is undertaken by the server side. The client side is responsible for displaying server current screen images and adding the initial region. The image segmentation process is as follows: Firstly, the server side loads a sinus CT image and sends current screenshot to all users. One user can set himself as current controller and clicks on the mobile device screen in the area of interest to generate a rectangular region border. This rectangular is the initial curve. Then this user sends it to the server side. Server side receives the marked image and iterates over the initial region using DRLSE in 215 times and then broadcasts the segmentation results to all users who are participating in the same meeting. The server side integrates image segmentation algorithm is as follow:

```
Algorithm 2. Segmentation interaction algorithm

Input:  CT image with initial marked curve
Output: server sends segmentation result image to cli-
        ents in broadcast
1 server loads sinus CT image data
2 server screenshots 10 frames
3 while server is listening client sides' entries do
4   if client side has been linked with server then
5     client side sends its client name and address
6     server sends current user to this client
7   else
8     server broadcasts 10 fps frames to client
9   end if
10  if one client side sets itself as current control-
       ler then
11    client side sends marked image to server side
12    server runs DRLSE algorithm
13    server broadcasts 10 fps frames to client sides
14   else
15     server broadcasts 10 fps frames to client sides
16   end if
17 end while
```

3 Results and Discussions

3.1 Segmentation Result

Since level-set was proposed, various improvements have been made. At present, level-set has been widely used in many fields such as recognition and image segmentation. DRLSE has been used in medical image segmentation of multiple organs, such as ventricular division, brain tumor analysis and kidney segmentation [24–26]. According to the reference, although the image data formats are different and organs are different, the segmentation result of DRLSE in these areas is reliable. Therefore, we choose to use DRLSE algorithm.

As to other CT image segmentation methods. The threshold-based method only considers the value of the pixel, without regard to the spatial feature of the image. The region-based method is also sensitive to noise in image, and it will cause the hole shapes or incoherent regions. These methods do not apply to paranasal sinus segmentation. However, we believe that neural network is a feasible research direction. It is the next research goal of our further work.

The following Fig. 3(a) shows one user sets himself as controller and draws two rectangular ROI as initial region. Fig. 3(b) shows the processed result image in iPad. After evolution, the curve is close to the target area's boundary. The iterative evolution of 215 times takes 2.7225 s. We tested 150 image data, and average computation time is 2.3732 s.

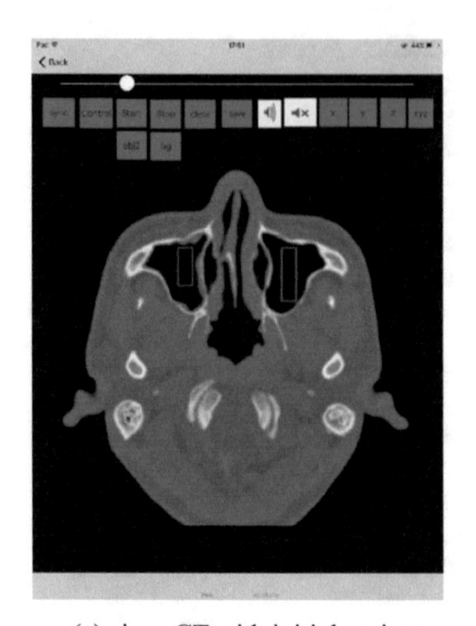

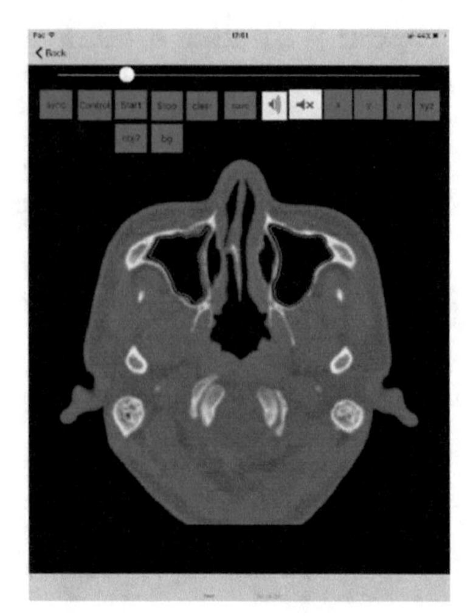

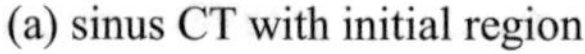

(a) sinus CT with initial region

(b) segmentation result

Fig. 3. Initial region and segmentation result in iPad

3.2 Platform Comprehensive Evaluation

Telemedicine platform has gained a lot of attention recently, and there were plenty of research achievements and practical applications. Each platform has its own advantages and disadvantages. We invited three medical experts to evaluate the characteristics of telemedicine platforms in the reference. In order to clearly demonstrate the differences of platforms, we use a table to sort out their attributes as follow:

Table 1. Comparison of platform characteristics

Reference	Device	Visualization	Collaboration	Segmentation	Plug-in
[2]	PC	√	√	√	×
[4]	PC	√	×	×	×
[5]	PC	√	×	×	×
[6]	PC	√	×	×	×
[10]	iPad	√	×	×	×
[11]	iPad	√	×	×	×
[13]	iPad	√	√	×	×
[14]	iPad	√	√	√	×
Our platform	iPad	√	√	√	√

As shown in Table 1, the existing methods [2–6] are only implemented in a PC, lacking the flexibility of mobile devices. These platforms can not eliminate space constraints. Székely, A. tested 102 applications on smartphone and tablets. The experiment result is that these easy-to-use devices equipped with excellent display may be used for diagnostic reading, reference, learning, consultation, and for communication with patients [20]. Therefore, the use of telemedicine platform in mobile devices is consistent with current trends and meets the needs of users.

Methods [8–11] are implemented in iPad, but collaboration between members cannot be achieved. Although they solve portability problems, they still can't simulate real meetings. Our platform ensures portability and collaboration, while adding segmentation function to improve practicability.

Plug-in in table measures scalability. Since our platform is a strict high cohesion and low coupling structure, segmentation module is an auxiliary function which is integrated to the system as a plug-in.

3.3 Bandwidth Test

Our platform is practical because the platform satisfies the needs of remote medical image analysis. To test the platform performance, we designed an experiment. The server side ran on an iMac with 3.2 GHz Intel Core i5 CPU, 16 GB RAM and a AMD Radeon R9 M390 (2 GB) graphics card, and the client side ran on an iPad Air2 with 1.5 GHz A8x CPU.

We evaluated the bandwidth occupied by the application in the mobile devices under 4G network. The bandwidth which platform needs for normal running maintained between 84.2 Kb/s and 91.4 Kb/s. The experiment result proved that our platform doesn't take up a lot of bandwidth, and general mobile devices can load such bandwidth.

4 Conclusion

In this paper, we introduced a collaborative medical processing telemedicine platform. Most telemedicine system attach importance to remote video conference and 3D image visualization. Although these platforms provide help for remote meetings, they still have no capacity for simulating the implementation of medical image analysis in real meetings. Our platform not only realizes telemedicine collaboration in real time, but also combines advantages of the classic medical multifunctional medical image analysis systems with telemedicine platforms in mobile devices. Besides, we focus on the treatment of patients with sinusitis. In this platform, we use DLRSE algorithm to realize automatic sinus CT ROI segmentation.

Our future work will focus on the reduction of bandwidth and adding more image processing plug-in. In the aspect of image segmentation, there is still an inexact problem of diseased tissue segmentation, and we plan to use the neural network to improve the segmentation veracity. Besides, we plan to solve 3D medical image segmentation in our platform.

References

1. Perednia, D.A., Brown, N.A.: Teledermatology: one application of telemedicine. Bull. Med. Libr. Assoc. **83**(1), 42–47 (1995)
2. Bernardes, P., et al.: KAMEDIN - teleconferencing and automatic image analysis for medical applications. In: Teixeira, J.C., Rix, J. (eds.) Modelling and Graphics in Science and Technology. Beiträge zur Graphischen Datenverarbeitung. Springer, Heidelberg (1996)
3. Moffitt, M.E., Richli, W.R., Carrasco, C.H., et al.: MDA-image: an environment of networked desktop computers for teleradiology/pathology. J. Med. Syst. **15**, 111–115 (1991)
4. Puech, P.A., Boussel, L., Belfkih, S., et al.: DicomWorks software for reviewing DICOM studies and promoting low-cost teleradiology. J. Digit Imaging **20**, 122 (2007)
5. Mohamed, A.S.A.: The use of teleradiology system linking a regional center of radiology to its district hospitals and clinics. In: ELemke, H.U., Inamura, K., Jaffe, C.C., Felix, R. (eds.) Computer Assisted Radiology/Computergestützte Radiologie, pp. 106–111. Springer, Heidelberg (1993)
6. Carson, G.C., Fath, S.J., Van Meter, T.A., et al.: IMAGEnet: a wide area teleradiology network. In: Emergency Radiology. LNCS, vol. 1, pp. 32–3. Springer, Heidelberg (1994)
7. Ivetic, D., Dragan, D.: Medical image on the go! J. Med. Syst. **35**(4), 499–516 (2011)
8. Park, J.B., Choi, H.J., Lee, J.H., Kang, B.S.: An assessment of the iPad 2 as a CT teleradiology tool using brain CT with subtle intracranial hemorrhage under conventional illumination. J. Digit. Imaging **26**(4), 683–690 (2013)
9. Pianykh, O.S.: DICOM and teleradiology digital imaging and communications in medicine (DICOM): a practical introduction and survival guide. J. Digit. Imaging, 281–317 (2012)
10. Zennaro, F., Grosso, D., Fascetta, R., Marini, M., Odoni, L., Di Carlo, V., Lazzerini, M., et al.: Teleradiology for remote consultation using iPad improves the use of health system human resources for paediatric fractures: prospective controlled study in a tertiary care hospital in Italy. BMC Health Serv. Res. **14**(1), 327 (2014)
11. Drnasin, I., Gogic, G., Tonkovic, S.: Interactive teleradiology. In: Dössel, O., Schlegel, W.C. (eds.) World Congress on Medical Physics and Biomedical Engineering, Munich, Germany. IFMBE Proceedings, vol. 25/5, pp. 290–294 . Springer, Heidelberg (2009)
12. Dreyer, K.J., Thrall, J.H., Hirschorn, D.S., Mehta, A.: PACS: a guide to the digital revolution. J. Med. Imaging Radiat. Oncol. **48**, 103 (2004)
13. Gortzis, L.G.: Clinical teleradiology: collaboration over the web during interventional radiology procedures. In: Kumar, S. (eds.) Teleradiology. Springer, Heidelberg (2008)
14. Schmid, J., Nijdam, N., Han, S., Kim, J., Magnenat-Thalmann, N.: Interactive segmentation of volumetric medical images for collaborative telemedicine. In: Magnenat-Thalmann, N. (ed.) 3DPH 2009. LNCS, vol. 5903, pp. 13–24. Springer, Heidelberg (2009)
15. Rosenfeld, R.M., Piccirillo, J.F., Chandrasekhar, S.S., Brook, I., Ashok Kumar, K., Kramper, M., Orlandi, R.R., Palmer, J.N., Patel, Z.M., Peters, A., Walsh, S.A., Corrigan, M. D.: Clinical practice guideline (update): adult sinusitis executive summary. Otolaryngol. Head Neck Surg. Off. J. Am. Acad. Otolaryngol. Head Neck Surg. **152**(4), 598–609 (2005)
16. Schroeder, W.J., Avila, L.S., Hoffman, W.: Visualizing with VTK: a tutorial. IEEE Comput. Graphics Appl. **20**(5), 20–27 (2000)
17. Haines, R.F., Chuang, S.L.: The effects of video compression on acceptability of images for monitoring life sciences experiments. In: EEE Computer Society Data Compression Conference. NASA-TP-3239, A-92040, NAS 1.60:3239 Snowbird, UT, United States (1992)
18. Schmalstieg, D.: The remote rendering pipeline: managing geometry and bandwidth in distributed virtual environments. Ph.D. Vienna University of Technology, Vienna (1997)

19. Li, C., Cu, C., Gui, C., et al.: Level set evolution without reinitialization: a new variational formulation. In: Processings of IEEE Computer Society Conference on Computer Vision and Pattern Recognition, pp. 430–436, San Deigo, USA (2005)
20. Székely, A., Talanow, R., Bágyi, P.: Smartphones, tablets and mobile applications for radiology. Eur. J. Radiol. **82**(5), 829–836 (2013)
21. Kass, M., Witkin, A., Terzopoulos, D.: Snakes: active contour models. Int. J. Comput. Vis. **1**, 1:321–1:331 (1988)
22. Wang, Y., Hong, F., Wu, E.: The implementation and design of medical image process sub-system based on VTK library. Comput. Eng. Appl. **2003**(08), 205–207 (2003)
23. Osher, S., Sethian, J.: Fronts propagating with curvature-dependent speed: algorithms based on Hamilton-Jacobi formulations. J. Comput. Phys. **79**(1), 12–49 (1988)
24. Liu, Y., Li, C., Guo, S., Song, Y., Zhao, Y.: A novel level set method for segmentation of left and right ventricles from cardiac MR images. In: 2014 36th Annual International Conference of the IEEE Engineering in Medicine and Biology Society, Chicago, IL, pp. 4719–4722 (2014)
25. Zabir, I., Paul, S., Rayhan, M.A., Sarker, T., Fattah, S.A., Shahnaz, C.: Automatic brain tumor detection and segmentation from multi-modal MRI images based on region growing and level set evolution. In: 2015 IEEE International WIE Conference on Electrical and Computer Engineering (WIECON-ECE), pp. 503–506, Dhaka, Bangladesh (2015)
26. Liang, R., Chen, X.J., Zhang, J.X.: Auto-Segmentation of lung in CT image series based on level set method with prior knowledge. In: 2017 3rd IEEE International Conference on Control Science and Systems Engineering (ICCSSE), Beijing, pp. 578–582 (2017)
27. Deng, Z., Chen, Y., Zhu, Z., Wang, Y., Wang, Y., Xu, M.: A collaborative and mobile platform for medical image analysis: a preliminary study. In: Chen, Y.W., Tanaka, S., Howlett, R., Jain, L. (eds.) Innovation in Medicine and Healthcare 2017. InMed 2017. Smart Innovation, Systems and Technologies, vol. 71, pp. 130–139. Springer, Cham (2018)
28. Deng, Z., et al.: Semi-automatic segmentation of paranasal sinuses from CT images using active contour with group similarity constraints. In: Chen, Y.W., Tanaka, S., Howlett, R., Jain, L. (eds.) Innovation in Medicine and Healthcare 2017. InMed 2017. Smart Innovation, Systems and Technologies, vol. 71, pp. 89–98. Springer, Cham (2018)

A Plug-In for Automating the Finite Element Modeling of Flatfoot

Zhongkui Wang[1(✉)], Shouta Yamae[1], Masamitsu Kido[2], Kan Imai[2], Kazuya Ikoma[2], and Shinichi Hirai[1]

[1] Department of Robotics, Ritsumeikan University, Shiga 525-8577, Japan
wangzk@fc.ritsumei.ac.jp, rr0037rp@ed.ritsumei.ac.jp, hirai@se.ritsumei.ac.jp

[2] Department of Orthopaedics, Kyoto Prefectural University of Medicine, Kyoto 602-8566, Japan
{masamits,kan-imai,kazuya}@koto.kpu-m.ac.jp
http://www.ritsumei.ac.jp/~hirai/index-e.html

Abstract. To automate the process of flatfoot finite element (FE) modeling, a software plug-in was developed and introduced in this paper. The plug-in was written in Python and based on the Abaqus Scripting Interface. It consists of three modules: script data, GUI (graphic user interface), and script command. The plug-in is integrated into Abaqus/CAE and can be easily adopted to reduce modeling time and efforts. The detailed procedures regarding FE modeling were automated by the proposed plug-in, and the users only have to determine and pick the corresponding nodes to represent the origin and insertion of ligaments, plantar fascias, and other small tissues of interests. By applying the proposed plug-in, the complicated modeling procedure can be simplified and sped up, and the users' workload can be dramatically alleviated.

Keywords: Automation · Plug-in · Flatfoot modeling · Finite element

1 Introduction

Despite being a common foot deformity, the flatfoot pathology is not fully understood and surgical decisions mainly rely on the foot surgeon's experiences. Cadaveric studies are limited by the lack of flatfoot donors, and research findings from a single cadaveric study are difficult to apply in a wide population due to the individual differences among flatfoot patients [1,2]. Computation model is another option for studying flatfoot pathology and tailoring patient-specific surgery. Spratley *et al.* proposed a 3D rigid-body model for simulating flatfoot deformity and MCO (medializing calcaneal osteotomy) surgery [3–5]. A finite element (FE) model was proposed by Lewis and was used to simulate different surgical corrections, such as talo-navicular arthrodesis, subtalar arthrodesis, and MCO [6]. The FE method has been frequently used to model human organs and tissues because of its continuum mechanics characteristics and its capability of

G. De Pietro et al. (Eds.): KES-IIMSS-18 2018, SIST 98, pp. 248–259, 2019.
https://doi.org/10.1007/978-3-319-92231-7_26

modeling irregular geometries and complex material properties. FE models of healthy foot and other foot deformities were also proposed for various purposes, such as for improving footwear design [7], studying the clawed hallux deformity [8], and developing ankle prostheses [9].

In our previous work, we developed FE flatfoot models using our own FE code [10] and commercial FE software [11]. The models showed potentials for reproducing flatfoot biomechanics and evaluating surgical procedures. However, the modeling process using FE software, such as Abaqus (Dassault System, MA), is relatively complicated and time consuming, and requires experiences with FE method. Because of the large daily workload of foot surgeons, it is hard to frequently implement the modeling process by foot surgeons for individual patient. Therefore, in this paper, we proposed a software plug-in to automate the modeling process and save valuable time of foot surgeons.

The paper is organized as follows. The methods involved with the development of the plug-in were presented in Sect. 2, followed by the results and

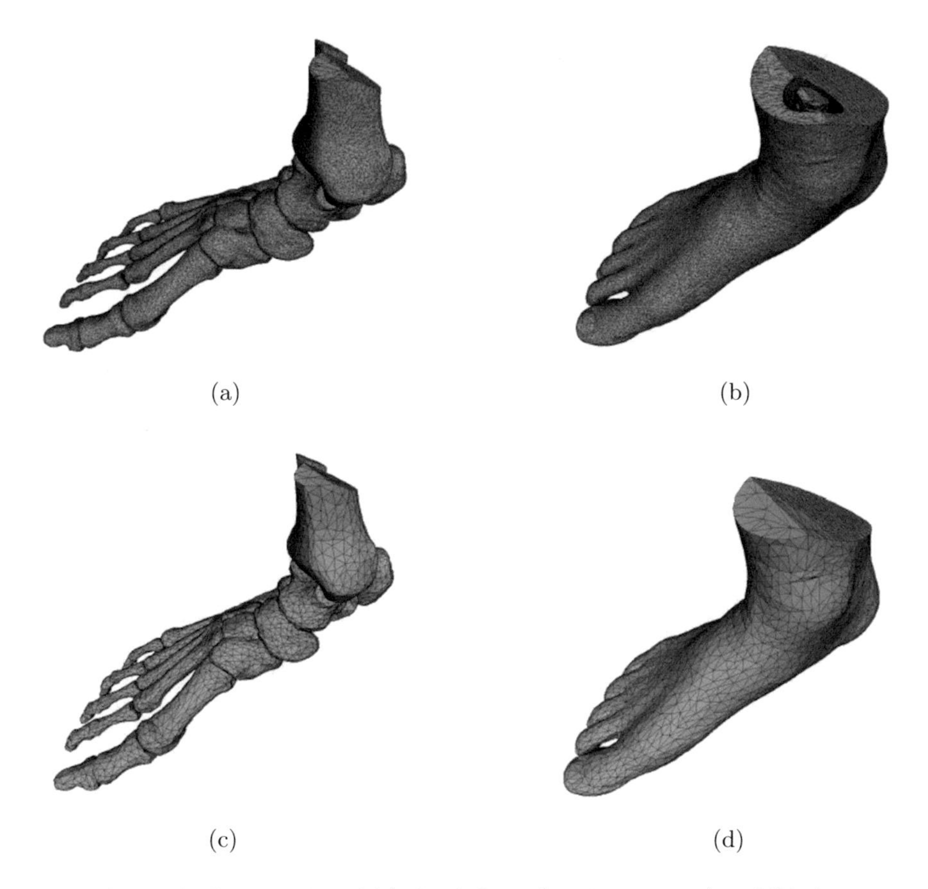

Fig. 1. The original geometries of (a) the skeleton (vertices: 55,551) and (b) the encapsulated soft tissue (vertices: 167,490), and the edited geometries of (c) the skeleton (vertices: 5,738) and (d) the soft tissue (vertices: 1,955).

performances in Sect. 3. The paper was concluded in Sect. 4 with suggestions of future work.

2 Methods

2.1 Preparation of Geometrical Data

First, CT images were taken for the flatfoot patient to examine the deformity. The boundaries of the skeleton and encapsulated tissue were then segmented from CT images using MIMICS (Materialise Inc., Leuven, Belgium) by foot surgeons. The segmented 3D geometries of the skeleton and encapsulated soft tissue were shown in Fig. 1a and b as examples. These original geometries are usually high-resolution and include internal structures, such as bone marrows, which are not suitable for FE modeling and simulation. Therefore, these original geometries were imported into MeshLab, an open source mesh editor [12], to reduce the node resolution and remove the internal structures. The final surface meshes of the skeleton and soft tissue were shown in Figs. 1c and d. The edited surface meshes are usually in a file format of 'STL' (Standard Triangulated Language), which unfortunately cannot be directly used in FE software Abaqus. To this end, the 'STL' files were converted into 'SAT' (Standard ACIS Text) files using a MATLAB file exchange function [13].

2.2 Modeling and Simulation of Flatfoot

The 'SAT' files of the skeleton and soft tissue were imported into FE software Abaqus for modeling. The small tissues, such as ligaments and plantar aponeuroses were not recognizable from CT images. Therefore, these small tissues were

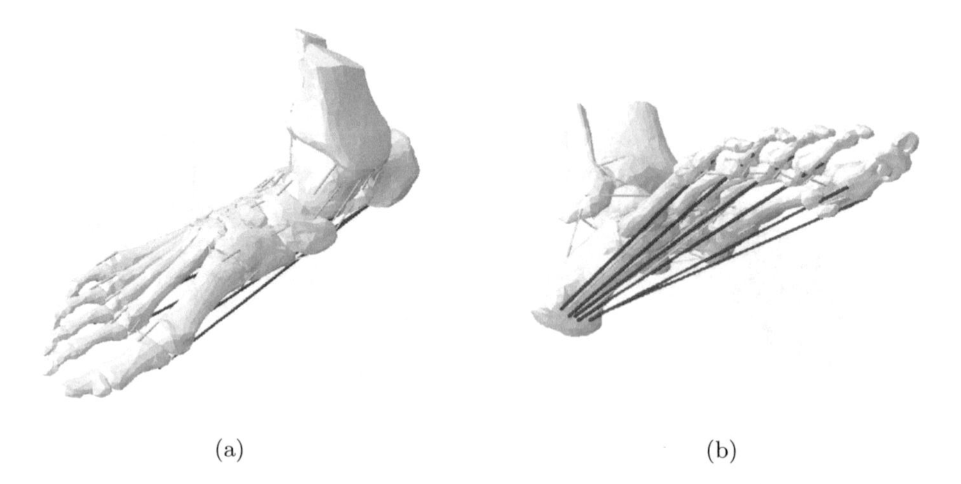

Fig. 2. The FE model of the skeleton (light gray), ligaments (yellow), and plantar aponeuroses (red) in (a) a dorsal view and (b) a plantar view. Soft tissue was hidden to show the ligaments and aponeuroses.

manually created by referring an anatomy book [14] and following the advice of foot surgeons. These tissues were modeled as wire parts and meshed with tension-only linear truss elements in Abaqus. The anatomical origin and insertion of the ligaments and aponeuroses were approximately located at the corresponding nodes on the skeleton surfaces (Fig. 2).

After having all the geometries of required skeleton and soft tissues, the following steps were performed in Abaqus before executing a simulation: (1) defining material properties, (2) defining element sections, (3) defining simulation steps, (4) creating instances and assembling instances, (5) setting sections and creating meshes for all parts, (6) defining contact properties, (7) creating surfaces and sets for contacts, constraints, and boundary conditions, (8) creating contacts, (9) creating constraints, (10) defining external loads, (11) defining boundary conditions, (12) creating a job for simulation. By submitting the job to Abaqus computation core, a simulation could be finally executed. For detailed information regarding material properties and contact conditions, one can refer to our previous work [11]. Examples of simulated stress distributions were shown in Fig. 3.

2.3 Automation Plug-In

To automate the modeling process, we employed the Abaqus Scripting Interface, which is an application programming interface (API), to access the functionality of Abaqus/CAE from Python scripts. We implemented the modeling automation functions as an Abaqus plug-in named 'Flatfoot' which can be easily accessed in the drop-down menu under 'Plug-ins' menu in Abaqus/CAE. The created plug-in consists of three separated modules located inside three subdirectories as shown in Fig. 4a.

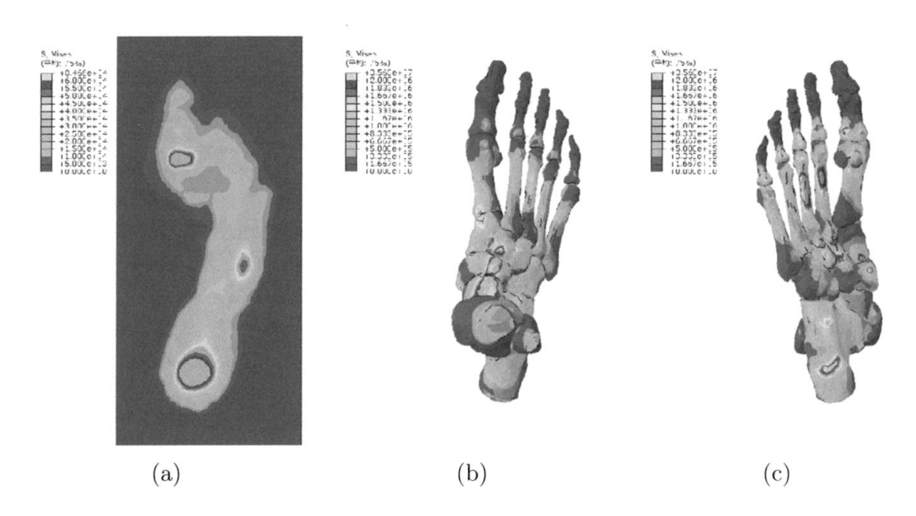

Fig. 3. The FE simulation results of stress distribution on (a) the ground support, (b) the skeleton in a dorsal view, and (c) the skeleton in a plantar view.

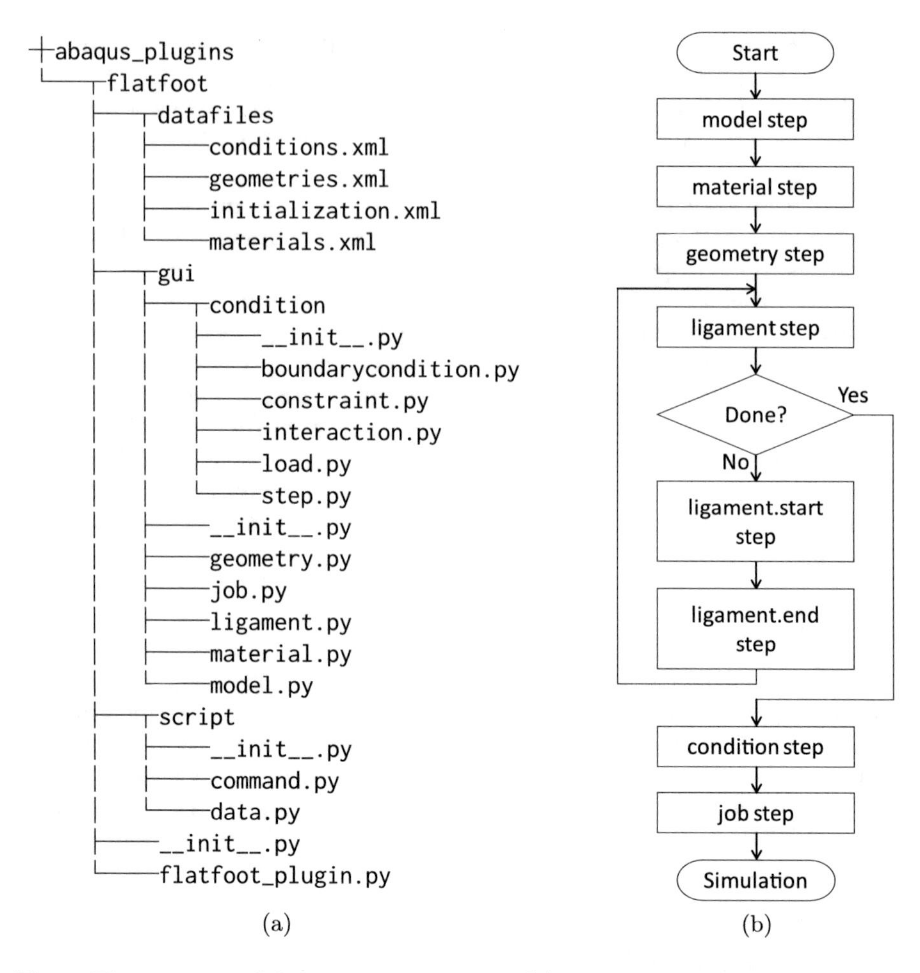

Fig. 4. The structure of the automation plug-in: (a) the directory tree consisting of all program files and (b) the order of the automation steps.

Script Data Module. The script data module consists of four programs located in the 'datafiles' subdirectory and they were executed to write data to the model database when the plug-in was invoked. The data included default settings for the FE model, such as the default condition settings in file 'conditions.xml', default geometrical files in 'geometries.xml', system units in file 'initialization.xml', and default material properties in file 'materials.xml', respectively.

GUI Module. The GUI module consists of all the programs defining the dialog boxes which were used to input data and output descriptions and information. The dialogs were programmed as Python classes and the contents of main input and output dialogs were summarized in Table 1.

Table 1. Input and output contents of main dialog boxes

Class name	Input content	Output content
ModelDialog	—	Model description
MaterialDialog	Property data of each material	—
GeometryDialog	Full path of geometry data (SAT files) and dimension scale of the geometry data	Information of tissue and ground generation, and meshing
LigamentDialog	Definition of ligaments and aponeuroses, selection of origin and insertion of ligaments and aponeuroses	—
ConditionDialog	Condition definitions, such as steps, interactions, constraints, etc.	Description of conditions
JobDialog	Desired CPU number	Job description

Table 2. Main functions and their descriptions

Function name	Operation
createBasicModel()	Model creation
createMaterialAndSection()	Creation of material and element properties
createBones()	To input bone geometries, create instances, sections, and meshes
createSkin()	To input skin geometry and create instance
createTissue()	To remove bones from skin and create soft tissue
createGround()	Ground creation
switchDisplayToLigament()	To display skeleton for creation of ligaments and aponeuroses
createLigament()	Creation of ligaments and aponeuroses
createStep()	Creation of simulation steps
createBonesContact()	Creation of interactions between neighboring bones
createGroundTissueContact()	Creation of interaction between tissue and ground
createConstraints()	Creation of constraints
createLoads()	Creation of external loads
createBoundaryConditions()	Creation of boundary conditions
submitBasicJob()	To define and submit a simulation job

Script Command Module. The script command module consists of programs defining all functions used to create the FE model based on the input information from user. Main functions were listed in Table 2.

When the plug-in was started, all modeling steps appeared one by one, as shown in Fig. 4b, and requested the user to input necessary information and guided the user through the modeling process. The following section graphically presented the performances of the developed plug-in for flatfoot modeling.

3 Results

By clicking the drop-down menu of 'Flatfoot' (Fig. 5a), the automation plug-in was started. The material dialog box appeared and requested the user to input

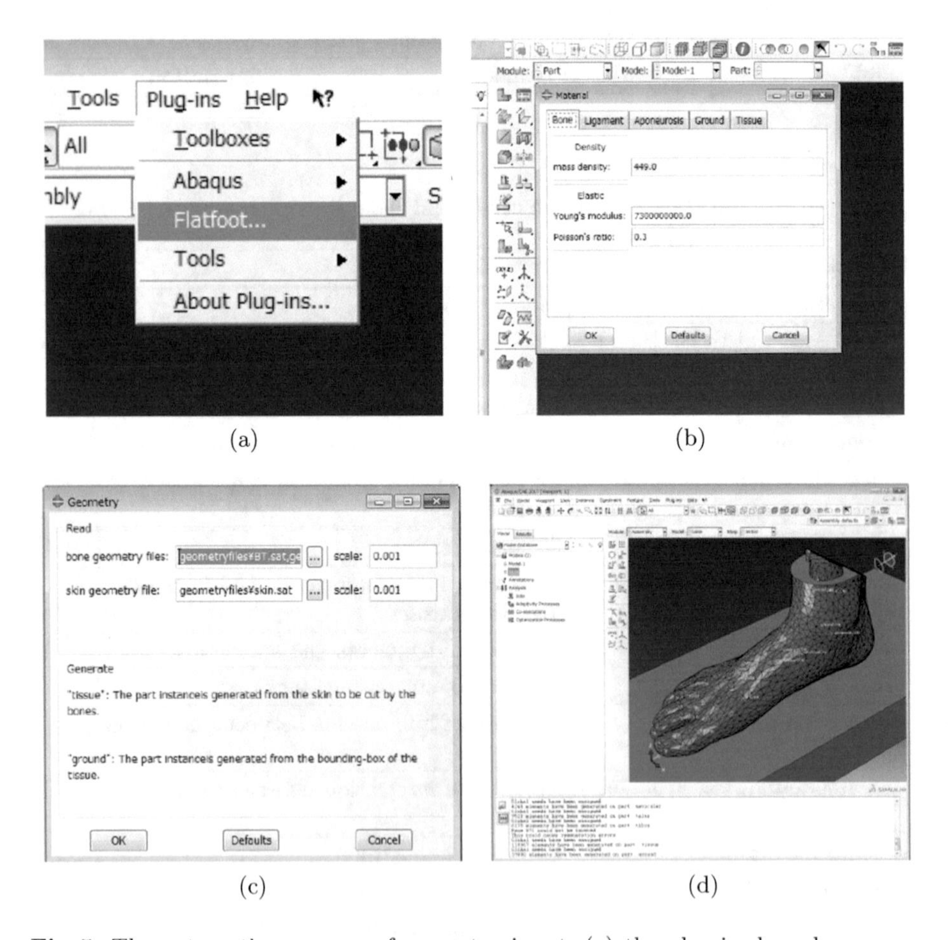

Fig. 5. The automation process of geometry input: (a) the plug-in drop-down menu 'Flatfoot', (b) the material dialog box, (c) the geometry dialog box, and (d) the input assembly of the flatfoot FE model.

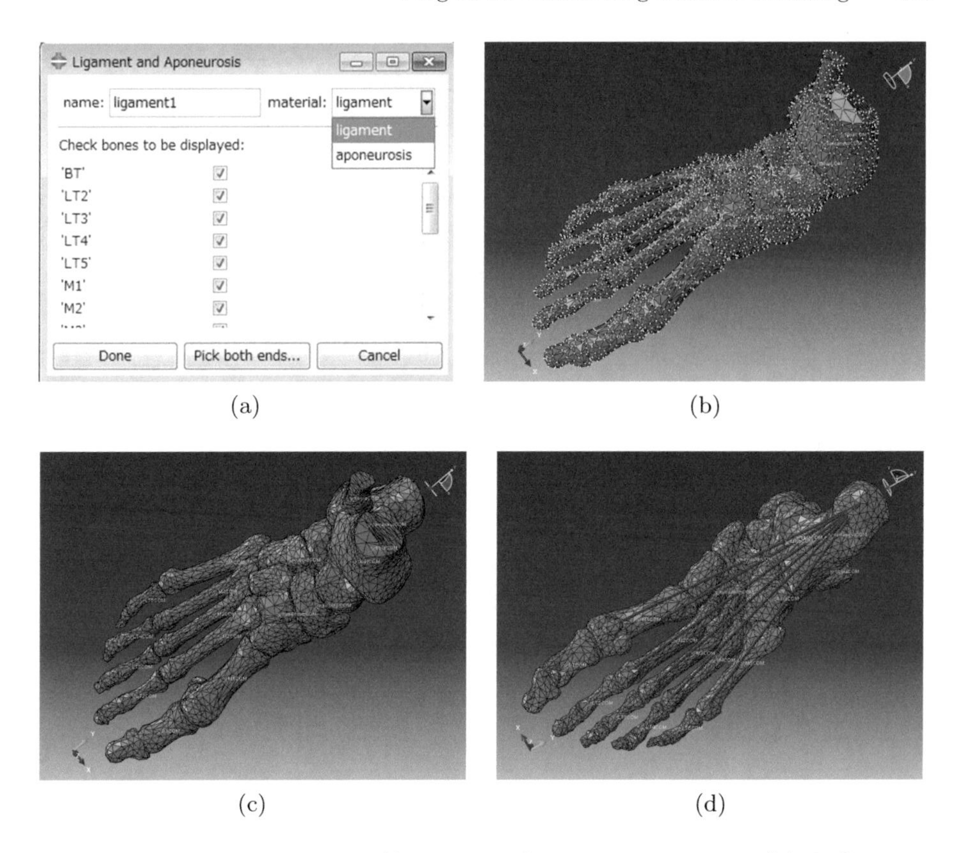

(a) (b) (c) (d)

Fig. 6. The automation process of ligament and aponeurosis creation: (a) the ligament dialog box, (b) the highlighted nodes on the skeleton for picking, (c) examples of some created ligaments, and (d) examples of some created aponeuroses.

the material properties of the involved materials in the FE model. Default values were predefined and the user can change them to the desired values. After setting the material properties, the geometry dialog appeared and asked the user to load the geometry data of the skeleton and skin, which should be prepared as validated 'SAT' files. The default system units were 'mm', 'MPa', 'N', 's', and 'tonne' for length, stress, force, time, and mass, respectively. Therefore, if the geometries were prepared in a length unit of 'm', the scale should be set to 0.001 to convert the unit to 'mm'. Once 'OK' was clicked, the geometries were imported into the FE model and the soft tissue was automatically generated by removing the skeleton out of the skin. In addition, a ground part was automatically generated beneath the flatfoot based on the position and size of the bounding-box of the flatfoot assembly. Subsequently, all imported parts were automatically meshed with 4-node linear tetrahedron elements. The generated flatfoot assembly was shown in Fig. 5d as an example.

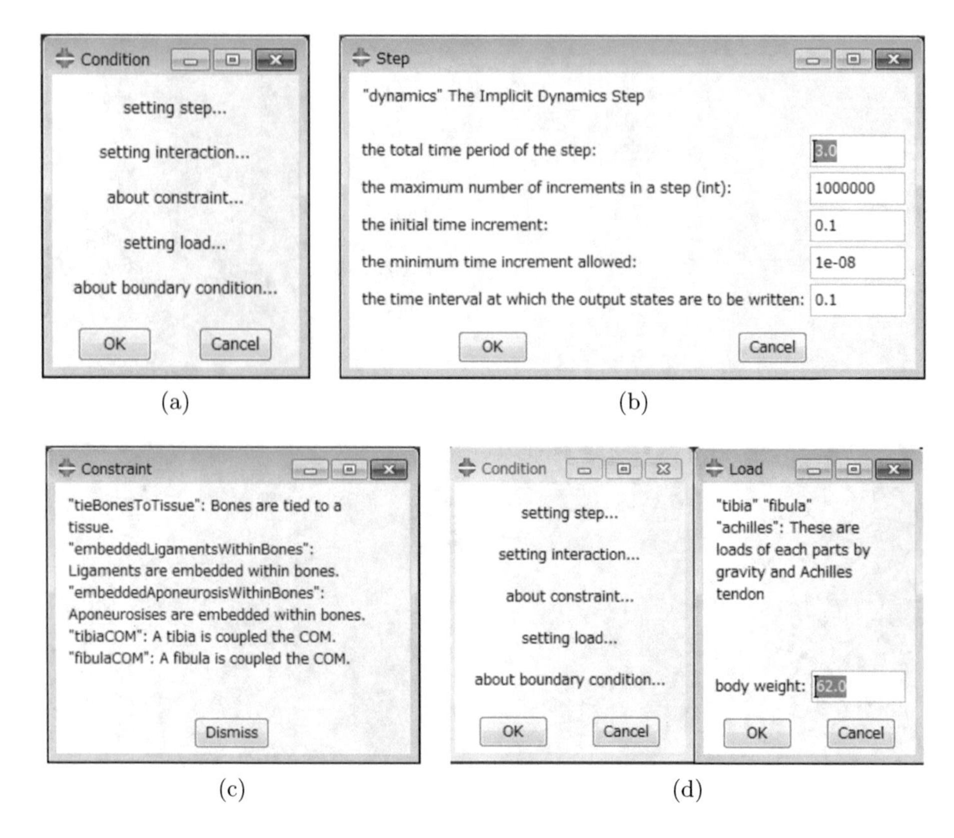

Fig. 7. The automation process of simulation condition settings: (a) the condition dialog box, (b) the simulation step settings, (c) the constraint descriptions, and (d) the external load setting.

Next, the ligament dialog appeared and assisted the user to create ligaments and aponeuroses. As shown in Fig. 6a, user could choose to create either ligament or aponeurosis and then gave it a proper name. For the convenience of picking desired nodes, the display of individual bones can be turned on or off by checking the check-box next to the corresponding bone instance. By clicking 'Pick both ends...' button, the model view was immediately switched to the skeleton assembly and the nodes for picking were highlighted, as shown in Fig. 6b. The user could then pick the desired nodes to represent the origin and insertion of the corresponding ligament or aponeurosis. This procedure can be repeated as many times as possible until all necessary ligaments and aponeuroses were created. Examples of some created ligaments and aponeuroses were shown in Fig. 6c and d, respectively.

After completing the creation of all ligaments and aponeuroses, the condition dialog box appeared and allowed the user to input and confirm simulation conditions as shown in Fig. 7a. For example, by clicking the 'setting step...' button, the step setting menu appeared (Fig. 7b) and the user could set the total

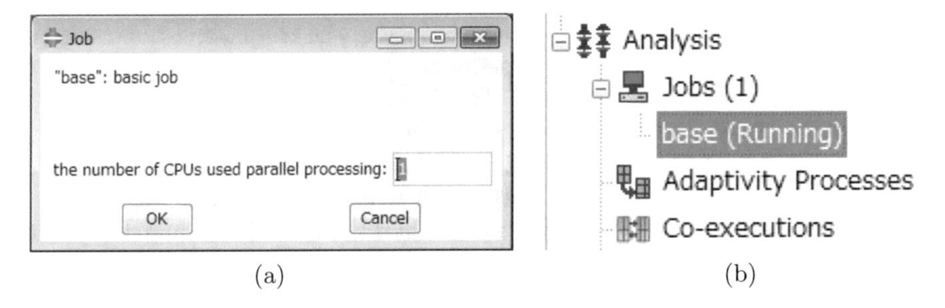

(a) (b)

Fig. 8. The creation of simulation job: (a) the job dialog box and (b) the confirmation of job running.

simulation time, the maximal number of increments, the initial time increment, and so on. The constraint information was shown in Fig. 7c which allowed the user to confirm the predefined constraints. These constraints were automatically created by the plug-in and did not require any user assists. The external loads were defined in the load menu as shown in Fig. 7d. In the case of simulating balanced standing, the only load was the body weight of the patient. The loads on tibia, fibula, and Achilles tendon were then automatically calculated based on the half body weight. Detailed calculations can be found in [15].

After completing the condition settings, the job dialog box appeared and requested the user to determine the CPU number for the current FE simulation as shown in Fig. 8a. Finally, simulation was started once the 'OK' button on the job dialog was clicked. At the bottom-left of the Abaqus/CAE model tree, the user could confirm that the job (named 'base') was successfully submitted and the simulation was running as shown in Fig. 8b. After the simulation was completed, the user could open the output database file ('.odb' file) in the working directory to examine the simulation results, such as the stress distributions shown in Fig. 3.

4 Conclusions

FE simulation provides a possible way to study flatfoot pathology and helps foot surgeons to tailor patient specific surgeries. However, the complicated modeling procedures and the requirement of FE experiences limited the wide applications of such simulations. This paper presented a software plug-in to automate the FE modeling process of flatfoot. Using the plug-in, the foot surgeons only have to determine and manually select the origin and insertion of ligaments and aponeuroses, which are of their interests, and the plug-in takes care of all the other settings regarding of FE simulation. The plug-in was based on Abaqus Scripting Interface and written in Python language. It was integrated in the Abaqus/CAE under the 'Plug-ins' menu. This plug-in can dramatically reduce the model development time and alleviates the workload of foot surgeon. Taking the flatfoot model shown in this paper as an example, it normally took at least one day (8 h) to complete the FE model following the traditional modeling procedure in Abaqus/CAE by an FE-experienced user. However, by introducing the developed plug-in, the modeling time was reduced to within 30 min even for an FE-inexperienced user. The developed plug-in is not limited to the automation of flatfoot modeling. It can be modified and applied to automatically model other foot deformities or other human body parts, such as arm and hand, as long as the geometrical data were properly prepared and the material properties, loading conditions, and boundary conditions were well determined.

Future work will extend the plug-in to automatically model and simulate flatfoot surgeries, such as the MCO and the lateral column lengthening (LCL), which require bone osteotomy, fragment shifting, and implant insertion.

Acknowledgments. This work was supported by the MEXT-Supported Program for the Strategic Research Foundation at Private Universities (2013–2017), and in part by JSPS KAKENHI Grant Numbers JP15H02230, and JP17K15806.

References

1. Niu, W., Yang, Y., Fan, Y., Ding, Z., Yu, G.: Experimental modeling and biomechanical measurement of flatfoot deformity. In: Proceedings of 7th Asian-Pacific Conference on Medical and Biological Engineering, pp. 133–138. Springer, Heidelberg (2008). https://doi.org/10.1007/978-3-540-79039-6_35
2. Blackman, A.J., Blevins, J.J., Sangeorzan, B.J., Ledoux, W.R.: Cadaveric flatfoot model: ligament attenuation and Achilles tendon overpull. J. Orthop. Res. **27**(12), 1547–1554 (2009). https://doi.org/10.1002/jor.20930
3. Spratley, E.M., Matheis, E.A., Curtis, W.H., Adelaar, R.S., Wayne, J.S.: Validation of a population of patient-specific adult acquired flatfoot deformity models. J. Orthop. Res. **31**(12), 1861–1868 (2013). https://doi.org/10.1002/jor.22471
4. Spratley, E.M., Matheis, E.A., Hayes, C.W., Adelaar, R.S., Wayne, J.S.: A population of patient-specific adult acquired flatfoot deformity models before and after surgery. Ann. Biomed. Eng. **42**(9), 1913–1922 (2014). https://doi.org/10.1007/s10439-014-1048-y
5. Spratley, E.M., Matheis, E.A., Hayes, C.W., Adelaar, R.S., Wayne, J.S.: Effects of degree of surgical correction for flatfoot deformity in patient-specific computational models. Ann. Biomed. Eng. **43**(8), 1947–1956 (2015). https://doi.org/10.1007/s10439-014-1195-1
6. Lewis, G.S.: Computational modeling of the mechanics of flatfoot deformity and its surgical corrections. Ph.D dissertation, Pennsylvania State University (2008)
7. Qiu, T., Teo, E., Yan, Y., Lei, W.: Finite element modeling of a 3D coupled foot-boot model. Med. Eng. Phys. **33**(10), 1228–1233 (2011). https://doi.org/10.1016/j.medengphy.2011.05.012
8. Isvilanonda, V., Dengler, E., Iaquinto, M., Sangeorzan, B.J., Ledoux, W.R.: Finite element analysis of the foot: model validation and comparison between two common treatments of the clawed hallux deformity. Clin. Biomech. **27**(8), 837–844 (2012). https://doi.org/10.1016/j.clinbiomech.2012.05.005
9. Ozen, M., Sayman, O., Havitcioglu, H.: Modeling and stress analyses of a normal foot-ankle and a prosthetic foot-ankle complex. Acta Bioeng. Biomech. **15**(3), 19–27 (2013). https://doi.org/10.5277/abb130303

10. Wang, Z., Imai, K., Kido, M., Ikoma, K., Hirai, S.: A finite element model of flatfoot (pes planus) for improving surgical plan. In: Proceedings of 36th Annual International Conference of the IEEE Engineering in Medicine and Biology Society, Chicago, pp. 844–847 (2014). https://doi.org/10.1109/EMBC.2014.6943723
11. Wang, Z., Imai, K., Kido, M., Ikoma, K., Hirai, S.: Study of surgical simulation of flatfoot using a finite element model. In: Chen, Y.W., Torro, C., Tanaka, S., Howlett, R., Jain, L.C. (eds.) Innovation in Medicine and Healthcare 2015. Smart Innovation, Systems and Technologies, vol. 45, pp. 353–363. Springer, Cham (2016). https://doi.org/10.1007/978-3-319-23024-5_32
12. Cignoni, P., Callieri, M., Corsini, M., Dellepiane, M., Ganovelli, F., Ranzuglia, G.: MeshLab: an Open-Source mesh processing tool. In: Sixth Eurographics Italian Chapter Conference, pp. 129–136 (2008). https://doi.org/10.2312/LocalChapterEvents/ItalChap/ItalianChapConf2008/129-136
13. STL to ACIS SAT conversion. https://jp.mathworks.com/matlabcentral/fileexchange/27174-stl-to-acis-sat-conversion Accessed 12 March
14. Netter, F.H.: Atlas of Human Anatomy, 5th edn, pp. 51–525. Elsevier, Amsterdam (2011)
15. Wang, Z., Kido, M., Imai, K., Ikoma, K., Hirai, S.: Towards patient-specific medializing calcaneal osteotomy for adult flatfoot: a finite element study. Comput. Method Biomec. **21**(4), 332–343 (2018)

Transparent Fused Visualization of Surface and Volume Based on Iso-Surface Highlighting

Miwa Miyawaki[1], Kyoko Hasegawa[2], Liang Li[2], and Satoshi Tanaka[2](✉)

[1] Graduate School of Information Science and Engineering, Ritsumeikan University, Kyoto, Japan
[2] College of Information Science and Engineering, Ritsumeikan University, Kyoto, Japan
stanaka@is.ritsumei.ac.jp

Abstract. Computer Graphics technology enables a three-dimensional representation of object's shape and inner structure. It is widely used in the field of visualization and simulation such as computer-aided design, scientific visualization, and medical simulation. Recent studies on implicit surface generation from shape measured three-dimensional point cloud data provide precise and refined surface visualization for complex objects from buildings and tangible heritages to the internal structure of the human body. However, to understand and analyze the structural characteristics of complex shapes, conventional methods, which visualize the whole object with one criterion, could not produce satisfactory results. A more comprehensive visualization method that extracts and highlights the edges and feature regions of a complex object is desired. In this paper, we propose a fused visualization method that extracts and highlights the shape characteristics of three-dimensional volume data of the human body. For the implicit surface generation, volume stochastic process sampling method is applied. The surface curvature is then calculated by projecting the mathematically well-defined curvature information at a point on the iso-surface to its tangent plane. The high curvature area is extracted as the feature region and transparently fused with the original volume data. The proposed method, which realizes three-dimensional transparent fusion of feature-highlighted iso-surface visualization and volume visualization, comprehensively visualizes global structure of the target medical data as well as emphasizes the structural characteristics in the feature region.

1 Introduction

Visualization of the three-dimensional (3D) shapes of the complex human body is important for understanding and handling the internal organs and structures in diagnosis and surgical decision making. Volume rendering and iso-surface rendering are two common methods for medical visualization. Volume rendering visualizes the global shape of the target, whereas iso-surface rendering is more

G. De Pietro et al. (Eds.): KES-IIMSS-18 2018, SIST 98, pp. 260–266, 2019.
https://doi.org/10.1007/978-3-319-92231-7_27

suitable for visualizing the local structural characteristics. In this paper, considering their respective advantages, we extracted the high curvature area for iso-surface rendering and transparently fused it with the volume rendering result. This enables us to recognize shapes of the iso-surface and it leads to easier understanding of 3D internal structure of a human body. In iso-surface rendering for medical volume data, a popular algorithm is marching cubes [1], which extracts an approximated polygonal mesh of an iso-surface from volume data. In this case, polygons that are frequently used in computer graphics can be used in rendering [2]. However, it is not suitable for high curvature area extraction by computing the differential of scalar fields because polygonal meshes are a set of small surfaces. In this paper, we construct continuous scalar fields by spline interpolation based on the volume data. The iso-surface is then precisely extracted as uniformly distributed 3D point clouds using the stochastic process sampling method that we proposed in our previous studies [3–8]. The curvature can be computed by differentiation of scalar fields. The point clouds that indicate the iso-surface with high curvature can be highlighted with certain colors and visualized by the Stochastic Point-based Rendering (SPBR) [2,9]. Furthermore, the original medical volume data are transformed into point cloud data and transparently fused and visualized with the highlighted high curvature area. Therefore, comprehensive visualization is realized, in which the whole shape of the target volume data can be globally perceived, whereas the local characteristics in the region of interest are emphasized with feature-highlighted iso-surface.

2 Rendering Method

Surface-volume fusion is realized by fusing both volume data and iso-surface which is created by spline interpolation of given volume data. In regard to render the volume data, according to the PBVR, the point distribution is proportionate to the opacity transfer function, and generated by Monte Carlo method [9]. Concerning the iso-surface data, uniformly distributed points are created using the theory of Brownian motion.

2.1 Volume Rendering

Particle-based Volume Rendering (PBVR) is used for volume rendering [9]. The PBVR is a novel method which enables high quality rendering of large scale data with opaque luminous particles. Luminous particles are created according to the density distribution function shown as below (Eq. 1).

$$\rho(\mathbf{x}) = -\frac{\ln(1-\alpha(\mathbf{x}))}{d^2 \Delta t}, \tag{1}$$

where, Δt is a parameter for sampling precision, and α is the opacity given by the transfer function. $\mathbf{x}$ is a position vector in the space where the volume data is defined. d is the diameter of the particle. The sectional area of the particle is approximated by a square with sides of d.

2.2 Surface Rendering

We adopt Stochastic Sampling Method (SSM) for surface rendering. The SSM is a method using stochastic differential equation. It is a simulation of Brownian motion on the implicit surface I_F, and the tracing of the Brownian particle is considered to be sample points. Figure 1 is a conceptual scheme of Brownian motion. In 3D euclidean space by position vector $\mathbf{q} = (q_1, q_2, q_3)$, the equation of the implicit surface is defined as follows:

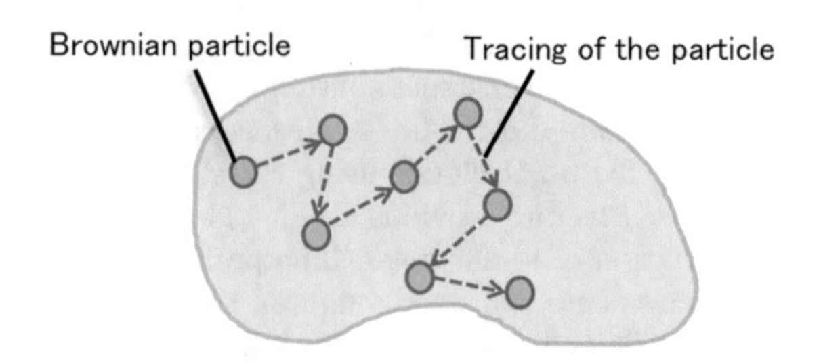

Fig. 1. Sampling iso-surfaces based on Brownian motion of a hypothetical particle

$$F(\mathbf{q}) = 0, \tag{2}$$

where, $\mathbf{q}$ is the position vector of a brown particle. Then, in order to use this equation, it is necessary to define $F(\mathbf{q})$ from a grid pattern volume data. Because it has scalar value, it is converted to a continuous function by spline interpolation. Then, an implicit surface $F(\mathbf{q}) = 0$ is defined as follows. C is iso-value and $S(\mathbf{q})$ is a scalar function.

$$F(\mathbf{q}) = S(\mathbf{q}) - C. \tag{3}$$

In the SSM, the Brownian motion on I_F is defined as the following stochastic differential equation:

$$dq_i(t) = dq_i^{(\mathrm{T})}(t) + dq_i^{(\mathrm{S})}(t), \tag{4}$$

where, t is the time step valuable and $dq_i^{(\mathrm{T})}$, $dq_i^{(\mathrm{S})}$ is defined as follows:

$$dq_i^{(\mathrm{T})} \equiv \sum_{j=1}^{d} P_{ij} dw_j, \tag{5}$$

$$dq_i^{(\mathrm{S})} \equiv -\frac{\alpha}{|\nabla F|^2} \left(\frac{\partial F}{\partial q_i} \right) \mathrm{Tr}\{\mathbf{H} \cdot \mathbf{P}\} \mathbf{dt}. \tag{6}$$

Here, $F(\mathbf{q})$ is a scalar function and dw_i is the Gaussian random variable with statistical properties,

$$\langle dw_i \rangle = 0, \qquad \langle dw_i dw_j \rangle = 2\delta_{ij} dt. \tag{7}$$

$dq_i^{(\mathrm{T})}$ is a random term that shows the movement of a Brownian particle which moves atomically from the implicit surface to a random place on 3D space.

P_{ij} is a projection operator which can extract the movement on tangent plane from random movement. $P_{ij} = n_i n_j$, where n_i is the i–th component of $\frac{\nabla F}{|\nabla F|}$. $dq_i^{(S)}$ is a stochastic correction term which corrects non-negligible jump from surface to I_F. By solving the stochastic differential equation numerically, we can generate uniformly-distributed points on I_F as schematically shown in Fig. 1.

2.3 Surface-Volume Fusion

Surface-volume fusion can be realized with the point data created by the PBVR and the SSM. These two point data are combined by point cloud fusion. It can be simply realized by merging the point clouds which have consistent positional information.

3 Feature Extraction of Iso-Surface

In order to highlight the feature regions, it is necessary to decide a definition of curvature and a threshold. Figure 2 is a conceptual schematic illustration of our transparent fused visualization of volume and iso-surface.

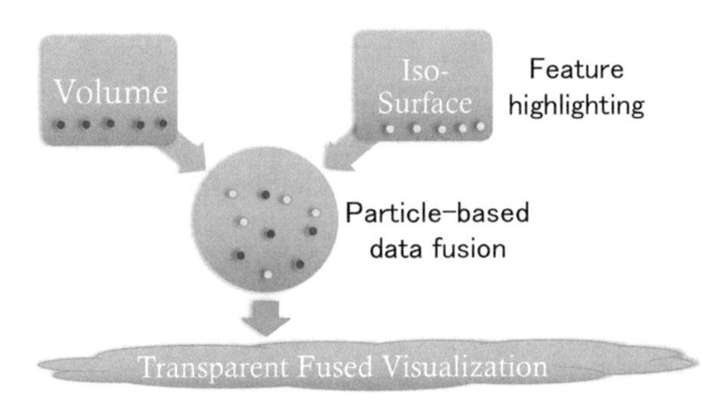

Fig. 2. Schematic illustration of our transparent fused visualization of volume and iso-surface.

3.1 Definition of Curvature

The curvature is defined by a method developed by ourselves, which is proposed in our previous study [10]. A typical definition of the curvature in two-dimensional space is the inverse of the radius of the circle with one tangent point on the curve. In this study, it is extended to adjust to an implicit surface and its equation is showed as follows:

$$K = \frac{\mathbf{Tr}\,[\mathbf{PH}]}{[\nabla F(\mathbf{q})]}, \tag{8}$$

where, Hessian's element H is defined as

$$\mathbf{H} = \begin{bmatrix} \frac{\partial^2 F(\mathbf{r})}{\partial x \partial x} & \frac{\partial^2 F(\mathbf{r})}{\partial x \partial y} & \frac{\partial^2 F(\mathbf{r})}{\partial x \partial z} \\ \frac{\partial^2 F(\mathbf{r})}{\partial y \partial x} & \frac{\partial^2 F(\mathbf{r})}{\partial y \partial y} & \frac{\partial^2 F(\mathbf{r})}{\partial y \partial z} \\ \frac{\partial^2 F(\mathbf{r})}{\partial z \partial x} & \frac{\partial^2 F(\mathbf{r})}{\partial z \partial y} & \frac{\partial^2 F(\mathbf{r})}{\partial z \partial z} \end{bmatrix}, \tag{9}$$

Hessian's elements have an information of second order differential of $F(\mathbf{q})$. Therefore, Hessian matrix shows the changing rate in all directions in the 3D space. However, we adopt the projection matrix of the Hessian matrix because only the changing rate in the tangential direction is required to compute the curvature. Thus, the curvature can be computed by only considering the tangential plane of the iso-surface.

3.2 Threshold for High Curvature Extraction

A threshold for defining high-curvature is decided based on the histogram of curvature. As the candidates for the threshold, average, mode and median value are available. In this study, average of the curvature is adopted because it can show high-curvature area as a feature better.

3.3 Color and Opacity Selection for Feature-Highlighting

In this study, high curvature area is considered to be the feature regions. Curvature K defined by Eq. 8 is used to decide the feature regions. For feature-highlighting, it is necessary to consider the proper color and opacity for highlighting the feature regions. Rainbow colors are suitable for visualizing the volume data. However, in surface-volume fusion, multiple colors may cause difficulties in feature area recognition because many colored lines and surfaces interfere with each other. Therefore, in this study, we colorize the volume data with gray scale and low opacity, whereas colorize the feature regions with highlighted color.

4 Visualization Results

The visualization result of high curvature area extraction is shown in Fig. 3. According to the result, the high curvature area in Fig. 3(b) is successfully extracted from the iso-surface data in Fig. 3(a).

Comparison of the conventional volume rendering [3] and the proposed transparent fusion of feature-highlight iso-surface visualization is shown in Fig. 4. According to the result, the proposed method successfully visualized the featured area as well as the whole shape of the volume data.

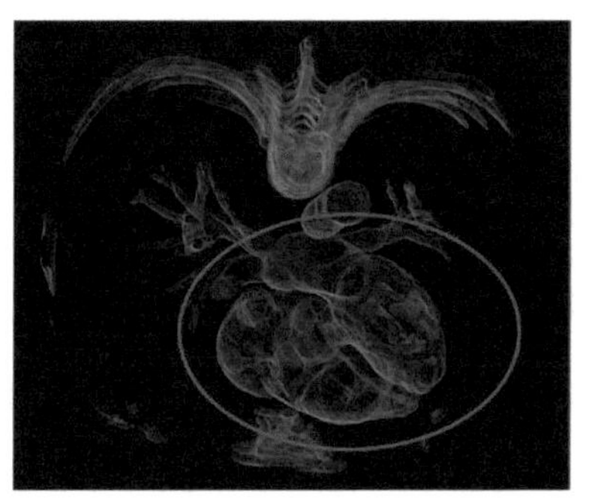

(a) Iso-surface rendering

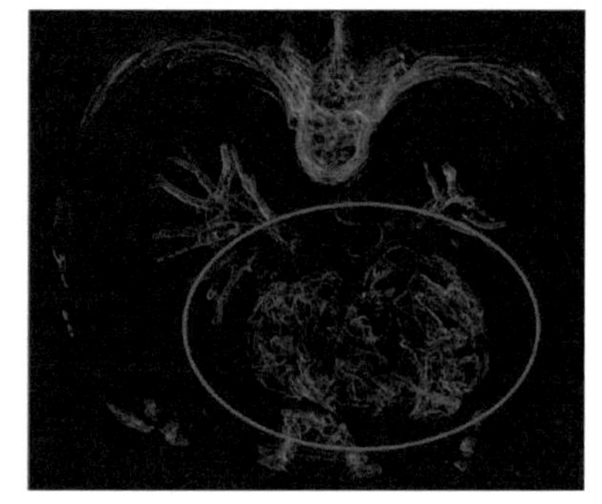

(b) High curvature area extracted from (a)

Fig. 3. Result of extracting high curvature surfaces

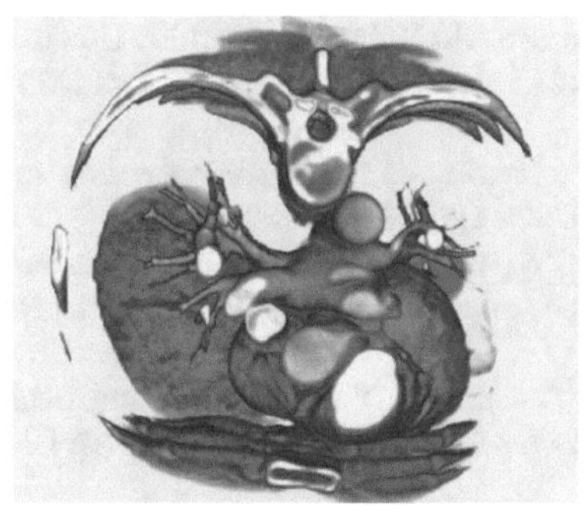

(a) the conventional volume rendering

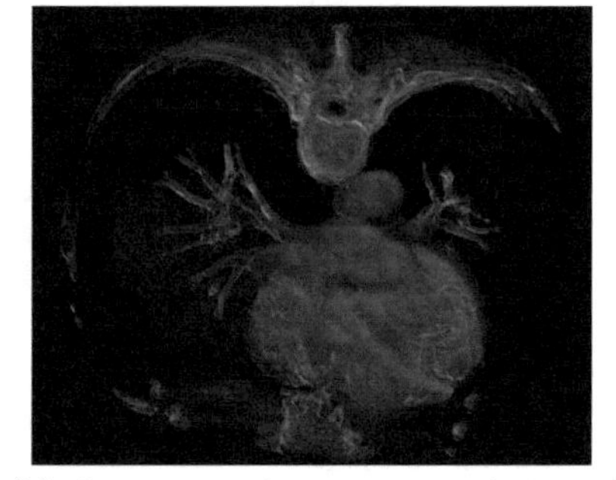

(b) the proposed transparent fusion of feature-highlighted iso-surface visualization

Fig. 4. Original volume data and the result of surface volume fusion

5 Conclusion

In this paper, we proposed the way to visualize the volume data and feature regions together. In order to make use of both features at the same time, transparent fused visualization was demonstrated. Then characteristic regions were visualized together with the volume data. In feature area visualization, the SSM, which can create complex and smooth surfaces, was used for iso-surface rendering. In addition, the volume data was visualized with gray scale and low opacity, whereas the feature area was emphasized with highlight color. In addition, the volume data was visualized with gray scale and low opacity, whereas the feature area was emphasized with highlight color. Application of the proposed method to stereoscopic vision is one of our future works. More accurate visual perception can be expected by introducing stereoscopic depth cues to our visualization method.

References

1. Lorenson, W.E., Cline, H.E.: Marching cubes: a high resolution 3D surface construction algorithm. In: SIGGRAPH 1987, pp. 163–169 (1987)
2. Tanaka, S., Hasegawa, K., Shimokubo, Y., Kaneko, T., Kawamura, T., Nakata, S., Ojima, S., Sakamoto, N., Tanaka, H.T., Koyamada, K.: Particle-based transparent rendering of implicit surfaces and its application to fused visualization. In: EuroVis 2012, pp. 25–29 (short paper), Vienna, Austria, 5–8 June 2012
3. Bloomenthal, J.: Polygonization of implicit surfaces. Comput. Aided Geom. Des. **5**, 341–355 (1988)
4. Tanaka, S., Morisaki, A., Nakata, S., Fukuda, Y., Yamamoto, H.: Sampling implicit surfaces based on stochastic differential equations with converging constraint. Comput. Graph. **24**(3), 419–431 (2000)
5. Tanaka, S., Shibata, A., Yamamoto, H., Kotsuru, H.: Generalized stochastic sampling method for visualization and investigation of implicit surfaces. Comput. Graph. Forum **20**(3), 359–367 (2001). Proceedings of Eurographics 2001
6. Tanaka, S., Nakamura, T., Ueda, M., Yamamoto, H., Shino, K.: Application of the stochastic sampling method to various implicit surfaces. Comput. Graph. **25**(3), 441–448 (2001)
7. Tanaka, S., Fukuda, Y., Yamamoto, H.: Stochastic algorithm for detecting intersection of implicit surfaces. Comput. Graph. **24**(4), 523–528 (2000)
8. Jo, Y., Oka, M., Kimura, A., Hasegawa, K., Saitoh, A., Nakata, S., Shibata, A., Tanaka, S.: Stochastic visualization of intersection curves of implicit surfaces. Comput. Graph. **31**(2), 230–242 (2007)
9. Sakamoto, N., Nonaka, J., Koyamada, K., Tanaka, S.: Volume rendering using a particles. In: IEEE International Symposium on Multimedia (ISM2006), San Diego, California, USA, 11–13 December 2006
10. Oka, M., Nakata, S., Tanaka, S.: Preprocessing for accelerating convergence of repulsive-particle systems for sampling implicit surfaces. In: IEEE SMI 2007 (Shape Modeling International 2007), Lyon, France, 13–15 June 2007

Joint Image Extraction Algorithm and Super-Resolution Algorithm for Rheumatoid Arthritis Medical Examinations

Tomio Goto[1(✉)], Yoshiki Sano[1], Takuma Mori[1], Masato Shimizu[1], and Koji Funahashi[2]

[1] Department of Computer Science and Engineering, Nagoya Institute of Technology, Gokiso-cho, Showa-ku, Nagoya 466-8555, Japan
t.goto@nitech.ac.jp, {shiki,mori,asimo}@splab.nitech.ac.jp
[2] Orthopaedic Surgery, Kariya Toyota General Hospital, 5-15 Sumiyoshi-cho, Kariya, Aichi 448-8505, Japan
http://www.splab.nitech.ac.jp/

Abstract. Super-resolution techniques have been widely used in fields such as television, aerospace imaging, and medical imaging. In medical imaging, X-rays commonly have low resolution and a significant amount of noise, because radiation levels are minimized to maintain patient safety. So, we proposed a novel super-resolution method for X-ray images, and a novel measurement algorithm for treatment of rheumatoid arthritis (RA) using X-ray images generated by our proposed super-resolution method. However, in our proposed system, there are several operations to do by doctors manually, and it is hard for them. By utilizing image recognition technology, it is possible to extract joint images from X-ray images automatically. In this paper, we will discuss an algorithm to extract joint images from X-ray images automatically. Experimental results show that correct joint images will be obtained for our proposed method. Therefore, our proposed measurement algorithm is effective for RA medical examinations.

Keywords: Image extraction · Joint space distance
Rheumatoid arthritis · Medical examinations

1 Introduction

X-ray images are widely used to diagnose a variety of diseases. However, to reduce the patient's exposure to radiation, X-ray dosage is minimized as much as possible. As a result, X-ray images contain a significant amount of noise and resolution is compromised. Thus, it is necessary to increase image resolution and reduce noise. We proposed a novel super-resolution system for X-ray images that consists of total variation (TV) regularization, a shock filter, and a median filter.

G. De Pietro et al. (Eds.): KES-IIMSS-18 2018, SIST 98, pp. 267–276, 2019.
https://doi.org/10.1007/978-3-319-92231-7_28

In addition, we proposed a novel measurement algorithm for the treatment of RA, using X-ray images generated by our proposed super-resolution system [1]. In this paper, we improve measurement accuracy to optimize parameters for super-resolution.

2 Super-Resolution System

Super-resolution is a technique for increasing the resolution of an enlarged image by generating new high-frequency components. This technique estimates and generates such components from the characteristics of the original signals. In recent years, various super-resolution techniques have been proposed, and most are classified as either reconstruction-based super-resolution [2] or learning-based super-resolution [3].

In resolution for X-ray images, which is used at medical operation, the resolution of one pixel will be 0.15 mm^2, and joint space distance will be about from 1.0 mm to 1.5 mm. Therefore, when the joint space distance is measured with one pixel error, the error will be from 10% to 15%, improving measurement accuracy is required.

It is possible to solve this problem by utilizing a super-resolution technique. By magnifying segmented images by a multiple of 4×4, a resolution of a pixel will be 0.0375 mm, thus the error can be suppressed from 2.5% to 3.75%. Also by utilizing the shock filter, clearer edges in images will be obtained, thus it is easy to select edges. However, the error still remains and it is not small to measure for RA medical examinations. In this paper, we improve measurement accuracy to set bigger magnification rate and optimal parameters for super-resolution.

A block diagram of our proposed super-resolution system is shown in Fig. 1. Each of the non-linear filters is explained in the following sections.

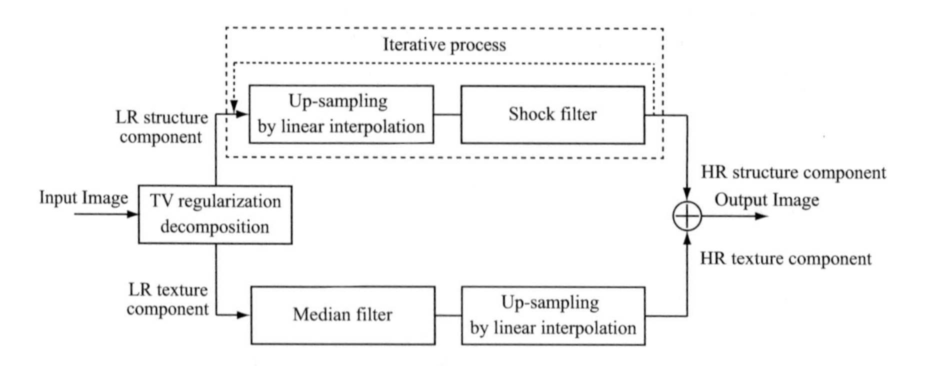

Fig. 1. Block diagram of our proposed super-resolution system.

2.1 Total Variation Regularization

As shown in Fig. 1, the TV regularization decomposition [4–8] is performed as follows. The structure component u is calculated to minimize the evaluation function $F(u)$ as shown in Eq. (1):

$$F(u) = \Sigma_{i,j}|\nabla u_{i,j}| + \lambda \Sigma_{i,j}|f_{i,j} - u_{i,j}|^2. \tag{1}$$

where f is a pixel value of the input image. The Chambolle's projection algorithm [9] is used to solve the minimization problem as shown in Eq. (2).

$$P_{i,j}^{(n+1)} = \frac{P_{i,j}^{(n)} + (\frac{\tau}{\lambda})\nabla(f + \lambda div P_{i,j}^{(n)})}{max\{1, |P_{i,j}^{(n)} + (\frac{\tau}{\lambda})\nabla(f + \lambda div P_{i,j}^{(n)})|\}}. \tag{2}$$

where P is a pixel value. The texture component v and the structure component u are obtained by using the equations in (3).

$$v = \lambda div P, \qquad u = f - v. \tag{3}$$

Figure 2 shows an example of the TV decomposition for an X-ray image, (a) is an original image, (b) is a structure component and (c) is a texture component.

2.2 Shock Filter

The shock filter is a nonlinear edge enhancement filter, which was proposed by Osher and Rudin [10] and Alvarez and Mazorra [11]. The process is achieved by utilizing Eq. (4):

$$u_{i,j}^{(n+1)} = u_{i,j}^{(n)} - sign\left(\Delta\left(K_\sigma * \Delta u_{i,j}^{(n)}\right)\right)\left|\nabla u_{i,j}^{(n)}\right| dt. \tag{4}$$

where u is a structure component, K is a smoothing filter. It is possible to reconstruct steep edges by calculating a simple operation; thus, this filter is suitable for high-speed processing. In addition, several artifacts generated during

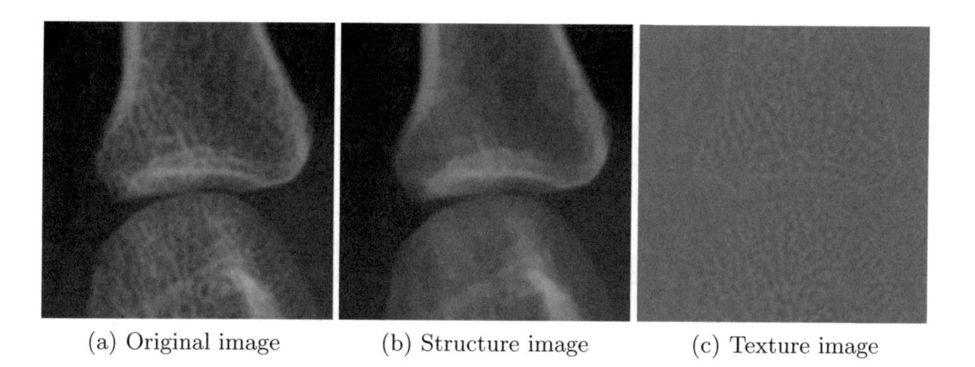

(a) Original image (b) Structure image (c) Texture image

Fig. 2. Example of total variation decomposition for X-ray image.

edge enhancement processing can be controlled successfully, i.e., ringing noise and jaggy noise. Figure 3 shows an input image and output images obtained by utilizing the shock filter.

2.3 Median Filter

Most noise is classified as a texture component by utilizing TV regularization. As mentioned previously, X-ray images contain significant noise. Therefore, we propose applying the median filter to the texture components of X-ray images. The median filter sorts nine pixel values in 3 × 3 pixels around the pixel of interest. Next, the filter replaces the fifth pixel value with a new pixel value of interest. This process is applied to all of the texture components.

3 Measurement Algorithm About Joint Space Distance

Rheumatoid arthritis (RA) is a disease that causes joint inflammation, and most commonly afflicts women between 30 and 50 years of age. As symptoms progress, the patient's joint space distance (JSD) will narrow. This change can be observed with X-ray images; however, at present, an accurate measurement method has not been established. Therefore, a more accurate JSD measurement technique is required. We proposed two JSD measurement algorithms [1]. Figure 4 shows an example of an output image. In our proposed method, we use an input image, which is magnified by utilizing a super-resolution method. Our algorithm for measurement is shown as follows:

1. Select several points on an edge of an upper bone by clicking mouse button, manually.
2. Set axes corresponding to a joint from selected points automatically.
3. Calculate coordinate values of selected points based on the axes.
4. Calculate a quadratic function by using the least squares method from the calculated fitting function.

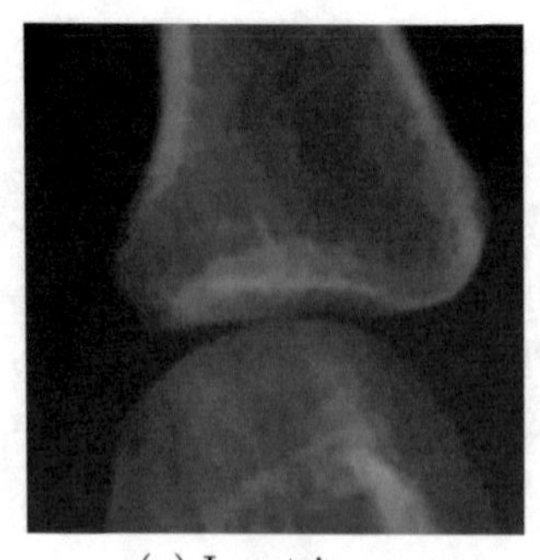

(a) Input image

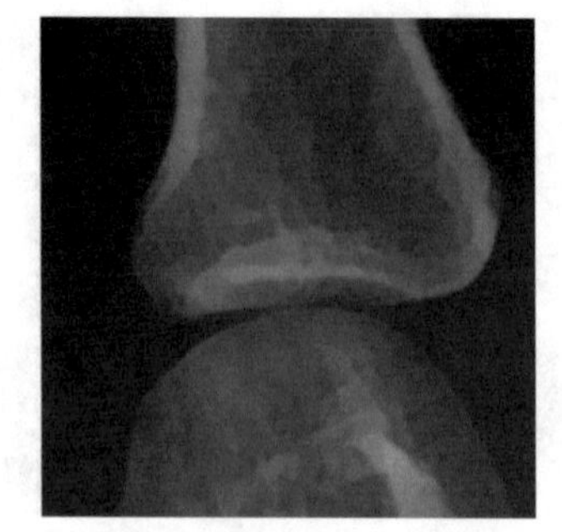

(b) Output image from shock filter

Fig. 3. Example of shock filter for X-ray image.

5. Calculate a quadratic function by selecting several points on an edge at a lower bone, similarly.
6. Measure the joint space distance from normal lines and integral operation.

We also propose another JSD measurement algorithm, which is calculated by using an area of JSD, and its algorithm for measurement is shown as follows:

1. An integral calculus range in a range of curve p is set, and to calculate the integral calculus S_R of differences between curve p and curve q.
2. An integral calculus range in a range of curve q is set, and to calculate the integral calculus S_B of differences between curve p and curve q.
3. The value W is defined as the distance between 2 points, which is automatically detected when the coordinate axis are set.
4. The JSD value D is calculated from two values: S_R and S_B and averages of the distances of the top and bottom as shown in Eq. (5).

$$D = \frac{S_R + S_B}{2W} \tag{5}$$

4 Proposed Extraction Method

By utilizing image recognition technology, we will extract parts of images around the third finger joints from many images, automatically. And by learning these images for correct images, it is possible to recognize joint space distance. We propose a novel method for extracting the third finger joints for correct images and the other images for incorrect images, which are used in the learning method.

Figure 6 shows our proposed extracting algorithm. First, to input the center coordinates in the third finger joints, then several points in the radius of the circle are selected, randomly. Then, $128\sqrt{2}$ square pixels are extracted because

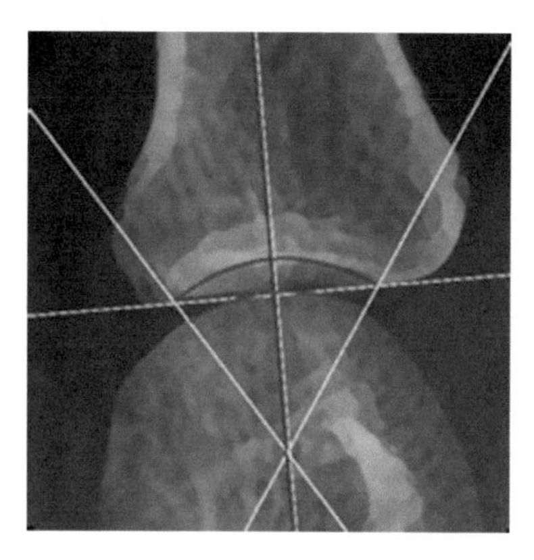

Fig. 4. Application for JSD calculation of X-ray image.

rotation will be processed in the post-processing and it is possible to prevent the purchase of noise by the rotation process. And those are turned in the range of ±20 degrees, then 128 square pixels in center are output as a correct image. Next, the Gaussian filter is processed to the X-ray image including the whole both hands, those are binarized, the isolated points are removed to do closing processing based on morphology operation, and bone regions are extracted as shown in Fig. 5.

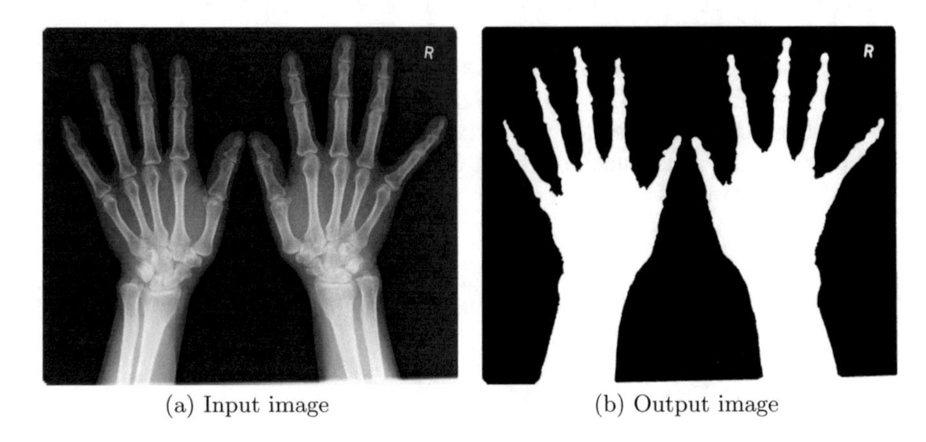

(a) Input image (b) Output image

Fig. 5. Extracted results for bone area

Finally, some points in the whole image are selected, randomly, and when the point is $128\sqrt{2}$ square pixels in bone regions and far from the center coordinates, 128 square pixels in center are output as an incorrect image. To select center image points randomly in the circle, in which the center coordinates draw, it is possible to control the gaps between the center point of joints and extracted images. Also, bone parts will be included in the center of the extracted images, which are far from the center of joints, and they are output as incorrect images, so that it is possible to output bone parts without the third joints of correct images.

5 Experimental Results

Learning images are made by using our proposed method, and machine learning with the Haar-Like feature by utilizing those images is performed.

5.1 Image Extraction for Machine Learning by Our Proposed Method

Figures 7 and 8 show the experimental results for correct and images, and Fig. 9 shows the experimental results for incorrect images. As a result, by utilizing

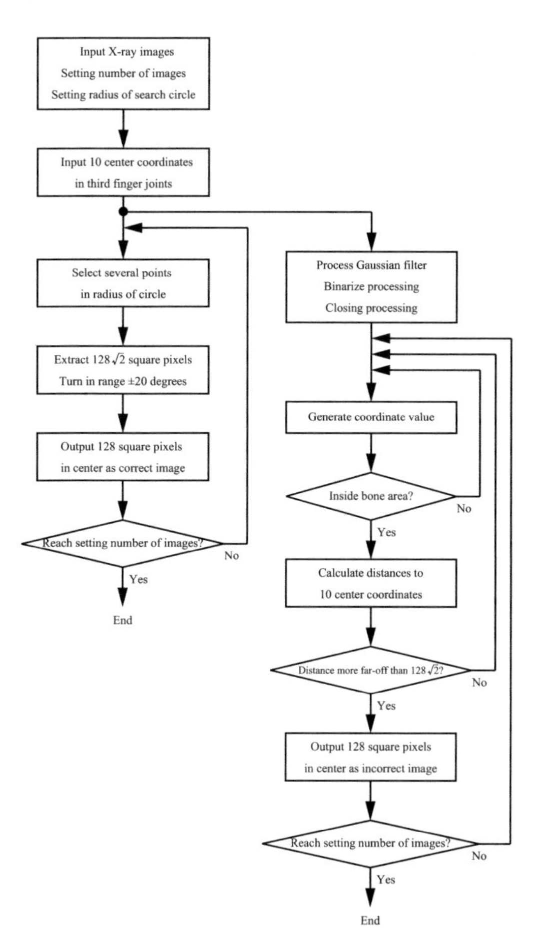

Fig. 6. Proposed extraction algorithm

the center points of joints, it is confirmed to extract correct images without greatly deviating from the center of those images. Also, to extract bone parts, it is effective method for features machine learning. And it is also confirmed to extract incorrect images with bone parts.

5.2 Feature Machine Learning Using Extracted Images

Machine learning is performed by using extracted images, which was made from 24 X-Ray images by using our proposed method, while changing the radius of the search circle with 0, 16 and 32 pixels. As a result, when the radius of the search circle was 0 pixel, it is possible to succeed machine learning and to make a detector. This means that our proposed method is effective for feature machine learning of the Haar-Like features.

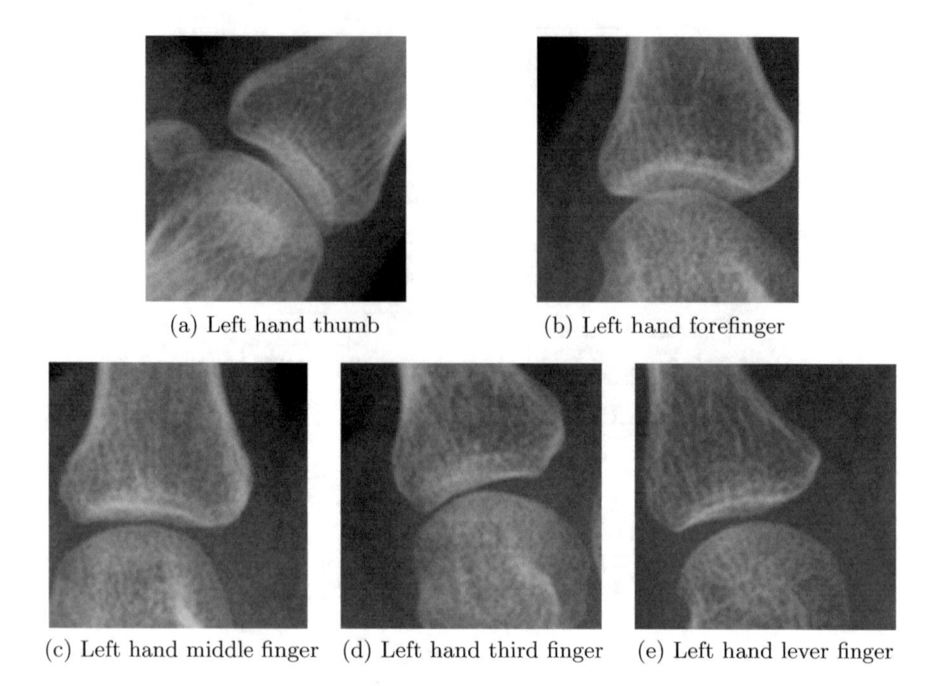

(a) Left hand thumb (b) Left hand forefinger

(c) Left hand middle finger (d) Left hand third finger (e) Left hand lever finger

Fig. 7. Output results for left hands as correct images

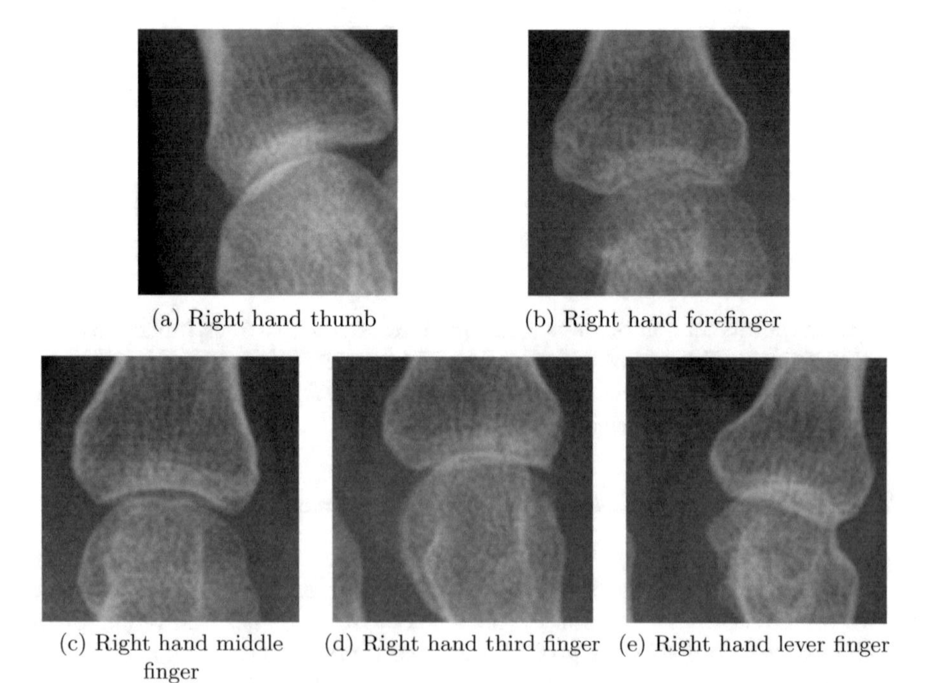

(a) Right hand thumb (b) Right hand forefinger

(c) Right hand middle finger (d) Right hand third finger (e) Right hand lever finger

Fig. 8. Output results for right hands as correct images

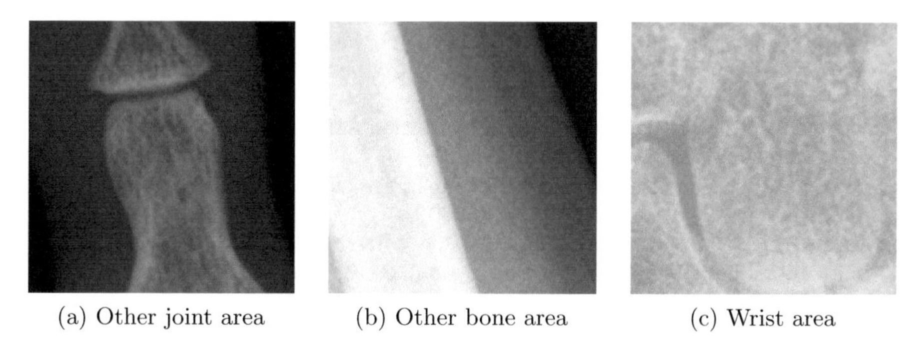

(a) Other joint area (b) Other bone area (c) Wrist area

Fig. 9. Output results as incorrect images

6 Conclusion

In this paper, an automation of measurement application for joint space distance is subject, and an algorithm for extracting images, which are used for feature learning at finger joints, has been proposed and its application has been implemented. By extracting images, which are used to the center of the third finger joints and bone parts as correct and incorrect images, respectively, the Haar-Like feature learning succeeds and it is possible to make its detector.

For further research, we intend to improve the detective accuracy to consider an enhancement of image features based on image processing and to consider the other features and effectiveness of our proposed method by using Deep-Learning.

References

1. Goto, T., Mori, T., Kariya, H., Shimizu, M., Sakurai, M., Funahashi, K.: Super-resolution technology for X-ray images and its application for rheumatoid arthritis medical examinations. In: International KES Conference on Innovation in Medicine and Healthcare (InMed), pp. 217–226, June 2016
2. Singh, A., Ahuja, N.: Single image super-resolution using adaptive domain transformation. In: IEEE International Conference on Image Processing (ICIP), pp. 947–951, September 2013
3. Cho, C., Jeon, J., Paik, J.: Example-based super-resolution using selfpatches and approximated constrained least squares filter. In: IEEE International Conference on Image Processing (ICIP), pp. 2140–2144, October 2014
4. Rudin, L., Ohser, S., Fetami, E.: Nonlinear total variation based noise removal algorithm. Phys. D **60**, 259–268 (1992)
5. Meyer, Y.: Oscillating patterns in image processing and non-linear evolution equation. In: The Fifteenth Dean Jacqueline B. Lewis Memorial Lectures. University Lecture Series, vol. 22. American Mathematical Society (1992)
6. Vese, L.A., Osher, S.J.: Modeling textures with total variation minimization and oscillating patterns in image processing. J. Sci. Comput. **19**, 553–572 (2003)
7. Aujol, J.-F., Aubert, G., Blanc-Feraud, L., Chambolle, A.: Image decomposition into a bounded variation component and an oscillating component. J. Math. Imaging Vis. **22**, 71–88 (2005)

8. Goto, K., Nagashima, F., Goto, T., Hirano, S., Sakurai, M.: Super-resolution for high resolution displays. In: IEEE Global Conference on Consumer Electronics (GCCE), pp. 309–310, October 2014
9. Chambolle, A.: An algorithm for total variation minimization and applications. J. Math. Imaging Vis. **20**(1), 89–97 (2004)
10. Osher, S.J., Rudin, L.I.: Feature-oriented Image enhancement using shock filters. SIAM J. Numer. Anal. **27**, 910–940 (1990)
11. Alvarez, L., Mazorra, L.: Signal and image restoration using shock filters and anisotropic diffusion. SIAM J. Numer. Anal. **31**(2), 590–605 (1994)

Smart Transportation Systems (KES-STS-18) Introduction

Smart Transportation Systems (KES-STS-18) Introduction

Introduction

Modern transportation systems have been transforming rapidly in recent years. The innovations and breakthroughs have motivated researchers on Smart Transportation Systems (STS) to expand the frontiers of relevant concepts and skillsets, which will affect the transport infrastructure modelling, safety analysis, freeway operations, intersection analysis, and other related leading edge topics. The papers in the STS section of this volume provide a snapshot of the latest developments in this area.

The topics includes Drone Deployment for Monitoring Air Pollution from Ships, multi-period performance of non-storable production with carry-over activities, Urban Rail Transit Demand Analysis and Prediction, Pricing of Shared-parking, estimation of the value of travel time, Stochastic Traffic Flow Fundamental Diagram, Passenger Car Equivalents, Road Functional Classification, New York Citi Bike System, Modeling and Analysis of Crash Severity for Electric Bicycle, signal timing strategies, Large Passenger Transfer Hub, Traffic impact analysis, Urban Distribution System, and connected and automated vehicles. The methods employed range from statistical analysis to discrete event modelling, from numerical data driven analysis to analytical investigation, from traffic flow theory to network modelling, and from urban transport to rural mobility corridors.

Xiabo Qu
Robert J. Howlett
Lakhmi C. Jain
Ljubo Vlacic

Organization

STS-18 International Programme Committee

A Modelling Framework of Drone Deployment for Monitoring Air Pollution from Ships

Jingxu Chen[1,2], Shuaian Wang[1(✉)], Xiaobo Qu[3], and Wen Yi[4]

[1] Department of Logistics and Maritime Studies,
The Hong Kong Polytechnic University, Hung Hom, Kowloon, Hong Kong
wangshuaian@gmail.com

[2] Jiangsu Key Laboratory of Urban ITS, Jiangsu Province Collaborative
Innovation Center of Modern Urban Traffic Technologies,
Southeast University, Nanjing, China
shenqiudeliming@163.com

[3] Department of Architecture and Civil Engineering,
Chalmers University of Technology, 412 96 Gothenburg, Sweden
xiaobo@chalmers.se

[4] School of Engineering and Advanced Technology, College of Sciences,
Massey University, Auckland, New Zealand
yiwen96@163.com

Abstract. Sulphur oxide (SOx) emissions impose a serious health threat to the residents and a substantial cost to the local environment. In many countries and regions, ocean-going vessels are mandated to use low-sulphur fuel when docking at emission control areas. Recently, drones have been identified as an efficient way to detect non-compliance of ships, as they offer the advantage of covering a wide range of surveillance areas. To date, the managerial perspective of the deployment of a fleet of drones to inspect air pollution from ships has not been addressed yet. In this paper, we propose a modelling framework of drone deployment. It contains three components: drone scheduling at the operational level, drone assignment at the tactical level and drone base station location at the strategic level.

1 Introduction

International shipping usually uses heavy fuel oil (HFO), whose sulphur content can be up to 3.5%, 3,500 times higher than levels of Euro V fuel used in automobiles. When ocean-going vessels approach container terminals, they emit a substantial amount of SOx. For example, in Hong Kong, about 3.8 million people live in close proximity to the Kwai Tsing container terminals and risk direct exposure to ship related SOx emissions, which account for 44% of the total SOx emissions in Hong Kong. In practice, sulphur oxide (SOx) emissions impose a serious health threat to the residents and a substantial cost to the local environment. To formulate environmentally conscious and sustainable policies, many countries and regions advocate fuel switch at berth. Ocean-going vessels are mandated to use low-sulphur fuel with maximum sulphur content (e.g. 0.5%) when docking at emission control areas (ECA).

G. De Pietro et al. (Eds.): KES-IIMSS-18 2018, SIST 98, pp. 281–288, 2019.
https://doi.org/10.1007/978-3-319-92231-7_29

Currently, there are three frequently used methods to verify whether ships have switched to low sulphur fuel in ECA. (i) The first one is fuel sampling. Fuel sampling for sulphur testing at laboratories is the most reliable way to verify the sulphur content of the fuel being used on board. However, fuel sample is costly and time consuming, and therefore is only carried out when there are clear signs of non-compliance (Fung 2016). (ii) The second one is on-board inspections of fuel-related documents, such as bunker delivery notes and log books. A major deficiency of this method is that these materials may not be sufficient for proving that compliant fuel has actually been used on board. (iii) The third one is fixed-point remote measurement, often using the sniffing method which can identify gross emitting ships from afar. Because of the higher uncertainty of remote measurement results, remote sensing results are used as a cost effective means of screening ships suspected of using non-compliant fuel. Suspected ships will have their fuel sampled at the next port of call for lab test. The fixed-point sniffing method can only inspect ships in small territorial water areas as the exhaust gases of ships must be close to the measurement equipment. Moreover, once ship operators learn the locations of the measurement equipment, they can adapt to it by switching to low sulphur fuel only when the ship is close to the measurement equipment (Fung 2016).

In recent years, airborne surveillance methods, typically drones equipped with sniffers, have been identified as an efficient way to pre-screen ships that are more likely to be non-compliant. Drones offer the advantage of covering a much wider range of surveillance areas (Ning 2016). This enables regulators to be able to effectively target suspected emitting ships for follow-up on-board inspections. Moreover, a drone could follow a ship, and transact multiple times across the plume in order to obtain a more reliable estimate. Another advantage of using drone technology is that it is much cheaper to deploy than traditional on-board inspection methods.

The usage of drones has been steadily gained attention from local government, industrial community and research community due to its high efficiency and convenience. In several countries, the local authorities have started piloting the use of drones to support regular ship inspection programs. For instance, Denmark announced it will spend about a million Euro to employ aerial sniffer drones to inspect the sulphur content of engine exhaust of ships (Collum 2015). Turkey announced it will use drones to complement existing enforcement mechanisms to monitor marine pollution, improve surveillance and enforce penalties (Roberts 2016). In industrial community, the Trident Alliance, which includes Maersk Line and 37 other companies, has given its endorsement to the idea of using drones for effective enforcement of ECA rules (Marine Electronics and Communication 2015). Furthermore, some studies focus on the technological aspects by developing sonar and optical flow sensors suitable for application to ship emission inspections. These studies have provided beneficial technical support of using drones to monitor ship emissions from the technological perspective. As the continuous increase of drone uses, efficient management of the drones will be vital to the enforcement of ECA rules. However, the managerial perspective of the deployment of a fleet of drones to inspect air pollution from ships has not been addressed particularly. From the perspective of academic research, although vehicle scheduling (Kuang et al. 2015; Qu et al. 2015, 2017; Liu et al. 2016, 2017a, b; Huang et al. 2016) and vessel scheduling (Zhen et al. 2017a, b, 2018) have been extensively studied, drone scheduling related research is scarce.

The optimal deployment of drones is a practical research topic which provides significant value in addressing climate change and environmental sustainability by reducing SOx emissions from shipping. To this end, this study aims to build a modelling framework of drone deployment (DP) for monitoring air pollution from ships. The DP modelling framework subsumes three main components, namely drone scheduling, drone assignment and drone base station location, which is elaborated in the next section.

2 Modelling Framework

In this section, we present the modelling framework for the deployment of drones. It can be decomposed into three components that span operational, tactical and strategic decisions: (1) Design the optimal schedule on a real-time basis for each drone, including which ships to be inspected by the drone and when to inspect each ship, at the operational level, (2) Identify the optimal assignment of drones to base stations on a daily basis at the tactical level, and (3) Evaluate the optimal locations of base stations for drones at the strategic level. The overall DP modelling framework is shown in Fig. 1.

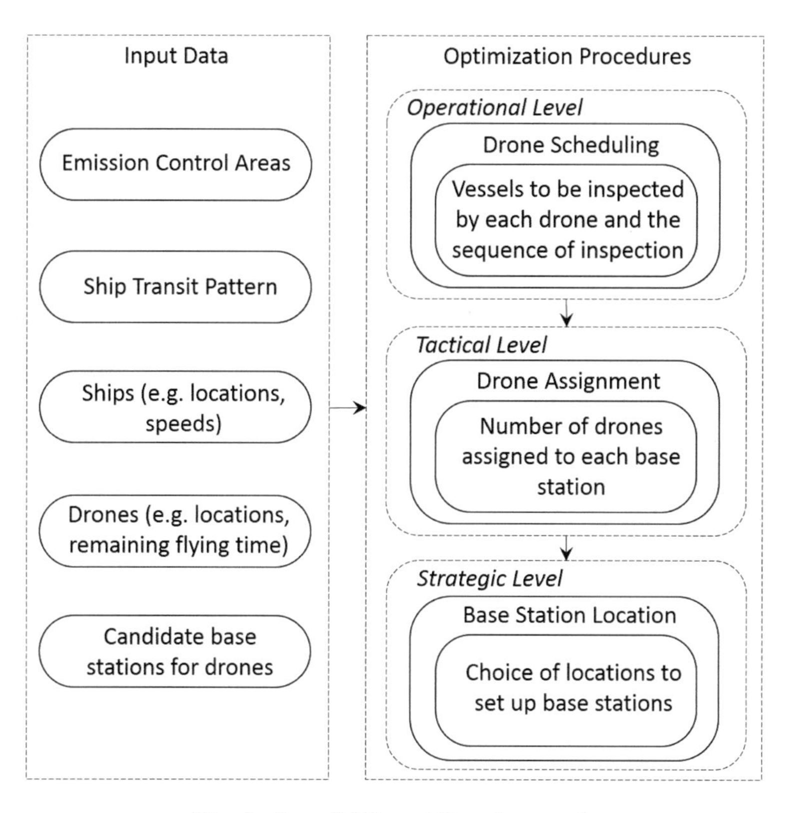

Fig. 1. Overall DP modelling framework

When drones leave a base station, they begin to follow ships in emission control areas, and transact multiple times across the plume in order to monitor ship emissions. Nevertheless, drones have flying time limitation. They can generally fly for up to two hours and associated flying distance is less than 100 km. After that, they have to return to a base station to replace depleted batteries with fully charged ones.

Meantime, to avoid ship collision, all ships are equipped with an automatic identification system (AIS). Information provided by the tracking system AIS also includes the identification of the ship, its position, sailing direction, and speed. According to AIS data (Marine Traffic 2017), we can obtain the real-time locations and speeds of all of the ships in the world.

2.1 Drone Scheduling

At the operational level, the optimal schedule needs to be determined for each drone at each station, including which ships to inspect by each drone and when to inspect each ship (or equivalently, the sequence of inspecting the ships). In addition, the schedule for a drone also includes information on when the drone should return to its base station for battery change.

In practice, there are a larger number of ships transiting the ECA and it may not be possible to inspect all of them. The purpose of drone scheduling is to maximize the total "weighted" number of ships that are inspected. The "weight" of a ship is the ship's chance of noncompliance. For example, ships with noncompliance history and ships whose operators have no prior experience of fuel switching should have a higher "weight", and ships that have already been inspected in other zones should have a lower "weight". Drones can be rescheduled every e.g. 10 min, by updating the AIS data of the ships, the inspection history of the ships (some ships have been inspected at dockside or in other zones), and the drone information (current location and remaining travelling distances before its battery is depleted).

The drone scheduling problem is related to the Vehicle Routing Problem with Time Windows (VRPTW) (Cordeau et al. 2000): a base station is a depot, a drone is a vehicle, a ship is a customer, and each ship has time window during which the ship should be inspected (otherwise the ship will sail out of ECA). The VRPTW has been proved to be strongly NP-hard, meaning that there is no polynomial-time or even pseudo-polynomial-time algorithm for it unless P = NP. A number of algorithms, such as column generation, branch and cut, branch and price, and meta-heuristics, have been proposed to identify optimal solutions for small-scale instances or good solutions for large-scale instances (Desrochers et al. 1992; Ho and Haugland 2004; Bräysy and Gendreau 2005; Kallehauge 2008).

However, the drone scheduling problem is much more difficult than the VRPTW because ships are sailing rather than stationary in the emission control area. In other words, the scheduling of drones should be carried out in a dynamic manner. Assuming the drone speed is fixed, the time required by a drone to fly from ship u to ship v, denoted by $d_{uv}(t)$, is dependent on the start flying time t from ship u. By contrast, in the VRPTW, $d_{uv}(t)$ is equal to a constant d_{uv} for all t. In this sense, droning scheduling is a three-dimensional VRPTW.

We can formulate a mathematical optimization model for the drone scheduling problem. The decision variables are as follows:

$z_v \in \{0, 1\}$: equal to 1 if ship v is inspected, and 0 otherwise;
$x_{uvk} \in \{0, 1\}$: equal to 1 if drone k inspects ship v immediately after inspecting ship u, and 0 otherwise;
$t_{vk} \geq 0$: start inspection time of ship v when inspected by drone k.

The inputs to the model mainly include the following:

w_v: the "weight" of ship v, i.e., the importance of inspecting ship v;
$[a_v, b_v]$: time window of ship v during which it should be inspected;
$d_{uv}(t)$: time required by a drone to fly from ship u time t to ship v;
Δ: Time required by a drone to inspect a ship;
Θ: Maximum flying duration of a drone without changing batteries.

Suppose that base station s has n_s drones. A basic version of the scheduling model for all of the n_s drones is as:

$$R_s(n_s) = \max \sum_v w_v z_v \tag{1}$$

subject to:

$$x_{uvk}(t_{uk} + \Delta + d_{uv}(t_{uk} + \Delta) - t_{vk}) \leq 0 \tag{2}$$

$$a_v \leq t_{vk} \leq b_v \tag{3}$$

$$\text{Total flying time of drone } k \text{ between two returns to base station} \leq \Theta,\ k = 1, 2, \ldots, n_s \tag{4}$$

and other constraints. The objective function (1) maximizes the total "weighted" number of ships that are inspected, denoted by $R_s(n_s)$. Constraint (2) ensures that there is sufficient time interval for drones to fly from one ship to the next. Constraint (3) guarantees that a ship must be inspected during its time window. Constraint (4) enforces the maximum flying duration of drones.

In Constraint (2), $d_{uv}(t)$ is generally a non-convex, non-differentiable function of t; moreover, $d_{uv}(t)$ may not even have an analytical expression. To overcome this difficulty, we plan to propose two solutions approaches. (i) For base stations with a small number of drones and ships to inspect, the time t can be discretized into small time intervals. After discretization, $d_{uv}(t)$ will be transformed into linear expressions with extra integer variables. Then, the above model can be transformed into a mixed-integer linear formulation, and we can develop exact algorithms to derive the optimal solution. (ii) If a base station has a large number of ships to inspect, the computational time of exact algorithms will be too long for practical application. We will hence design heuristics or meta-heuristics to obtain near-optimal solutions in reasonable time (Wang et al. 2013a; Liu et al. 2014; Zhen et al. 2016a).

2.2 Drone Assignment

Suppose that there are S base stations. To facilitate drone management, the ECA can be divided into S zones, one for each base station. Given a fleet of N drones, we need to decide how many drones to assign to each base station. This mainly depends on how many ships will transit each zone. Since the ship transit patterns on different days of a week (as well as on holidays) is different, drones may need to be repositioned every day from one base station to another. The drone assignment problem decides, for each day, based on the ship transit pattern, the number of drones n_s to assign to each base station $s = 1, 2, \ldots, S$. In Sect. 2.1, Eq. (1) provides the maximum “weighted” number of ships that can be inspected by n_s drones assigned to base station s. The drone assignment model, which maximizes the “weighted” number of ships that can be inspected by all of the N drones, can be formulated as:

$$\max_{n_s \text{nonnegative integer}} \sum_{s=1,2,\ldots,S} R_s(n_s) \tag{5}$$

subject to:

$$\sum_{s=1,2,\ldots,S} n_s \leq N \tag{6}$$

The above drone assignment model is an integer linear program. If $R_s(n_s)$ is concave in n_s, then this problem can be solved in polynomial time using the greedy algorithm proposed in Wang and Wang (2016). If $R_s(n_s)$ is not concave, then we can use dynamic programming to identify the optimal solution (Zhen et al. 2017c). In the dynamic programming procedure, there are S stages, each of which represents a base station; at each stage s, there are at most N states, each of which representing the total number of drones that have already been assigned to stages $1, 2, \ldots, s-1$. The running time of the dynamic programming solution is $O(SN)$, and hence the optimal solution can be obtained efficiently.

2.3 Base Station Location

In ECA, there can be many possible locations for establishing base stations for drones. A total of S base stations should be chosen from the candidate locations in order to maximize the value in Eq. (5). If there are W candidate locations, then the number of feasible solutions is $O(2^W)$. Moreover, to evaluate the quality of each solution, we need to solve the drone assignment problem, embedded with the drone scheduling problem. As a result, the base station location problem is a three-stage decision process that is extremely computationally intensive (Zhen et al. 2016b).

We can use a multi-start local-search algorithm to identify the locations to establish the base stations. The algorithm will start from many initial solutions. A local search procedure is applied to each initial solution, meaning that the solution will be repeatedly compared with its “neighboring” solutions, and be replaced by the best neighbor, until there is no neighbor that is better than it.

3 Conclusions

Drones have been identified as an efficient approach to monitor ship air pollution. Previous studies mainly focused on technical issues of using drones to monitor ship emissions from the technological perspective. Yet, the managerial perspective of the deployment of a fleet of drones is limited. This study proposed a modelling framework of drone deployment drones to inspect air pollution from ships. The framework contained three components which span operational, tactical and strategic decisions. At the operational level, the optimal schedule was designed on a real-time basis for each drone, including which ships to be inspected by the drone and when to inspect each ship. At the tactical level, the optimal assignment of drones to base stations was conducted on a daily basis. At the strategic level, we evaluated the optimal locations of base stations for drones. The proposed optimization models and algorithms would be conducive to enhancing the efficiency of drone deployment and then reducing SOx emissions from ships.

Acknowledgment. This research is sponsored by Environment and Conservation Fund Project 92/2017 and the Youth Program (No. 71501038), General Project (No. 71771050), Key Projects (No. 51638004) of the National Natural Science Foundation of China, and the Natural Science Foundation of Jiangsu Province in China (BK20150603). The views and opinions expressed in this article are those of the authors and do not necessarily reflect the official policy or position of any agency of government.

References

Fung, F.: Enforcement of fuel switching regulations – practices adopted in the US, EU and other regions, and lessons learned for China (2016). http://www.nrdc.cn/information/informationinfo?id=168. Accessed April 2017

Ning, Z.: "Drone" technology update in port and ship emission monitoring and management (2016). https://www.polyu.edu.hk/cee/MOVE-2016/4b-12-NING.pdf. Accessed April 2017

Collum, J.: Meeting the SECA Challenge (2015). http://www.maritime-executive.com/magazine/meeting-the-seca-challenge. Accessed April 2017

Roberts, L.: Drone detection of marine fuel sulphur emissions (2016). http://www.clydeco.com/insight/article/drone-detection-of-marine-fuel-sulphur-emissions. Accessed April 2017

Marine Electronics and Communication. Drones lead the way in emissions compliance (2015). http://www.marinemec.com/news. Accessed April 2017

Kuang, Y., Qu, X., Wang, S.: A tree-structured crash surrogate measure for freeways. Accid. Anal. Prev. **77**, 137–148 (2015)

Qu, X., Wang, S., Zhang, J.: On the fundamental diagram for freeway traffic: a novel calibration approach for single-regime models. Transp. Res. Part B: Methodological **73**, 91–102 (2015)

Qu, X., Zhang, J., Wang, S.: On the stochastic fundamental diagram for freeway traffic: model development, analytical properties, validation, and extensive applications. Transp. Res. Part B: Methodological **104**, 256–271 (2017)

Liu, Z., Wang, S., Chen, W., Zheng, Y.: Willingness to board: a novel concept for modeling queuing up passengers. Transp. Res. Part B **90**, 70–82 (2016)

Liu, Z., Yi, W., Wang, S., Chen, J.: On the uniqueness of user equilibrium flow with speed limit. Netw. Spat. Econ. **17**(3), 763–775 (2017a). https://doi.org/10.1007/s11067-017-9343-4

Liu, Z., Wang, S., Zhou, B., Cheng, Q.: Robust optimization of distance-based tolls in a network considering stochastic day to day dynamics. Transp. Res. Part C **79**, 58–72 (2017b)
Huang, D., Liu, Z., Liu, P., Chen, J.: Optimal transit fare and service frequency of a nonlinear origin destination based fare structure. Transp. Res. Part E **96**, 1–19 (2016)
Zhen, L., Wang, S., Zhuge, D.: Analysis of three container routing strategies. Int. J. Prod. Econ. **193**, 259–271 (2017a)
Zhen, L., Liang, Z., Zhuge, D., Lee, L.H., Chew, E.P.: Daily berth planning in a tidal port with channel flow control. Transp. Res. Part B: Methodological **106**, 193–217 (2017b)
Zhen, L., Wang, K., Wang, S., Qu, X.: Tug scheduling for hinterland barge transport: a branch-and-price approach. Eur. J. Oper. Res. **265**(1), 119–132 (2018)
Marine Traffic. https://www.marinetraffic.com/. Accessed April 2017
Cordeau, J.F., Desaulniers, G., Desrosiers, J., Solomon, M.M., Soumis, F.: The VRP with Time Windows. Montréal: Groupe d'études et de recherche en analyse des décisions (2000)
Desrochers, M., Desrosiers, J., Solomon, M.: A new optimization algorithm for the vehicle routing problem with time windows. Oper. Res. **40**(2), 342–354 (1992)
Ho, S.C., Haugland, D.: A tabu search heuristic for the vehicle routing problem with time windows and split deliveries. Comput. Oper. Res. **31**(12), 1947–1964 (2004)
Bräysy, O., Gendreau, M.: Vehicle routing problem with time windows, Part I: route construction and local search algorithms. Transp. Sci. **39**(1), 104–118 (2005)
Kallehauge, B.: Formulations and exact algorithms for the vehicle routing problem with time windows. Comput. Oper. Res. **35**(7), 2307–2330 (2008)
Wang, S., Liu, Z., Meng, Q.: Systematic network design for liner shipping services. Trans. Res. Rec. J. Transp. Res. Board **2330**, 16-23. (2013a)
Liu, Z., Meng, Q., Wang, S., Sun, Z.: Global intermodal liner shipping network design. Transp. Res. Part E: Logistics Transp. Rev. **61**, 28–39 (2014)
Zhen, L., Wang, S., Wang, K.: Terminal allocation problem in a transshipment hub considering bunker consumption. Naval Res. Logistics **63**(7), 529–548 (2016a)
Wang, S., Wang, X.: A polynomial-time algorithm for sailing speed optimization with containership resource sharing. Transp. Res. Part B Methodological **93**, 394–405 (2016)
Zhen, L., Wang, S., Zhuge, D.: Dynamic programming for optimal ship refueling decision. Transp. Res. Part E: Logistics Transp. Rev. **100**, 63–74 (2017c)
Zhen, L., Zhuge, D., Zhu, S.L.: Production stage allocation problem in large corporations. Omega **73**, 60–78 (2016b)

A Novel Approach to Evaluating Multi-period Performance of Non-storable Production with Carry-Over Activities

Barbara T. H. Yen[1(✉)], Lawrence W. Lan[2], and Yu-Chiun Chiou[2]

[1] School of Engineering and Built Environment, Griffith University, Gold Coast Campus, Southport 4222, Australia
t.yen@griffith.edu.au
[2] Department of Transportation and Logistics Management, National Chiao Tung University, Hsinchu, Taiwan, ROC

Abstract. This paper proposes a novel approach, called dynamic integrated slack-based measure (DISBM) modeling approach, to evaluating multi-period non-radial slacks of non-storable production characterized with carry-over activities. The proposed modeling approach incorporates conventional dynamic slack-based measure (DSBM) technical efficiency and service effectiveness into data envelopment analysis (DEA) modeling such that multi-period input excesses, output shortages and consumption gaps can be simultaneously determined. Some important properties of the proposed DISBM modeling are explored. A case study on the efficiency and effectiveness of Taiwan's intercity bus transport during 2007–2010 is presented. The results indicate that the proposed DISBM modeling is superior to conventional DSBM modeling in terms of benchmarking power, and that the non-radial slacks associated with input, output and consumption variables do provide rational information to rectify the inefficient and/or ineffective units throughout the production process.

Keywords: Data envelopment analysis (DEA)
Dynamic integrated slack-based measure (DISBM) · Technical efficiency
Service effectiveness · Bus transport

1 Introduction

In service industries, some products are perishable or non-storable and with lagged-productive or carry-over effects (Yu and Lee 2009; Hsieh and Lin 2010; Lan et al. 2013; Wu et al. 2013). The transport and lodging services are two typical examples of non-storable production systems because one can never stockpile the surplus services in low-demand periods (off-peak hours) to be used in high-demand periods (peak hours). When such non-storable services are produced and a portion of which are not concurrently consumed, the technical effectiveness (a joint effect of both technical efficiency and service effectiveness) would certainly be less than its technical efficiency. In the airline and lodging examples, all the unoccupied seats (rooms) at any

G. De Pietro et al. (Eds.): KES-IIMSS-18 2018, SIST 98, pp. 289–299, 2019.
https://doi.org/10.1007/978-3-319-92231-7_30

flight (night) are gone forever and its revenues for that flight (night) will be lost and can never be retrieved. Essentially, technical efficiency and service effectiveness for non-storable production systems represent two distinct measurements (Fielding 1987; Lan and Lin 2003; Chiou et al. 2010). To elucidate this concept, Fig. 1 manifests three distinctive performance measurements for the scheduled transport carriers' production, including technical efficiency, service effectiveness and technical effectiveness. Conventional benchmarking methods, however, mainly measure the technical efficiency without jointly accounting for the service effectiveness. Here arises an important issue regarding the holistic measurement of efficiency and effectiveness with consideration of non-storable effects.

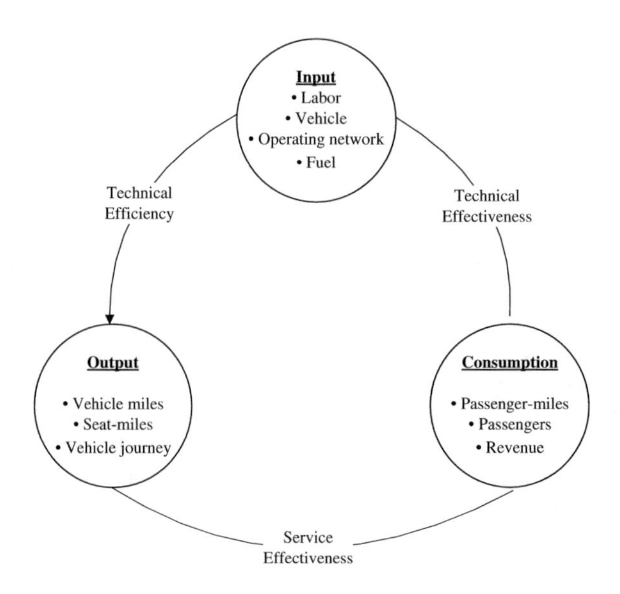

Fig. 1. Distinctive performance measurements for transport service

To tackle the aforementioned difficulty, this study proposes a novel DEA approach, the dynamic integrated slack-based measure (DISBM) modelling, which not only accounts for three distinctive performance measures of non-storable production but also considers carry-over effects in multi-period contexts. Specifically, the proposed novel approach will measure the non-radial slacks to adequately represent the input excesses, output shortages and consumption gaps for non-storable production with carry-over activities existing in the study horizon.

The rest of this paper is organized as follows. Section 2 formulates the proposed DISBM modeling. An empirical study is carried out in Sect. 3. Finally, conclusions and avenues of future research are addressed.

2 The Proposed Models

2.1 The Fundamentals

Tone and Tsutusi (2010) propose a dynamic slack-based measure (DSBM) model which directly evaluates the input excesses and the output shortages of a DMU in multi-period contexts. The DSBM model can be expressed by a fractional programming in λ^t, s_{it}^-, s_{it}^+, s_{it}^{good}, s_{it}^{bad}, and s_{it}^{free} with input-, output- and non-oriented formulations. Follow Tone and Tsutusi (2010), we further symbolize the four category links as z^{good}, z^{bad}, z^{free} and z^{fix}. The continuity of link flows (i.e., carry-overs) between periods t and $t+1$ can be guaranteed by the following condition:

$$\sum_{j=1}^{J} z_{ijt}^{\alpha} \lambda_j^t = \sum_{j=1}^{J} z_{ijt}^{\alpha} \lambda_j^{t+1} (\forall i;\ t = 1, \ldots, T-1). \tag{1}$$

where α stands for *good, bad, free* and *fixed*. This constraint is critical for the dynamic model, since it connects activities in periods t and $t+1$. We can further express $DMU_o(o = 1, \ldots, J)$ as follows:

$$x_{iot} = \sum_{j=1}^{J} x_{ijt} \lambda_j^t + s_{it}^- (i = 1, \ldots, m;\ t = 1, \ldots, T), \tag{2}$$

$$x_{iot}^{fix} = \sum_{j=1}^{J} x_{ijt}^{fix} \lambda_j^t (i = 1, \ldots, p;\ t = 1, \ldots, T), \tag{3}$$

$$y_{iot} = \sum_{j=1}^{J} y_{ijt} \lambda_j^t - s_{it}^+ (i = 1, \ldots, s;\ t = 1, \ldots, T) \tag{4}$$

$$y_{iot}^{fix} = \sum_{j=1}^{J} y_{ijt}^{fix} \lambda_j^t (i = 1, \ldots, r;\ t = 1, \ldots, T) \tag{5}$$

$$q_{iot} = \sum_{j=1}^{J} q_{ijt} \lambda_j^t - s_{it}^+ (i = 1, \ldots, l;\ t = 1, \ldots, T) \tag{6}$$

$$q_{iot}^{fix} = \sum_{j=1}^{J} q_{ijt}^{fix} \lambda_j^t (i = 1, \ldots, n;\ t = 1, \ldots, T), \tag{7}$$

$$z_{iot}^{good} = \sum_{j=1}^{J} z_{ijt}^{good} \lambda_j^t - s_{it}^{good} (i = 1, \ldots, ngood;\ t = 1, \ldots, T), \tag{8}$$

$$z_{it}^{bad} = \sum_{j=1}^{J} z_{ijt}^{bad} \lambda_j^t + s_{it}^{bad} (i = 1, \ldots, nbad;\ t = 1, \ldots, T), \tag{9}$$

$$z_{iot}^{free} = \sum_{j=1}^{J} z_{ijt}^{free} \lambda_j^t + s_{it}^{free} (i = 1, \ldots, nfree;\ t = 1, \ldots, T), \tag{10}$$

$$z_{iot}^{fix} = \sum_{j=1}^{J} z_{ijt}^{fix} \lambda_j^t (i = 1, \ldots, nfix;\ t = 1, \ldots, T), \tag{11}$$

$$\sum_{j=1}^{J} \lambda_j^t = 1 (t = 1, \ldots, T), \tag{12}$$

$$\lambda_j^t,\ s_{it}^-,\ s_{it}^+,\ s_{it}^{good},\ s_{it}^{bad} \geq 0 \text{ and } s_{it}^{free} : free(\forall i, t). \tag{13}$$

where $\lambda^t \in R^n (t = 1, \ldots, T)$ is the intensity vector for period t, and *nbad*, *nfree* and *nfix* are the number of bad, free and fixed links, respectively. The last constraint, Eq. (12), corresponds to the variable-returns-to-scale (VRS) scenario (Cooper et al. 2007). If this constraint is not included in the formulation, the model would become a constant-returns-to-scale (CRS) scenario. $s_{it}^-, s_{it}^+, s_{it}^{good}, s_{it}^{bad}$ and s_{it}^{free} are slack variables denoting, respectively, input excess, output shortage/consumption gap, link shortfall, link excess and link deviation.

The input-, output- and non-oriented DSBM models evaluate the overall efficiency of DMU_o with variables $\lambda_j^t, s_{it}^-, s_{it}^+, s_{it}^{good}, s_{it}^{bad}$ and s_{it}^{free}. Essentially, the objective functions of input- and output-oriented DSBM models are denoted as [DSBM$_I$] and [DSBM$_O$], in which their corresponding efficiency scores, ρ_I and ρ_O, are obtained from the following:

$$[\text{DSBM}_I]\ Min\ \rho_I = \frac{1}{T} \sum_{t=1}^{T} w^t \left[1 - \frac{1}{m + nbad} \left(\sum_{i=1}^{m} \frac{w_i^- s_{it}^-}{x_{iot}} + \sum_{i=1}^{nbad} \frac{s_{it}^{bad}}{z_{iot}^{bad}} \right) \right] \tag{14}$$

$$[\text{DSBM}_O]\ Min\ \rho_O = \frac{1}{T} \sum_{t=1}^{T} w^t \left[1 / \left(1 + \frac{1}{s + ngood} \sum_{i=1}^{s} \frac{w_i^+ s_{it}^+}{y_{iot}} + \sum_{i}^{good} \frac{s_{it}^{good}}{z_{iot}^{good}} \right) \right] \tag{15}$$

[DSBM$_I$] and [DSBM$_O$] are subject to Eqs. (1)–(5) and (8)–(13), where w^t, w_i^- and w_i^+ are weights to period t, input i and output i, respectively, which are supplied exogenously according to their importance, and which also satisfy the following conditions:

$$\sum_{t=1}^{T} w^t = T \sum_{i=1}^{m} w_i^- = m \text{ and } \sum_{i=1}^{s} w_i^+ = s. \tag{16}$$

If all weights are even, then w^t, w_i^- and w_i^+ can be set $w^t = 1(\forall t)$, $w_i^- = 1(\forall i)$ and $w_i^+ = 1(\forall i)$.

2.2 The Proposed DISBM Models

Basically, the proposed dynamic integrated slack-based measure (DISBM) models use the above [DSBM$_\text{I}$] to measure technical efficiency, denoted as [DSBM$_\text{I}$-TE], and the above [DSBM$_\text{O}$] to measure service effectiveness, denoted as [DSBM$_\text{O}$-SE]. The technical efficiency score ρ_{TE} and the service effectiveness score ρ_{SE} can be respectively obtained from the following:

$$[\mathrm{DSBM_I - TE}]\ Min\ \rho_{TE} = \frac{1}{T}\sum_{t=1}^{T} w^t\left[1 - \frac{1}{m + nbad}\left(\sum_{i=1}^{m}\frac{w_i^- s_{it}^-}{x_{iot}} + \sum_{i=1}^{nbad}\frac{s_{it}^{bad}}{z_{iot}^{bad}}\right)\right] \tag{17}$$

subject to Eqs. (1)–(5), (8)–(13) and to the conditions in Eq. (16).

$$[\mathrm{DSBM_O - SE}]\ Min\ \rho_{SE} = 1/\left\{\frac{1}{T}\sum_{t=1}^{T} w^t\left[1 + \frac{1}{l + ngood}\left(\sum_{i=1}^{l}\frac{w_i^+ s_{it}^+}{q_{iot}} + \sum_{i}^{good}\frac{s_{it}^{good}}{z_{iot}^{good}}\right)\right]\right\} \tag{18}$$

subject to Eqs. (1) and (4)–(13), where w^t and w_i^+ are weights to period t and consumption variable i which are supplied exogenously according to their importance, and which also satisfy the following conditions:

$$\sum_{t=1}^{T} w^t = T \text{ and } \sum_{i=1}^{l} w_i^+ = l \tag{19}$$

Under CRS context, we integrate the above [DSBM$_\text{I}$-TE] and [DSBM$_\text{O}$-SE] into the DEA modeling, denoted as [DISBM-CRS]. The integrated efficiency values ρ_{TE-SE} can therefore be obtained from the following:

$$[\mathrm{DISBM - CRS}]\ Min\ \rho_{TE-SE} = \frac{\frac{1}{T}\sum_{t=1}^{T} w^t\left[1 - \frac{1}{m+nbad}\left(\sum_{i=1}^{m}\frac{w_i^- s_{it}^-}{x_{iot}} + \sum_{i=1}^{nbad}\frac{s_{it}^{bad}}{z_{iot}^{bad}}\right)\right]}{\frac{1}{T}\sum_{t=1}^{T} w^t\left[1 + \frac{1}{l+ngood}\left(\sum_{i=1}^{l}\frac{w_i^+ s_{it}^+}{q_{iot}} + \sum_{i}^{good}\frac{s_{it}^{good}}{z_{iot}^{good}}\right)\right]} \tag{20}$$

subject to Eqs. (1)–(11), (13) and also satisfying the conditions in Eqs. (16) and (19), where ρ_{TE-SE} denotes the integrated efficiency score of DMU_o. If Eq. (20) subject to Eqs. (1)–(13) can also satisfy the conditions in Eqs. (16) and (19), [DISBM-CRS] would become [DISBM-VRS]. Following Tone (2001), both [DISBM-CRS] and [DISBM-VRS] can be solved as a linear programming problem.

The overall efficiency score for each DMU can be determined by the above [DISBM-CRS] and [DISBM-VRS] formulations. The proposed DISBM modelling approach would have the same properties as the non-oriented DSBM model in Tone and Tsutsui (2010). However, the superiority of proposed modelling approach is that it can be used to evaluate the efficiency for multiple departments under multi-period contexts.

If further applied to Q-department production process as shown in Fig. 2, more generalized specifications can be formulated. Let Q denote the number of departments, N_q denote the number of variables in the qth department, J denote the number of DMUs. For the non-storable commodities in this paper, Q would be 3; N_1 denote the number of input variables; N_2 denote the number of output variables; N_3 denote the number of consumption variables. The generalized [DISBM-CRS] problem with Q-department production process, denoted as [GDISBM-CRS], can be formulated as follows:

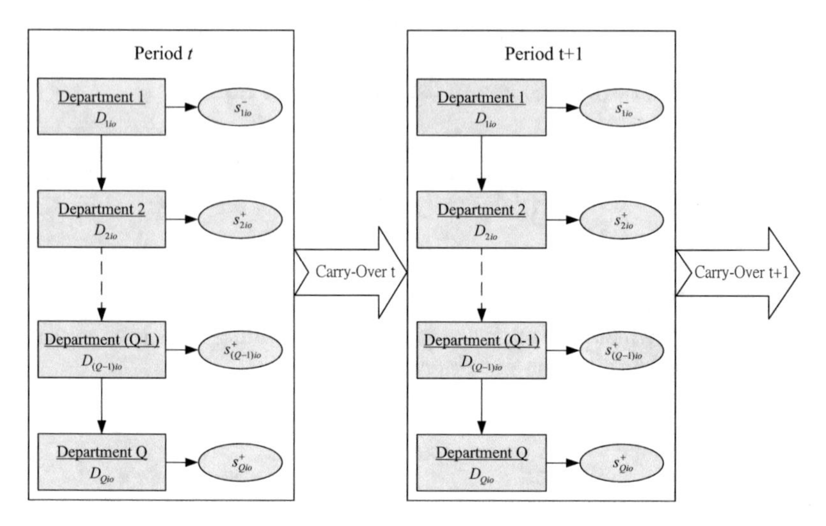

Fig. 2. A dynamic structure for Q-department production process

3 Case Study

To demonstrate the proposed DISBM models, a case study of intercity bus transport is presented. This case study evaluates the technical efficiency and service effectiveness of 33 intercity bus companies in Taiwan over 5 periods: 2007–2011. The variables are arranged as follows: two input variables—number of buses (NB) and operating network (ON), two output variables—number of bus runs (NBR) and bus-km (B-km), three consumption variables—number of passengers (NP), passenger-km (P-km) and average number of on-board passengers per run (AOP), and one "carry-over" variable—operating revenue (OR). The data are excerpted from Annual Report published by the Ministry of Transportation and Communications. Table 1 presents the descriptive statistics over the study horizon.

We employ the proposed [DISBM-CRS] model to jointly measure the overall efficiency scores of each bus company under CRS with equal weight for each period.

Table 1. Descriptive statistics for 33 Taiwan's intercity bus carriers (2007–2011)

Year	Item	Input		Output		Consumption			Carry-over
		NB	ON (km)	NBR	B-km	NP	P-km	AOP	OR (NT$)
2007	Mean	96	744	139,712	18,386,535	2,154,120	238,273,351	12	336,265,423
	Std. Dev.	213	1,838	263,626	45,130,129	4,823,277	602,765,086	7	823,159,026
	Max.	950	7,748	903,181	205,837,912	11,988,249	2,700,030,241	13	3,337,133,985
	Min.	4	65	5,050	329,260	74,539	2,791,207	8	5,185,209
2008	Mean	98	763	122,465	14,138,313	2,126,299	205,021,705	12	278,527,123
	Std. Dev.	218	1,834	234,293	33,319,389	4,698,601	554,095,918	5	685,077,548
	Max.	984	7,754	669,094	149,780,836	11,404,110	2,627,437,002	18	2,958,772,172
	Min.	4	65	4,423	288,380	63,342	2,252,757	8	4,639,645
2009	Mean	92	701	122,498	14,066,109	2,184,718	212,989,660	13	278,042,113
	Std. Dev.	199	1,825	239,241	32,426,242	4,653,203	559,339,746	5	670,046,987
	Max.	898	7,754	626,808	139,180,568	11,356,019	2,575,575,086	19	2,851,468,376
	Min.	4	155	1,340	110,434	18,784	743,544	7	1,774,544
2010	Mean	95	710	132,435	14,265,515	2,261,038	207,547,398	12	286,078,619
	Std. Dev.	202	1,834	244,384	32,532,300	4,661,122	510,925,714	5	673,008,106
	Max.	867	7,752	603,848	133,432,051	9,499,374	2,153,016,957	16	2,595,274,445
	Min.	4	155	1,460	119,643	19,108	697,851	6	1,667,873
2011	Mean	94	872	138,063	14,636,130	2,406,328	220,154,265	13	303,953,552
	Std. Dev.	200	2,004	254,276	33,902,692	5,007,787	537,882,008	5	700,503,283
	Max.	831	7,690	1,322,583	123,899,304	27,090,186	2,059,963,115	17	2,886,772,136
	Min.	1	65	3,254	212,161	36,623	1,377,022	6	2,779,518

Table 2. Results of [DISBM-CRS] model in 2007–2011

DMU	Overall efficiency		2007		2008		2009		2010		2011	
	TE	SE	TE	SE	TE	SE	TE	SE	TE	SE	TE	SE
1	0.891	0.847	0.724	0.470	0.730	0.767	1.000	1.000	1.000	1.000	1.000	1.000
2	0.849	0.266	0.721	0.061	0.786	0.101	1.000	1.000	0.737	0.101	1.000	0.067
3	0.999	0.752	1.000	1.000	0.997	0.339	1.000	1.000	1.000	1.000	1.000	0.421
4	0.873	0.688	0.867	0.424	0.792	0.739	1.000	1.000	0.757	0.741	0.947	0.535
5	0.807	0.478	0.645	0.472	0.742	0.673	0.902	0.099	1.000	0.610	0.746	0.534
6	0.844	0.641	0.756	0.552	0.898	0.744	0.866	0.584	0.876	0.656	0.826	0.671
7	1.000	1.000	1.000	1.000	1.000	1.000	1.000	1.000	1.000	1.000	1.000	1.000
8	0.886	0.108	0.795	0.069	0.853	0.063	1.000	0.119	0.915	0.125	0.865	0.165
9	1.000	1.000	1.000	1.000	1.000	1.000	1.000	1.000	1.000	1.000	1.000	1.000
10	1.000	1.000	1.000	1.000	1.000	1.000	1.000	1.000	1.000	1.000	1.000	1.000
11	0.996	0.211	1.000	0.129	0.991	0.160	1.000	0.329	1.000	0.208	0.987	0.228
12	0.989	0.221	1.000	0.159	1.000	0.430	0.945	0.350	1.000	0.075	1.000	0.089
13	0.960	0.958	1.000	1.000	1.000	1.000	1.000	1.000	1.000	1.000	0.800	0.788
14	0.884	0.292	0.842	0.217	0.885	0.466	1.000	0.207	1.000	0.311	0.691	0.260
15	0.977	0.843	0.887	0.216	1.000	1.000	1.000	1.000	1.000	1.000	1.000	1.000
16	0.926	0.701	0.777	0.535	0.998	0.757	0.904	0.772	0.950	0.754	1.000	0.688
17	0.950	0.966	1.000	1.000	1.000	1.000	1.000	1.000	1.000	1.000	0.750	0.831
18	0.737	0.769	0.654	0.625	0.709	0.839	0.781	0.750	0.775	0.807	0.767	0.823
19	0.937	0.878	1.000	1.000	1.000	1.000	1.000	1.000	0.787	0.728	0.900	0.661
20	0.617	1.000	0.994	1.000	1.000	1.000	0.013	1.000	0.078	1.000	1.000	1.000
21	0.773	0.724	0.602	0.724	0.759	0.748	0.757	0.673	0.749	0.713	1.000	0.762
22	0.836	0.883	0.331	0.840	1.000	1.000	1.000	1.000	1.000	1.000	0.850	0.577
23	0.863	0.799	0.510	0.767	1.000	1.000	1.000	1.000	1.000	1.000	0.803	0.229
24	0.533	0.119	0.597	0.065	0.338	0.152	0.546	0.156	0.543	0.179	0.642	0.041
25	0.644	0.284	0.667	0.224	0.500	0.418	0.767	0.209	0.640	0.302	0.646	0.269
26	0.745	0.545	0.779	0.723	0.802	0.637	0.752	0.486	0.703	0.452	0.688	0.428
27	0.593	0.196	0.523	0.128	0.603	0.190	0.607	0.290	0.605	0.220	0.629	0.154
28	0.554	0.432	0.602	0.523	0.321	0.417	0.564	0.361	0.639	0.457	0.644	0.401
29	1.000	1.000	1.000	1.000	1.000	1.000	1.000	1.000	1.000	1.000	1.000	1.000
30	0.895	0.877	0.834	0.836	0.927	0.788	0.714	0.760	1.000	1.000	1.000	1.000
31	0.658	0.472	0.553	0.273	0.572	0.301	0.543	0.335	0.623	0.452	1.000	1.000
32	1.000	1.000	1.000	1.000	1.000	1.000	1.000	1.000	1.000	1.000	1.000	1.000
33	0.962	0.774	1.000	1.000	1.000	1.000	0.983	0.725	0.827	0.673	1.000	0.472

Table 2 presents the overall and 5-period efficiencies. For comparison, we also employ separated DSBM models which measure the non-oriented technical efficiency and service effectiveness for each company. Table 3 reports the results of separated DSBM models.

Table 3. Results of separated DSBM models in 2007–2011

DMU	Overall efficiency		2007		2008		2009		2010		2011	
	TE	SE	TE	SE	TE	SE	TE	SE	TE	SE	TE	SE
1	1.000	1.000	1.000	1.000	1.000	1.000	1.000	1.000	1.000	1.000	1.000	1.000
2	1.000	1.000	1.000	1.000	1.000	1.000	1.000	1.000	1.000	1.000	1.000	1.000
3	1.000	1.000	1.000	1.000	1.000	1.000	1.000	1.000	1.000	1.000	1.000	1.000
4	0.964	1.000	1.000	1.000	1.000	1.000	1.000	1.000	0.822	1.000	1.000	1.000
5	0.845	0.663	0.752	0.858	0.803	0.841	0.896	0.261	1.000	0.725	0.772	0.629
6	0.965	0.855	0.863	0.966	0.965	0.909	1.000	0.752	1.000	0.829	1.000	0.819
7	1.000	1.000	1.000	1.000	1.000	1.000	1.000	1.000	1.000	1.000	1.000	1.000
8	0.738	0.154	0.457	0.121	0.453	0.185	0.863	0.163	0.917	0.139	1.000	0.162
9	1.000	1.000	1.000	1.000	1.000	1.000	1.000	1.000	1.000	1.000	1.000	1.000
10	1.000	1.000	1.000	1.000	1.000	1.000	1.000	1.000	1.000	1.000	1.000	1.000
11	0.867	0.296	0.692	0.289	0.788	0.355	1.000	0.329	0.949	0.230	0.905	0.277
12	1.000	0.993	1.000	0.997	1.000	0.982	1.000	0.997	0.999	0.994	0.999	0.997
13	1.000	1.000	1.000	1.000	1.000	1.000	1.000	1.000	1.000	1.000	1.000	1.000
14	0.758	0.449	0.706	0.309	0.885	0.474	0.810	0.381	0.860	0.449	0.530	0.632
15	1.000	1.000	1.000	1.000	1.000	1.000	1.000	1.000	1.000	1.000	1.000	1.000
16	0.958	0.739	0.889	0.533	1.000	0.806	0.947	0.778	0.957	0.813	1.000	0.763
17	0.943	1.000	1.000	1.000	1.000	1.000	1.000	1.000	1.000	1.000	0.714	1.000
18	0.838	0.766	0.774	0.651	0.765	0.895	0.925	0.688	0.887	0.784	0.837	0.811
19	1.000	1.000	1.000	1.000	1.000	1.000	1.000	1.000	1.000	1.000	1.000	1.000
20	0.617	1.000	0.994	1.000	1.000	1.000	0.013	1.000	0.078	1.000	1.000	1.000
21	0.781	0.708	0.617	0.719	0.776	0.745	0.777	0.650	0.779	0.695	0.958	0.730
22	0.914	1.000	0.570	1.000	1.000	1.000	1.000	1.000	1.000	1.000	1.000	1.000
23	1.000	1.000	1.000	1.000	1.000	1.000	1.000	1.000	1.000	1.000	1.000	1.000
24	1.000	1.000	1.000	1.000	1.000	1.000	1.000	1.000	1.000	1.000	1.000	1.000
25	0.620	0.324	0.593	0.298	0.558	0.349	0.686	0.293	0.614	0.360	0.650	0.320
26	1.000	1.000	1.000	1.000	1.000	1.000	1.000	1.000	1.000	1.000	1.000	1.000
27	0.699	0.488	0.605	0.435	0.721	0.511	0.733	0.510	0.787	0.581	0.651	0.402
28	0.643	0.483	0.760	0.631	0.339	0.445	0.683	0.373	0.717	0.523	0.717	0.445
29	1.000	1.000	1.000	1.000	1.000	1.000	1.000	1.000	1.000	1.000	1.000	1.000
30	1.000	1.000	1.000	1.000	1.000	1.000	1.000	1.000	1.000	1.000	1.000	1.000
31	1.000	1.000	1.000	1.000	1.000	1.000	1.000	1.000	1.000	1.000	1.000	1.000
32	1.000	1.000	1.000	1.000	1.000	1.000	1.000	1.000	1.000	1.000	1.000	1.000
33	1.000	1.000	1.000	1.000	1.000	1.000	1.000	1.000	1.000	1.000	1.000	1.000

Figures 3 and 4 compare the overall technical efficiency and service effectiveness of DISBM to those of DSBM, respectively. From Figs. 3 and 4, obviously, the benchmarking power of DISBM is much superior to that of DSBM. It should be mentioned that the efficiency scores cannot be compared directly because the problem structures are different—DISBM has an aggregated structure but DSBM has an independent one. Besides, DSBM over emphasizes the importance of production sector

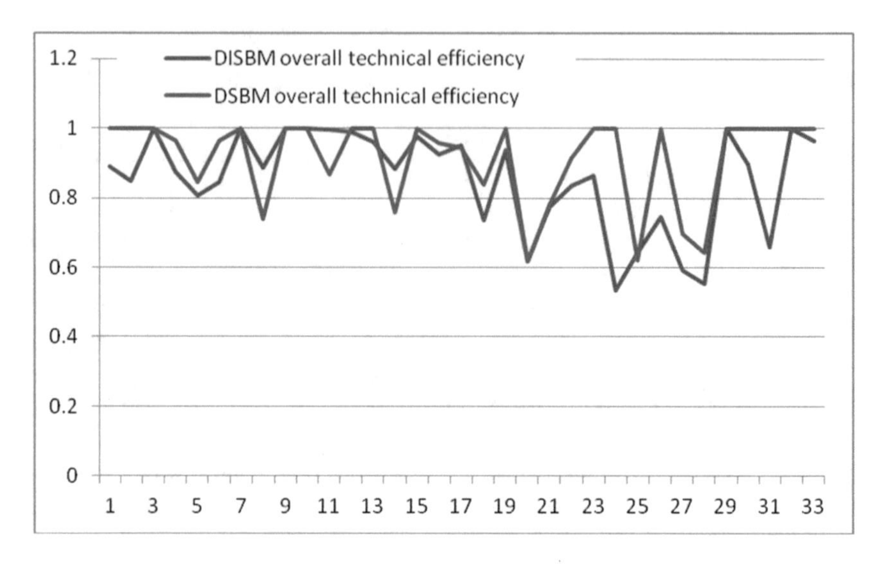

Fig. 3. Technical efficiency comparison of DISBM and DSBM models.

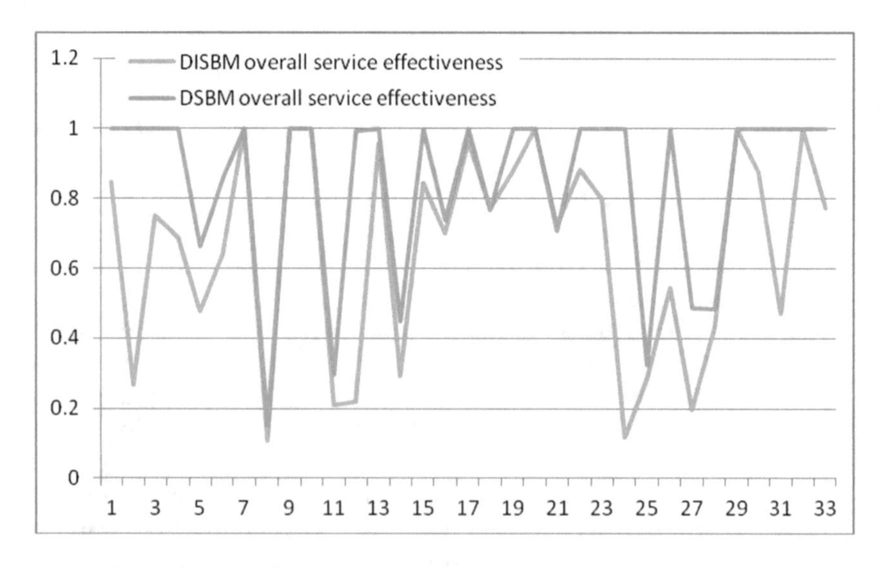

Fig. 4. Service effectiveness comparison of DISBM and DSBM models.

because of its independent structure—the production sector has been evaluated twice with completely opposite position. The production sector is viewed as output sector for technical efficiency and output shortages are measured. On the contrary, it is treated as input sector for service effectiveness and input excesses are measured. This difference explains why some DMUs are evaluated with efficiency (effectiveness) in DSBM but inefficiency (ineffectiveness) in DISBM.

4 Conclusions

This paper has contributed to DEA literature in several ways. In theory, the novel DISBM approach takes into account the non-storable production properties with carry-over effects by integrating both DSBM technical efficiency and service effectiveness into the DEA modelling such that the non-radial slack values for input, output and consumption can be jointly determined. Therefore, the proposed DISBM modelling can precisely account for input excesses, output shortages and consumption gaps. In practice, the case study has demonstrated that DISBM is superior to DSBM in terms of benchmarking power.

The proposed DISBM models are specified with integration of input-oriented technical efficiency and output-oriented service effectiveness. Other specifications or even multi-objective specifications deserve further exploration. This study only illustrates a case study (bus transit). It is a challenging issue to apply the proposed DISBM models to an enterprise with many branches or departments that are vertically and/or horizontally interrelated, such as the supply chain systems within an enterprise, the postal mail operation (pickup, processing and delivery), among others.

Acknowledgement. This study was sponsored by National Science Council of the Republic of China (NSC101-2221-E-233-010 and NSC102-2221-E-233-004).

References

Chiou, Y.C., Lan, L.W., Yen, T.H.: A joint measurement of efficiency and effectiveness for non-storable commodities: integrated data envelopment analysis approaches. Eur. J. Oper. Res. **201**, 477–489 (2010)

Fielding, G.J.: Managing Public Transit Strategically. Jossey-Bass Inc., San Francisco (1987)

Tone, K., Tsutsui, M.: Dynamic DEA: a slacks-based measure approach. Omega **38**, 145–156 (2010)

Wu, W.W., Lan, L.W., Lee, Y.T.: Benchmarking hotel industry in a multi-period context with DEA approaches: a case study. Benchmarking Int. J. **20**, 152–168 (2013)

Yu, M.M., Lee, B.C.Y.: Efficiency and effectiveness of service business: evidence from international tourist hotels in Taiwan. Tour. Manag. **30**(4), 571–580 (2009)

Hsieh, L.F., Lin, L.H.: A performance evaluation model for international tourist hotels in Taiwan: an application of the relational network DEA. Int. J. Hospitality Manag. **29**(1), 14–24 (2010)

Lan, L.W., Wu, W.W., Lee, Y.T.: Strategic benchmarking for tourist hotels in Taiwan. East-Asia Rev. **480**, 35–52 (2013)

Lan, L.W., Lin, E.T.J.: Technical efficiency and service effectiveness for railways industry: DEA approaches. J. Eastern Asia Soc. Transp. Stud. **5**, 2932–2947 (2003)

Tone, K.: A slacks-based measure of efficiency in data envelopment analysis. Eur. J. Oper. Res. **130**(3), 498–509 (2001)

Cooper, W.W., Huang, Z., Li, S.X., Parker, B.R., Pastor, J.T.: Efficiency aggregation with enhanced Russell measures in data envelopment analysis. Socio-Economic Plann. Sci. **41**(1), 1–21 (2007)

Urban Rail Transit Demand Analysis and Prediction: A Review of Recent Studies

Zhiyan Fang, Qixiu Cheng(✉), Ruo Jia, and Zhiyuan Liu

Jiangsu Key Laboratory of Urban ITS, Jiangsu Province Collaborative Innovation Center of Modern Urban Traffic Technologies, School of Transportation, Southeast University, Nanjing, China
qixiu.cheng@seu.edu.cn

Abstract. Urban rail transit demand analysis and forecasting is an essential prerequisite for daily operations and management. This paper categorizes the proposed demand forecasting methods, and focuses on traditional models, statistical models and machine learning approaches, according to their features and fields. Especially, influential and widely-used methods including the four-stage model, land use models, time series methods, Logit regression, Artificial Neural Networks (ANNs) and other referring methods are all taken into discussion.

1 Introduction

Demand forecasting is an essential process for the rational planning of transport system facilities and infrastructures [1]. Urban rail transit is a key component of public transportation, which provides fast, convenient and relatively long-distance public transport service [2]. Its demand prediction has become an essential part of traffic flow prediction for its widely use in dense urban cities.

Influential factors including fares, willingness to board, network planning and assessment are all important elements in rail transit system, where elaborate studies contributes to the improvement of demand forecasting in a direct or indirect way. In this paper, those elements are introduced briefly, while mainly focusing on recent studies in the field of passenger demand forecasting, for its importance and influence in urban rail transit planning, construction and administration. Generally, main hurdles in inaccurate forecasting results often result from [3] the in-conformity of urban and transportation planning, the lack of investigation data, the unawareness of influential factors, the deficiency of metro network and the problems existing in demand forecasting models and applications. With regard to the above-mentioned weakness mostly referring to classic, relatively macroscopical methods, the development of dynamic and day-to-day prediction also receives much attention in the recent years. The widely applicant of big data, as a by-product in the information era, contributes to the rapid development of forecasting models as well.

In the following sections, Sect. 2 presents a brief analysis of influential factors in rail transit system; in Sect. 3, a classification of its demand forecasting methodology is proposed, elaborated in traditional models, statistical models and machine learning

G. De Pietro et al. (Eds.): KES-IIMSS-18 2018, SIST 98, pp. 300–309, 2019.
https://doi.org/10.1007/978-3-319-92231-7_31

approaches, which gives a literature review of the current works regarding to rail transit demand and relating analysis; Sect. 4 gives a conclusion of this paper.

2 Transit Flow Analysis

Urban rail transit is a large, complicated system, where large numbers of elements have to be taken into account before prediction process, also for the sake of further improving the operation efficiency and service level. The system analysis ranges from fare structure, queuing problem handling, station choice making, feeder modes for metro transit and so on. All the studies provides a different view in considering urban rail transit system and promotes its demand forecasting, directly or indirectly.

For most passengers involving public transport, fare is one of the most important factors in route decision. It guides the passenger route choice and boarding time choice in several ways [4]. Approaches have been brought forward to alleviate passenger congestion, including a new nonlinear distance-based transit fare structure, measured by origin and destination Euclidean distance [5], giving passenger greater freedom to make efficient trip plan and achieving systematic optimized solution. Some fare considerations even help in demand prediction [6].

Passenger queuing reflects the willingness to board and the level of service. Queue parameter can be added as a variable in metro network design [7], and the quantitative investigation of board willingness [8] provides a new insight into demand prediction problems. Particularly, the methodology of those studies focus more on mathematical model construction and solving algorithms, which is also a relatively blank area in metro flow forecasting.

With a lot of forecasting studies stay at the station level, station choice for passengers plays an important part as well, for it is a key component in regression models for forecasting. The choice analysis itself, on the other hand, can also be achieved by regression models [9], or by mathematical model at user equilibrium [10].

Transit feeder modes, another influential factor in demand prediction analysis, are also taken into consideration. As a reflection of land use patterns, which is a significant category in traditional forecasting methods, feeder modes represent a considerable barrier to ridership remaining to be lightened. Luckily, it can be predicted by taking weather effects into consideration [11], or modelled involving both observed and unobserved factors for Transit-Oriented-Development (TOD) application [12].

Generally, all those considerations give a relatively comprehensive and elaborate view of the planning and administrative details, proposed from the view of all transport authority, transit company and passengers. Moreover, some studies about Stop-skipping scheme [13] and transit network design [14], or other analysis involving route performance evaluation [15] and fundamental diagram [16] of other public transports can also be used for reference, combining unique features of rail transit itself. It is reasonable to believe that the transit flow analysis provides a new, broader view to demand forecasting, and deep its consideration in relative factors.

3 Models for Rail Transit Demand Prediction

In general, traffic flow prediction models can be divided to two broad categories: qualitative methods, such as Delphi method, economic surveys and analogical method [17]; and quantitative methods [1]. However, to the best of our knowledge, there is little study applying qualitative methods in urban rail demand forecasting, potentially partly because of the modernity.

Among large varieties of quantitative methods, it can be primarily divided into four large categories: traditional models, including classic four-stage aggregate model and models regarding land use; statistical models including time series methods and regression models; machine learning approaches including various kinds of artificial neural networks and deep learning method; and finally simulation software packages, suitable for macro-level, meso-level and micro-level forecasting, depending on the functions (Fig. 1).

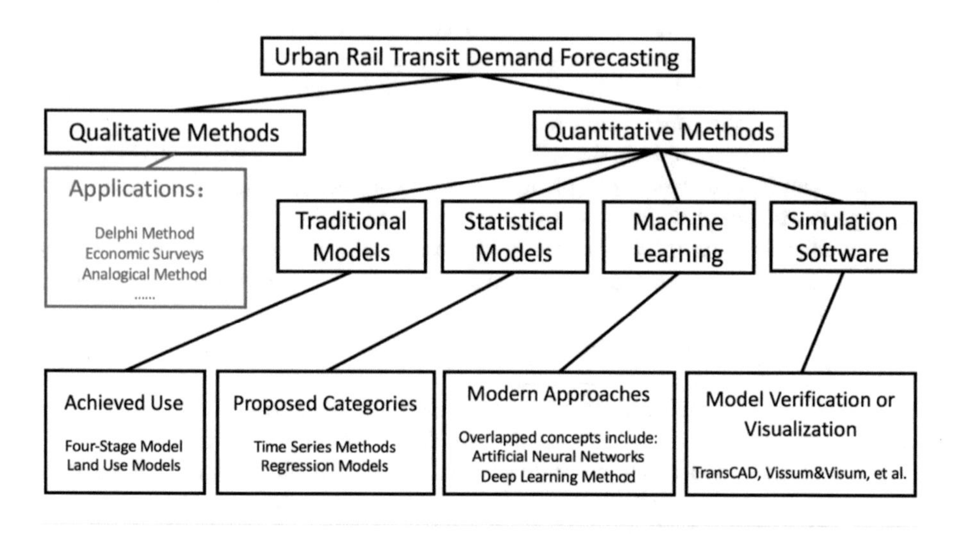

Fig. 1. Classification chart of urban rail transit demand prediction methods

Specifically, time series methods are based on historical data, including exponential smoothing, seasonal coefficient method and gray correlation analysis [17]; regression models aim at constructing regression equations revealing relationship between passenger flow and influential factors, including linear regression model, non-linear regression model [17] and important discrete models like Logit regression applied in disaggregate model. Artificial Neural Networks (ANNs) range from Feedback Neural Network (FNN) to Back Propagation Neural Network Model (BPNN), or other algorithms and methods.

Simulation software packages, such as TransCAD, Visum and Vissim, CUBE and others, are used for model verification in planning or as visualization tools more often, which has outstanding merits with regard to macroscopic project assessment and comparison.

The reason mostly relies on it providing a relatively cheap and convenient way when mathematical model construction is too complicated to access. However, the shortcomings are still obvious: it gives an approximate estimation of forecasting methods like four-stage model, without the ability of combining statistical methods and Artificial Neural Networks for further increasing accuracy.

3.1 Traditional Models

Traditional models, unlike time series and machine learning aiming at pure statistical analysis and simplified variable input, provide the demand forecasting in a more macroscopic way. Furthermore, it can be further divided into the four-stage model and land use models [18]. Four-stage model was firstly proposed in Chicago Area Transportation Study in 1962, based on the survey among the trips of citizens, where the four stages refer to trip generation or attraction, trip distribution, modal split and traffic assignment [19]. On the other hand, land use models are direct ridership forecast models (DRMs) [20], concentrating on the analysis of trip generation to get prediction results, and are more applicable in newly constructed metro lines.

Typically, the four-stage model functions to provide aggregate prediction, with inborn limitations of precision, only accurate to the magnitudes of passenger demand in the urban transportation network [21], because it was originally developed for city traffic distribution [20]. In the level of metro lines and stations, other methods or considerations, such as disaggregate models, are usually combined. It was introduced in urban rail travel demand analysis as early as 1985 [22], and also achieves its application in data investigation [23]. In the study, a nonlinear mathematical programming model is put up at the same time, to better understand traffic pattern transferring between urban metro and intercity railway. However, its wider use remains to be discussed for the difficulty of calibrating model parameters.

On the other hand, land use models are sometimes combined with the four-stage model or statistical models, as a group of variables or influential factors. Nevertheless, compared to the four-stage model that receives more attention and is basically mature and perfect, land use models are closer to one way of thinking and dealing; thus a wide range of methodology can be categorized as land use analysis, as long as it seldom combines with other typical methods, such as a decision tree based forecasting method with Canonical Correlation Analysis (CCA) to identify key land use variables [24]. It is also an essential part in new-line-oriented models, for the input data of those models is relatively in short supply. In the recent studies, its prediction can also be accurate to 15 min, achieving the accuracy level of a machine learning method, by a multi-factor model [25]. However, its precision has to be further improved with the influencing mechanism analysis [26], where induced passenger flow is taken into account, and is proved to have severe effect on land use and existing metro lines.

3.2 Statistical Models

Statistical models achieve the widely use and large categories in traffic demand forecasting, and are often applied congregated with other models. Within the field, time series methods and regression models are used most often.

Time series methods consider only the historical data and the change pattern, ignoring the impact of outside factors, which is convenient but unstable [27]; while regression models take lots of relative factors into consideration, making it effort-costing in model construction process. And logit model family, a set of discrete [28] regression model, has especially frequent use in urban rail transit demand forecasting. Generally, it comprises Multinomial Logit Model (MNL), Nest Logit Model (NL), Cross-Nested Model (CNL), Mixed Logit Model and so on.

Among all the time series models, Autoregressive Integrated Moving Average (ARIMA) model receives most attention in traffic flow forecasting, and it particularly provides a reliable prediction of peak-time spreading in 2016 [29]. Different kinds of regression models are also brought up and compared, including linear regression, logarithmic regression, exponential regression and Power regression [30]. However, even the regression parameter can be adjusted to a satisfying level, the lack of accuracy is still a problem. For other methods of greater precision, like K nearest neighbors (KNN) nonparametric regression [31], or other dynamic regression models combined time series model and regression model [32], the iterative comparison and analysis based on real-world but historical data are usually time-costing and effort-consuming, thus not suitable for real-time forecasting. Especially for the prediction during events, where the data for irregular holidays and unexpected events is hard or impossible to process, the effect of the forecasting is therefore heavily weakened.

Apart from that, Logit model is of significance in rail transit prediction and traffic analysis as well. The utilization of Logit model considers uncertainty [33], thus better increase the model practicability [34, 35]; although the nature of Logit model improve efficiency and effectiveness, treating data as nominal also ignores important ordering information [36]. It is worth noting that Logit model achieves a good performance combining both traffic zone and individual traveler features in a multi-level model [37], bringing aggregate and disaggregate model, but still having difficulty in data collecting regarding to several influential factors. Its application also involves demand forecasting in preliminary design phase [21], covering the gap between traditional theories about land use and statistical models.

Moreover, other methods, involving K-means clustering analysis [38] and fractal theory [39], can also be categorized as statistic models. K-means clustering analysis has its limitation of only completing the first step of origin-destination distribution; although the fractal theory has a less than 7% maximum error, it remains to be improved when regarding to parameter estimation. Therefore, despite the large number of categories and wide use of statistic models, its application towards practical demand forecasting is still limited and needs optimization.

3.3 Machine Learning Approaches

Compared with traditional models, which mainly focus on the reason of trip generation and analyzing relative factors, modern models utilizing machine learning are sometimes easier to achieve higher accuracy and quick response. Among machine learning methods, Artificial Neural Networks (ANNs) receive a wider use for its outperformance in different fields. ANNs, or connectionist systems, are a rising computer system inspired by biological neural networks, which have the ability to learn (or progressively improve performance on) certain tasks, without the need of task-specific programming [40].

Benefiting from the broad categories of neural network methods and other deep learning methods, relatively large amount of adjusted models were brought up in the last decade.

Some machine learning methods, such as Support Vector Machine (SVM) [41] or spatial weight Least Square-Support Vector Machine (LS-SVM) [20] proposed and focused on relatively earlier than artificial neural networks, give a preliminary try in rail transit demand forecasting. The SVM model assures high accuracy and good adaptability with ensured stability at the same time, but its fatal weakness falls upon its macroscopic prediction only accurate to years, thus limiting its application; LS-SVM is good at simplified data size, in spite of its low accuracy, mainly used as supplementary models. Therefore, up-dated models are needed, mainly leading to the introduce of Artificial Neural Networks.

Among all the ANNs, error Back Propagation Neural Network (BPNN) model achieves a most widely use in urban rail transit forecasting [42–44], where Back Propagation (BP) algorithm is also an essential algorithm applied in other networks [45]. BPNN is a multilayer mapping network minimizing an error backward while information transmitted forward [46]. Relatively, according to the studies, BPNN lacks stability in metro flow forecasting; although it achieves short-term prediction with satisfying accuracy in some cases, with flexible and efficient reaction to the change of influential factors, the time interval is not short enough. Even so, its convenience can still not be denied for its ability of basically approximating any nonlinear function with any accuracy [46].

To overcome the shortcomings of BPNN while giving full play of its strength, improved ANNs are introduced, for instance, Elman Neural Network (ENN). It is a typical type belonging to Fuzzy Neural Network (FNN), able to forecast in short-term with an average relative error of 1.9% [47], which is an obvious improvement of BPNN; however, the minimized error may refers to an over-fitting problem. Moreover, Wavelet Neural Network (WNN) [48], combing classic sigmoid neural networks and the wavelet analysis [49], has high prediction accuracy, stability and feasibility at the same time, indicating a applicable approach in daily-life administration.

Furthermore, there are also neural networks interweaving non-machine-learning methods, targeting at developing strength of both categories: the accurate and automatic prediction of neural networks and the practical feature analysis in statistic or other methods. ANN binding regression model has been applied in peak-hour cross-section passenger volume (PCPV) forecasting [50], achieving good characters of convergence and rapidity, while the prediction is only specific to days. Besides, the correlative factors remain to be adjusted in the study, indicating that some disadvantages can not be ignored even if the advantages are combined. A multi-level model combing both an ANN and Kalman filter has precise prediction of subway transport hub [51] as well, taking historical error into account, but it requires separate flow prediction in different stations. Thus we can see the bright future of neural network forecasting, or combined neural networks, although the further improvement and adjustment for practical use still remains.

4 Conclusion

Urban rail transit system, one of the essential public transports, relies on precise prediction to improve its planning, construction, management and maintenance.

In this paper, a literature review regarding urban rail transit forecasting categorizes its methodology, with a focus on traditional models, statistical models and machine learning approaches. Traditional models comprise the four-stage model and land use models, relatively macroscopic and generation-based; statistical models include time series methods and regression models; machine learning approaches mainly consist of Artificial Neural Networks and SVM, and methods combing ANNs and regression models. All the methods have their merits and limits.

In the future, more studies will be focused on improvement of the proposed methods, regarding to accuracy, efficiency and convenience in operation, as well as efforts including mapping out influencial factors, foreseeing social and economic effects, studying model utility, et al.; meanwhile, microscopical forecasting [52] will receive more attention, with a easy access to large amount of real-world data and quick computer processing algorithm. However, traffic features will also be combined with proposed models, as an adjustment, making future calculation both accurate, characteristic and automatic.

Acknowledgement. This study is supported by the General Projects (No. 71771050) and Key Projects (No. 51638004) of the National Natural Science Foundation of China, and the Natural Science Foundation of Jiangsu Province in China (BK20150603).

References

1. Milenkovic, M., Nebojsa, B.: Railway Demand Forecasting (2016)
2. Los vehículos, aeronaves. Urban Rail Transit. Betascript Publishing, London (2011)
3. Xiao, J.: Demand Forecasting Method of Inter-city Rail Transit, Urban Mass Transit (2006)
4. Wang, Z., Li, X., Chen, F.: Impact evaluation of a mass transit fare change on demand and revenue utilizing smart card data. Transp. Res. Part A Policy Pract. **77**, 213–224 (2015)
5. Huang, D., Liu, Z., Liu, P., Chen, J.: Optimal transit fare and service frequency of a nonlinear origin-destination based fare structure. Transp. Res. Part E Logistics Transp. Rev. **96**, 1–19 (2016)
6. Börjesson, M.: Forecasting demand for high speed rail. Transp. Res. Part A Policy Pract. **70**, 81–92 (2014)
7. Zhu, J., Hu, L., Jiang, Y., Khattak, A.: Circulation network design for urban rail transit station using a PH(n)/PH(n)/C/C queuing network model. Eur. J. Oper. Res. **260**, 1043–1068 (2017)
8. Liu, Z., Wang, S., Chen, W., Zheng, Y.: Willingness to board: A novel concept for modeling queuing up passengers. Transp. Res. Part B Methodological **90**, 70–82 (2016)
9. Shao, C., Xia, J.C., Lin, T.G., Goulias, K.G., Chen, C.: Logistic regression models for the nearest train station choice: a comparison of captive and non-captive stations. Case Stud. Transp. Policy **3**, 382–391 (2015)
10. Wang, S., Qu, X.: Station choice for Australian commuter rail lines: equilibrium and optimal fare design. Eur. J. Oper. Res. **258**, 144–154 (2017)

11. Gong, X., Currie, G., Liu, Z., Guo, X.: A disaggregate study of urban rail transit feeder transfer penalties including weather effects. In: Transportation (2017)
12. Puello, L.C.L.P., Geurs, K.T.: Modelling observed and unobserved factors in cycling to railway stations: application to transit-oriented-developments in the Netherlands. Eur. J. Trans. Infrastruct. Res. **15**, 27–50 (2015)
13. Liu, Z., Yan, Y., Qu, X., Zhang, Y.: Bus stop-skipping scheme with random travel time. Transp. Res. Part C Emerg. Technol. **35**, 46–56 (2013)
14. Yan, Y., Liu, Z., Meng, Q., Jiang, Y.: Robust optimization model of bus transit network design with stochastic travel time. J. Transp. Eng. **139**, 625–634 (2013)
15. Yan, Y., Liu, Z., Bie, Y.: Performance evaluation of bus routes using automatic vehicle location data. J. Transp. Eng. **142**, 04016029 (2016)
16. Qu, X., Zhang, J., Wang, S.: On the stochastic fundamental diagram for freeway traffic: Model development, analytical properties, validation, and extensive applications. Transp. Res. Part B Methodological **104**, 256–271 (2017)
17. Bai, L.: Urban rail transit normal and abnormal short-term passenger flow forecasting method. J. Transp. Syst. Eng. Inf. Technol. **17**, 127–135 (2016)
18. Guo, J., Liu, X.: An Analysis of Forecast of the Passenger Flow of Urban Rail Transit, Shanxi Science & Technology (2017)
19. Baidubaike. Four-Stage Model. https://baike.baidu.com/item/%E5%9B%9B%E9%98%B6%E6%AE%B5%E6%B3%95/3012753. Accessed Jan 2018
20. Zhou, J.Z., Zhang, D.Y.: Direct ridership forecast model of urban rail transit stations based on spatial weighted LS-SVM. J. Chin. Railway Soc. **36**, 1–7 (2014)
21. Zhang, Z.N., Cheng, Y., Yi-Lin, M.A., Sun, F.L.: The forecast method of urban rail transit passenger volume in the preliminary design phase. J. Transp. Eng. (2017)
22. Kato, H., Kaneko, Y.: Choice of Travel Demand Forecast Models: Comparative Analysis in Urban Rail Route Choice (2007)
23. Zhou, C.: Research on demand forecast of passenger transfer of rail transit and intercity railway-taking Beijing as an example. In: Presented at the International Conference on Mechatronics, Materials, Chemistry and Computer Engineering, August 2017
24. Li, X., Liu, Y., Gao, Z., Liu, D.: Decision tree based station-level rail transit ridership forecasting. J. Urban Plann. Dev. **142**, 04016011 (2016)
25. He, Z., Huang, J., Du, Y., Wang, B., Yu, H.: The prediction of passenger flow distribution for urban rail transit based on multi-factor model. In: IEEE International Conference on Intelligent Transportation Engineering (2016)
26. Wang, Z.: Passenger flow prediction model of the newly constructed urban rail transit line. In: International Conference of Logistics Engineering and Management, pp. 1301–1306 (2014)
27. Yang, R., Wu, B.: Short-term passenger flow forecast of urban rail transit based on BP neural network. In: Intelligent Control and Automation, pp. 4574–4577 (2010)
28. Hensher, D.A., Greene, W.H.: The mixed logit model: the state of practice. Transportation **30**(2), 133–176 (2003)
29. Li, R., Rushton, L., Jones, M.: Peak spreading forecast in urban rail transit demand. In: Presented at the Australasian Transport Research Forum (ATRF), 38th, 2016, Melbourne, Victoria, Australia November (2016)
30. Shang, B., Zhang, X.N.: Passengers flow forecasting model of urban rail transit based on the macro-factors. Adv. Eng. Forum **6–7**, 688–693 (2012)
31. Liu, M., Jiao, P., Sun, T.: On Short-term Forecasting Model of Passenger Flow in Urban Rail Transit, Urban Mass Transit (2015)

32. Li, B.: Research on the computer algorithm application in urban rail transit holiday passenger flow prediction. In: International Conference on Network and Information Systems for Computers (2017)
33. Liu, Z., Wang, S., Meng, Q.: Optimal joint distance and time toll for cordon-based congestion pricing. Transp. Res. Part B Methodological **69**, 81–97 (2014)
34. Liu, Z., Wang, S., Meng, Q.: Toll pricing framework under logit-based stochastic user equilibrium constraints. J. Adv. Transp. **48**, 1121–1137 (2014)
35. Liu, Z., Wang, S., Zhou, B., Cheng, Q.: Robust optimization of distance-based tolls in a network considering stochastic day to day dynamics. Transp. Res. Part C Emerg. Technol. **79**, 58–72 (2017)
36. Zheng, Z., Liu, Z., Liu, C., Shiwakoti, N.: Understanding public response to a congestion charge: a random-effects ordered logit approach. Transp. Res. Part A Policy Pract. **70**, 117–13 (2014)
37. Wu, L., Yang, Y.: Research of multilevel models for demand forecast of urban rail transit. In: International Conference on Electric Technology and Civil Engineering, pp. 1444–1447 (2011)
38. He, Z., Wang, B., Huang, J., Du, Y.: Station passenger flow forecast for urban rail transit based on station attributes. In: IEEE International Conference on Cloud Computing and Intelligence Systems, pp. 410–414 (2015)
39. Zhang, L., Jia, Y., Yin, X., Niu, Z.H.: The Arrival Passenger Flow Short-Term Forecasting of Urban Rail Transit Based on the Fractal Theory. Springer, Heidelberg (2014)
40. Wikipedia. Artificial Neural Network. https://en.wikipedia.org/wiki/Artificial_neural_network. Accessed Jan 2018
41. Li, Z., Zhang, Q., Wang, L.: Flow prediction research of urban rail transit based on support vector machine. In: International Conference on Transportation Information and Safety, pp. 2276–2282 (2011)
42. Li, J., Ye, X., Ma, J.: Forecasting method of urban rail transit ridership at station-level on the basis of back propagation neural network. In: Transportation Research Board Annual Meeting (2015)
43. Hou, Y., Dong, H., Jia, L.: A study on the forecast method of urban rail transit. In: Proceedings of the 2015 International Conference on Electrical and Information Technologies for Rail Transportation, pp. 365–372. Springer, Heidelberg (2016)
44. Li, Q., Qin, Y., Wang, Z., Zhan, M., Liu, Y., Zhao, Z., Li, Z.: The research of urban rail transit sectional passenger flow prediction method. J. Intell. Learn. Syst. Appl. **5**(4), 227–231 (2013)
45. Li, J., Cheng, J.H., Shi, J.Y., Huang, F.: Brief Introduction of Back Propagation (BP) Neural Network Algorithm and Its Improvement, vol. 169, pp. 553–558 (2012)
46. Wang, L., Zeng, Y., Chen, T.: Back propagation neural network with adaptive differential evolution algorithm for time series forecasting. Expert Syst. Appl. **42**, 855–863 (2015)
47. Li, Q., Qin, Y., Wang, Z.Y., Zhao, Z.X., Zhan, M.H., Liu, Y.: Prediction of urban rail transit sectional passenger flow based on elman neural network. Appl. Mech. Mater. **505–506**, 1023–1027 (2014)
48. Yue, X., Zheng, Y., Lin, J.: Urban rail transit passenger flow prediction based on improved WNN. In: Computer Engineering & Applications (2016)
49. Alexandridis, A.K., Zapranis, A.D.: Wavelet neural networks: a practical guide. Neural Netw. **42**, 1–27 (2013)

50. Zhu, G., Yang, C., Huang, D., Zhang, P.: A Combined Forecasting Model of Urban Rail Transit Peak-Hour Cross-Section Passenger Volume (2015)
51. Li, S.W.: Passenger flow forecast algorithm for urban rail transit. Telkomnika Indonesian J. Electr. Eng. **12** (2013)
52. Zhou, M., Qu, X., Li, X.: A recurrent neural network based microscopic car following model to predict traffic oscillation. Transp. Res. Part C **84**, 245–264 (2017)

Pricing of Shared-Parking Lot: An Application of Hotelling Model

Wei Zhang[1] and Shuaian Wang[1,2](✉)

[1] Department of Logistics and Maritime Studies,
The Hong Kong Polytechnic University, Hung Hom, Kowloon, Hong Kong
wei.sz.zhang@connect.polyu.hk, wangshuaian@gmail.com
[2] The Hong Kong Polytechnic University Shenzhen Research Institute,
Nanshan District, Shenzhen, China

Abstract. Shared-parking lot brings utilization improvement, but also has its disadvantage compared with traditional parking lot while they are competing for public users. In the market including both shared-parking lot and traditional parking lot, parking lot operators need to know how to deal with parking price to be competitive in the market. The Hotelling model is applied in this paper to study the product differentiation of traditional parking lot and shared-parking lot, with some equilibrium analyses to figure out equilibrium parking prices of both parking lots while considering their competition in the market. Two points of indifferent consumers exist in the competition of the traditional parking lot and the shared- parking lot.

1 Introduction

Sharing economy has become very popular in the last few years. Multiple sharing economy platforms are booming, such as Uber, Airbnb, Relay Rides and so on, which enable owners to rent out their durable goods while not using them. New Internet-based markets occur correspondingly, then several issues related to ownership, rental rates, quantities, prices, and surplus generated are heatedly discussed both in the industry and academic area. People claim that shared economy brings efficiency, opportunity and sociability. This opinion can be well represented in the aspect of urban logistics.

In the progress of city modernization, the number of vehicles increases as more people are attracted to the metropolis. Gradually, urban areas are saturated commercial tall buildings but lack of parking lots. One solution is public transport (Liu et al. 2013, 2016; Liu and Meng 2014), however, private cars are attractive for various reasons. Parking of private cars has become a confusing and conflicting problem in many areas which occurs every day. Such problem cannot be solved only by constructing more parking lots to fulfill the parking requirement of vehicles, because the land resources are too limited, especially metropolis such as Hong Kong. Instead of such conventional method, some new solution is in need to improve the efficiency and capacity of existing parking lots, with considering the cost and feasibility.

'Shared-parking strategy' is a relatively new method to improve parking efficiency suggested recently by Shao et al. (2016). This idea comes from the phenomenon that a

G. De Pietro et al. (Eds.): KES-IIMSS-18 2018, SIST 98, pp. 310–317, 2019.
https://doi.org/10.1007/978-3-319-92231-7_32

large number of private parking spots in residential communities are empty when their owners go to work during the day, and are only occupied at night when their owners go back home. (These owners live in one location and work in another, and their travel pattern can be predicted very well daily, weekly or monthly.) If the owners are willing to offer their personal parking spots to other people, so-called public users, when the parking spots are not occupied in the daytime, the shortage of parking spaces nearby could be alleviated to a great extent and parking efficiency could also be improved.

There is some existing literature on shared-parking lots, most related to parking lot allocation. These studies often assume that 'demand' from public users and 'supply' from parking spot owners are both given before allocation, and the parking fee is fixed and preset. For example, Shao et al. (2016) build binary integer linear programming models to make an optimum allocation, aiming to maximize the profit under constraints of parking space and time duration. Parking pricing, so far, has received relatively little attention in the literature. However, a better understanding of strategic parking pricing is important, not only because adjusting parking fees can regulate parking demand (Cassandras and Geng 2013) and relieve the heavy parking pressure (Teodorovic and Lucic 2006), but also because drivers are always searching for cheaper parking lots (Chou et al. 2008). The competition between the traditional parking lot and shared-parking lot, so far, has received little attention, and parking lot operators should know how to deal with parking price to compete in the market, especially for shared-parking operator owing to the inconvenience nature of the shared-parking lot.

Inconvenience of the shared-parking lot is that, public users have to request parking (do reservations beforehand) for integrate time intervals, which means that if a public user wants to park during 9:30–11:45 am, she will be suggested to request for three-time interval 9:00–10:00 am, 10:00–11:00 am and 11:00–12:00 am. The reason behind is that, each parking spot has different available time gaps (because the parking spot owners have different departure time and coming back time), only collecting such parking requests with integrate time intervals, the shared-parking company can allocate certain public user to specific parking spot, so that to make the shared-parking lot as fully occupied as possible to make more profit. For shared-parking lot operator to do the optimum allocation, public users have to request parking for integrate time intervals. But for the traditional parking lot, the public user can start parking at any time as preference, which is 9:30 am in this example. She does not need to do reservation which starts from 9:00 am, actually earlier than her real parking start time. She also does not need to delay the original schedule to start parking at 10:00 am. This is the main differentiation of traditional parking lots and shared-parking lots.

For studying product differentiation in markets with multiple competitors, the Hotelling model is probably the most well-known one. Hotelling's classic paper studied two firms, which are located at the two endpoints of a linear city with length one. Firm one is located at $x = 0$, firm two is located at $x = 1$. Consumers are uniformly distributed with density 1 along this linear city and incur a transportation cost t per unit of distance. Each consumer has unit demand and will buy from the firm which has a cheaper generalized price (price plus transportation cost) and does not exceed surplus s. Let p_A and p_B denote the price by firm A and firm B, the demand for firm A is given by $D_A(p_A, p_B) = \tilde{x}$ where $p_A + t\tilde{x} = p_B + t(1 - \tilde{x})$, and the demand for firm B is given by $D_B(p_A, p_B) = 1 - D_A(p_A, p_B)$, as can be seen in Fig. 1. The customer located

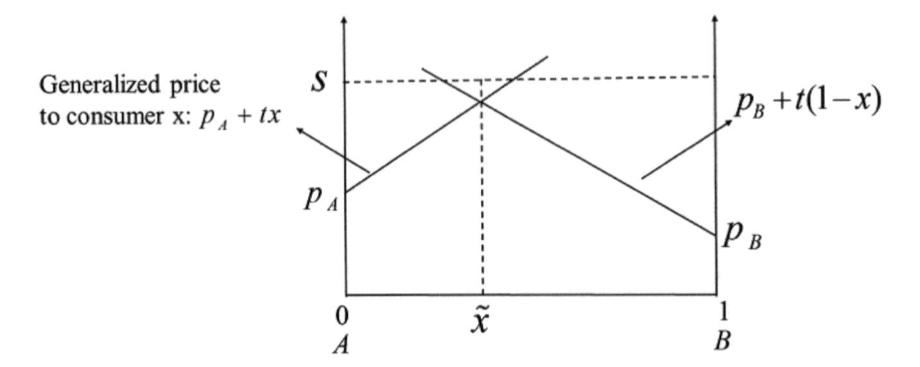

Fig. 1. The equilibrium of the classic Hotelling model

at $\tilde{x}$ is called an indifferent consumer who is just indifferent between firm A and firm B. Once we know the indifferent consumer and then define the demand function of firm A and B, the payoff functions of both firms can be derived, therefore the equilibrium prices can be determined.

Although the hotelling model originally framed in the context of location differentiation along a linear market, it has various possible interpretations. In this case of two kinds of parking lots, we may intuitively think that, the parking price of the shared-parking lot could be lower than the traditional parking lot to attract more drivers. Although shared-parking lot has more inconvenience compared with traditional parking lot, under lower parking price motivation of shared-parking lot, public user who wants to park at 10:30 am may move up her original schedule to start parking at 10:00 am or to delay the original schedule to start parking until 11:00 am so as to park in the shared parking lot. Then, Hotelling's 'space' becomes 'parking time' for public users, and 'transport costs' between consumers and suppliers become 'early arrival cost' and 'late arrival cost', with considering the cost of being late is usually higher than the cost of being early (Small 1982). In this paper, we propose an application of the Hotelling model that has exactly this interpretation, with some equilibrium analysis to figure out equilibrium parking prices of the traditional parking lot and the shared-parking lot while considering the competition between them.

2 Problem Description and Model Formulation

2.1 Problem Description

Indifferent public users and two kinds of parking lots, traditional parking lot A and shared-parking lot B, are considered as participants in the market. To simplify the problem, indifferent public users (parking demand) are all generated from the one-time interval. In this time interval, the arrival of public users follows a Poisson distribution, as usually considered in previous literature (Richardson 1974; Cleveland 1963; Blunden 1971). Suppose each time interval has one-hour length. For example, our modeling time interval is from 9:00 to 10:00, then 9:00 indicates 0 and 10:00 indicates

1 in the Hotelling model, and public users are uniformly distributed along this interval $[0, 1]$. Compared with the classic Hotelling model that a linear city has length one, here a time interval has length one, that is to say, we consider time distance instead of space distance in this problem.

Both parking lots are in the same residential area other than scattered, and if the public user wants to park in this area, she has no preference for specific parking lot in the aspect of geographic reason. The only differentiation of two parking lots is that, for the traditional parking lot A, public users can park at any time once they arrive at the parking lot; for the shared-parking lot B, public users whose prefer starting parking time is in the middle of a time interval need to decide to move up the schedule to the beginning of this time interval, or to delay the schedule until the beginning of the next time interval. Each public user has unit demand, either choosing to park in the traditional parking lot A or the shared-parking lot B.

Suppose all the public users generated in this time interval want to park for a whole daytime. This makes sense for public users who work nearby the parking lot so that they need to park for a whole daytime. What is more, the start time of their work always concentrates on one specific time interval, for example, from 8:00 to 9:00, which is to be considered as our modeling period. As the ends of the working hour in the afternoon also do not have big differences, if we charge the parking fee per hour which is the same as the practical situation, the parking revenue earning from each public user of our interests just has a small variance limited in one or two hour's revenue. Compared to the whole day profit, this small part can be neglected in order to simplify the question. So, the parking fee can be set for one unit consumption of parking spot, that is to say, traditional parking lot p_A and shared-parking lot p_B are both considered as a whole day parking fee. Both traditional parking lot A and shared-parking lot B have the marginal cost as they sell one parking spot, which is defined as c_A and c_B, being considered as the operation cost for the companies to maintain daily operation.

The inconvenience of the shared-parking lot will be captured by early arrival cost coefficient t_1 and late arrival cost coefficient t_2. Both costs are up to public users to pay, indicating the unwillingness of public users to change their schedules. For the early arrival cost coefficient t_1, although the public user can still arrive at the shared-parking lot according to her schedule, she has to pay for the integrate time interval, so that the early arrival cost can also be interpreted as the unhappy emotion (the public user may think that she pays more parking fee than she actually deserves to be charged). For the late arrival cost coefficient t_2, it represents the bad outcome for the public user if she is late, for example, to go to work after the regulated working start time, late to arrive her office. As the cost of being late is usually higher than the cost of being early (Small 1982), the late arrival cost coefficient t_2 is larger than the early arrival cost coefficient t_1.

2.2 Model Formulation

In order to derive the demands of the traditional parking lot A and the shared-parking lot B, we need to derive the public user with the preference arrival time x that is just indifferent between parking in the traditional parking lot A and the shared-parking lot B. Suppose a public user actually wants to park her car at a time duration of x from the beginning of the time interval (or $(1 - x)$ to the end of the time interval). The early arrival cost is $t_1 x$ if she chooses to arrive at the shared-parking lot earlier than her original time schedule, the late arrival cost is $t_2(1 - x)$ if she chooses to arrive at the shared-parking lot later than her original time schedule. So, the total cost for the public user parking in the traditional parking lot is p_A, but the total cost for the public user parking in the shared parking lot is $p_B + t_1 x$ or $p_B + t_2(1 - x)$. The Nash equilibrium can be expressed by Eqs. (1) and (2), then x_1 and x_2 can be derived as Eqs. (3) and (4).

$$p_A = p_B + t_1 x_1 \tag{1}$$

$$p_A = p_B + t_2(1 - x_2) \tag{2}$$

$$x_1 = \frac{p_A - p_B}{t_1} \tag{3}$$

$$x_2 = \frac{p_B - p_A + t_2}{t_2} = \frac{p_B - p_A}{t_2} + 1 \tag{4}$$

Figure 2 can be derived by above equations, there exist two time points defined as x_1 and x_2. For the public user who has the parking preference at the time x_1, it is indifferent for her to park at the traditional parking lot A with a higher parking fee p_A or park at the shared-parking lot B with a lower parking fee p_B but a higher inconvenience to move up her schedule. For the public user who has the parking preference at the time x_2, it is indifferent for her to park at the traditional parking lot A with a higher parking fee p_A, or park at the shared-parking lot B with a lower parking fee p_B but a higher inconvenience to delay her schedule.

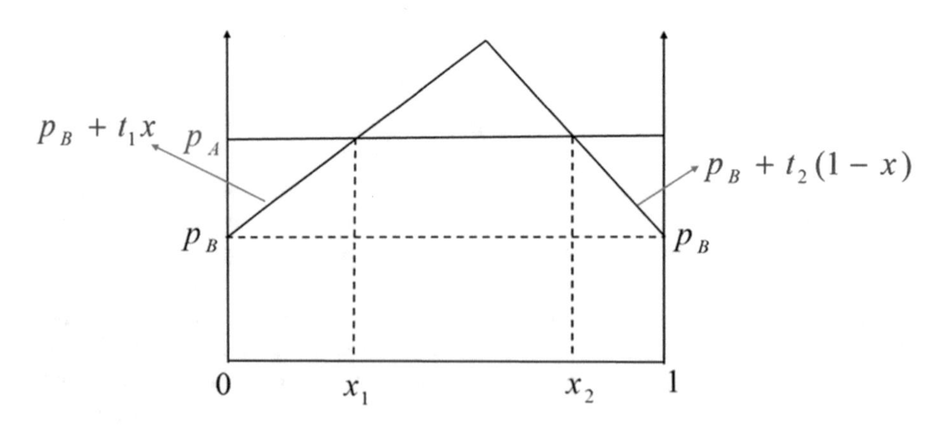

Fig. 2. The equilibrium of the traditional parking lot and shared-parking lot in $[0, 1]$

Once we know the indifferent public user, we may define the demand function of the traditional parking lot A by Eq. (5) and the demand function of the shared-parking lot B by Eq. (6). It can be seen in Fig. 2 that parts of the market are covered by the shared-parking lot B, which are the public users who have the parking preferences in the time duration of $[0, x_1]$ and $[x_2, 1]$. Residual market is covered by the traditional parking lot A, which are the public users who have the parking preferences in the time duration of $[x_1, x_2]$. This result is intuitive since, for the public users with the parking preferences in the middle of the time interval, either to move up their schedules or to delay their schedules is not a wise choice.

$$D_A(p_A, p_B) = \int_{x_1}^{x_2} 1dz = z|_{x_1}^{x_2} = x_2 - x_1 = 1 + \frac{p_B - p_A}{t_2} - \frac{p_A - p_B}{t_1} \tag{5}$$

$$D_B(p_A, p_B) = \int_{0}^{x_1} 1dz + \int_{x_2}^{1} 1dz = z|_0^{x_1} + z|_{x_2}^{1} = x_1 + (1 - x_2) = \frac{p_A - p_B}{t_1} - \frac{p_B - p_A}{t_2} \tag{6}$$

The maximization problem of the traditional parking lot A is

$$\underset{p_A}{Max}\, \Pi^A(p_A, p_B) = (p_A - c_A) D_A(p_A, p_B) = (p_A - c_A)\left(1 + \frac{p_B - p_A}{t_2} - \frac{p_A - p_B}{t_1}\right) \tag{7}$$

By deriving its first-order condition $\frac{d\prod^A}{dp_A} = 0$, we have

$$\begin{aligned} &\left(1 + \frac{p_B - p_A}{t_2} - \frac{p_A - p_B}{t_1}\right) - \frac{p_A - c_A}{t_2} - \frac{p_A - c_A}{t_1} = 0 \\ &p_A = \left(\frac{t_1 t_2}{t_1 + t_2} + p_B + c_A\right) \Big/ 2 \end{aligned} \tag{8}$$

The maximization problem of the shared-parking lot B is

$$\underset{p_B}{Max}\, \Pi^B(p_A, p_B) = (p_B - c_B) D_B(p_A, p_B) = (p_B - c_B)\left(\frac{p_A - p_B}{t_1} - \frac{p_B - p_A}{t_2}\right) \tag{9}$$

By deriving its first-order condition $\frac{d\prod^B}{dp_B} = 0$, we have

$$\begin{aligned} &\frac{p_A - p_B}{t_1} - \frac{p_B - p_A}{t_2} - \frac{1}{t_1}(p_B - c_B) - \frac{1}{t_2}(p_B - c_B) = 0 \\ &p_B = \frac{p_A + c_B}{2} \end{aligned} \tag{10}$$

Then, by Eqs. (8) and (10), the equilibrium prices are determined as follow

$$p_A^* = \left(\frac{2t_1 t_2}{t_1 + t_2} + 2c_A + c_B\right) \Big/ 3 \tag{11}$$

$$p_B^* = \left(\frac{t_1 t_2}{t_1 + t_2} + c_A + 2c_B\right) \Big/ 3 \tag{12}$$

3 Conclusion and Future Research

The classic Hotelling model originally framed in the context of location differentiation along a linear market, but in this paper another interpretation is considered that timing is a relevant measure of product differentiation. We study two competing parking lots, a shared-parking lot and a traditional parking lot. Hotelling's 'space' becomes'parking time' for public users, and 'transport costs' of consumers become 'early arrival costs' and 'late arrival costs' of parking lot public users. A general formulation that has exactly this interpretation is proposed in this paper with equilibrium analyses, to figure out equilibrium parking prices of two kinds of parking lots. Another major difference from the classic Hotelling model is that, we have two points of indifferent consumers in the competition of the traditional parking lot and the shared- parking lot, instead of only one point of the indifferent consumer in the classic Hotelling's case.

In the future, the proposed model can be extended. In this paper, the parking fee of the traditional parking lot p_A and of the shared-parking lot p_B are set as whole day parking fees. This assumption can be relaxed, for example, parking fees are charged per hour and are different between each hour based on the parking demand in the specific time interval. Then, the distribution of parking duration of public users should also be considered (the parking duration cannot be a whole daytime as considered in this paper), that is to say, the situation of the current time interval may make influence on the situations of the following time intervals, the demand for each time interval is not independent and elastic demand or even stochastic demand (Meng and Qu 2012; Qu et al. 2017; Zhou et al. 2017) can be considered to make the problem more realistic.

Acknowledgment. This research is sponsored by the National Natural Science Foundation of China (No. 71771050).

References

Blunden, W.R.: The Land-Use/Transport System. Analysis and Synthesis (1971)
Chou, S., Lin, S., Li, C.: Dynamic parking negotiation and guidance using an agent-based platform. Expert Syst. Appl. **35**(3), 805–817 (2008)
Cleveland, D.E.: Accuracy of the periodic check parking study. Traffic Eng. **33**(12), 14–17 (1963)
Geng, Y., Cassandras, C.: New "smart parking" system based on resource allocation and reservations. IEEE Trans. Intell. Transp. Syst. **14**(3), 1129–1139 (2013)
Liu, Z., Meng, Q.: Bus-based park-and-ride system: a stochastic model on multimodal network with congestion pricing schemes. Int. J. Syst. Sci. **45**(5), 994–1006 (2014)
Liu, Z., Wang, S., Chen, W., Zheng, Y.: Willingness to board: a novel concept for modeling queuing up passengers. Transp. Res. Part B Methodol. **90**, 70–82 (2016)
Liu, Z., Yan, Y., Qu, X., Zhang, Y.: Bus stop-skipping scheme with random travel time. Transp. Res. Part C Emerg. Technol. **35**, 46–56 (2013)

Meng, Q., Qu, X.: Estimation of rear-end vehicle crash frequencies in urban road tunnels. Accid. Anal. Prev. **48**, 254–263 (2012)

Qu, X., Zhang, J., Wang, S.: On the stochastic fundamental diagram for freeway traffic: model development, analytical properties, validation, and extensive applications. Transp. Res. Part B Methodol. **104**, 256–271 (2017)

Richardson, A.J.: An improved parking duration study method. In: Proceedings of the 7th Australian Road Research Board Conference, Adelaide, Australia, pp. 397–413 (1974)

Shao, C., Yang, H., Zhang, Y., Ke, J.: A simple reservation and allocation model of shared parking lots. Transp. Res. Part C Emerg. Technol. **71**, 303–312 (2016)

Small, K.A.: The scheduling of consumer activities: work trips. Am. Econ. Rev. **72**(3), 467–479 (1982)

Teodorovic, D., Lucic, P.: Intelligent parking systems. Eur. J. Oper. Res. **175**(3), 1666–1681 (2006)

Zhou, M., Qu, X., Li, X.: A recurrent neural network based microscopic car following model to predict traffic oscillation. Transp. Res. Part C Emerg. Technol. **84**, 245–264 (2017)

Estimating the Value of Travel Time Using Mixed Logit Model: A Practical Survey in Nanjing

Cheng Lv[1], Ling Dai[1], Kai Huang[1,2](✉), and Zhiyuan Liu[1]

[1] Jiangsu Province Collaborative Innovation Center of Modern Urban Traffic Technologies, Southeast University, Nanjing 210096, China
lvcheng1996@126.com, 979370348@qq.com, leakeliu@163.com

[2] Institute of Transport Studies, Department of Civil Engineering, Monash University, Melbourne 3800, Australia
h.uangkai@foxmail.com

Abstract. Value of travel time (VOTT) is a fundamental index in transportation economics, which represents the trip cost that the travelers are willing to pay for saving their travel time, working as a communicator between money and time. VOTT is of great importance in urban transportation planning and traffic forecasting as VOTT of travelers will impact on their traveling mode, frequency and routes. VOTT is calculated by estimating the ratio of the utility function on travel time and cost. The model is specified through logit regression analysis. In this paper, a stated preference survey on the VOTT of travelers is conducted in Nanjing, China. We adopt mixed logit model, which is a flexible model that allows the parameters to vary across the population and is not restricted by the IIA property, to perform the regression, and analyze the distribution of VOTT.

1 Introduction

Value of travel time (VOTT) is gaining increasing attention in transportation economics, as the economic instrument is now being an indispensable component of transportation planning, like the congestion tolling, where total travel cost need to be taken as the objective of the optimal toll design problem [1–3]. VOTT represents the trip cost that the travelers are willing to pay for saving an additional unit of travel time, working as a communicator between money and time, which is crucial to analyze the travelers' travel mode choice [4–7]. VOTT varies among residents with different ages, living standards, salaries, personal tastes, travel purposes, and so on. It implicitly reflects the value that the travelers create within the same time as they spent on journey, as well as the subjective preference of them [8]. Therefore, there are generally two perspectives on the estimating of VOTT, the first being an indirect method considering GDP and regional average household income, the other being a direct observation by calculating the marginal rate of substitution of money for travel time [9].

The concept of VOTT was first put forward by Becker using microeconomics theories in 1965 [10]. Johnson and De Serpa later explained and extended Becker's theory by taking the value of non-working time and the value of saving time into

G. De Pietro et al. (Eds.): KES-IIMSS-18 2018, SIST 98, pp. 318–328, 2019.
https://doi.org/10.1007/978-3-319-92231-7_33

account respectively [11, 12]. Cesario's paper examines the role of time costs—both on-site and travel—in models describing recreation behavior [13]. Discrete choice modelling based on random utility theory was first introduced to the study of VOTT in the 1970s [14, 15]. The advancement of computer continued to facilitate such analysis in the upcoming decades. Boyd and Mellman and Cardell and Dunbar first applied mixed logit model in modeling automobile demand [16, 17]. Ben-Akiva et al. applied mixed logit model on customer-level data [18]. Greene extended the mixed logit model to account for the variance of the random parameter distribution [19]. Train specified a logit formula for the mixing distribution [20].

The structure of this paper is as follows. Section 2 explains the methodology used in this paper to estimate the value of VOTT, including the basic concept of mixed logit model. Section 3 summarizes the data collected, presenting the main structure of the data, and how the survey was designed. Section 4 covers the details of the analysis, including the specification of the model, the calculation of VOTT and how the VOTT distributes. Section 5 provides conclusion of this paper and recommendations.

2 Methodology

2.1 Estimating the Value of VOTT

Based on the utility maximization theory, travelers tend to choose the mode whose utility top among all the alternatives. The basic form of utility function where the utility of alternative i is evaluated by respondent n can be expressed as

$$U_{ni} = V_{ni} + \varepsilon_{ni} \tag{1}$$

$$V_{ni} = \beta_{0_{ni}} + \beta_{C_{ni}} C_{ni} + \beta_{T_{ni}} T_{ni} \tag{2}$$

where ε is the stochastic error term, C_{ni} denotes the travel cost, and T_{ni} denotes the travel time.

VOTT demonstrates the willingness of travelers to pay more for saving their travel time. It is defined as the marginal rate of substitution of money for travel time. So, VOTT can be obtained by dividing the estimated marginal utility of travel time with the estimated marginal utility of travel cost.

$$\text{VOTT} = \frac{\beta_{T_{ni}}}{\beta_{C_{ni}}} \tag{3}$$

2.2 Mixed Logit Model

Usually, binary logit model or multimodal logit model is adopted in the VOTT estimation. While BL or MNL model is easy to perform, they cannot account for the various preference among population.

Mixed logit model is a flexible model that allows for both potential observed and unobserved heterogeneity of individuals in the models [19, 21]. The parameters in the model are not necessarily fixed, they can vary across the population with density $f(\beta|\theta)$, where θ is the true parameter of the distribution [22]. The conditional probability that individual n chooses alternative i will be:

$$P_{ni}(\theta) = \int L_{ni}(\beta_n) f(\beta|\theta) \mathrm{d}\beta \tag{4}$$

$$L_{ni}(\beta_n) = \frac{e^{\beta_n' X_{ni}}}{\sum_i e^{\beta_n' X_{ni}}} \tag{5}$$

The parameters can follow various distribution such as normal distribution and lognormal distribution.

The estimation of VOTT generated from the MXL model can be obtained by calculating the ratio between the expectations of $\beta_{T_{ni}}$ and $\beta_{C_{ni}}$.

$$\mathrm{VOTT} = \frac{E(\beta_{T_{ni}})}{E(\beta_{C_{ni}})} \tag{6}$$

3 Data

To get the travel data of residents in Nanjing, China, a stated preference (SP) survey is conducted in the six main districts of Nanjing, China. Tablet devices are used in the survey to facilitate the collection of data, keeping the data accurate and organized.

SP survey is widely used when people's preferences need to be characterized. In an SP survey, a set of hypothetical scenarios are presented to respondents, thus their preference as well as the factors behind those choices can be acquired through analysis on the survey data.

In this survey, 2436 observations of working purpose and 2454 observations of non-working purpose are finally collected. The gender ratio is 1:0.904. Detailed demographic information on the respondents is given below (Table 1).

Table 1. Demographic information

		Freq.	Percent
Purpose	Non-working	2,454	50.18
	Working	2,436	49.82
Gender	Female	2,568	52.52
	Male	2,322	47.48
Age	Under 18	540	11.04
	19–30	1,230	25.15
	31–50	2,052	41.96
	Above 51	1,068	21.84

(continued)

Table 1. (*continued*)

		Freq.	Percent
Income	Under 1500	1,014	20.74
	1500–3000	1,014	20.74
	3000–5000	1,230	25.15
	5000–7000	732	14.97
	7000–9000	264	5.4
	9000–11000	48	0.98
	11000–15000	294	6.01
	Above 15000	294	6.01
Car	No	2,316	47.36
	Own	2,574	52.64
Total		4890	

The questionnaire is divided into two parts. The first part collects the basic information of the respondents, including gender, age, income, occupation and car ownership.

The second part is the SP questions, which require respondents to make multiple choices between car and public transit with different travel cost and time (Table 2).

Table 2. Variables and descriptions

Variable	Type	Description
Gender	Discrete variable	Number 0 and 1 represent the gender of respondents, which denote male and female respectively
Age	Discrete variable	Number 1, 2, 3 and 4 represent the age of respondent the age of respondents, which denote under 18, 19–30, 31–50 and above 50 respectively
Income	Discrete variable	A set of ranked numbers are used to represent the monthly income of respondents, namely 750, 2250, 4000, 6000, 8000, 11000, 13000 and 17000 (yuan)
Occupation	Discrete variable	Dummy variables named after occup1, occup2, …, occup10 represent the occupation of respondents, which denote civil service, professional/technical personnel, clerk, service industry, worker/farmer, retiree, student, business owner, jobless and others
Car ownership	Discrete variable	Number 1 and 0 represent the car ownership of respondents, which denote the status of own and not own respectively
Travel purpose	Discrete variable	Number 1 and 0 represent the travel purpose of respondents, which denote working purpose and non-working purpose respectively
Travel cost	Continuous variable	The cost that a respondent spends in a trip

(*continued*)

Table 2. (*continued*)

Variable	Type	Description
In-vehicle time	Continuous variable	The travel time that a respondent spends in vehicle in a trip
Waiting time	Continuous variable	The time that a respondent spends on waiting for the vehicle or walking to the station
Travel mode	Discrete variable	Number 0 and 1 represent two different travel modes, which denote car and public transit

Twelve SP questions are presented as trade-offs to the respondents in the SP survey, with six questions for working purpose and the other six for non-working purpose. In each trade-off, the travel cost, time in vehicle, time for waiting and travel purpose of two modes are listed.

The arrangement of SP questions follows uniform design. In most cases, orthogonal design is implemented for consistency and reliability. However, to ensure that the question set is orderly and comparable, orthogonal design requires every combination of value is presented, which might create many redundant questions. The longer the questionnaire is, the less patient respondents will be, leading to less reliable data. Uniform design can effectively cut down the number of questions and a significant amount of information can be obtained [23] (Table 3).

Table 3. The question sets of SP survey

Public transit			Car		
Time in vehicle	Time for waiting	Travel cost	Time in vehicle	Time for waiting	Travel cost
35	4	2	11	4	9
60	6	3.5	23	4	18
80	8	1.5	35	4	6
20	10	3	5	4	15
45	12	1	17	4	3
70	14	2.5	29	4	12

4 Analysis

Before performing the estimation, mixed logit model requires that the distribution of all random parameters is defined. Waiting time, in-vehicle time and travel cost are assumed to be random parameters.

Let the utility function take this form:

$$U_{ni} = V_{ni} + \varepsilon_{ni} \tag{7}$$

$$\begin{aligned} V_{ni} = \beta_0 + \beta_1 C_{ni} + \beta_2 Tin_{ni} + \beta_3 Twait_{ni} + \beta_4 P_{ni} \\ + \beta_5 G_n + \beta_6 A_n + \beta_7 I_n + \beta_8 CO_n + \beta_9 O_n \end{aligned} \tag{8}$$

where C denotes travel cost, Tin denotes in-vehicle travel time, $Twait$ denotes waiting time, P denotes travel purpose, G denotes the gender of respondent, A denotes the age of respondent, I denotes the monthly income of respondent, CO denotes the car ownership of respondent, O denotes the occupation of respondent.

The random parameters in the mixed logit models are assumed to follow normal distribution. Assume the variance of random parameters to be σ, then the specification of the models will define $\beta_i = b + n\sigma_i$, where n is the random variable.

Considering each individual's preference might be correlated over choice situations [24], the utility will be:

$$\begin{aligned} U_{ni} &= \beta_{ni} X_{ni} + \varepsilon_{ni}, \\ \beta_{ni} &= b + \tilde{\beta}_{ni}, \\ \tilde{\beta}_{ni} &= \rho \tilde{\beta}_{ni-1} + \mu_{ni} \end{aligned} \tag{9}$$

In the first two models, all three random coefficients are treated as independent, while those of the last one is treated as correlated. Insignificant variables (at 95% confidence level) are dropped in the second and the third model.

The models are fitted by using maximum likelihood estimation, and 500 Halton draws are used for the simulation in the MXL model fitting.

The parameter estimates are given in Table 4. The signs of all three variables concerning time and cost are negative as expected, indicating the inclination of travelers for shorter travel time and lower cost.

The gender of travelers doesn't show significant impact on the choice of modes; however, the sign indicates that men have a higher tendency to take public transit than women. The negative income variable and car variable show that travelers who have higher income and own cars are more likely to travel by car. As to occupation, clerk, service industry personnel, worker/farmer, retiree, student show significant positive result, indicating they have preference for public transit.

The pdf of coefficients of in-vehicle time and cost are given in Figs. 1 and 3. As stated in Section Methodology, the VOTT can be calculated by dividing the expectation of travel time and cost.

By calculating the VOTT in terms of waiting time and in-vehicle time separately, the value of waiting time is much higher than the value of in-vehicle travel time when the random variables are specified as independent in model 2, while it shows an opposite result in model 3. Besides, the value of in-vehicle travel time calculated with model 3 is fairly higher than model 2.

Since the distributions of travel time and travel cost are assumed to be both normal distributed, the density function of the ratio for them can be derived with following methods.

Table 4. Fitting results

	MXL-1		MXL-2		MXL-3	
	Independent random variables		Independent random variables		Correlated random variables	
	Coeff.	Std. Err.	Coeff.	Std. Err.	Coeff.	Std. Err.
Random parameter mean						
Twait	−0.601488	0.387914	−0.229464	0.027991	−0.283464	0.094541
Tin	−0.190557	0.108299	−0.090441	0.015256	−0.332790	0.128513
Cost	−0.409931	0.247421	−0.167934	0.021507	−0.473610	0.178956
Random parameter spread						
Twait	0.515017	0.378428				
Tin	0.132803	0.097509	−0.031771	0.015963	0.378083	0.149709
Cost	−0.133749	0.081279	0.053799	0.009482	0.346549	0.133890
Fixed parameter mean						
Cons.	−3.740195	3.227165	−0.922042	0.414774	−4.093648	2.004217
Purpose	1.459034	0.960903	0.573027	0.149170	1.545897	0.657596
Gender	−0.248122	0.366990				
Age	0.988642	0.758775	0.320385	0.093034	0.726042	0.337448
Income	−0.000319	0.000188	−0.000138	0.000025	−0.000250	0.000091
Car	−1.176451	0.842220	−0.440384	0.169100	−1.090855	0.503751
Occup1	0.180971	0.734168				
Occup2	−0.591264	0.748903				
Occup3	1.168006	1.023242	0.605694	0.176888	1.341028	0.559809
Occup4	1.328594	0.954646	0.791448	0.237921	2.253904	0.873869
Occup5	2.684016	1.987757	1.196046	0.368190	3.205311	1.416476
Occup6	2.257389	1.535544	1.335631	0.354423	4.812361	2.006421
Occup7	2.184064	1.755448	0.881984	0.227638	2.065349	0.875564
Occup8	0.031870	1.052584				
Occup9	−2.706807	2.759435				
LL	−1908.2557		−1911.7960		−1867.6314	
Prob > chi2	0.0000		0.0000		0.0000	
AIC	3860.209		3853.592		3767.263	
BIC	4011.159		3961.413		3882.272	
VOTT_wait	88.037		81.983		35.911	
VOTT_in	27.891		32.313		42.160	

Given two normally distributed random variables X and Y, where $X \sim N(\mu_X, \sigma_X)$, $Y \sim N(\mu_Y, \sigma_Y)$. Define two new variables U and V, let $U = X/\sigma_X$, $V = Y/\sigma_Y$, then $U \sim N(\mu_X/\sigma_X, 1)$, $V \sim N(\mu_Y/\sigma_Y, 1)$. Then, the density function of $W = U/V$ is:

$$g(W) = \frac{1}{2\pi} Q e^{M} \tag{10}$$

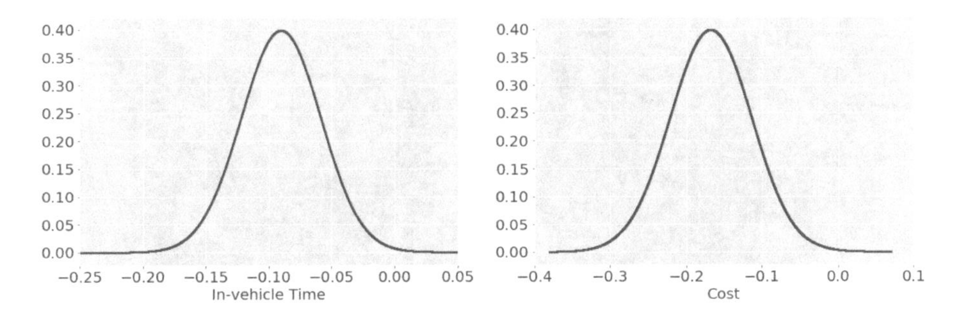

Fig. 1. PDF plot of coefficients of in-vehicle time and cost (Model 2)

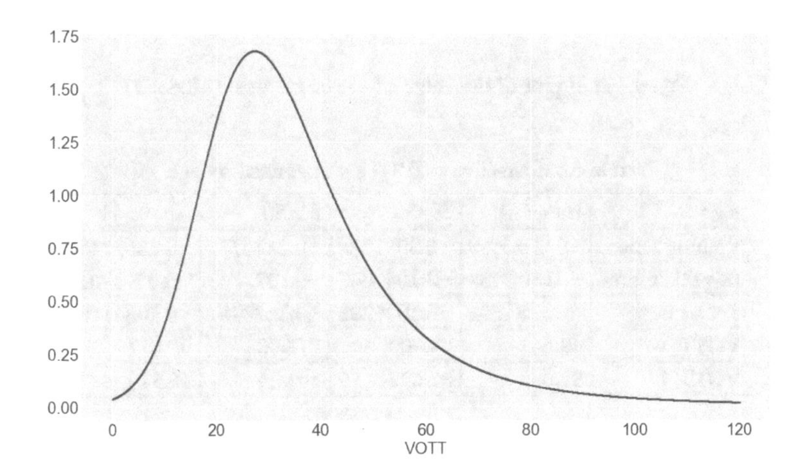

Fig. 2. PDF plot of the value of in-vehicle time (Model 2)

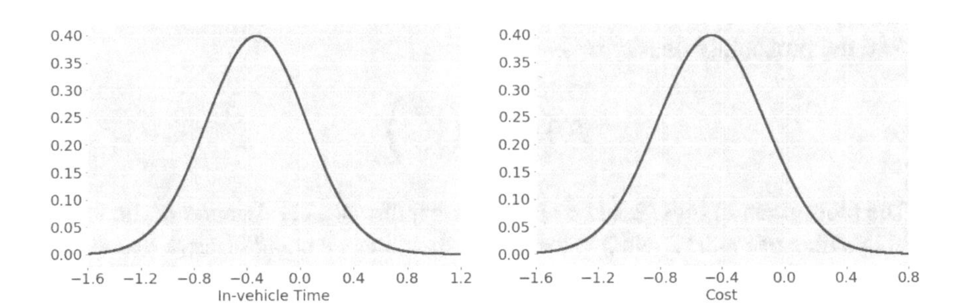

Fig. 3. PDF plot of coefficients of in-vehicle time and cost (Model 3)

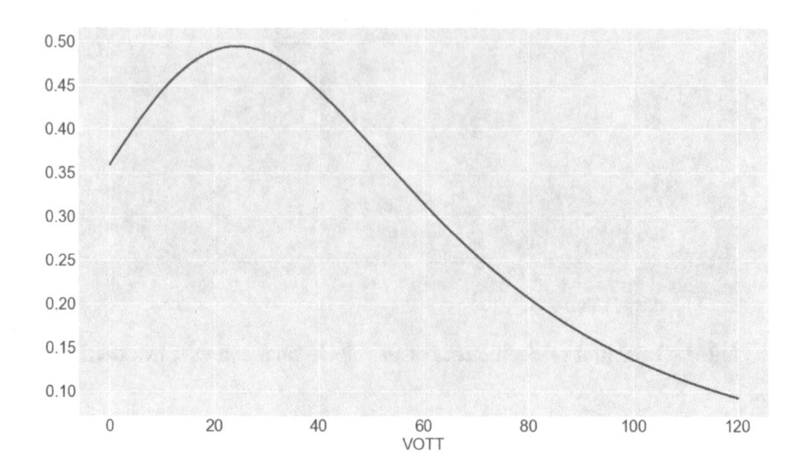

Fig. 4. PDF plot of the value of in-vehicle time (Model 3)

Table 5. Estimation of VOTT categorized by age

Age	Under 18	19–30	31–50	Above 51
Waiting time	−0.0956488	−0.3085028	−0.2116755	−0.2869740
In-vehicle time	−0.0612468	−0.1041047	−0.1327863	−0.0498866
Travel cost	−0.1181795	−0.1975288	−0.1628487	−0.1834140
VOTT_w	48.561	93.709	77.990	93.877
VOTT_in	31.095	31.622	48.924	16.319

Where $M = -\frac{1}{2}\left(\frac{\mu_Y}{\sigma_Y}w - \frac{\mu_X}{\sigma_X}\right)^2 s^2$, $Q = ks\sqrt{2\pi}[1 - 2\Phi(-k/s)] + 2s^2e^{-k^2/2s^2}$, $s = \frac{1}{\sqrt{1+w^2}}$, $k = \left(\frac{\mu_X}{\sigma_X}w + \frac{\mu_Y}{\sigma_Y}\right)s^2$, Φ denotes the standard normal cumulative distribution function [25].

And the probability density of $Z = X/Y$ is:

$$f(t) = \frac{\sigma_Y}{\sigma_X} g\left(\frac{\sigma_Y}{\sigma_X}t\right) \tag{11}$$

The plots given in Figs. 2 and 4 are the probability density function of the value of in-vehicle time of model 2 and 3 respectively, in which we could observe that the most probable value of VOTT both lie around RMB25/h. Besides, the variance of Fig. 3 is significantly greater than that of Fig. 2, hence contributing to too many zero value of VOTT in Fig. 3, which is impossible in real life.

Table 5 lists the estimation of VOTT categorized by age. Travelers between 31–50 years old, who make up the bulk of the working-age population, show the highest value of in-vehicle time. Old travelers have the lowest value of in-vehicle time, as citizen above 60 years old can enjoy the welfare policy of taking public transit with discount or for free.

5 Conclusion

This paper estimates the value of travel time in Nanjing using a stated preference survey. Six stated preference questions are presented to the respondents for each travel purpose. Mixed logit models, which allows for the heterogeneity among travelers, are established whereby the VOTT is obtained by calculating the ratio of the coefficients of travel time and travel cost, and the influencing factors are analyzed. According to the density function plots, Model 2 gives a more reasonable result, suggesting that the value of in-vehicle time is RMB32.313/h and the value of waiting time is RMB81.983/h. Therefore, adequate emphasis should be placed on forming a reliable timetable for the running of public transit system, in order to avoid too long waiting time and ensure its attraction towards travelers.

Acknowledgment. This study is supported by the Projects of International Cooperation and Exchange (No. 51561135003), Key Projects (No. 51638004) and Youth Projects (No. 71501038) of the National Natural Science Foundation of China.

References

1. Liu, Z., Meng, Q., Wang, S.: Speed-based toll design for cordon-based congestion pricing scheme. Transp. Res. Part C **31**, 83–98 (2013)
2. Liu, Z., Wang, S., Meng, Q.: Optimal joint distance and time toll for cordon-based congestion pricing. Transp. Res. Part B **69**, 81–97 (2014)
3. Liu, Z., Wang, S., Zhou, B., Cheng, Q.: Robust optimization of distance-based tolls in a network considering stochastic day to day dynamics. Transp. Res. Part C **79**, 58–72 (2017)
4. Chen, J., Liu, Z., Zhu, S., Wang, W.: Design of limited-stop bus service with capacity constraint and stochastic travel time. Transp. Res. Part E **83**, 1–15 (2015)
5. Liu, Z., Yan, Y., Qu, X., Zhang, Y.: Bus stop-skipping scheme with random travel time. Transp. Res. Part C **35**, 46–56 (2013)
6. Liu, Z., Wang, S., Chen, W., Zheng, Y.: Willingness to board: a novel concept for modeling queuing up passengers. Transp. Res. Part B **90**, 70–82 (2016)
7. Athira, I.C., Muneera, C.P., Krishnamurthy, K., Anjaneyulu, M.V.L.R.: Estimation of value of travel time for work trips. Transp. Res. Procedia **17**, 116–123 (2016)
8. Wang, F.: Study on behavior value of time based on stated preference survey (Master's thesis). Beijing University of Technology (2005)
9. Kockelman, D.K., Chen, T.D., Larsen, D.K., Nichols, B.: The economics of transportation systems: a reference for practitioners. University of Texas (2013)
10. Becker, G.S.: A theory of the allocation of time. Econ. J. **75**, 493–517 (1965)
11. Johnson, M.B.: Travel time and the price of leisure. Econ. Inq. **4**(2), 135–145 (1966)
12. De Serpa, A.: A theory of the economics of time. Econ. J. **81**, 828–846 (1971)
13. Cesario, F.J.: Value of time in recreation benefit studies. Land Econ. **52**(1), 32 (1976)
14. Daly, A.J., Zachary, S.: Commuters' value of time. Transportation Research Record (1975)
15. Mcfadden, D.L.: Conditional logit analysis of qualitative choice behavior. In: Frontiers in Econometrics (1972)
16. Boyd, J.H., Mellman, R.E.: The effect of fuel economy standards on the US automotive market: an hedonic demand analysis. Transp. Res. Part A **14**(5–6), 367–378 (1980)

17. Cardell, N.S., Dunbar, F.C.: Measuring the societal impacts of automobile downsizing. Transp. Res. Part A **14**(5–6), 423–434 (1980)
18. Ben-Akiva, M., Bolduc, D., Bradley, M.: Estimation of travel model choice models with randomly distributed values of time. Transp. Res. Rec. **1413**, 88–97 (1993)
19. Greene, W.H., Hensher, D.A., Rose, J.: Accounting for heterogeneity in the variance of unobserved effects in mixed logit models. Transp. Res. Part B **40**, 75–92 (2006)
20. Train, K.: Mixed logit with a flexible mixing distribution. J. Choice Model. **19**(2000), 40–53 (2016)
21. Devarasetty, P.C., Burris, M., Shaw, W.D.: The value of travel time and reliability-evidence from a stated preference survey and actual usage. Transp. Res. Part A **46**(8), 1227–1240 (2012)
22. Algers, S., Bergström, P., Dahlberg, M., Dillén, J.L.: Mixed logit estimation of the value of travel time. Working Paper, Department of Economics, Uppsala University (1998)
23. Li, R., Lin, D.K., Chen, Y.: Uniform design: design, analysis and applications. Int. J. Mater. Prod. Technol. **20**(1), 101–114 (2004)
24. Train, K.E.: Discrete Choice Methods with Simulation. Cambridge University Press, Cambridge (2009)
25. Qiao, C.G., Wood, G.R., Lai, C.D., Luo, D.W.: Comparison of two common estimators of the ratio of the means of independent normal variables in agricultural research. Adv. Decis. Sci. **2006**, 14 (2006)

Validation of an Optimization Model Based Stochastic Traffic Flow Fundamental Diagram

Jin Zhang(✉) and Xu Wang

Griffith School of Engineering, Griffith University,
Gold Coast, QLD 4222, Australia
{jin.zhang4,michael.wang}@griffithuni.edu.au

Abstract. The fundamental diagram is used to represent the graphical layout and determine the mathematical relationships among traffic flow, speed and density. Based on the observational speed-density database, the distribution of speed is scattered in any given traffic state. In order to address the stochasticity of traffic flow, a new calibration approach has been proposed to generate stochastic traffic flow fundamental diagrams. With this proposed stochastic fundamental diagram, the residual and stochasticity of the performance of calibrated fundamental diagrams can be evaluated. As previous work only shows the validation of one model, in this paper, we will use field data to validate other stochastic models. Greenshields model, Greenberg model, and Newell model are chosen to evaluate the performance of the proposed stochastic model. Results show that the proposed methodology fits field data well.

1 Introduction

In order to analyze a macroscopic traffic model involving traffic parameters – flow, speed and density, a fundamental diagram has been applied to represent relationships between these traffic characteristics to distinguish vehicular traffic flow from other kinds of flow [1–3]. For the speed-density relationship, since the fundamental diagram of the speed-density relationship was first introduced by Greenshields in 1935 [4], the development of this relationship has been greatly increased and there are various models to represent the relationship now: e.g. Greenberg (1959), Underwood (1961), Newell (1961), Kerner and Konhäuser (1994), Del Castillo (1995a&b), Li and Zhang (2001), MacNicolas (2008), Ji et al. (2010), Wang et al. (2011), Keyvan-Ekbatani et al. (2012, 2013) [5–16]. Note that these models are all deterministic in nature which essentially describes average system behaviors from a statistical perspective. However, as the freeway traffic is more complex than deterministic, predictable, and homogeneous fluids governed by physical laws, the collected database scattered throughout the entire range of traffic flow [17, 18]. Hence, stochastic models are more reasonable and accurate to represent the speed-density relationship.

From the previous literature, a few stochastic models have been proposed to represent the traffic flow relationship. Soyster and Wilson [19] proposed a simple stochastic model for traffic on hills using a Poisson process, and this model is too

G. De Pietro et al. (Eds.): KES-IIMSS-18 2018, SIST 98, pp. 329–337, 2019.
https://doi.org/10.1007/978-3-319-92231-7_34

specific and ideal. Kerner [20] model only provided the range of speed at a given density, so the distribution of speed was missing. Other probabilistic models like Muralidharan [21], Fan and Seibold [22], Jabari and Liu [23] models can generate the distribution of speed/flow as a function of given density, but they are all based on specific flow models: Muralidharan et al. model based on the triangular deterministic model; Jabari and Liu model based on Newell's simplified microscopic car-following model; Fan and Seibold introduce a new varying parameter to reflect the randomness in the Aw-Rascle-Zhang (ARZ) model. Qu et al. [12] propose a generic approach to generate a stochastic fundamental diagram which has wide applicability to all speed-density models, but they only use Underwood model for validation purpose. The objective of this paper is to evaluate other four well known models in Table 1.

Table 1. Well-known speed-density models

Models	Function	
Greenshields et al. (1935)	$v = v_f\left(1 - \frac{k}{k_j}\right)$	v_f, k_j
Greenberg (1959)	$v = v_o \ln\left(\frac{k_j}{k}\right)$	v_o, k_j
Newell (1961)	$v = v_f\left\{1 - \exp\left[-\frac{\lambda}{v_f}\left(\frac{1}{k} - \frac{1}{k_j}\right)\right]\right\}$	v_f, k_j, λ

2 Data

In this paper, we use the GA400 data. This GA400 dataset includes a total of 47,815 observations of speeds and densities from 76 stations were collected on the Georgia State Route 400 by snapshot, and the data is frequently used for calibrating the speed-density relation in the literature (e.g. Wang et al. (2011) [1]). Each station provides average observations at every 20 s and we have one-year of continuous observations in 2003 which is long enough to describe the fundamental diagram. In this project, the original data is aggregated every 5 min when used to generate the fundamental diagrams. Note that Qu et al. (2015) and Qu et al. (2017) both applied this dataset for their case studies.

3 Methodology

Having obtained the calibrated deterministic speed-density models, we apply an optimization model to calibrate a family of percentile-based speed-density curves by introducing another parameter α in the model [M']. This new model [M'] is to calibrate a $100\alpha^{\text{th}}$ percentile based speed-density curve such that the ratio between weighted residual of observations below the calibrated curve and the total residual is α. The mathematical representation of α is,

$$\frac{\sum_{i=1}^{m} \varpi_i |g_\alpha(k_i, v_i)|}{\sum_{i=1}^{m} \varpi_i |v_i - v_\alpha(k_i)|} = \alpha \quad (1)$$

and

$$g_\alpha(k_i, v_i) = \begin{cases} v_\alpha(k_i) - v_i, & \text{if } v_i - v_\alpha(k_i) < 0 \\ 0, & \text{otherwise} \end{cases} \quad (2)$$

In Qu et al. (2017), they further prove that Eq. (1) is satisfied at the optimal solution to the following optimization model [M'] and therefore we can solve [M'] to calibrate the $100\alpha^{th}$ percentile based speed-density line:

$$[\mathrm{M}'] \quad \min C(k_{jam}, v_f) = \sum_{i=1}^{m} (1 - 2\alpha)\varpi_i (g_\alpha(k_i, v_i))^2 + \sum_{i=1}^{m} \alpha \varpi_i (v_i - v_\alpha(k_i))^2 \quad (3)$$

subject to:

$$v_\alpha(k_i) = f(k_i, \beta), i = 1, 2, \ldots, m \quad (4)$$

$$g_\alpha(k_i, v_i) = \begin{cases} v_\alpha(k_i) - v_i, & \text{if } v_i - v_\alpha(k_i) < 0 \\ 0, & \text{otherwise} \end{cases} \quad (5)$$

This procedure means that, the fundamental diagram calibrated by Model [M'] is the $100\alpha^{th}$ percentile based speed-density curve based on speed-density model - $v = f(k, \beta)$. By changing α between 0 and 1, the stochastic fundamental diagram is obtainable. Again, here if we replace Eq. (4) by other functions in Table 1, the corresponding optimization models can be established to generate the $100\alpha^{th}$ percentile based speed-density curve with respect to other speed-density models. The results are shown in Fig. 1.

4 Results

According to the Fig. 1, we can have all the corresponding percentile based speeds for any given density. In other words, the cumulative distribution function (CDF) and probability density function (PDF) of speeds at any given density are obtainable. Then, we use each model in Table to generate the CDFs and PDFs at a given density. In order to validate the performance of the generated CDFs and PDFs, we also generate the empirical CDFs and PDFs with respect to the given densities (k = 10, 40, 60 veh/km). As mentioned above, we use Greenshields model, Greenberg model, and Newell model to validate the proposed methodology by Qu et al. (2017). We compare the empirical

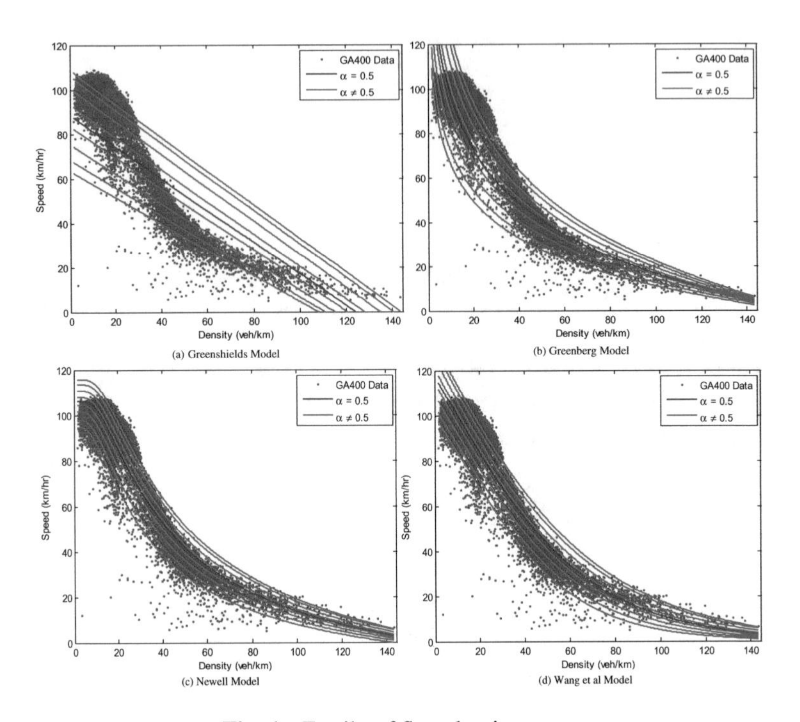

Fig. 1. Family of flow-density curves

CDFs/PDFs and generated CDFs/PDFs by the proposed methodology in combination with the three well known deterministic models. The results are presented in Figs. 2, 3 and 4.

As can be seen in Figs. 2, 3 and 4, most of the generated CDFs and PDFs perfectly re-establish the empirical CDFs and PDFs. Hypothesis tests also included to compare with the PDFs. It should be noted that the empirical speed PDF may have zigzag sections through the whole density range. For example, when density at 60 veh/km, there is a zigzag section from 15 km/h to 25 km/h. However, most of the proposed model based approach can still capture this zigzag pattern and practically re-establish the pattern of speed-density data points. Moreover, the data points at higher densities may not have enough data points to generate complete histograms for validation. For example, when density equals 60 vehs/km, we have no data when speed is from 18 km/h to 22 km/h. Although the histograms are incomplete, the generated PDFs and CDFs still reasonably re-establish the empirical ones. The evaluation results are similar to Underwood Model in Qu et al. (2017).

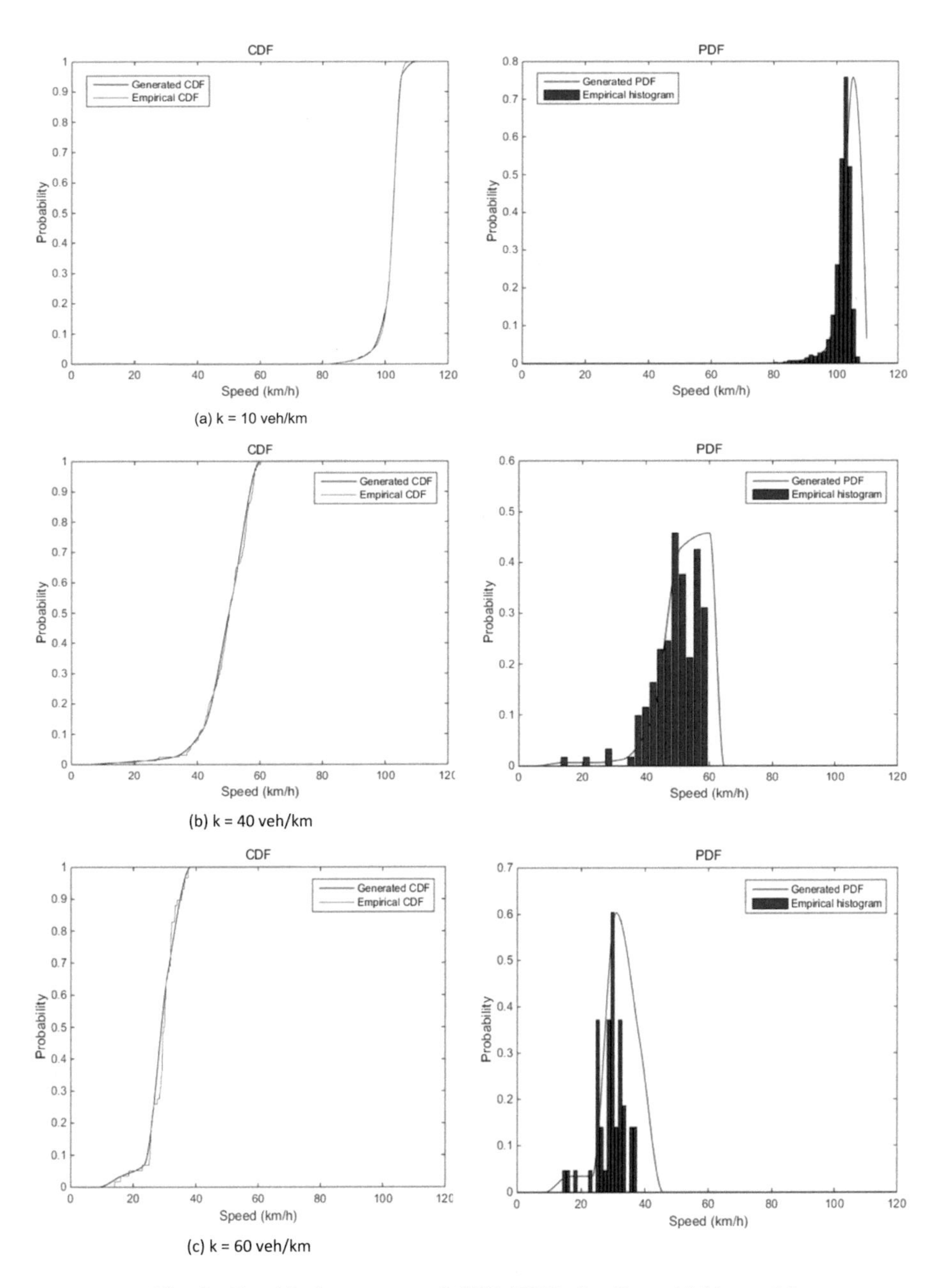

Fig. 2. Empirical vs generated CDFs/PDFs for Greenshields model

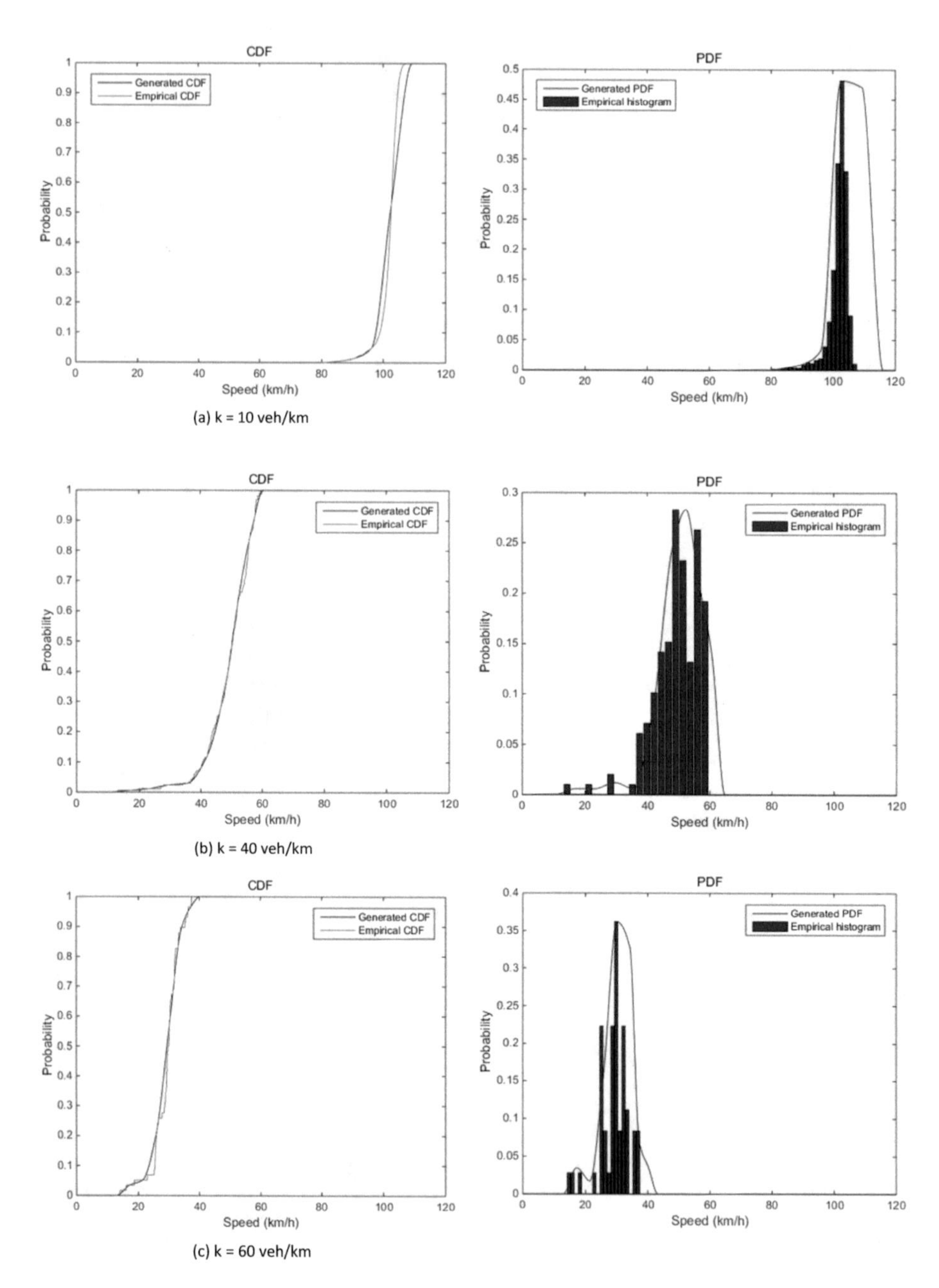

Fig. 3. Empirical vs generated CDFs/PDFs for Greenberg model

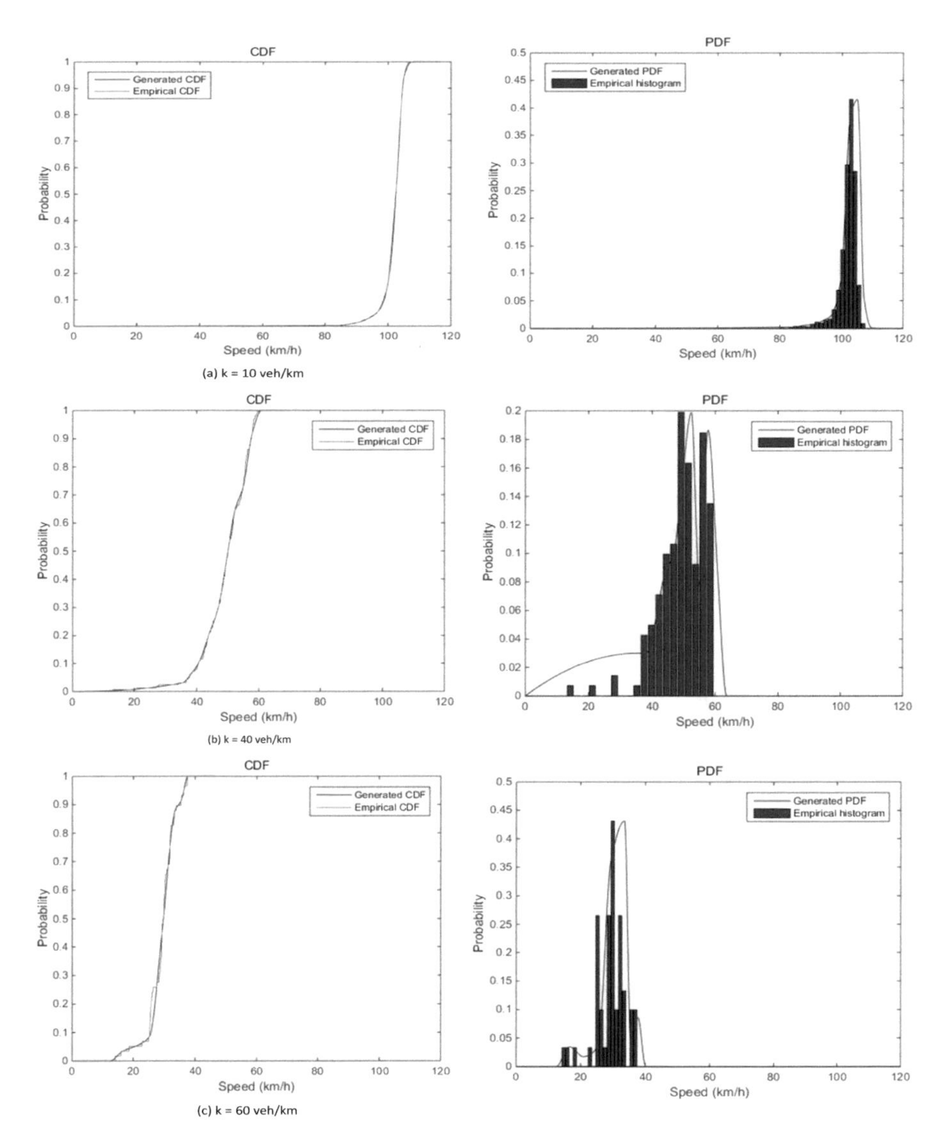

Fig. 4. Empirical vs generated CDFs/PDFs for Newell model

5 Conclusion

This research is an extension of Qu et al. (2017), in which the authors propose an optimization model based on the theorem of probability to generate stochastic fundamental diagrams. However, for the model validation section, they only use Underwood model. In this paper, we use other three well-known models: Greenshields model, Greenberg model, and Newell model to evaluate the performance of the proposed stochastic model. Based on our results, the proposed methodology can be applied in combination with Greenshields model, Greenberg model, and Newell model. The model can be useful for the

References

1. Wang, H., Li, H., Chen, Q., Ni, D.: Logistic modeling of the equilibrium speed–density relationship. Transp. Res. Part A **45**, 554–566 (2011)
2. Qu, X., Wang, S., Zhang, J.: On the fundamental diagram for freeway traffic: a novel calibration approach for single-regime models. Transp. Res. Part B **73**, 91–102 (2015)
3. Jin, S., Qu, X., Zhou, D., Xu, C., Ma, D., Wang, D.: Estimating cycleway capacity and bicycle equivalent units for electric bicycles. Transp. Res. Part A **77**, 225–248 (2015)
4. Greenshields, B.D., Bibbins, J.R., Channing, W.S., Miller, H.H.: A study of traffic capacity. Highw. Res. Board Proc. **14**, 448–477 (1935)
5. Greenberg, H.: An analysis of traffic flow. Oper. Res. **7**, 255–275 (1959)
6. Underwood, R.T.: Speed, volume, and density relationship: quality and theory of traffic flow. In: Yale Bureau of Highway Traffic, pp. 141–188 (1961)
7. Newell, G.F.: Nonlinear effects in the dynamics of car following. Oper. Res. **9**(2), 209–229 (1961)
8. Kerner, B.S., Konhäuser, P.: Structure and parameters of clusters in traffic flow. Phys. Rev. E **50**(1), 54–83 (1994)
9. Del Castillo, J.M., Benítez, F.G.: On the functional form of the speed-density relationship–I: general theory. Transp. Res. Part B **29**(5), 373–389 (1995)
10. Del Castillo, J.M., Benítez, F.G.: On the functional form of the speed-density relationship–II: empirical investigation. Transp. Res. Part B **29**(5), 391–406 (1995)
11. Li, J., Zhang, H.M.: Fundamental diagram of traffic flow: new identification scheme and further evidence from empirical data. Transp. Res. Rec. **2011**, 50–59 (2001)
12. Qu, X., Zhang, J., Wang, S.: On the stochastic fundamental diagram for freeway traffic: model development, analytical properties, validation, and extensive applications. Transp. Res. Part B **104**, 256–271 (2017)
13. MacNicholas, M.J.: A simple and pragmatic representation of traffic flow. In: Symposium on The Fundamental Diagram: 75 Years, Transportation Introduction Research Board, Woods Hole, MA (2008)
14. Ji, Y., Daamen, W., Hoogendoorn, S., Hoogendoorn-Lanser, S., Qian, X.: Investigating the shape of the macroscopic fundamental diagram using simulation data. Transp. Res. Rec. **2161**, 40–48 (2010)
15. Keyvan-Ekbatani, M., Kouvelas, A., Papamichail, I., Papageorgiou, M.: Exploiting the fundamental diagram of urban networks for feedback-based gating. Transp. Res. Part B **46**(10), 1393–1403 (2012)
16. Keyvan-Ekbatani, M., Papageorgiou, M., Papamichail, I.: Urban congestion gating control based on reduced operational network fundamental diagrams. Transp. Res. Part C **33**, 74–87 (2013)
17. Zhou, M., Qu, X., Li, X.: A recurrent neural network based microscopic car following model to predict traffic oscillation. Transp. Res. Part C **84**, 245–264 (2017)
18. Kuang, Y., Qu, X., Wang, S.: A tree-structured crash surrogate measure for freeways. Accid. Anal. Prev. **77**, 137–148 (2015)
19. Soyster, A.L., Wilson, G.R.: A stochastic model of flow versus concentration applied to traffic on hills. Highw. Res. Board **456**, 28–39 (1973)

20. Kerner, B.S.: Experimental features of self-organization in traffic flow. Phys. Rev. Lett. **81**(17), 3797 (1998)
21. Muralidharan, A., Dervisoglu, G., Horowitz, R.: Probabilistic graphical models of fundamental diagram parameters for simulations of freeway traffic. Transp. Res. Rec. **2249**, 78–85 (2011)
22. Fan, S., Seibold, B.: Data-fitted first-order traffic models and their second-order generalisations: comparison by trajectory and sensor data. Transp. Res. Rec. **2391**, 32–43 (2013)
23. Jabari, S.E., Liu, H.X.: A probabilistic stationary speed–density relation based on Newell's simplified car-following model. Transp. Res. Part B **68**, 205–223 (2014)
24. Liu, Z., Yan, Y., Qu, X., Zhang, Y.: Bus stop-skipping scheme with random travel time. Transp. Res. Part C **35**, 46–56 (2013)

Estimation of Heavy Vehicle Passenger Car Equivalents for On-Ramp Adjacent Zones Under Different Traffic Volumes: A Case Study

Xu Wang[1], Weiwei Qi[2(✉)], and Mina Ghanbarikarekani[3]

[1] Griffith School of Engineering, Griffith University, Gold Coast, QLD 4222, Australia
[2] School of Civil Engineering and Transportation, South China University of Technology, Guangzhou, China
ctwwqi@scut.edu.cn
[3] School of Civil and Environmental Engineering, University of Technology Sydney, 2007 Sydney, Australia

Abstract. Due to the difference in operational characteristic, heavy vehicles have been viewed as a hindrance in traffic flow and capacity analysis. The emergence of passenger car equivalents (PCE) can assist traffic agencies in better understanding the impact of heavy vehicles on passenger vehicles in the mixed traffic stream, by converting a heavy vehicle of a subject class into the equivalent number of passenger cars. However, according to existing literature, most researchers have devoted to the estimation of PCE for basic freeway sections. Therefore, in this study, we explore the variation of heavy vehicle PCE for on-ramp adjacent zones under varying traffic volume. A one-lane on-ramp in Queensland, Australia, is selected for a case study and four existing PCE approaches are applied in the calculation of PCE. They are homogenization based method, time headway based method, traffic flow based method, and multiple regression method, respectively. The final PCE values are compared to those derived from VISSIM simulation model. The following conclusions are drawn: (1) homogenization based method cannot reveal the variation trend of PCE factors over traffic volume; (2) the results obtained through time headway and traffic flow based methods are more consistent with outcome from simulation model.

Keywords: Passenger car equivalents · Passenger car units · Freeway capacity VISSIM simulation

1 Introduction

To balance transport costs and time, increasing motor vehicles have swarmed into freeways. Accordingly, capacity analysis has been of vital importance in planning, design and operation of freeways [1–3]. However, variations in traffic composition complicate the process. Varying vehicles differ in size, speed, acceleration and braking capacity, so they cannot be treated as identical. As a result, converting heavy vehicles (HVs) to the equivalent number of passenger cars (PCs) is required. The term,

G. De Pietro et al. (Eds.): KES-IIMSS-18 2018, SIST 98, pp. 338–346, 2019.
https://doi.org/10.1007/978-3-319-92231-7_35

passenger car equivalents (PCE) or passenger car units (PCU), was first introduced in the highway capacity manual (HCM) in 1965 and defined as the number of PCs that are displaced by a HV of subject category, under the prevailing roadway and traffic conditions [4]. It has been treated as a significant indicator to reveal the adverse impact of HVs on the quality of traffic flow on freeways [5–12]. As per the most recently issued HCM [13], the PCE value was applied to approximate HV adjustment factors (f_{HV}) for basic freeway sections through the following equation:

$$f_{HV} = \frac{1}{1 + P_T(E_T - 1) + P_R(E_R - 1)} \tag{1}$$

where P_T and P_R are percent trucks and percent creational vehicles. E_T and E_R are PCE values for trucks and creational vehicles, respectively.

According to the extensive literature, PCE values can be derived from various traffic flow characteristics such as speeds, time headways, volumes and percentages of HVs as well as vehicular and freeway characteristics [14–19]. All researchers have believed that HVs and PCs in any vehicle-mixed stream should be separately analysed because of the following points: (1) HVs significantly differ from standard PCs in speed and operational characteristics (maneuverability, acceleration, braking, etc.); (2) Compared to PCs, HVs typically need more space and larger time headway to keep safe; (3) an increase in the percentage of HVs in a mixed traffic flow has an effect on traffic operation and efficiency. However, the PCE estimates are generally based on homogenous traffic conditions and basic freeway sections in developed countries. There was few research focusing on the PCE estimations for HVs on on- and off-ramps and weaving zones. Therefore, the objective of the paper is to quantify PCE values of HVs on freeway on-ramps. To achieve it, we explore the variation trend in PCE values under varying traffic volume. Five existing methods are applied for the estimation.

The remainder of the paper is organized as follow. Section 2 reviews five existing approaches for PCE factors. Section 3 describes the research segment and the procedure of data collection and processing. Section 4 presents five sets of PCE values and compares them with those provided in 2010 HCM [13]. Section 5 concludes.

2 Existing Methods for PCE Estimates

This section presents PCE calculation methods which are commonly used by researchers and traffic agents. Among them, time headway based method, traffic flow based method, multiple regression method, and simulation method are more common [15].

2.1 Homogenization Based Methods

Built on the homogenization method (HM) proposed in 1965, Chandra and Sikdar [12] considered the impact of HVs on the transverse space of PCs. They revised the original HM by replacing the length of vehicles with the horizontal projected area of vehicles, mathematically,

$$PCE_i = \frac{A_i/V_i}{A_c/V_c} = \frac{A_i}{A_c} \times \frac{V_c}{V_i} \quad (2)$$

where A_i and A_c are the rectangular projected area of type i HVs and PCs. However, this method has a limitation in use. It is more appropriate for the quantification of PCEs for homogeneous traffic in developed countries, as its proposal was based on the traffic condition where the composition of vehicles is relatively singularized and most motorists strictly follow the lane discipline of the road.

2.2 Time Headway Based Methods

Based on the initial time headway based method (THM), Krammes and Crowley [20] further analysed time headway for four different car-following, which reflects headway differences between HVs and PCs. In addition, the effect of non-subject HVs on a mixed-traffic stream was also taken into account. The equation is given as follows,

$$PCE_s = \frac{(1-p)\left(h_{ps} + h_{sp} - h_{pp}\right) + ph_{ss}}{h_{pp}} \quad (3)$$

where p is the proportion of HVs in a mixed traffic stream. h_{ps} is mean headway for PCs following HVs of the subject class, h_{sp} is mean headway for HVs of the subject class following PCs, h_{pp} is mean headway for PCs following PCs, h_{ss} is mean headway for HVs of the subject class following HVs of the subject class. The data collection needs to focus on the four kinds of time headways.

2.3 Traffic Flow Based Method

Summer et al. [21] developed the traffic flow based method (TFM) to estimate PCE values based on a flow vs density curve generated by simulation model. It took into account macroscopic parameters such as traffic flow rates and percent HVs, which reduces the difficulty in collecting microscopic parameters. The equation is given as follows,

$$PCE_S = \frac{1}{\Delta p}\left[\frac{q_B}{q_S} - \frac{q_B}{q_M}\right] + 1 \quad (4)$$

where Δp is the percentage of the subject HVs displacing an equal number of PCs in a mixed traffic flow. Webster and Elefteriadou [15] recommended it be 5%. q_B is the base vehicle flow rate at a constant traffic density where only PCs are involved. q_M is the mixed vehicle flow rate at a constant traffic density where typical percent HVs are involved. 7.3% of HVs was used in this study. q_S is the subject HV flow rate at a constant traffic density where a certain number of PCs in the mixed traffic stream is replaced with an equal number of the subject PCs. In this case, q_B and q_S were indirectly obtained through flow vs density curve.

2.4 Multiple Linear Regression Model

The multiple linear regression model (MLRM) has been widely used to calibrate coefficients in a relationship [22–24]. It is built on the conception that different classes of vehicles significantly differ in speed reduction potentials. The speed reduction coefficients for any category of vehicles can be quantified by the relationship between predictor variables (the number of different types of vehicles) and the response variable (the average traffic stream speed under certain traffic volume). MLRM is mathematically given as follows,

$$MS = u_f + C_1 N_p + C_2 N_h + C_3 N_b + C_4 N_o \tag{5}$$

where MS is mean traffic stream speed. u_f is free flow speed. N_p, N_h, N_b and N_o are the number of PCs, HVs, buses and other vehicles in traffic stream, respectively. C_1, C_2, C_3 and C_4 are marginal effects of the number of different types of vehicles on mean traffic stream speed. PCE values for type i vehicles can be calculated based on the ratio of C_i to C_1. However, the MLRM is built on premise that the average traffic speed is a linear function of traffic volume. Therefore, it is not appropriate for the estimation of PCE under all traffic scenarios.

3 Data Collection and Processing

3.1 Site Description

The one-hour based traffic data under varying traffic volumes were obtained from the zone adjacent to the one-lane zip-merging on-ramp in Queensland, which is shown in Fig. 1.

Fig. 1. The location of the data collection point and study site

3.2 Data Processing

Considering the homogeneity of traffic in Queensland, vehicles were categorized into four groups: (1) trucks with the length of 12.5 m are viewed as HVs; (2) buses with 12.5 m are standard buses; (3) 5-m short cars are regarded as reference PCs; (4) the rest of vehicles in the mixed traffic stream are treated as other vehicles due to the marginal percent composition. The one-hour based data were collected from nine different time periods, which reflects actual variations in traffic volumes. The speeds for 50 HVs and 100 PCs were retrieved from each video and averaged in order to estimate PCE values using the HM. The parameters for the HM are shown in Table 1. The proportions of HVs under nine traffic volumes were derived from videos. The headways for four car-following scenarios were obtained based on 20 corresponding car-following events under each scenario, which is given in Table 2.

Table 1. The parameters used for the HM

Traffic flow (vehs/hr/ln)	Time periods	Average speed for HVs (m/s)	Average speed for PCs (m/s)
413	21:00–22:00	26.7	29.8
609	20:00–21:00	25.0	28.1
786	19:00–20:00	24.3	27.0
1008	12:00–13:00	24.0	26.4
1221	05:00–06:00	23.7	25.9
1389	16:00–17:00	23.6	25.9
1610	09:00–10:00	23.5	25.8
1793	17:00–18:00	23.4	25.8
2016	06:00–07:00	23.6	25.8

Table 2. The parameters used for the THM

Traffic flow (vehs/hr/ln)	Proportion of HGVs	Headway for pp (s)	Headway for ss (s)	Headway for sp (s)	Headway for ps (s)
413	0.054	6.09	6.28	6.25	6.11
609	0.063	4.01	4.71	4.45	4.39
786	0.075	2.93	3.84	3.79	3.41
1008	0.081	2.28	3.19	3.44	2.89
1221	0.079	1.80	2.75	2.89	2.46
1389	0.089	1.54	2.46	2.74	1.99
1610	0.081	1.20	2.11	2.38	1.59
1793	0.079	1.08	1.92	2.13	1.46
2016	0.083	1.06	1.95	2.21	1.39

The parameters for the TFM were derived according to simulation models. In this study, VISSIM was used to simulate and generate three kinds of traffic flow rates, as the interaction between PCs and HVs during overtaking and merging and the behaviour of car drivers in the neighbourhood of HVs can be handled internally by VISSIM. It is necessary that VISSIM model need to be calibrated with traffic flow parameters (e.g. time headway, traffic volume, percent HVs, etc.) and be validated with traffic trajectory data extracted from videos recorded on site before use. The parameters used for the TFM are shown in Table 3. To estimate PCE values using MLRM, speeds for 100 vehicles were retrieved from each video and averaged. The number of each category of vehicles under nine traffic flow conditions was counted and listed in Table 4.

Table 3. The parameters used for TFM

Traffic flow (vehs/hr/ln)	Flow rate for fixed traffic stream (vehs/hr/ln)	Flow rate for base vehicle stream (vehs/hr/ln)	Flow rate for subject HV stream (vehs/hr/ln)
413	421	433	418
609	611	625	605
786	795	811	780
1008	1023	1042	989
1221	1203	1229	1150
1389	1411	1446	1322
1610	1598	1635	1490
1793	1784	1798	1674
2016	1989	2035	1837

Table 4. The parameter used for MLRM

Traffic flow (vehs/hr/ln)	Average traffic stream speed (km/hr)	No. of PCs	No. of HVs	No. of buses	No. of other vehicles
413	102.7	391	15	5	2
609	98.5	571	26	7	5
786	94.8	727	44	10	5
1008	91.9	927	61	14	6
1221	88.7	1125	81	11	4
1389	86.6	1266	96	18	9
1610	85.4	1480	109	11	10
1793	84.6	1652	123	10	8
2016	83.1	1849	137	19	11

Due to the lack of a judgement criterion for PCE values, the simulation method (SM) was used, as a control group, to compare with PCE values derived from other four methods. In this study, VISSIM simulation model was used to estimate PCE values simply based on freeway merging capacity theory. We let the subject HV only stream and PC only stream respectively merge into the study site. The maximum numbers of the subject HVs and PCs merging into the freeway mainline under each traffic volume were obtained through multiple simulation runs. The ratio of the maximum number of PCs merging into the freeway to that of the subject HVs was finally viewed as the PCE value.

4 A Comparative Analysis

The preliminary results for four different PCE estimates and the SM were calculated based on data provided in Sect. 3 and tabulated in Table 5, which reflects the variation trend of PCE values over the increase in traffic volumes.

Table 5. The results for different PCE methods

Traffic flow (vehs/hr/ln)	HM	THM	TFM	MLRM	SM
413	2.79	1.03	1.15	1.58	1.20
609	2.81	1.20	1.20	1.58	1.26
786	2.78	1.45	1.39	1.58	1.45
1008	2.75	1.75	1.70	1.58	1.81
1221	2.73	1.94	1.94	1.58	2.29
1389	2.74	2.03	2.38	1.58	2.45
1610	2.74	2.26	2.48	1.58	2.57
1793	2.76	2.28	2.32	1.58	2.64
2016	2.73	2.35	2.69	1.58	2.70

As can be seen in Table 5, PCE values calculated by the HM slightly fluctuate with the increase in traffic volumes. As the projected area for the subject HVs and reference PCs has been fixed, PCE values for the HM only replies on the ratio of average speed for PCs to that for HVs. The final results reflect the speed is less sensitive to the change in traffic volume, which is consistent with other researchers' conclusions. Accordingly, the HM is not a proper indicator to predict the variation in traffic flow. For the THM, time headway and percent HVs, were taken into account. As a result, PCE values estimated by the THM are relatively consistent with those simulated through the SM, but it slightly underestimate PCE values at high traffic volume conditions. The density-flow curve and simulation based TFM performs slightly better than the THM, since its predictor variables are exactly three types of traffic stream rates. Therefore, it is an acceptable indicator to predict the variation in PCE values with the increase in traffic volumes. The MLRM requires a mass of data based on each traffic scenario. Due to the limitation of data, we simply derive a fixed PCE value under nine traffic scenarios.

5 Conclusion

The objective of this study is to estimate the variation in PCE values of the subject heavy vehicles (HVs), with the increase in traffic volumes, for an on-ramp adjacent zone. To this end, the zone, in the neighbourhood of a one-lane zip-merging on-ramp in

Queensland, Australia, was selected as the study site. Based on data collected from the study site and generated by VISSIM, PCE values were calculated through four existing PCE methods: (1) homogenization based method (HM); (2) time headway based method (THM); (3) traffic flow based method (TFM); (4) multiple linear regression model (MLRM). Besides, the simulation method (SM) was applied to generate a set of PCE reference values which is a criterion to judge the most appropriate PCE method. Through a comparative analysis, the following conclusions can be drawn: (1) the HM cannot properly predict the variation trend of PCE values over traffic volumes due to the low sensitivity of the speed to the change in traffic volumes; (2) both the THM and TFM can derive the results which are relatively consistent with outcomes from VISSIM simulation model. The developed PCE is very useful for further traffic engineering studies with mixed traffic flow [23, 24].

References

1. Adnan, M.: Passenger car equivalent factors in heterogenous traffic environment-are we using the right numbers? Procedia Eng. **77**, 106–113 (2014)
2. Qu, X., Zhang, J., Wang, S.: On the stochastic fundamental diagram for freeway traffic: model development, analytical properties, validation, and extensive applications. Transp. Res. Part B **104**, 256–271 (2017)
3. Qu, X., Wang, S., Zhang, J.: On the fundamental diagram for freeway traffic: a novel calibration approach for single-regime models. Transp. Res. Part B **73**, 91–102 (2015)
4. TRB (Transportation Research Board): Special report 209: Highway Capacity Manual, third ed. Transportation Research Board, Washington, DC (1994)
5. De Luca, M., Dell'Acqua, G.: Calibrating the passenger car equivalent on Italian two line highways: a case study. Transport **29**(4), 449–456 (2013)
6. Elefteriadou, L., Torbic, D., Webster, N.: Development of passenger car equivalents for freeways, two-lane highways, and arterials. Transp. Res. Rec. J. Transp. Res. Board **1572**, 51–58 (1997)
7. Fan, H.: Passenger car equivalents for vehicles on Singapore expressways. Transp. Res. Part A Gen. **24**(5), 391–396 (1990)
8. Okura, I., Sthapit, N.: Passenger car equivalents of heavy vehicles for uncongested motorway traffic from macroscopic approach. Doboku Gakkai Ronbunshu **512**, 73–82 (1995)
9. Al-Kaisy, A., Jung, Y., Rakha, H.: Developing passenger car equivalency factors for heavy vehicles during congestion. J. Transp. Eng. **131**(7), 514–523 (2005)
10. Al-Obaedi, J.: Estimation of passenger car equivalents for basic freeway sections at different traffic conditions. World J. Eng. Technol. **04**(02), 153–159 (2016)
11. Benekohal, R., Zhao, W.: Delay-based passenger car equivalents for trucks at signalized intersections. Transp. Res. Part A Policy Pract. **34**(6), 437–457 (2000)
12. Chandra, S., Sikdar, P.K.: Factors affecting PCU in mixed traffic situations on urban roads. Road Transp. Res. **9**(3), 40–50 (2000)
13. TRB (Transportation Research Board): Highway Capacity Manual. Transportation Research Board, Washington, DC (2010)
14. Huber, M.J.: Estimation of passenger car equivalents of trucks in traffic stream. Transp. Res. Board **869**, 60–70 (1982)

15. Webster, N., Elefteriadou, L.: A simulation study of truck passenger car equivalents (PCE) on basic freeway sections. Transp. Res. Part B Methodol. **33**(5), 323–336 (1999)
16. Tiwari, G., Fazio, J., Pavitravas, S.: Passenger car units for heterogeneous traffic using a modified density method. In: Transportation Research Circular E-C018: 4th International Symposium on Highway Capacity, Maui, Hawaii, USA (2000)
17. Kuang, Y., Qu, X., Wang, S.: A tree-structured crash surrogate measure for freeways. Accid. Anal. Prev. **77**, 137–148 (2015)
18. Qu, X., Yang, Y., Liu, Z., Weng, J., Jin, S.: Potential crash risks of expressway on-ramps and off-ramps: a case study in Beijing, China. Saf. Sci. **70**, 58–62 (2014)
19. Jin, S., Qu, X., Zhou, D., Xu, C., Ma, D., Wang, D.: Estimating cycleway capacity and bicycle equivalent units for electric bicycles. Transp. Res. Part A **77**, 225–248 (2015)
20. Krammes, R.A., Crowley, K.W.: Passenger car equivalents for trucks on level freeway segments. Transp. Res. Rec. J. Transp. Res. Board **1091**, 10–17 (1986)
21. Sumner, R., Hill, D., Shapiro, S.: Segment passenger car equivalent values for cost allocation on urban arterial roads. Transp. Res. Part A Gen. **18**(5–6), 399–406 (1984)
22. Meng, Q., Qu, X.: Estimation of vehicle crash frequencies in road tunnels. Accid. Anal. Prev. **48**, 254–263 (2012)
23. Easa, S.M., Qu, X., Dabbour, E.: Improved pedestrian sight distance needs at railroad-highway grade crossings. J. Transp. Eng. Part A Syst. **143**(7), 04017027 (2017)
24. Wang, S., Meng, Q., Liu, Z.: Fundamental properties of volume-capacity ratio of a private toll road in general networks. Transp. Res. Part B **47**, 77–86 (2013)

Development of Road Functional Classification in China: An Overview and Critical Remarks

Yadan Yan(✉), Yang Li, and Pei Tong

School of Civil Engineering, Zhengzhou University, Zhengzhou 450001, China
yanyadan@zzu.edu.cn, 1372594218@qq.com,
958298975@qq.com

Abstract. Road functional classification is of significance for achieving high transportation efficiency. The road network planning and design mainly depend on road functions. This concept is used in almost all countries throughout the world. This paper presents a review on the development of road functional classification systems in codes and standards in China over the past decades, whose importance received limited attention to date. Functional classification systems include the urban road functional classification system and the highway functional classification system. The review includes major components of the road functional classification, consisting of context analysis, modal accommodations and detailed categories. In addition to these concerns, some ideas for future development, especially for context definitions, functional classification concepts and criteria, and quantitative functional evaluation, are discussed.

1 Introduction

Functional classification is the process by which streets and highways are grouped into classes, or systems, according to the character of service they are intended to provide [1]. It defines the role that a particular road segment plays in serving traffic flow through the network [2]. Functional classifications have impacts on land use both directly and indirectly [3]; meantime, urban layout and land use significantly affect road function. Inappropriate land use hinders roads from functioning as they are intended [4]. Functional classifications also have important impacts on many aspects of road planning and management, including the structure of road network, traffic safety, ranking of construction works, and investments in infrastructure [4–6].

A road network performs most efficiently and safely from both a traffic/road operation and a road safety perspective if roads are designed, operated and maintained to serve their intended purposes [7, 8]. In recent years, the "urbanized highway" phenomenon and the street concept have caused the rethinking of road functions. In this regard, objectives of the paper are to present an overview on the development of road functional classification systems in national codes and standards in China over the past decades. Furthermore, discussions and some critical remarks are developed.

G. De Pietro et al. (Eds.): KES-IIMSS-18 2018, SIST 98, pp. 347–355, 2019.
https://doi.org/10.1007/978-3-319-92231-7_36

2 Context Analysis

The context category decision is a starting point that leads to geometric design choices, as they will be influenced by the road type [3]. Then the context and road type will define modes that need to be considered. Two contexts have been used in the national standards. According to "Code for design of urban road" [9], urban roads and highways are divided by the boundary of the urban planning area in China. Roads that serve transportation needs within the urban area belong to the urban road category. Users of urban roads are vehicles, bicycles, pedestrians and buses, and they are administrated by the municipal department; while users of highways are mainly vehicles, and they are administrated by the transportation department. Highways usually do not set exclusive lanes for non-motorized modes.

Until 2013, the context of urbanization region was proposed in "Technical Standard of Urbanization Regional Highway Engineering" [10]. This local standard is issued by Shanghai Housing and Urban-Rural Development Commission. Urbanization regional highways are highways that are located within the suburban scope, whose land use plan is the urban construction land use or industrial land use.

3 Urban Road Functional Classification

3.1 Urban Road Types

For decades, the urban road network structure in China has relied on four urban road types: expressways, major arterials, minor arterials and branches [9, 11, 12]. This basic taxonomy has evolved since it is proposed in "Code for urban road design" [9] in 1990, shown as follows:

1. Expressways: providing service for large, long distance and fast traffic.
2. Major arterials: providing traffic function and citywide connections.
3. Minor arterials: mainly providing collecting and distributing functions. They also has the serving function.
4. Branches: providing connections between minor arterials and internal roads of residential districts, industrial districts and traffic facilities. The main function should be serving function and dedicated to serving the traffic of local areas.

It can be found that though functions of urban roads are provided, related descriptions are vague and difficult to apply in practices. The first national guidance involving the definition of road function is "Code for planning of intersection on urban roads" issued in 2011 [13]. In this code, the term "traffic function" is defined as the role that traffic infrastructure plays in the transportation system, and the traffic service provided for travellers. This code also describes functions and basic design requirements for each type of intersection.

In "Code for design of urban road engineering" [12] issued in 2012, the description of expressways is modified as "achieving the function of fast and continuous passing"; while other types of urban roads remain their former descriptions. Road types determine specific requirements for their design. Functional classifications in both "Code for planning of intersection on urban roads" [13] and "Code for design of urban road engineering" [12] act as input for the range selection of geometric design elements of urban road intersections and urban roads, such as design speed, capacity, level of service, alignment or intersection control type.

As far as the entire project development procedure is concerned, planning phase is before the design phase. A common concept in urban road network planning is the structuring of urban roads depending on the function of the urban road [14]. As usual, a function is assigned to an urban road which then influences its design. However, the "Code for transportation planning on urban road" [11] issued in 1995, which is currently in use, does not involve road functional classifications. It simply divides urban roads into four types (i.e., expressways, major arterials, minor arterials and branches) and provides planning indicators such as road network density, the number of lanes and road width for each type of road.

Hence, in 2016, a draft version of "Code for urban comprehensive transport system planning" [15] for comments was released. More detailed function description are given in the draft. According to trip characteristics, urban roads are divided into arterials, locals, and collectors. Collectors provide connections between arterials and locals. These three categories are further subdivided into four types, i.e., expressways, major arterials, minor arterials and branches, which are consistent with the conventional urban road types. Four urban road types are then subclassified into eight classes, as shown in Table 1.

Table 1. Functional classifications in "Code for Urban Comprehensive Transport System Planning" (draft, 2016).

Category	Type	Class	Function description
Arterial	Expressway	Class I	Providing fast and efficient traffic service for long distance trips of motor vehicles
		Class II	Providing fast traffic service for medium and long distance trips of motor vehicles
	Major arterial	Class I	Providing connections between major activity areas, and connections to entry and exit of the urban area
		Class II	Providing connections between major districts and sub-areas
		Class III	Providing connections between major districts and linkage within the district, and offering service for land use along the road
Collector	Minor arterial	–	Providing connections between arterials and locals
Local	Branch	Public	Roads for local circulation and access
		Non-public	Roads for local circulation and access but not administrated by the municipal department

3.2 Modal Accommodations

The functional classification system based on four types of urban roads is helpful and convenient for designers to use. Nevertheless, existing codes focus on motor vehicles and severely limit design choices for an urban road network which intends to meet multimodal needs. In recent years, the concept of street design has been introduced in China, which concentrates on public spaces and accommodating modes such as bicycles and pedestrians [16].

According to "Shanghai street design guidelines" [17] issued in 2016, streets in Shanghai are categorized considering activities along the road and characteristics of the located context. Five street types are addressed, including commercial streets, living and service streets, landscape and leisure streets, traffic streets and comprehensive streets. Road types and street types can be cross-classified. For instance, a commercial street can also be either a major arterial or a minor arterial. The "Code for urban comprehensive transport system planning" [15] also takes into account multimodal needs. Modal priorities are addressed for eight classes of urban roads, as shown in Table 2.

Table 2. Multimodal accommodation in "Code for Urban Comprehensive Transport System Planning" (draft, 2016).

Class	Bus transit	Bicycle and pedestrian
Class I expressway	Arterial and general bus routes	–
Class II expressway		–
Class I major arterial	Arterial and general bus routes	Priority
Class II major arterial		Priority
Class III major arterial		Priority
Minor arterial		Priority
Public branch	Feeder bus routes	Priority
Non-public branch	–	Priority

4 Highway Functional Classification

In the first and officially issued version of "Technical standard of highway engineering" in 1981 [18], highways are categorized as motorways, class I highways, Class II highways, class III highways and class IV highways, according to their technical characteristics. After two revisions in the following fifteen years, road function was finally paid more attention in the 1997 edition of "Technical standard of highway engineering" [19]. There are three categories including arterials, collectors and locals. These three categories are further divided into five classifications, which have evolved and still in use now, indicated as follows:

1. Freeways are the highest classification of highways. They are exclusively used by vehicles, and are designed as access controlled facilities with limited locations at which vehicles can enter or exit the road (typically via on- or off- ramps).

2. Class I highways provide connections between freeways and suburb or low density (few people) areas of large cities. Freeways and class I highways are arterial highways.
3. Class II highway are arterial highways serving medium cities and large mining districts or port districts.
4. Class III highways are local highways providing connections for towns or villages. Their main function is the collector function.
5. Class IV highways are local highways.

In the following 2003 edition of "Technical standard of highway engineering" [20], the collector function of class III highways are assigned to class I highways. As a result of vague and subjective provisions about functional classifications, designers prefer using traffic volumes as the basis of highway type designation instead of using the function. Context characteristics and road network structures are also rarely considered in engineering practices. With regard these issues, the term "highway function" is first defined in the 2014 version of "Technical standard of highway engineering" [21]. It is the ability to provide traffic services of passing through, collecting and distributing, and access. The importance of function is prominently highlighted. Different from the previous functional classification system, highway functions are classified into three categories. Principal arterials and minor arterials provide the passing through function. Major collectors and minor collectors serve collecting and distributing functions. Local highways provide the access function. Highways are still divided into five types, but selection of highway types are based on expected functions:

1. For principal arterials, type choices should be freeways.
2. For minor arterials, type choices should be freeways, class I or class II highways.
3. For major collectors, type choices could be class I or class II highways.
4. For minor collectors, type choices could be class II or Class III highways.
5. For locals, type choices could be Class III or class IV highways highways.

5 Discussions and Ideas for Future Development

After decades of development, road functional classification in China has been gaining importance. Despite provisions formulated in existing codes and standards, significant work remains. Suggestions and needs for future work are suggested here.

5.1 Context Definitions

For several decades in China, whether a road is an urban road or a highway is determined by the urban area boundary. The context is limited by the binary urban/rural classification. Two functional classification systems are developed correspondingly, as clarified in previous sections. However, with the rapid development of urbanization, new context categories such as suburban areas and rural communities emerge quickly. The urbanization rate is 57.35% in 2016. Phenomena of "urbanized highway" have pervasively existed, resulting in the traffic safety problem. There is also no recognition of the road or street concept for villages and rural communities [22]. Designs such as speed limit sign, sidewalks and signal controls, which are used in the urban area, may be required where a rural highway enters a village.

Other than China, this issue has also been raised in U.S. [23, 24]. In order to deal with this problem, considering factors of density, land uses and building setbacks, five context definitions are proposed in the "Expanded functional classification system for highways and streets" by TRB's National Cooperative Highway Research Program (NCHRP) in 2017 [25]. The context categories are rural, rural town, suburban, urban and urban core. In order to achieve the integration of urban and rural development, care must be also taken in establishing an integrated function classification system in China. A set of contexts according to the engineering practices and characteristics of administrative divisions in Chinese cities need to be established. It should be also noted that a road may transition into different contexts over its length and this will be reflected in the design considerations and cross sections [26]. However, neither the urban road functional classification system nor the highway functional system so far has provided any design guidance for transitional segments between an urban road and a highway.

5.2 Functional Classification Concepts and Criteria

The relationship between road functional classifications is ambiguous. Classifying a road within a specific category occurs during planning stages of the project based on network-wide needs and thus is well in advance of the geometric design stage [23]. However, the national standard formulation lags behind engineering needs. The road function concept is actually introduced in the planning after it being used in geometric designs. Standards on road network planning currently used by transportation agencies, engineers and planners has not involved the concept of functional classifications yet [11]. The lack of a specific code or standard has resulted in a situation that road functional analysis is considered insufficient for acceptance.

Definitions of road functions are not well studied and precisely clarified. Terms of road functions currently used in codes and standards are inconsistent. For instance, what are differences between the traffic function and the traffic service function? There are also the same problems with the living function and the serving function. Besides, the serving function in some literature is interpreted as accommodating non-motorized modes, i.e., bicycles and pedestrians [26]; while it is regarded as accessibility or providing many opportunities for entry and exit in some other literature [27].

The concept of functional classification should define the role that a particular road segment plays in serving this flow of traffic through the network [2]. Hence, besides traffic volumes, influencing factors including travel needs, efficiency of travel, collecting and distributing and access points et al. should be also considered when functional classifications are assigned to roads. Among these factors, travel needs ought to be the most fundamental and the most important element. Furthermore, although the current international road standards promote the classification of roads according to these influencing factors, the procedures are yet strongly qualitative and based on the planner or designer's judgment [2, 7]. Recent work in [28] used data mining techniques, based on the knowledge of a data set previously collected, to quantify the membership degree of a road to a given functional class.

5.3 Quantitative Functional Evaluation

Functional classifications play an important role at stages of road planning and design. But performance is commonly evaluated based on the concept of capacity and level of service [12, 21]. Hence, there is a question about the relationship between road functions and level of service. For example, capacity vs. connection function, which is more important? If functional evaluation is conducted, what indicators are appropriate?

In addition, despite road functional classifications have been proven to be useful, operational performance of a road may not meet the target function defined in planning or design stages. It is thus necessary to diagnose causes for not realizing the expected function, so that corrective actions can be implemented. Some researchers have begun addressing this issue [4, 29]. A misuse of the road, compared with its proper characteristics, in fact, worsens users' safety [28, 30, 31] and leads to an increase in social costs and possible inefficiencies in the programming of maintenance [32].

6 Conclusions

Functional classifications are important for the network planning and road design. In this paper, we reviewed road functional classification systems in the national codes and standards in the past decades in China. It is suggested that the urban road functional classification system and the highway functional classification system are two isolated systems, resulting from the binary urban/rural context. Under the background of urbanization and multimodal accommodation, an integrated functional classification system is urgently needed. The inconsistence problem also exists between functional classifications in the planning stage and functional classifications in the design stage. Refinements may be made to organize the relationship. Moreover, the complete set of function concepts, classification criteria, and the procedure for determining the function of a road could be addressed and paid more attention.

Acknowledgements. This research work is supported by the National Natural Science Foundation of China (No. 51678535).

References

1. U.S. Department of Transportation, Federal Highway Administration. Highway Functional Classification Guidelines: Concepts, Criteria and Procedures (1989). http://www.co.marquette.mi.us/departments/road_commission/docs/FHWA_Functional_Classification_Guildlines.pdf. Accessed Jan 2018
2. U.S. Department of Transportation, Federal Highway Administration. Highway Functional Classification Concepts, Criteria and Procedures (2013). https://www.fhwa.dot.gov/planning/processes/statewide/related/highway_functional_classifications/fcauab.pdf. Accessed Jan 2018

3. Litman, T.: Evaluating transportation land use impacts: considering the impacts, benefits and costs of different land use development patterns (2012). www.vtpi.org/landuse.pdf. Accessed Jan 2018
4. Dong, J.X., Cheng, T., Xu, J.J., Wu, J.P.: Quantitative assessment of urban road network hierarchy planning. Town Plan. Rev. **84**(4), 445–472 (2013)
5. Qu, X., Yang, Y., Liu, Z., Weng, J., Jin, S.: Potential crash risks of expressway on-ramps and off-ramps: a case study in Beijing, China. Saf. Sci. **70**, 58–62 (2014)
6. Kuang, Y., Qu, X., Wang, S.: A tree-structured crash surrogate measure for freeways. Accid. Anal. Prev. **77**, 137–148 (2015)
7. Toronto Transportation Services. City of Toronto Road Classification System (2013). https://www1.toronto.ca/City%20Of%20Toronto/Transportation%20Services/Road%20Classification%20System/Files/pdf/2012/rc_document.pdf. Accessed Jan 2018
8. Qu, X., Wang, S., Zhang, J.: On the fundamental diagram for freeway traffic: a novel calibration approach for single-regime models. Transp. Res. B-Meth. **73**, 91–102 (2015)
9. CJJ 37-90. Code for urban road design (1990)
10. Shanghai Housing and Urban-Rural Development Commission. Technical Standard of Urbanization Regional Highway Engineering (2013)
11. GB 50220-95. Code for transportation planning on urban road (1995)
12. CJJ 37-2012. Code for design of urban road engineering (2012)
13. GB 50647-2011. Code for planning of intersection on urban roads (2011)
14. Friedricha, M.: Functional structuring of road networks. Transp. Res. Procedia **25**, 568–581 (2017)
15. Ministry of Housing and Urban-Rural Development of the People's Republic of China and General Administration of Quality Supervision, Inspection and Quarantine of the People's Republic of China. Code for Urban Comprehensive Transport System Planning (draft) (2016)
16. Jin, S., Qu, X., Zhou, D., Xu, C., Ma, D., Wang, D.: Estimating cycleway capacity and bicycle equivalent units for electric bicycles. Trans. Res. A-Pol. **77**, 225–248 (2015)
17. Shanghai Municipal Government of Planning and Land Resources. Shanghai Street Design Guidelines (2016)
18. JTJ 01-81. Technical Standard of Highway Engineering (1981)
19. JTJ 001-97. Technical Standard of Highway Engineering (1997)
20. JTG B01-2003. Technical Standard of Highway Engineering (2003)
21. JTG B01-2014. Technical Standard of Highway Engineering (2014)
22. Guo, M., Xu, Y.C.: China's road classification system should adapt to the needs of the times (2017). http://www.thepaper.cn/newsDetail_forward_1707610. Accessed Jan 2018
23. Stamatiadis, N., Kirk, A., King, M., Chellman, R.: Development of a context sensitive multimodal functional classification system. Presented at the Transportation Research Board 96th Annual Meeting, Washington D.C. (2017)
24. Stamatiadis, N., Kirk, A., Jasper, J., Wright, S.: Functional classification system to aid contextual design. Transport Res. Rec., No. 2638, pp. 18–25 (2017)
25. Stamatiadis, N., Kirk, A., Hartman, D., Jasper, J., Wright, S., King, M., Chellman, R.: An expanded functional classification system for highways and streets. Prepublication draft of NCHRP Research Report 855. Transportation Research Board, Washington D.C. (2017)
26. Liu, B., Yan, L., Wang, Z.: Reclassification of urban road system: integrating three dimensions of mobility, activity and mode priority. Transp. Res. Procedia **25**, 627–638 (2017)
27. Cai, J.: Study on the importance of access roads. City Plann. Rev. **29**(3), 84–88 (2005)
28. D'Andrea, A., Cappadona, C., La Rosa, G., Pellegrino, O.: A functional road classification with data mining techniques. Transport **29**(4), 419–430 (2014)

29. Yan, Y.D., Ning, Z.Q., Li, H., Wang, D.W.: Urban expressway functional reliability. J. Highw. Transp. Res. Develop. **10**(2), 68–72 (2016). (English Edition)
30. Kockelman, K.M.: Modeling traffic's flow-density relation: accommodation of multiple flow regimes and traveler types. Transportation **28**(4), 363–374 (2001)
31. Karlaftis, M.G., Golias, I.: Effects of road geometry and traffic volumes on rural roadway accident rates. Accid. Anal. Prev. **34**(3), 357–365 (2002)
32. Cafiso, S., Lamm, R., La Cava, G.: Fuzzy model for safety evaluation process of new and old roads. Transport Res. Rec., No. 1881, pp. 54–62 (2004)

Unlock a Bike, Unlock New York: A Study of the New York Citi Bike System

Yinghao Chen, Zhiyuan Liu, and Di Huang(✉)

Jiangsu Key Laboratory of Urban ITS, Jiangsu Province Collaborative Innovation Center of Modern Urban Traffic Technologies, School of Transportation, Southeast University, Nanjing, China
vic_dhuang@163.com

Abstract. As urban populations grow, there is a growing need for efficient and sustainable modes, such as bicycling. The shortage of bicycle demand data is a barrier to design, planning, and research efforts in bicycle transportation before. In July 2013, the New York City implements the bike-sharing system, Citi Bike, and makes their data available for analysis. Data used in this study includes the information about active stations, average bicycles available, total annual membership, maintenance issues, events of vandalism and calls and emails to system center. Through statistic description, partial correlation analysis and principle component analysis, final variables are obtained. Finally, a Poisson regression model was adopted for the analysis. The analysis results are useful for understanding the influential factors including temperatures and weathers, which reflected by seasons generally, and supplements associated with rules or policy of bike-sharing system. In addition, the inferential results of these models provide guidance on future planning of station and bike supplement.

1 Introduction

Many benefits of bicycle sharing systems (BBS) have led to the rapid growth of these systems around the world in the recent years. In fact, over 1000 cities have already started or are considering the initiation of a BBS (Meddink 2015). These systems include docking stations where users can rent bicycles and travel to a different docking station where the bicycle is returned, providing individuals increased flexibility to bicycle without the traditional burdens of owning a bicycle (such as the need to secure their bicycles or perform regular maintenance). Proponents see BBS as a way to increase the usefulness of public transportation by providing connectivity to and from transit stops. Indeed, the dispersed land-use patterns in many cities, and even in the urban core of some cities, make this 'last-mile' connectivity particularly troublesome for transportation planners, and low-cost, convenient bicycle rentals have the potential to increase the 'reach' of transit greatly (Midgley 2009, Liu et al. 2017).

Given the growing attention towards bicycle-sharing systems, it is important to examine the current performance of BSS operation to improve the effectiveness of BSS schemes (Fishman et al. 2013). Further, understanding factors influencing BSS demand will allow us to better coordinate the installation of new systems or modify existing

G. De Pietro et al. (Eds.): KES-IIMSS-18 2018, SIST 98, pp. 356–366, 2019.
https://doi.org/10.1007/978-3-319-92231-7_37

systems. Focuses have been put on the examination of the impact of various attributes on BSS usage, which usually estimated, by arrivals and departures. The biggest difference between these studies are difference variables considered, including BSS infrastructure (such as number of BSS stations and stations' capacity), transportation network infrastructure (such as length of bicycle facilities, streets and major roads), land use (such as population and job density), point of interests (such as presence of subway stations, restaurants, businesses and universities), and meteorological and temporal attributes (such as temperature and time of day). Moreover, plenty of analyses on BBS from multiple perspectives have been carried previously, providing useful insights on the BSS usage patterns as well. For example, gender gap between the users of BSS is found to be an issue where the majority of BSS users are male (Faghih-Imani and Eluru 2015, Murphy and Usher 2015). Further, research efforts demonstrated that BSS users prefer to use the existing bicycle facilities such as bicycle lanes (Faghih-Imani and Eluru 2015, González et al. 2015). Related methods used in previous researches can be found in Table 1 whereas some typical findings are shown in Table 2.

Table 1. Summary of estimating usage pattern methods in previous studies

Author	Year	Method
Kaltenbrunner et al.	2010	Method based on time series analysis considering temporal and meteorological variables
Borgnat et al.	2011	Method based on time series analysis considering temporal and meteorological variables
Han et al.	2014	Method based on time series analysis considering temporal and meteorological variables
Giot and Cherrier	2014	Method based on time series analysis considering temporal and meteorological variables
Rudloff and Lackner	2014	Method based on count models considering dummy variables whether a station is full or empty for the three closest stations
Faghih-Imani et al.	2014	Method based on a linear mixed model analyzing arrival and departure rates

However, the earlier research efforts have neglected to adequately consider the initial factors of people's willingness of choosing BBS, both inside and outside, such as weather, gender, age, quality of service and so on. Though some studies have considered these factors, they did not provide a quantitative expression. Thus, the objective of my study is to find relative certain relationship between these factors and users' attitudes. However, attitude is an non-observable variable and it is hard to be estimated directly. Therefore, we use the change of total membership to reflect people's preference since the mechanism of the membership system that we study features monthly renewal. Poisson regression is implemented using STATA to explore the relationship.

Table 2. Summaries of findings on BBS in previous studies

Author	Year	Factors
Fuller et al.	2011	Convenience of BSS as well as the presence of a BSS station closer to home location
Lathia et al.	2012	the differences between BSS short-term users and BSS annual members' preferences
O'Brien et al.	2014	Adoption of BSS for commuting purposes on weekdays
Gebhart and Noland	2014	The impact of weather information (such as temperature and humidity) on BSS usage
Faghih-Imani and Eluru	2015	Gender gap between the users of BSS
Faghih-Imani and Eluru	2015	Bicycle facilities such as bicycle lanes

2 Data Sources

2.1 Data Background

New York City implemented their bike-sharing system, Citi Bike, in July 2013, and makes their data publically available for analysis. Citi Bike is New York City's bike share system, and the largest in the nation. Citi Bike launched in May 2013 and has become an essential part of our transportation network. The data used in our research was obtained from Citi Bike website (https://www.citibikenyc.com/system-data). The Citi Bike website provides trip dataset for every month of operation since July 2013. The trip dataset includes information about origin and destination stations start time and end time of trips, user types i.e. whether the user was a customer with an annual membership pass or a temporary pass, and the age and gender for members' trips only. Additionally, the stations' capacity and coordinates as well as trip duration are also provided in the dataset. The built environment attributes such as bicycle routes and subway stations are derived from New York City open data (https://nycopendata.socrata.com/) while the socio-demographic characteristics are gathered from US 2010 census and the weather information was obtained for the Central Park station from National Climatic Data Center.

2.2 Data Preprocessing

Motivate, the operator of the Citi Bike program, provides monthly reports to the NYC Department of Transportation. Monthly data from June 2013 to February 2017, excluding several month is not available. Then, we deleted some of them by choosing representative variables preliminarily. For example, I chose average riders per day instead of trips by annual members and trips by casual members. So, the initial task for variables is finished.

Next, we used number 1 to 4 to represent four seasons since that we thought it is important to analyze the effect of season to the number of membership because of weather, temperatures. Then, we turned it to four dummy variables to satisfy the

Table 3. Final variables

Variable	Variable explanation
X1	Time
X2	Year
X3	Season: February, March, April-1; May, June, July-2; August, September, October-3; November, December, January-4
X4	Spring-1; Others-0
X5	Summer-1; Others-0
X6	Autumn-1; Others-0
X7	Winter-1; Others-0
X8	Active stations
X9	An average of bicycles available or in use
X10	Total annual membership
X11	Overall ridership trips
X12	Overall miles
X13	Average rides per day
X14	Average distance(miles)
X15	Times of each bicycle used per day
X16	Maintenance issues
X17	Vandalism
X18	Calls to call center
X19	Emails to center

requirement of methods of regression on count data. The final variables are shown in Table 3.

2.3 Statistic Description

SPSS gives a result of statistic description for X9, X10, X11, X12, X13, X14, X15, X16, X17, X18, and X19, which are shown in Table 4. These variables are all associated with operation of system and will be affected by external conditions, which is obviously different from rest variables such as time and seasons that are inherent comparatively and active stations referred to policy decision.

It's clearly shown that all the variables analyzed commonly own large ranges and standard deviations, indicating that the value of these variables change together by certain objective rules. In order to study factors of total annual membership which can be considered to be a characterization of citizens' attitude towards BBS, we did a line chart (Fig. 1) to observe the changes of total annual membership. We can find that the line of monthly total annual membership shows a rising trend generally. However, when the abscissa values between two years, usually in November, December and January, a sharp gap appeared. Therefore, it's understandable that total annual membership experienced a sudden down in a monthly-renewed membership system during wintertime. That's why we classified time into four seasons before.

Table 4. Descriptive statistics

	Minimum	Maximum	Mean	Std. deviation
An average of bicycles available or in use	3318	9557	5950.97	1375.408
Total annual membership	52130	121592	93199.85	13844.980
Overall ridership trips	209795	1573653	850583.69	345957.931
Overall miles	276296	3432044	1583585.15	789617.335
Average rides per day	7493	52990	28164.97	11724.438
Average distance(miles)	.90000000	2.7200000	1.790512820	.3619752006
Times of each bicycle used per day	1.50	7.32	4.6610	1.67918
Maintenance issues	968	19879	5192.51	4910.959
Vandalism	4	447	72.33	95.612
Calls to call center	4646	65798	22918.59	15165.566
Emails to center	313	52044	4288.10	8185.289
Valid N (list wise)				

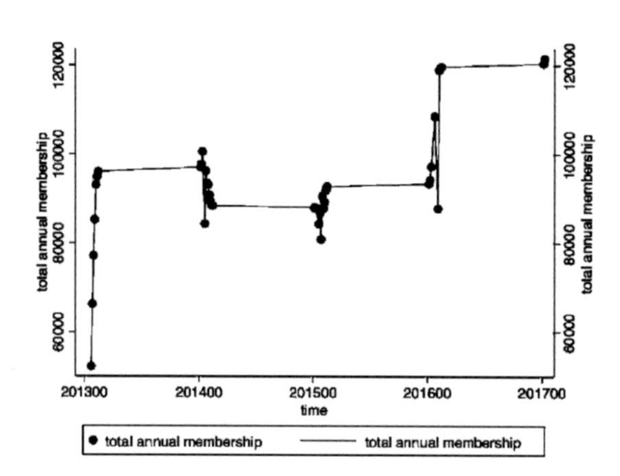

Fig. 1. Line chart of total annual membership

3 Methodology

3.1 Poisson Regression

To help illustrate the principle elements of a Poisson regression model (Meng and Qu 2012), consider the number of accidents occurring per year at various intersections in a city. In a Poisson regression model, the probability of intersection i having y_i accidents per year (where y_i is a nonnegative integer) is given by

$$P(y_i) = \frac{e^{-\lambda_i}\lambda_i^{y_i}}{y_i!} \tag{1}$$

This model is estimable by standard maximum likelihood methods, with the likelihood function given as

$$L(\beta) = \prod \frac{EXP[-EXP(\beta X_i)][EXP(\beta X_i)]^{y_i}}{y_i!} \tag{2}$$

The log of the likelihood function is simpler to manipulate and more appropriate for estimation, and is given as

$$LN(\beta) = \sum_{i=1}^{n} [-EXP(\beta X_i) + y_i \beta X_i - LN(y_i!)] \tag{3}$$

3.2 Principle Components Analysis

Principal components analysis has two primary objectives: to reduce a relatively large multivariate data set, and to interpret data (Johnson and Wichern 1992). These objectives are accomplished by explaining the variance–covariance structure using a few linear combinations of the originally measured variables. Through this process a more parsimonious description of the data is provided—reducing or explaining the variance of many variables with fewer well-chosen combinations of variables. If, for example, a large proportion (70%–90%) of the total population variance is attributed to a few uncorrelated principal components, then these components can replace the original variables without much loss of information, and also describe different dimensions in the data. Principal components analysis relies on the correlation matrix of variables—so the method is suitable for variables measured on the interval and ratio scales.

4 Data Analysis and Results

4.1 Partial Correlation

As Table 5 shows below, we can easily notice that X8, X9, X16 have all have significant correlation with X1 (time). Therefore, it's reasonable to assume that X8, X9, X16 have something to do with time. If we look up the variables table, we can find that these variables are all associated with policy of BBS more or less, since the number of active stations and supply of bikes were decided by the operator of system based on something like market research, which is not related to this study. So, these variables must have relationships inside.

Table 5. Partial correlation of variables

Control variables			X1	X8	X9	X16	X17	X18	X19
X10	X1	Correlation Significance (2-tailed) df.	1.000 . 0	.645 .000 36	.247 .134 36	.440 .006 36	−.492 .002 36	−.262 .112 36	.025 .880 36
	X8	Correlation Significance (2-tailed) df.	.645 .000 36	1.000 . 0	.693 .000 36	.694 .000 36	−.501 .001 36	.124 .460 36	.488 .002 36
	X9	Correlation Significance (2-tailed) df.	.247 .134 36	.693 .000 36	1.000 . 0	.630 .000 36	−.262 .112 36	.382 .018 36	.403 .012 36
	X16	Correlation Significance (2-tailed) df.	.440 .006 36	.694 .000 36	.630 .000 36	1.000 . 0	−.358 .027 36	.356 .028 36	.562 .000 36
	X17	Correlation Significance (2-tailed) df.	−.492 .002 36	−.501 .001 36	−.262 .112 36	−.358 .027 36	1.000 . 0	.139 .405 36	−.127 .448 36
	X18	Correlation Significance (2-tailed) df.	−.262 .112 36	.124 .460 36	.382 .018 36	.356 .028 36	.139 .405 36	1.000 . 0	.367 .023 36
	X19	Correlation Significance (2-tailed) df.	.025 .880 36	.488 .002 36	.403 .012 36	.562 .000 36	−.127 .448 36	.367 .023 36	1.000 . 0

4.2 Principal Component Analysis

Based on my conclusion before, we tried to do a factor analysis on X8, X9, X16. The result is shown in Tables 6, 7.

Table 6. Total variance explained

Component	Initial eigenvalues			Extraction sums of squared loadings		
	Total	% of Variance	Cumulative %	Total	% of Variance	Cumulative %
1	2.638	87.944	87.944	2.638	87.944	87.944
2	.232	7.741	95.685			
3	.129	4.315	100.000			
Extraction Method: Principle Component Analysis.						

From tables above, the value of cumulative is 87.944%, which means that it's sufficient for component1 to represent the combination of X8, X9 and X10. When it comes to the practical meaning of Component1, we would like to conclude it as strategy made by operator of Bicycle-Sharing System. There are some reasons for this conclusion. First, we can notice that X8(the number of active stations) increase by time from 322 to 598, it is caused by the development and expansion of BBS's operating scale. Obviously, it's the same to X9(the average number of bikes available at each

Table 7. Component matrix

	Component
	1
X8	.953
X9	.919
X16	.941
Extraction Method: Principal Component Analysis.	
a. 1 components extracted.	

station). Refer to X16(reported Maintenance issues), it's a kind of service provided by BBS company for users, and in this degree, the essence in it is the same as well. Therefore, the result of component analysis is both mathematically and practically meaningful. And Component1 is given as

$$\mathrm{Component1} = 0.953 * \mathrm{X8} + 0.919 * X9 + 0.941 * X16 \tag{4}$$

In addition, component1 increases with increases in X8, X9, X16.

4.3 Poisson Regression

Table 8 shows the parameter estimates of a Poisson regression estimated on the monthly total annual membership data. Component1 appears to have a smaller influence than season-effect, contrary to what is expected. Additionally, the number of monthly total annual membership increases with Vandalism increases. It indicated the increasing attention by public regardless of that the effect of this kind of attention is positive or negative, but it virtually reflected the important influence of BBS. Moreover, the positive correlation between Vandalism and Total annual membership is

Table 8. Results of poisson regression

Poisson regression						Number of obs = 39
LR chi2(7) = 69173.78		Log likelihood = –5435.4103		Pseudo R2 = 0.8649		
X10	Coef.	Std.Err.	z	P > z	95% Conf.Interval	
X1	–0.0000786	7.79e–06	–10.09	0.000	–.0000939	–.000633
X4	.0579179	.0016761	34.55	0.000	.0546328	.0612031
X5	–.0143767	.0015509	–9.27	0.000	–.0174164	–.0113371
X7	.0694784	.0013492	51.50	0.000	.066834	.0721228
X17	.0001259	6.61e–06	19.03	0.000	.0001129	.0001389
X19	–.0000102	9.55e–08	–107.10	0.000	–.0000104	–.00001
component	.0000185	1.28e–07	144.35	0.000	.0000183	.0000188
_cons	27.07603	1.567789	17.27	0.000	24.00322	30.14884

understandable. So, we can conclude that the signs of the estimated parameters are in line with expectation.

The mathematical expression for this Poisson regression model is as follows,

$$\begin{aligned} E(y_i) = LN(\lambda) = & \ 27.07603 + 0.0000185 Component1 \\ & - 0.0000102X19 + 0.0001259X17 + 0.0694784X7 \\ & - 0.0143767X5 + 0.0579179X4 - 0.0000786X1 \end{aligned} \tag{5}$$

where Component1 is the new variable received through component analysis before, indicating the change of operators' strategy, X19 is the number of emails to center every month, X17 is the events of Vandalism happened in a month and X4, X5, X7 are dummy variables about four seasons based on the effect off autumn to X10.

5 Summary and Conclusions

With the growing installation of bike-sharing system infrastructure there is substantial interest in identifying factors contributing to the demand of these systems. Earlier research efforts mostly have put focus on the arrivals and departures rates of one BBS station to indicate the demand of these systems. Others have done some researches on the characteristics of different users that have correlation with the demand more or less. However, either of the two kinds study provide some mathematical expression to estimate citizens' willingness or attitude to use these systems, they neglected to adequately consider in psychological aspect. It is possible that the total annual membership that renewed every month in the Citi Bike in New York can indicate people's idea to some degree.

This paper first presents statistic descriptions by SPSS to observe the characteristics of initial data, and we assume that it is possible that seasons have a big effect on the annual membership of BBS, so we turn data associated with seasons into four dummy variables. Then, a partial correlation analysis was done to select a series of variables closely related. Afterwards, a principle component analysis was used to reduce a relatively large multivariate data set, we received Component1 representing a kind of strategy maybe by operators in this case. Finally, a Poisson regression is achieved and we get the mathematical expression of relationship between monthly total annual membership and other factors. In addition, this model was in line with what we expected.

Results are useful for understanding the factors associated with factors like temperatures and weathers that reflected by seasons generally and supplements associated with rules or policy of bike-sharing system. Therefore, the inferential result of these models provides guidance on future planning of station and bike supplement.

The bike-sharing data used for this study offers many additional possibilities for understanding the generation of trips. There are many other ways that one can evaluate different users, such as by gender, daytime and nighttime trips, and linked station pairs. More detailed information on urban design and land use characteristics in the immediate vicinity of each station may also shed light on features that attract riders or contribute riders to become a member of BBS. As a result, future research will undoubtedly explore these issues to achieve a more precise model and try to do some related prediction.

References

Faghih-Imani, A., Eluru, N.: Incorporating the impact of spatio-temporal interactions on bicycle sharing system demand: a case study of New York CitiBike system. J. Transp. Geogr. **54**, 218–227 (2016)

Borgnat, P., Abry, P., Flandrin, P., Robardet, C., Rouquier, J.-B., Fleury, E.: Shared bicy- cles in a city: a signal processing and data analysis perspective. Adv. Complex Syst. **14**(03), 415–438 (2011)

Fishman, E.: Bikeshare: a review of recent literature. Transp. Rev. **36**, 92–113 (2015)

Fuller, D., Gauvin, L., Kestens, Y., Daniel, M., Fournier, M., Morency, P., Drouin, L.: Use of a new public bicycle share program in Montreal. Canada. Am. J. Prev. Med. **41**(1), 80–83 (2011)

Faghih-Imani, A., Eluru, N.: Analyzing bicycle sharing system user destination choice preferences: an investigation of Chicago's Divvy system. J. Transp. Geogr. **44**, 53–64 (2015)

Faghih-Imani, A., Eluru, N., El-Geneidy, A., Rabbat, M., Haq, U.: How land-use and urban form impact bicycle flows: evidence from the bicycle-sharing system (BIXI) in Montreal. J. Transp. Geogr. **41**, 306–314 (2014)

Giot, R., Cherrier, R.: Predicting bikeshare system usage up to one day ahead. In: IEEE Symposium Series in Computational Intelligence 2014. Workshop on Computational Intelligence in Vehicles and Transportation Systems, pp. 1–8 (2014)

González, F., Melo-Riquelme, C., de Grange, L.: A combined destination and route choice model for a bicycle sharing system. Transportation **43**, 407–423 (2015)

Han, Y., Côme, E., Oukhellou, L.: Towards bicycle demand prediction of large-scale bicycle sharing system. Presented at Transportation Research Board 93rd Annual Meeting, No. 14–2637 (2014)

Jin, S., Qu, X., Zhou, D., Xu, C., Ma, D., Wang, D.: Estimating cycleway capacity and bicycle equivalent units for electric bicycles. Transp. Res. Part A **77**, 225–248 (2015)

Kaltenbrunner, A., Meza, R., Grivolla, J., Codina, J., Banchs, R.: Urban cycles and mobility patterns: exploring and predicting trends in a bicycle-based public transport system. Pervasive Mob. Comput. **6**(4), 455–466 (2010)

Liu, Z., Yan, Y., Qu, X., Zhang, Y.: Bus stop-skipping scheme with random travel time. Transp. Res. Part C **35**, 46–56 (2013)

Meddin, R., DeMaio, P.: The Bike-Sharing World Map. http://www.bikesharingworld.com. Accessed 20 July 2015

Meng, Q., Qu, X.: Estimation of vehicle crash frequencies in road tunnels. Accid. Anal. Prev. **48**, 254–263 (2012)

Midgley, P.: The role of smart bike-sharing systems in urban mobility. Journeys **2**, 23–31 (2009)

Murphy, E., Usher, J.: The role of bicycle-sharing in the city: analysis of the Irish ex-perience. Int. J. Sustain. Transp. **9**, 116–125 (2015)

O'Brien, O., Cheshire, J., Batty, M.: Mining bicycle sharing data for generating in-sights into sustainable transport systems. J. Transp. Geogr. **34**, 262–273 (2014)

Qu, X., Zhang, J., Wang, S.: On the stochastic fundamental diagram for freeway traffic: model development, analytical properties, validation, and extensive applications. Transp. Res. Part B **104**, 256–271 (2017)

Noland, R.B., Smart, M.J., Guo, Z.: Bikeshare trip generation in New York city. Transp. Res. Part A **94**, 164–181 (2016)

Rudloff, C., Lackner, B.: Modeling demand for bikesharing systems: neighboring sta-tions as source for demand and reason for structural breaks. Transp. Res. Rec. **2430**, 1–11 (2014)

Modeling and Analysis of Crash Severity for Electric Bicycle

Cheng Xu[1,2(✉)] and Xiaonan Yu[1]

[1] Department of Traffic Management Engineering,
Zhejiang Police College, Hangzhou 310053, China
xucheng@zjjcxy.cn
[2] College of Civil Engineering and Architecture,
Zhejiang University, Hangzhou 310058, China

Abstract. Electric bicycle (E-bike) traffic crashes have become an important traffic safety problem in many Chinese cities. Based on the traffic crash data of E-bikes from Xintang region in Hangzhou, China, the time distribution, spatial distribution, and influencing factors for electric bicycles-related traffic crashes were analyzed, and the main factors that affect the traffic crash of electric bicycles were obtained. On this basis, a logistic model of the influencing factors on the severity of traffic crashes for electric bicycles was set up. The key factors affecting the severity of traffic crashes on electric bicycles were obtained, which provided the basis for the prevention and safety management of traffic crashes on electric bicycles.

Keywords: Electric bicycle · Traffic crash · Management countermeasure

1 Introduction

With the rapid social and economic development, the cities of China have been expanding. Urban commuter travel radius also continues to increase. Take Hangzhou as an example, the average travel distance of urban residents increased from 4.4 km in 2010 to 5.6 km in 2015. With the increase of travel distance, the traditional human bicycle has been unable to meet the daily travel needs. The E-bike due to have the advantages of low cost, flexible operation, easy riding, has become an important way of travel in many cities in China. On the one hand, electric bicycles effectively alleviate the travel demand brought by the serious shortage of urban public transport services in our country, further enhance the travel efficiency of urban residents, especially low-income travelers. On the other hand, electric bicyclists, due to their fast riding speed, weak safety awareness, poor awareness of self-protection and prominent violations. In recent years, it has become a difficult point of urban traffic management. The traffic crashes involving electric bicycles also show a tendency of increasing year by year. In Zhejiang Province, for example, in 2016 there were 1110 fatalities involving electric bikes. An increase of 5.1% over the same period of last year, accounting for 26.5% of the total number of traffic crashes in the province. Among them, 924 people, who drove electric cyclists, died in traffic crashes, up 9.1%. Therefore, it is important to analyze the basic characteristics and development trend of E-bike crash in detail,

G. De Pietro et al. (Eds.): KES-IIMSS-18 2018, SIST 98, pp. 367–375, 2019.
https://doi.org/10.1007/978-3-319-92231-7_38

explore the space-time characteristics of traffic crash deeply, and quantitatively establish the influencing factors of electric bicycle traffic crash so as to provide strong data support for E-bikes traffic management and crash prevention theoretical and practical significance.

In recent years, research on safety risks and crashes of E-bikes have been carried out widely (Jin et al. 2015; Xu et al. 2016; Qu et al. 2017). The first bicycle safety performance functions (SPFs) were presented and applied to Boulder, Colorado by Nordback et al. (2014). Such functions provide a basis for both future investigations into safety treatment efficacy and for prioritizing intersections to better allocate scarce funds for bicycle safety improvements. Carter et al. (2007) developed a macro-level bicycle intersection safety index (Bike ISI) that would allow engineers, planners, and other practitioners to use known intersection characteristics to prioritize intersection approaches with respect to bicycle safety proactively. Weber et al. (2014) analyzed the age distribution, the use of helmets, the degree of injury, the type of crash, and the type of crash as well as the multi-factor correlations based on the chi-square test for cases involving electric bicycle crashes in Switzerland between 2011 and 2012. Langford et al. (2015) based on GPS technology analysis of hybrid bicycle safety behavior in four cases and found that electric bicycles and bicycles do not drive on a regular basis and violations of signal signs marking traffic violations of these two types of violations rates are high. Based on the questionnaire survey of e-bike users distributed in Denmark, Haustein and Moller (2016) concluded that the riding style and attitude of E-bike drivers play a crucial role in major traffic crashes.

The above research conducted an in-depth analysis on the safety behavior of electric bicycles and some crash data, However, due to the large difference in driving environment at home and abroad, foreign countries lack a lot of data on the operation and behavior of electric bicycles, making the relevant conclusions not applicable to the road traffic conditions in China. Therefore, this paper takes the data of traffic crashes involving electric bicycles occurring in Xintang Street, Xiaoshan District, Hangzhou, Zhejiang Province from June 2015 to May 2016 as a sample, analyzes the spatiotemporal distribution characteristics and basic statistical characteristics of traffic crashes involving electric bicycles. Based on this, the key factors affecting the severity of electric bicycle crash are analyzed by generalized linear regression model. According to the result of data analysis, this paper puts forward the countermeasures for the traffic management of electric bicycles and provides an empirical basis for the traffic safety management and crash prevention of electric bicycles.

2 Statistics and Analysis of Electric Bicycle Crash

2.1 Crash Data Source

China's road traffic crashes are generally divided into summary procedures to deal with crashes and general procedures to deal with crashes. Simple incident handling refers to the case of a simple or general case, only causing minor damage to vehicles or minor injuries, the parties to the crash facts and responsibilities identified uncontested traffic crash. Since a large number of crashes involving electric bicycles are handled by simple procedures, it is often the case that many crashes caused by electric bicycles are missed by analyzing only general procedures for dealing with crashes. Therefore, the text adopts the summary procedure to deal with crash and general procedure to deal with crash data from 0:00:20, Jun 1, 2015 to 24:00, May 31, 2016 in the area of Xintang Sub-district, Xiaoshan District, Hangzhou. Among them, the summary program to deal with 1091 crashes, the general procedure to deal with 21 crashes. In the meantime, during the statistical period, the number of traffic lights involved in electric bicycle traffic crashes in Xintang was 4427, accounting for 26% of the total number of border police officers involved, indicating that the traffic crashes involving electric bicycles was in a high status.

The crash database category includes the crash occurrence time, weather, location, cause of the incident, personal injury and property damage, degree of loss, crash liability, crash type, age of electric bicycle driver, gender and other related data. Through the above analysis of various factors, we can conduct in-depth analysis on the causes of traffic crashes involving electric bicycles and provide reference for the prevention of traffic crashes.

2.2 Spatial and Temporal Distribution of Crashes

E-bike traffic crash analysis area is Xintang streets. The total area of 34.8 km^2, resident population of 65000, more than 140,000 non-local population. As the area is located at the junction of urban and rural areas, there are a large number of migrant workers, high E-bike ownership in the area, and heavy demand for commuting during peak periods. The traffic conflicts between electric bicycles and other vehicles are serious and the traffic safety situation is severe.

From the time point of view, electric bicycle traffic crashes show the phenomenon of time is not balanced. Figures 1 and 2 show the time distribution of traffic crashes in different months and different time periods respectively. In the distribution of the month, the number of electric bicycle crashes in a year except for January-February due to the Spring Festival and other reasons there is a clear downward trend, as well as a marked increase in March, the number of traffic crashes in other months was basically stable. This is mainly related to the return of migrant workers to the Spring Festival and the concentration of migrant workers back to Hangzhou. In the period distribution, the distribution of electric bicycle crashes showed obvious saddle type, basically the same with the time distribution characteristics of traffic demand, and the rapid increase of traffic brought the rapid growth of traffic crashes.

From the perspective of space, Fig. 3 is a thermodynamic diagram of the crash drawn on the basis of all the data concerning traffic crashes of electric bicycles in a year within the scope of Xintang Squadron. As can be seen from the figure, the traffic crashes involving electric bicycles are mostly concentrated in the trunk roads of the area center, among them, the three major arterial roads, namely Tongmei Line, Xiao-Ming Line and Nanxiu Road, accounted for 15.16%, 12.25% and 11.24% of the total number of crashes, respectively. Traffic crashes showed a more obvious clustering effect.

At the same time, due to the large number of intersection points, traffic demand, electric bicycles running red lights highlight the illegal acts, the intersection is also

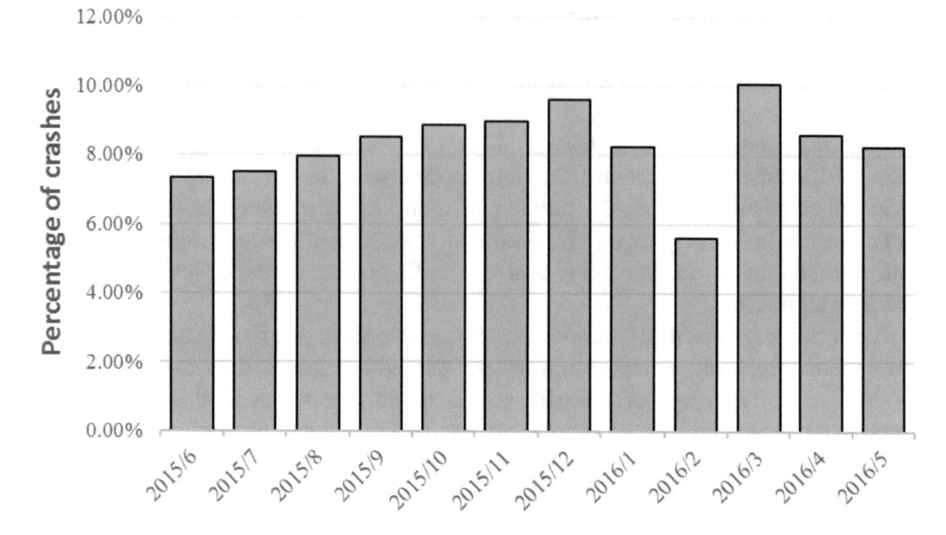

Fig. 1. Electric bicycles in different months the proportion of crashes

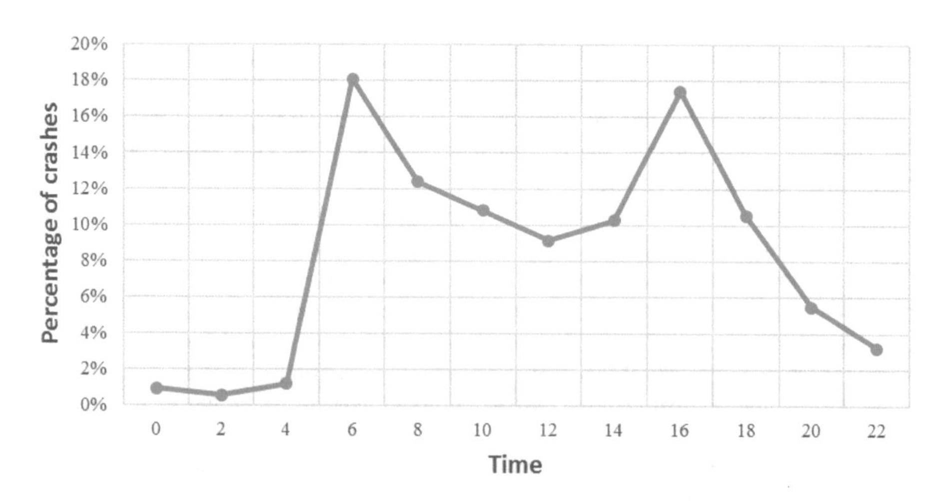

Fig. 2. Electric bicycles in different periods accounted for the proportion of crashes

involved in electric bicycle crash prone areas. Among them, there were more than 20 traffic crashes in five intersections each year.

2.3 Statistical Characteristics of Crashes

This section analyzes the statistical characteristics of traffic crashes involving electric bicycles from the four aspects of age, gender, crash type and illegal behavior. In terms of age, it can be seen from the data that the highest percentage of cyclists in the 40–50 age group should be closely related to the number of travelers in this group. Meanwhile,

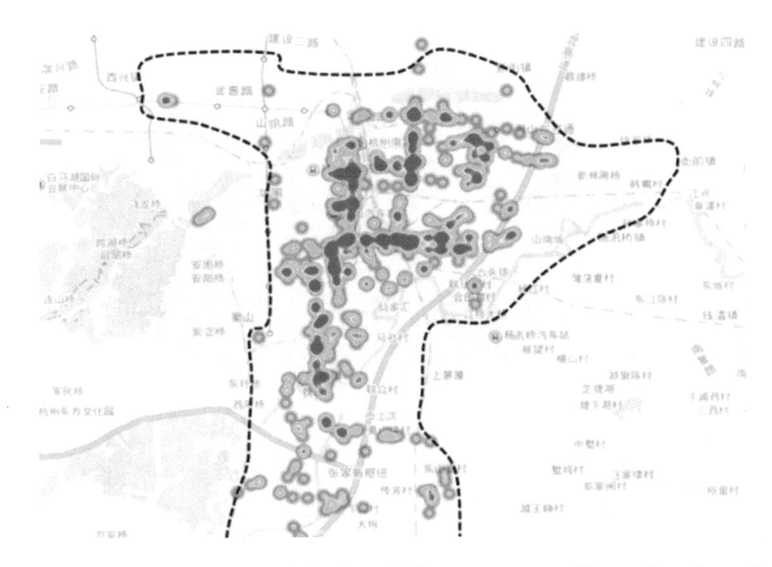

Fig. 3. Electricity within the jurisdiction of Xintang streets traffic crashes thermal map

riders in the 60–80 age group showed a downward trend due to reaction time, cognitive ability and operation ability, resulting in a large proportion of casualties. In terms of gender, male pilots accounted for 58.39% of all incidents and summary pilots handled 41.61% of female pilots, with a ratio of men to women of about 3:2. In the general crash handling process, male pilots accounted for 85.71%, female pilots accounted for 14.29%, male to female ratio of about 6:1. The main reason for this difference should be that men's cyclists are more risk-taking and have a higher percentage of serious violations, resulting in greater casualties. In terms of types of crashes, Fig. 4 shows the number of different types of crashes, with motor vehicles and electric bicycle crashes and electric bicycle crashes accounting for 79.84% and 17.42%, accounting for the absolute majority. These two types of crashes are also the main causes of injuries to electric bicyclists.

3 Impact Analysis on the Severity of Electric Bicycle Crash

The severity of traffic crashes is one of the main indicators of the severity of the crash. Therefore, in-depth analysis of the severity of the crash is helpful to identify the main factors causing the crash injury and provide the data support for further reducing the crash injury. In order to further analyze which factors will affect the severity of electric bicycle traffic crashes, we classify the statistics of the rider's gender, age, crash time, month, weather, road form, crash type and illegal behavior as shown in Table 1.

It can be seen that the incidence of electric bicycle and motor vehicle crashes in electric bicycle traffic crashes accounts for nearly 80%, Including motor vehicle crashes accounted for 51.3% of the total crash, indicating motor vehicle illegal behavior is still the main factor in traffic crashes caused by electric bicycles. In order to further analyze

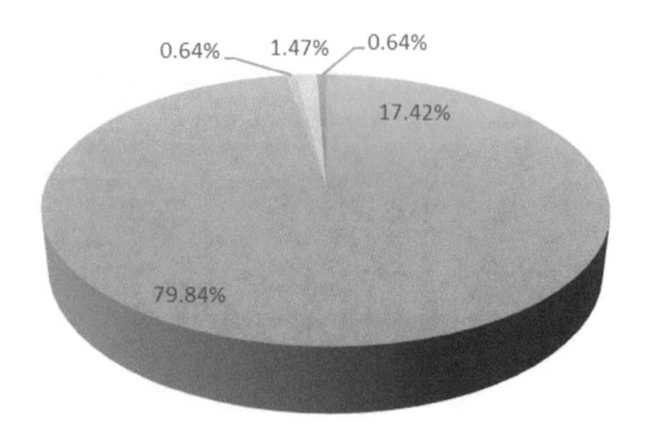

Fig. 4. Crash type scale diagram

Table 1. Electric bicycle crash discretization result

Crash variable	Type description	Discretization value	Number of crashes (ratio)
Gender	Male	1	674 (60.6%)
	Female	2	438 (39.4%)
Age	Youth (0–40)	1	477 (42.9%)
	Middle aged (40–60)	2	516 (46.4%)
	Elderly (60 and above)	3	119 (10.7%)
Time of occurrence	0–6 o'clock	1	32 (2.9%)
	6–10 o'clock	2	334 (30.0%)
	10–16 o'clock	3	342 (30.8%)
	16–20 o'clock	4	308 (27.7%)
	20–24 o'clock	5	96 (8.6%)
Occurrence of the month	1–2	1	154 (13.8%)
	3–4	2	206 (18.5%)
	5–6	3	178 (16.0%)
	7–8	4	171 (15.4%)
	9–10	5	192 (17.3%)
	11–12	6	211 (19.0%)
Weather	Sunny day	1	515 (46.3%)
	Cloudy day	2	365 (32.8%)
	Rain and snow	3	232 (20.9%)
Road form	Intersection	1	474(42.6%)
	Road section	2	638 (57.4%)

(*continued*)

Table 1. (*continued*)

Crash variable	Type description	Discretization value	Number of crashes (ratio)
Illegal behavior	Motor vehicle illegal	1	571 (51.3%)
	Non-motor vehicle driving in reverse	2	126 (11.3%)
	Driving electric bicycles in violation of the provisions of manned	3	95 (8.5%)
	Non-motor vehicles are not driving in the non-motor vehicle lane	4	66 (5.9%)
	Non-motor vehicles do not drive on the right side of the road	5	61 (5.5%)
	Turn non-motorized vehicles to keep going straight	6	46 (4.1%)
	Non-motor vehicles violate traffic signal regulations	7	45 (4.1%)
	Other non-motor vehicle offenses	8	102 (9.3%)
Crash type	Electric bicycles and electric bicycles	1	190 (17.1%)
	Electric bicycles and motor vehicles	2	889 (79.9%)
	Electric bicycles and bicycles	3	7 (0.6%)
	Electric bicycles and pedestrian	4	18 (1.7%)
	Electric bicycle unilateral crash	5	8 (0.7%)
Degree of damage	No/mild damage	0	711 (63.9%)
	Moderate to severe injury (including death)	1	401 (36.1%)

the impact of electric bicycle traffic crash damage factors, this paper divided the electric bicycle traffic crashes into mild injury and severe injury two categories, the multivariate logistic model was used to model the influencing factors as follows:

$$\ln\left(\frac{p}{1-p}\right) = g(x) = \beta_0 + \beta_1 x_1 + \cdots + \beta_8 x_8 \tag{1}$$

where, p is the probability of light injury caused by electric bicycle traffic crash, x_1 to x_8 are the rider's gender, age, crash time, month, weather, road form, crash type and illegal act variables. Among them, the age and crash time are continuous variables, others are discrete variables, and β_0 to β_8 are the parameters to be calibrated.

Suppose there are m observation samples, and the observed values are $y_1, y_2, \ldots, y_m$, respectively, let $p_i = P(y_i = 1|x_i)$ be the probability of y_i= 1 for a given condition. Similarly, the probability of y_i = 0 is $P(y_i = 1|x_i) = 1 - p_i$, so the probability of getting an observation is $P(y_i) = p_i^{y_i}(1 - p_i)^{1-y_i}$.

Since each observation sample is independent of each other, their joint distribution is the product of each edge distribution. In this way, the likelihood function is obtained as:

$$L(\beta) = \prod_{i=1}^{m} (\pi(x_i))^{y_i} (1 - \pi(x_i))^{1-y_i} \tag{2}$$

The parameter estimate that maximizes the likelihood function is the parameter of the logistic model (Zhou et al. 2017). Take the logarithm of L(β):

$$\ln L(\beta) = \sum_{i=1}^{m} \{y_i \ln[\pi(x_i)] + (1 - y_i) \ln[1 - \pi(x_i)]\} \tag{3}$$

Continuing to find partial derivatives of these $n + 1$ β_i respectively, we get $n + 1$ equations as follows:

$$\frac{\partial \ln L(\beta_k)}{\partial \beta_k} = \sum_{i=1}^{m} x_{ik}[y_i - \pi(x_i)] = 0 \quad k = 0, 1, 2, \ldots, n \tag{4}$$

The solution to the system of equations can be obtained using a gradient ascent algorithm or a Newton iterative algorithm. The solution to the above equations is the estimated parameter of the logistic regression model. Table 2 shows the logistic regression results. From the p-value results, it can be seen that the coefficients β_2, β_7 and β_8 are significant, that is, the age of the rider, the offense of the crash, the degree of injury of the crash type and the electric bicycle traffic crash are significant, Among them, the types of violations and incidents more affect the degree of damage caused by the crash. Other factors such as time, month, weather and road sections do not significantly affect the extent of the crash.

Table 2. Logistic model regression analysis of parameter results

	Estimate the value of the parameter	*t* statistics	*P* value
β_0	106.334	0.000	1.000
β_1	0.606	1.509	0.131
β_2	0.029	2.061	**0.039**
β_3	−0.042	−1.175	0.240
β_4	−0.064	−1.283	0.200
β_5	−0.130	−0.544	0.587
β_6	−111.298	0.000	1.000
β_7	−0.216	−3.060	**0.002**
β_8	3.810	9.677	**0.000**

4 Conclusion

Electric bicycles have always been one of the most important hidden dangers of urban road traffic safety in our country. On this basis, the paper analyzes the influencing factors of eight kinds of electric bicycle traffic crashes. Constructed a logistic regression model to analyze the quantitative relationship between the influencing factors and the extent of accidental injury. Revealed the main factors affecting the severity of electric bicycle crash damage and laid the data support for traffic safety management and crash prevention of electric bicycle.

Acknowledgements. This work was supported by the Zhejiang Provincial Natural Science Foundation of China (LQ17E080001), and the China Postdoctoral Science Foundation.

References

Carter, D.L., Hunter, W.W., Zegeer, C.V., Stewart, J.R., Huang, H.: Bicyclist intersection safety index. Transp. Res. Rec. J. Transp. Res. Board **2031**, 18–24 (2007)

Haustein, S., Moller, M.: E-bike safety: individual-level factors and incident characteristics. J. Trans. Health **3**(3), 386–394 (2016)

Jin, S., Qu, X., Zhou, D., Xu, C., Ma, D., Wang, D.: Estimating cycleway capacity and bicycle equivalent units for electric bicycles. Transp. Res. Part A **77**, 225–248 (2015)

Langford, B.C., Chen, J., Cherry, C.R.: Risky riding: naturalistic methods comparing safety behavior from conventional bicycle riders and electric bike riders. Accid. Anal. Prev. **82**, 220–226 (2015)

Nordback, K., Marshall, W.E., Janson, B.N.: Bicyclist safety performance functions for a U.S. city. Accid. Anal. Prev. **65**, 114–122 (2014)

Qu, X., Zhang, J., Wang, S.: On the stochastic fundamental diagram for freeway traffic: model development, analytical properties, validation, and extensive applications. Transp. Res. Part B **104**, 256–271 (2017)

Weber, T., Scaramuzza, G., Schmitt, K.-U.: Evaluation of E-bike accidents in Switzerland. Accid. Anal. Prev. **73**, 47–52 (2014)

Xu, C., Yang, Y., Jin, S., Qu, Z., Hou, L.: Potential risk and its influencing factors for separated bicycle paths. Accid. Anal. Prev. **87**, 59–67 (2016)

Zhou, M., Qu, X., Jin, S.: On the impact of cooperative autonomous vehicles in improving freeway merging: a modified intelligent driver model based approach. IEEE Trans. Intell. Transp. Syst. **18**(6), 1422–1428 (2017)

A Comparative Analysis for Signal Timing Strategies Under Different Weather Conditions

Weiwei Qi(✉) and Huiying Wen

School of Civil Engineering and Transportation,
South China University of Technology, Guangzhou 510641, China
qwwhit@163.com, hywen@scut.edu.cn

Abstract. This paper provides design and evaluation of two signal timing strategies with five cases respectively under different degree of weather condition. The main aim of this project is finding out the most suitable signal timing strategies depending on the weather condition by comparing the analysis results. The procedure mainly includes two parts in traffic signal design, which are signal timing plans design and performance analysis. Methods of design and performance analysis include Webster's signal timing method, Akcelik's signal timing method and HCM delay method. It is clear that the weather factor has huge impact on performance of signalized intersection. Apparently, compared to strategy B, strategy A is more suitable for bad weather condition.

Keywords: Signal timing strategy · Weather condition · Level of service

1 Introduction

As one of the most uncontrollable disturbances that affect the operational performance of the signalized intersection, environment disturbances such as rainfall, snow, ice, fog, high wind and flooding can have destructive impact on the intersection [1]. Adverse weather condition will cause severe traffic paralysis by reducing traffic flow speed, increasing vehicles delay and boosting possibility of road crash, especially in urban area [2]. Queensland has been the most vulnerable state in Australia suffers flooding taking population into account. Records covering 1788 to 1996 indicate that at least 2213 persons have been killed in floods in Australia [3]. Considering the significant impact of weather factors on the signalized intersections, there are two important aspects that need to be accomplished, one is to compare the different cases under different degree of weather condition in same signal timing strategy, and the other is to find out the optimal signal timing strategy that has the best performance (i.e., smallest delay, optimal v/c ratio [4] and highest level of service [5]) corresponding to the weather condition.

The signal timing strategies designed for normal weather condition may not suitable to bad weather conditions. However, it is possible to design another signal timing strategy by changing the optimization aim, which may have a relative large delay under normal weather condition [6], but a much smaller delay under adverse weather condition. As a result, it is significant to study the impact of bad weather on signal timing

G. De Pietro et al. (Eds.): KES-IIMSS-18 2018, SIST 98, pp. 376–385, 2019.
https://doi.org/10.1007/978-3-319-92231-7_39

strategies and maintain the signal timing plans in an efficient level of operation based on weather condition [7].

2 Information of the Selected Intersection

The study site is choose as the intersection beside harbor town shopping center, located on the cross-point of Gold Coast Highway and Oxley Drive. As a typical urban intersection in coastal city, Gold Coast, it is an ideal object for this project. All the data include intersection geometric, traffic flow rates and time headways were collected from this intersection.

As can been seen from the Fig. 1, this is a four-leg intersection that is controlled by the signal lights. Each approach has four approximately 3 metres-wide lanes, one left-turn lane, two through lanes and one right-turn lane. All lanes on the entrance driveways are numbered, from 1 to 16.

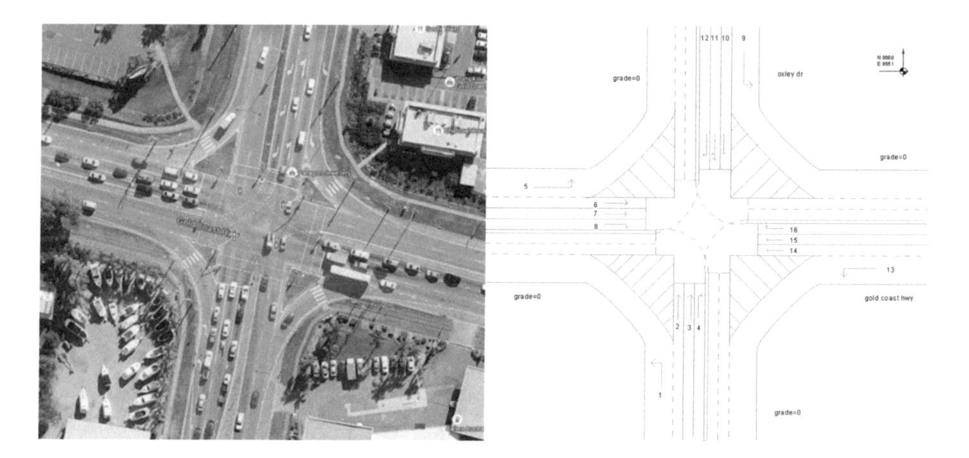

Fig. 1. Intersection of gold coast HWY and Oxley Dr.

3 Comparison of Estimated Headway and Real Headway

There are three main disturbances that affect the operational performance of the signalized intersection: physical factors of the roadway, traffic factors and environmental disturbances [8]. The environmental factor is most uncontrollable one among them, so the preparation in advance could be very significant [9]. As the most common environmental disturbances in the coastal city, Gold Coast, the impact analysis of rainfall will be discussed. In order to collect the real data, the actual headways in 32 cycles were measured in this intersection that are placed in Table 1.

The headway refers to the time interval between two vehicles. The first headway is the time interval between the initiation of the green signal and the instant vehicle crossing the curb line. The second headway is the time interval between the first and

Table 1. The constant headway value in 32 cycles

Cycles	1	2	3	4	5	6	7	8
h (s)	2.281	1.813	1.961	1.843	1.826	1.662	1.922	2.113
Cycles	9	10	11	12	13	14	15	16
h (s)	1.913	1.925	1.753	1.763	1.812	1.931	2.291	1.858
Cycles	17	18	19	20	21	22	23	24
h (s)	1.993	1.703	2.010	1.614	2.043	1.798	2.062	1.767
Cycles	25	26	27	28	29	30	31	32
h (s)	1.887	1.926	1.807	2.051	1.756	1.912	1.884	2.168

second vehicle crossing the stop line. The headway will be constant after several vehicles passing [10].

As can be easily seen from Table 1, the maximum headway is 2.2914 s, even smaller than the average headway (2.3285 s) under the normal weather condition. In order to analyze these data in a more intuitive way, Table 1 is re-drawn as a PDF figure.

The headways of 32 cycles are divided into 9 groups with a class interval 0.113 s. From the Fig. 2, it can be found that the real average headway is around 1.9077 s, much less than the normal weather condition (2.3285 s). Using the relationship between headway and saturation flow rates, a value of 1887 veh/h for saturation flow rate can be determined. Details of grouping and frequency are presented in Table 2.

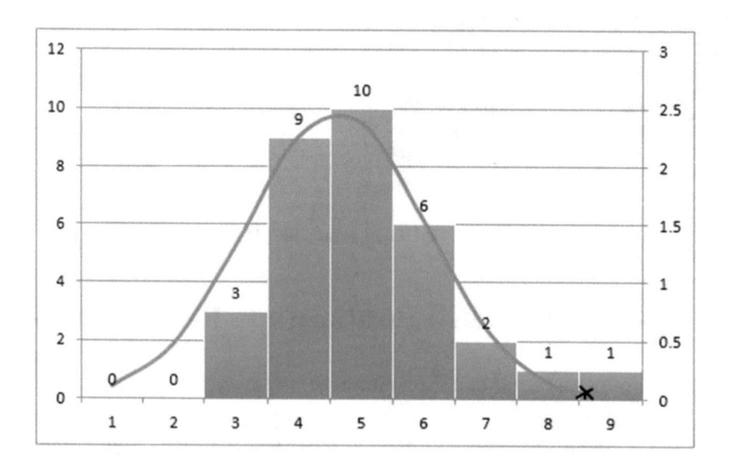

Fig. 2. Probability density function of headway

Group 1, group 2, group 8 and group 9 have a very low frequency, which means most headways are between the intervals of (1.613 s, 2.178 s). The headway estimated under normal weather condition (2.328 s) is greater than 2.178 s, locates outside of the interval, so the signal timing strategies design and performance analysis based on this conservative headway estimation will be worse than the real situation.

Table 2. Details of grouping

No. of group	Grouping	Frequency
1	<1.5 s	0
2	1.5 s ~ 1.613 s	0
3	1.613 s ~ 1.726 s	3
4	1.726 s ~ 1.839 s	9
5	1.839 s ~ 1.952 s	10
6	1.952 s ~ 2.065 s	6
7	2.065 s ~ 2.178 s	2
8	2.178 s ~ 2.291 s	1
9	>2.291 s	1

4 Impacts of Bad Weather on Timing Strategy A

In this section, several cases under different density of rain condition will be compared by using the exactly the same methodology. In addition, the signal phase scheme is same as well. Then, the saturation flow rates for each weather conditions are presented in the Table 3.

Table 3. Saturation flow rates for each weather conditions

Lanes	v_i (veh/h)	s_1^* (veh/h)	s_2^* (veh/h)	s_3^* (veh/h)	s_4^* (veh/h)	s_5^* (veh/h)
1	162.56	1426	1140.8	998.2	855.6	1887
2	168.3	1650	1320	1155	990	1887
3	198	1650	1320	1155	990	1887
4	147.31	1364	1091.2	954.8	818.4	1887
5	136.896	1426	1140.8	998.2	855.6	1887
6	89.1	1650	1320	1155	990	1887
7	99	1650	1320	1155	990	1887
8	178.56	1488	1190.4	1041.6	892.8	1887
9	196.788	1426	1140.8	998.2	855.6	1887
10	178.2	1650	1320	1155	990	1887
11	69.3	1650	1320	1155	990	1887
12	111.6	1550	1240	1085	930	1887
13	111.23	1426	1140.8	998.2	855.6	1887
14	158.4	1650	1320	1155	990	1887
15	138.6	1650	1320	1155	990	1887
16	195.3	1550	1240	1085	930	1887

Note:
s_1 = saturation flow rates under normal weather condition,
s_2 = saturation flow rates with a 20% drop due to certain density of rain,
s_3 = saturation flow rates with a 30% drop due to certain density of rain,
s_4 = saturation flow rates with a 40% drop due to certain density of rain, and
s_5 = saturation flow rates based on the real headway measured in-filed.

Obviously, the demand flow rates for each lanes are fixed value. A series of density of rainfall cause different saturation flow rate drops, because the impacts of rainfall on traffic operation mostly reflect in saturation flow rates drops. The next step is to determine the flow ratios and summation of the flow ratios for each case. All data have been integrated in the following Table 4.

Table 4. Flow ratios of each lanes for each cases

Lanes	y_1^*	y_2^*	y_3^*	y_4^*	y_5^*
1	0.1140	0.1425	0.1629	0.1900	0.0864
2	0.1020	0.1275	0.1457	0.1700	0.0890
3	0.1200	0.1500	0.1714	0.2000	0.1049
4	0.1080	0.1350	0.1543	0.1800	0.0779
5	0.0960	0.1200	0.1371	0.1600	0.0726
6	0.0540	0.0675	0.0771	0.0900	0.0472
7	0.0600	0.0750	0.0857	0.1000	0.0525
8	0.1200	0.1500	0.1714	0.2000	0.0949
9	0.1380	0.1725	0.1971	0.2300	0.1044
10	0.1080	0.1350	0.1543	0.1800	0.0943
11	0.0420	0.0525	0.0600	0.0700	0.0366
12	0.0720	0.0900	0.1029	0.1200	0.0594
13	0.0780	0.0975	0.1114	0.1300	0.0588
14	0.0960	0.1200	0.1371	0.1600	0.0837
15	0.0840	0.1050	0.1200	0.1400	0.0737
16	0.1260	0.1575	0.1800	0.2100	0.1033

Note:
y_1 = flow ratios of each lanes under normal weather condition,
y_2 = flow ratios of each lanes based on 20% dropped saturation flow rates,
y_3 = flow ratios of each lanes based on 30% dropped saturation flow rates,
y_4 = flow ratios of each lanes based on 40% dropped saturation flow rates, and
y_5 = flow ratios of each lanes based on real saturation flow rates.

From Table 5, the summation flow ratios are 0.45, 0.5625, 0.6428, 0.75 and 0.3863 corresponding to fully saturation flow rates, 20% dropped saturation flow rates, 30% dropped saturation flow rates, 40% dropped saturation flow rates and real saturation flow rates respectively. All summation flow ratios are less than 0.9 which means the same intersection geometric design and signal phase design adopted for all cases are acceptable. Having determined the summation flow ratios that are acceptable, signal cycle length should be computed now. The final results of minimum cycle length and the optimal cycle length for all cases are placed in Table 6.

From case 1 to case 4, it can be found that the summation of flow ratios and the cycle lengths have upward trends. In other words, the more adverse the weather conditions are, the longer the cycle lengths are. The final step for signal timing strategies design is to determine the effective green times and effective green ratios, which is shown in Table 7.

Table 5. Flow ratios of each signal phases and summation flow ratios for each cases

Phases	$y^*_{1\ phase\ i}$	$y^*_{2\ phase\ i}$	$y^*_{3\ phase\ i}$	$y^*_{4\ phase\ i}$	$y^*_{5\ phase\ i}$
A	0.0960	0.1200	0.1371	0.1600	0.0837
B	0.1080	0.1350	0.1543	0.1800	0.0943
C	0.1200	0.1500	0.1714	0.2000	0.1049
D	0.1260	0.1575	0.1800	0.2100	0.1033
	Y^*_k				
	0.4500	0.5625	0.6428	0.7500	0.3863

Note:
$y_{1\ phase}$ = flow ratios of each signal phases under normal weather condition,
$y_{2\ phase}$ = flow ratios of each signal phases based on 20% dropped saturation flow rates,
$y_{3\ phase}$ = flow ratios of each signal phases based on 30% dropped saturation flow rates,
$y_{4\ phase}$ = flow ratios of each signal phases based on 40% dropped saturation flow rates,
$y_{5\ phase}$ = flow ratios of each signal phases based on real saturation flow rates, and
Y_k = summation flow ratios for each phases successively.

Table 6. The minimum cycle length and the optimal cycle length for all cases

	Case 1*	Case 2*	Case 3*	Case 4*	Case 5*
Y	0.45	0.563	0.643	0.75	0.386
C_m (s)	36.364	45.767	56.022	80.000	32.573
C_0 (s)	63.636	80.092	98.039	140.000	57.003

Note:
Case 1 = the case under normal weather condition,
Case 2 = the case based on 20% dropped saturation flow rates,
Case 3 = the case based on 30% dropped saturation flow rates,
Case 4 = the case based on 40% dropped saturation flow rates, and
Case 5 = the case based on the real saturation flow rate.

Table 7. Effective green times and effective green ratios for all cases

	Phase	Case 1	Case 2	Case 3	Case 4	Case 5
t_{EG} (s)	A	9.31	12.81	16.64	25.60	8.05
	B	10.47	14.41	18.71	28.80	9.01
	C	11.64	16.01	20.89	32.00	10.06
	D	12.22	16.81	21.86	33.60	9.87
t_G (s)	A	9.31	12.81	16.64	25.60	8.05
	B	10.47	14.41	18.71	28.80	9.01
	C	11.64	16.01	20.89	32.00	10.06
	D	12.22	16.81	21.86	33.60	9.87
λ	A	0.1463	0.1599	0.1696	0.1829	0.1413
	B	0.1646	0.1799	0.1907	0.2057	0.1581
	C	0.1829	0.1999	0.2130	0.2286	0.1766
	D	0.1920	0.2099	0.2229	0.2400	0.1732

Till now, the signal timing plans of all cases under different weather condition have been obtained. It is easy to determine all the values. To sum up the important information and find out the trend of delay based on the different weather conditions, Table 8 and Fig. 17 are made.

It can be easily found from Fig. 3 that the average control delays for the whole intersection are going up, because the weather condition is becoming worse. This would support the point of weather impacts on the intersection operations, the worse the weather conditions are, the lower the level of services are, which are presented in the Table 8 as well.

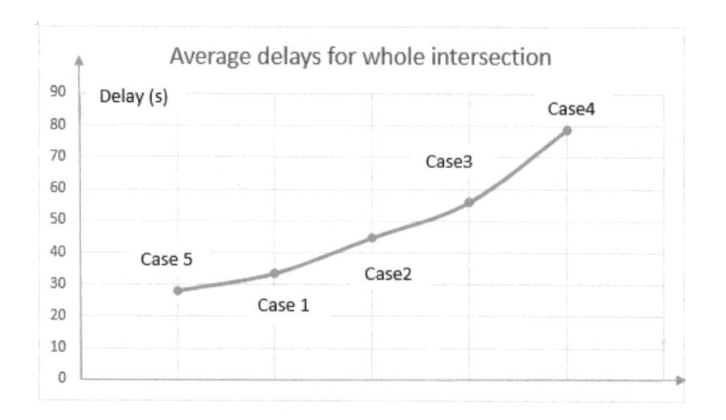

Fig. 3. Trend of the delay for all cases

Table 8. Delay for the whole intersection and LOS for all cases

Cases	d_I (s)	LOS
1	33.59	C
2	44.76	D
3	55.93	D
4	78.63	E
5	28.12	C

5 Impacts of Bad Weather on Timing Strategy B

In this section, in order to find the most optimal timing strategies for all kinds of weather conditions, all the cases discussed in last section will be discussed again, however, using another signal timing strategy that aims on both average stop frequency and average delay time. For this purpose, the Akcelik's optimal cycle length equation will be used. In addition, the basic known terms, such as intersection geometric, demand flow rates, signal phase scheme, saturation flow rates under different weather situations and flow ratios, are completely same as timing strategy A, which will be repeated here as $C_0 = \frac{(1.4+K)L+6}{1-Y}$.

The factor K is stopping compensation factor. The smallest value of K helps to get a shortest cycle length that ensures a small average delay and queue length. So, in this case, take K value as –0.3. For example, under normal weather condition, the value of optimal cycle length equals to ((1.4 – 0.3) * 20 + 6)/(1 – 0.45) = 50.91 s. The final results of signal timing strategies for all cases are presented in Table 9.

Table 9. Final results for timing strategy B

	Phase	Case1	Case2	Case3	Case4	Case5
C_0 (s)		50.19	64	78.41	112	45.62
t_{EG} (s)	A	6.594	9.387	12.447	19.63	6.25
	B	7.418	10.56	14.019	22.08	6.96
	C	8.243	11.73	15.57	24.53	6.85
	D	8.655	12.32	16.354	25.76	9.87
t_G (s)	A	6.594	9.387	12.447	19.63	6.25
	B	7.418	10.56	14.019	22.08	6.96
	C	8.243	11.73	15.57	24.53	6.85
	D	8.655	12.32	16.354	25.76	9.87
λ	A	0.1295	0.1467	0.1587	0.1728	0.1217
	B	0.1457	0.165	0.1788	0.1971	0.137
	C	0.1619	0.1833	0.1986	0.219	0.1526
	D	0.17	0.1925	0.2086	0.23	0.1502

Table 10. Delay for the whole intersection and LOS for all cases

Cases	d_I (s)	LOS
1	34.313	C
2	55.298	D
3	82.073	F
4	131.131	F
5	27.86	C

Then, the performance analysis results, delay and LOS, will be placed below. All the important data are integrated in Table 10 and Fig. 4. Similar to signal timing strategy A, signal timing strategy B has an up-trend of average delays as the weather condition is getting worse. However, the growth rates of these two strategies are different, which decide the applicability of the signal timing strategies.

6 Comparative Analysis

Two signal timing strategies of the same intersection using the different cycle length equation lead to distinct results that one is suitable for good weather condition and the other applies to bad weather condition. The results of two performance analysis will be compared in the Table 11 and Fig. 5.

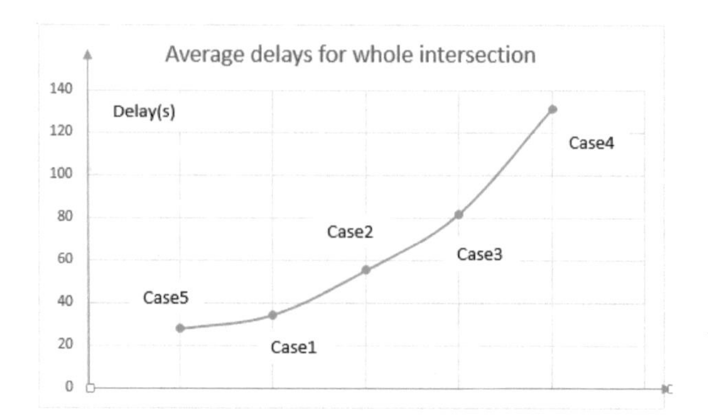

Fig. 4. Trend of the delay for all cases

Table 11. Average control delay of two signal timing strategies

Cases	Delay of strategy A (s)	Delay of strategy B (s)
5	28.112	27.856
1	33.59	34.313
2	44.76	55.298
3	55.93	82.073
4	78.63	131.131

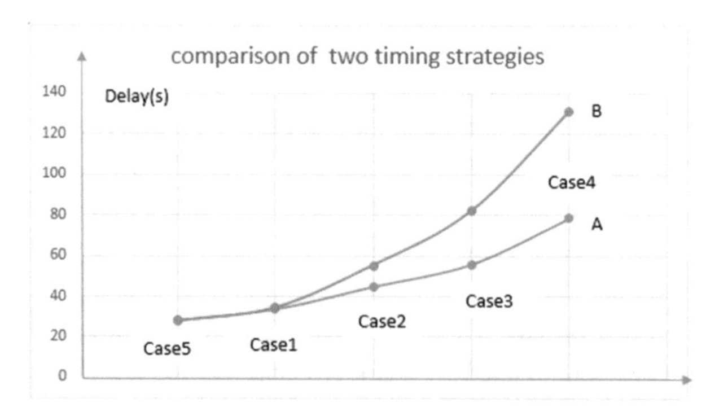

Fig. 5. Comparison of two timing strategies

From Fig. 5, it can be easily found that delays for two signal timing strategies both go up when the weather conditions are getting worse. However, the trend of delays for signal timing strategy B has a greater gradient, which means with the same environment change, delays for strategy B increases more.

The values of delay for strategy B is 27.856 s, smaller than the value of delay for strategy A (28.112 s) in case 5 that based on the real headway. As a result, the signal timing plan may adopts the signal timing strategy B as it has a better performance in normal weather condition. As comparison, in case 4 which has the worst weather condition (with a 40% drop of saturation flow rate), the value of delays are 78.63 s and 131.131 s for strategy A and B respectively. In other words, in this case, signal timing strategy A is much more suitable for this intersection as it has a much smaller delay.

Acknowledgments. The study is supported by the National Natural Science Foundation of China (NO. 71701070), the Natural Science Foundation of Guangdong Province (NO. 2016A030310427).

References

1. Keyvan-Ekbatani, M., Papageorgiou, M., Knoop, V.L.: Controller design for gating traffic control in presence of time-delay in urban road networks. Transp. Res. Part C **59**, 308–322 (2015)
2. Goodwin, L.C., Pisano, P.A.: Weather-responsive traffic signal control. ITE J. **74**(6), 28–33 (2004)
3. Coates, L.: Flood fatalities in Australia, 1788–1996. Aust. Geogr. **30**(3), 391–408 (1999)
4. Wang, S., Meng, Q., Liu, Z.: Fundamental properties of volume-capacity ratio of a private toll road in general networks. Transp. Res. Part B **47**, 77–86 (2013)
5. Zhou, M., Qu, X., Li, X.: A recurrent neural network based microscopic car following model to predict traffic oscillation. Transp. Res. Part C **84**, 245–264 (2017)
6. Qu, X., Wang, S., Zhang, J.: On the fundamental diagram for freeway traffic: a novel calibration approach for single-regime models. Transp. Res. Part B **73**, 91–102 (2015)
7. Sharma, A., Bullock, D., Peeta, S., Krogmeier, J.: Detection of inclement weather conditions at a signalized intersection using a video image processing algorithm. In: 4th IEEE Digital Signal Processing Workshop and 12th Signal Processing Education Workshop, pp. 150–155 (2006)
8. Shin, C.H., Choi, K.: Saturation flow rate estimation under rainy weather conditions for on-line traffic control purpose. KSCE J. Civ. Eng. **2**(3), 211–222 (1998)
9. Al-Kaisy, A., Freedman, Z.: Weather - responsive signal timing: practical guidelines. Transp. Res. Rec. J. Transp. Res. Board **1978**(1), 49–60 (2006)
10. Qu, X., Zhang, J., Wang, S.: On the stochastic fundamental diagram for freeway traffic: Model development, analytical properties, validation, and extensive applications. Transp. Res. Part B **104**, 256–271 (2017)

Study on Macroscopical Layout Optimization Model of Large Passenger Transfer Hub Facilities Based on NSGA-II

Chengyuan Mao[1(✉)], Yiming Bie[2], Kan Zhou[3], and Weiwei Qi[4]

[1] College of Engineering, Zhejiang Normal University, Jinhua, China
maocy@zjnu.cn
[2] School of Traffic Science and Engineering, Harbin Institution of Technology, Harbin, China
byimingbie@126.com
[3] Shenzhen Municipal Design and Research Institute Co. Ltd., Shenzhen, China
63638192@qq.com
[4] School of Civil Engineering and Transportation, South China University of Technology, Guangzhou 510641, China
qwwhit@163.com

Abstract. In general, large passenger hubs contain a variety of transportation functional areas. The facility configuration requirements of different functional areas are not the same, and a rational collocation of spatial positions and the number of functional areas will greatly improve the level of service. By analyzing the connotations of the functional area, an optimization objective function was built, based on the average expected passenger walking time, average cross-collision delays [average delay due to collisions between passengers moving in opposing directions] of passengers, and the cost of the hub, with constraint conditions on functional acreage, shape and location, as well as a model of macroscopical layout optimization of hub facilities. A variety of optimization problem-solving methods were compared, and given the features of this model, the genetic algorithm NSGA-II was selected. Finally, this paper gives an example of a three-storey high-speed rail hub, evaluates the hub's layout scheme, and proves that the model is effective and feasible. Therefore, the research results can be used to evaluate the layout plans of passenger transfer facilities for large hubs, and provide a theoretical foundation for perfecting hub transfer facilities.

1 Introduction

In recent years, many high-speed rail hubs have been built in China's cities. As people have used them, some layout problems of the functional areas have gradually emerged. In the final analysis, such problems turn out to be due to the lack of an effective method for setting out the functional areas and a lack of evaluation methods. The existing research results have mostly focused on the macroscopical layout and siting problems (Bie et al. 2012), traffic analysis and prediction (Mao et al. 2015; Qu et al. 2014; Qu and Wang 2015), and qualitative analysis of facility configurations. Feng (2010) built a dual-objective optimization model of a comprehensive transfer hub, considering

G. De Pietro et al. (Eds.): KES-IIMSS-18 2018, SIST 98, pp. 386–396, 2019.
https://doi.org/10.1007/978-3-319-92231-7_40

the hub's land (space) condition based on the aims of minimizing the construction costs and transfer delays, but he did not give the specific solution to the model. Qiu and Gu (2006) analyzed the characteristics of the planar and three-dimensional layouts of some typical international integrated transport hubs, providing a potential reference for the facility configuration optimization of China's high-speed rail hubs.

Foreign studies of facility configuration and layout originated from industry manufacturing (Kusiak and Heragu 1987). With the development of optimization theory, various configurations of optimization problems have emerged, including the configuration of building facilities, the layout of facilities etc., which could provide references for the design of transfer facilities at high-speed transport hubs. Lee et al. (2005) built an optimization model of functional zones, which took into account the constraints of internal walls and corridors, and took the maximum satisfaction with the functional zones as the objective. He also used an improved genetic algorithm to calculate the model. Aiello et al. (2012) put forward the flat space division method of building facilities, and utilized the genetic algorithm to calculate a multi-objective optimization problem of building facilities layout. McKendall et al. (2006) used a simulated annealing algorithm to settle the dynamic building facility layout optimization problem.

In conclusion, meaningful research has been carried out studying traffic delays and layouts of transport hubs, but the evaluation method for transfer facility layouts and functions is not clear enough. Therefore, this paper present a macroscopical layout optimization model of passenger transfer facilities based on passenger transfer expected time and delays. By analysing and evaluating the rationality of the transfer facilities of large-scale passenger hubs, this model could provide a theoretical basis for the construction of such hubs.

2 Establishment of the Macroscopical Layout Model of Passenger Transfer Facilities

2.1 Objective Function

The 1st Objective Function: Expected Traveling Time

Different layouts of functional areas directly determine the distance between transfer facilities, which affects the expected traveling time. Therefore, in this paper the minimal average expected traveling time is selected as one of the objective functions, as shown in Eq. (1).

$$\text{Min} \quad t_{des} = \frac{\sum_{j=1}^{M}\sum_{i=1}^{M} f_{ij}(dH_{ij}/v_{hd} + dV_{ij}/v_{vd})}{\sum_{j=1}^{M}\sum_{i=1}^{M} f_{ij}} \tag{1}$$

where

t_{des} – the average desired traveling time (s);

f_{ij} – the number of passengers moving from functional area i to functional area j per unit of time (p/h);
dH_{ij} – the horizontal distance of the link channel between functional area i and functional area j (m);
dV_{ij} – the vertical distance of the link channel between functional area i and functional area j (m);
v_{hd} – the desired speed of passengers traveling horizontally, 1.362 m/s;
v_{vd} – the desired speed of passengers traveling vertically: upstream 0.391 m/s, downstream 0.289 m/s;
M – the number of functional areas.

The 2nd Objective Function: Cross-collision Delays

Cross-collision delays, average delay due to collisions between passengers moving in opposing directions, have the greatest impact on passengers. Therefore, the minimal average Cross-collision delay is considered as one of the objective functions, as shown in Eq. (3).

$$\text{Min} \quad d_c = \frac{\sum_{l \in L} K_l d_c^l Q_l}{\sum_{j=1}^{M} \sum_{i=1}^{M} f_{ij}} \tag{3}$$

Where

d_c – the average cross-collision delay for passengers(s);
L – a set, which contains all link channels between functional areas;
K_l – binary variable, $K_l = 1$ if passengers crossed, and $K_l = 0$ otherwise;
d_c^l – the average cross-collision delay for passengers on the lth link channel combination(s), calculated by Eq. (4);
Q_l – the number of passengers transferred by the lth link channel combination per unit of time(p/h).

The total passenger arrival rate, the arrival rate difference absolute value of the passenger from different directions and the angle of conflict are the three major factors to decide the average intersection delay. According to the relationship between the three factors, the model of the average conflict delay is built, and it can be optimized by the Levenberg-Marquardt, as shown in Eq. (4)

$$d_c = d_c^{\theta} + (p_1 x + p_2 y + p_3 xy)\exp\left(p_4 \theta^5 + p_6\right) \tag{4}$$

Where

x – the arrival rate difference absolute value of the passenger from different directions (p/s);
y – the arrival rate of passengers (p/s);

d_c^{θ} – the average cross-collision delay for passengers, when the angle of conflict is $\theta(s)$;
p_i – correction coefficient

The 3rd Objective Function: The Cost of the Transfer Hub
When designing the layout of a large passenger transfer hub, construction cost savings are very important, and the smaller the total investment, the better. Therefore, the minimal cost of the large passenger transfer hub is considered as one of the objective functions, as shown in Eq. (5).

$$\text{Min} \quad Cost = \sum_{n=0}^{N-1}\sum_{i=1}^{M} K_{ni}A_iP_{ni} + \sum_{j=1}^{M}\sum_{i=1}^{M}(dH_{ij}WH_{ij}\text{PH}_{\text{link}} + dV_{ij}WV_{ij}\text{PV}_{\text{link}}) \quad (5)$$

Where
$Cost$ – the cost of the hub(yuan);
N – the number of layers in the hub;
A_i – the area of functional area i(m^2);
P_{ni} – the construction cost per unit area of functional area i in the nth layer (yuan/m^2);
K_{ni} – binary variable, $K_{ni} = 1$ when functional area i is located inlayer n, and $K_{ni} = 0$ otherwise;
HP_{link} – the per unit area construction costs of the horizontal link channel between functional areas (yuan/m^2);
PV_{link} – the per unit area construction costs of the vertical link channel between functional areas (yuan/m^2);
WH_{ij} – the width of the horizontal link channel between functional area i and functional area j (m);
WV_{ij} – the width of the vertical link channel between functional area i and functional area j (m).

2.2 Establishment of the Constraints

Functional Areas. In order to achieve the requirements, the various functional areas must be of the right size to match passengers' needs. Also, the sum of the functional areas are bounded by the total area of the building. These constraints on the functional areas can be described as in Eqs. (8) to (10).

$$g_1 = A_i - A_i^{\max} \leq 0, \quad \forall i \quad (8)$$

$$g_2 = A_i^{\min} - A_i \leq 0, \quad \forall i \quad (9)$$

$$g_3 = \sum_{i=1}^{M} K_{ni}A_i - A_n^{total}, \quad \forall n \tag{10}$$

Where

A_i^{max} – the maximum permitted building area of functional area $i(\mathrm{m}^2)$;

A_i^{min} – the minimum permitted building area of functional area $i(\mathrm{m}^2)$;

A_n^{totol} – the total building area inlayer $n(\mathrm{m}^2)$.

Shape of Functional Areas. When the various functional areas meet the size requirements, they also need to avoid thin strips, which would cause micro design inconvenience in the internal transfer facility. Since it is assumed that functional areas are rectangular, the aspect ratio is used to restrain the functional areas' shape, as shown in Eqs. (11) and (12).

$$g_4 = \alpha_i - \alpha_i^{\max} \leq 0, \quad \forall i \tag{11}$$

$$g_5 = \alpha_i^{\min} - \alpha_i \leq 0, \quad \forall i \tag{12}$$

where

α_i – the aspect ratio of functional area i, calculated by Eq. (13);

α_i^{max} – the maximum allowable aspect ratio of functional area i;

α_i^{min} – the minimum allowable aspect ratio of functional area i.

$$\alpha_i = \frac{\max\{h_i, w_i\}}{\min\{h_i, w_i\}} \tag{13}$$

Where

h_i – the height of functional area $i(\mathrm{m})$;

w_i – the width of functional area i (m).

Location Constraint. The location constraint means that functional areas on the same floor cannot overlap each other, and some functional areas must or cannot be built in particular locations. These constraints can be described by Eqs. (14) and (15).

$$g_6 = |x_i - x_j| - \frac{W_i + W_j}{2} \geq 0 \cap g_7 = |y_i - y_j| - \frac{H_i + H_j}{2} \geq 0 \tag{14}$$

$$g_8 = (x_i, y_i, z_i) = (\mathrm{x}\ ,\mathrm{y}\ ,\mathrm{z}) \tag{15}$$

Where

W_i – the width of functional area $i(\mathrm{m})$;

H_i – the height of functional area $i(\mathrm{m})$;

(x, y, z) – the coordinates of a particular location.

Model Development. To sum up, we can get the macroscopical layout model of passenger transfer facilities, as shown in Eqs. (16) and (17).

$$\text{Min} \quad F(x) = [t_{des}, d_c, Cost]^{\mathrm{T}} \tag{16}$$

$$\text{s.t.} \quad G(x) = [g_1, g_2, g_3, g_4, g_5, g_6, g_7, g_8]^{\mathrm{T}} \tag{17}$$

3 Solving the Large Passenger Hub Transfer Facility Macroscopic Distribution Optimization Model Based on NSGA-II

3.1 Selection of Method for Solving the Model

The conventional methods for solving multi-objective optimization problems mainly include the weight coefficient method, the constraint method, the simulated annealing algorithm, the ant colony optimization algorithm, particle swarm optimization and the genetic algorithm. In view of the model's features, all six of these multi-objective optimization problem solution methods were comprehensively considered. The features of the solution methods include operating convenience, solution accuracy, global optimization capability, convergence rate, and maturity constraints. The genetic algorithm was chosen for solving the model.

Genetic algorithms mainly comprise three methods: the Simple Genetic Algorithm (SGA), Non-dominated Sorting Genetic Algorithm (NSGA) and the Non-dominated Sorting Genetic Algorithm-II (NSGA-II). Compared with SGA and NSGA, NSGA-II has distinct advantages in terms of global optimization capability, calculation efficiency and solutions that maintain diversity. Many scholars have verified the validity of the algorithm through actual optimization studies. Therefore, NSGA-IIwas chosen as the solution method for this paper.

3.2 NSGA-II Solution Steps

According to the basic operating procedures of NSGA-II, the solution procedure can be divided into the following six steps:

(1) Variable coding. In order to reduce the length of the coding, a strip structure is used in this paper to state the functional area layout issues. Each floor of the large passenger hub is divided into a variable-width strip region. Each strip region in the longitudinal direction is further divided into a variable-height rectangle.
(2) Selection operation. A certain amount of initial coding is generated randomly, which represents the multiple solutions to optimizing the layout model. The initial coding group is conducted in anon-dominated sequence, and the non-dominated sequence and degree of congestion are calculated. According to the non-dominated

sequence and the degree of congestion, N parent codings are chosen. After N parent coding crossovers and mutations, N progeny codings are produced.

(3) Crossover operation. The crossover operation is only conducted for code section 1, code section 2 and code section 3. The cross rate is the probability of mutual pairing codings being selected for crossover operations. The cross-way is a two-point crossover, which randomly disposes two junctions in the encoding section to exchange data between two intersections, producing the progeny encoding.

(4) Mutation operation. The mutation operation is only conducted for code section 1 and code section 2. The mutation rate is the probability of a mutual pairing coding being selected for mutation operations. The mutation method is a uniform mutation, which mutates the data of code section 1 and code section 2 in the parent with a small probability, producing the child code.

(5) Elitist selection. After the crossover and mutation operation of the parent code sections, N child code sections are produced. The child code sections and N parent code sections are merged, and then this 2N coding is sorted, using a non-dominated method. Then the N code sections are elected by the elitist selection of NSGA-II. After that, the procedure moves to the second step for a new round of genetic manipulation.

(6) Processing of constraints. The genetic research work focused on unconstrained optimization problems, including the search space limit method, the penalty function method and the constraints transform. Considering the advantages and disadvantages of the three processing methods, and the nature and complex extent of the macroscopical layout optimization model of large passenger transfer hub facilities, the search space limit method is used to treat the constraints on area and location. The penalty function method is used to treat the constraint on the total area of functional areas on the same floor. When the sum of the area of functional areas on the same floor is more than the maximum allowable building area, then the non-dominated sequence rank is increased, the degree of congestion is defined as infinite and the probability inherited by the next generation is reduced.

4 Case Study

Macroscopical layout optimization of a three-storey high-speed rail hub is used as an example to analyze the effectiveness of NSGA-II. Suppose there are 10 transport functional areas within the high-speed rail hub, each functional area corresponding to a number shown in Table 1. The costs per unit area of the functional areas are shown in Table 2, the construction costs of the horizontal link channel between functional areas are 1800 yuan/m^2, and the construction costs of the horizontal link channel between functional areas are 2000 yuan/m^2.

The maximum allowable building range of each floor is a rectangular area with length 150 m and width 120 m. The required areas of the functional areas and the aspect ratio of each functional area are shown in Table 3. The per hour transfers of traffic between the various functional areas are shown in Table 4.

Table 1. Functional areas

Functional area	No.	Functional area	No.	Functional area	No.	Functional area	No.
Bus parking	1	Ticketing	4	Bus transfer	7	Taxi drop-off	9
Outbound	2	Car drop-off	5	Taxi pick-up	8	Orbit transfer	10
Passenger waiting	3	Car parking	6				

Table 2. Unit costs of functional areas

Area No.	Unit cost (yuan/m^2)			Area No.	Unit cost (yuan/m^2)		
	B1 Floor	1st Floor	2nd Floor		B1 Floor	1st Floor	2nd Floor
1	—	3000	—	6	5000	2000	6000
2	4000	—	—	7	5000	3000	6000
3	4000	—	3500	8	5000	1500	5000
4	4000	2000	3000	9	5000	1500	5000
5	5000	1500	5000	10	15000	9000	35000

Note: "-" indicates that the functional area is not allowed to be located on the corresponding floor.

Table 3. Area & Aspect ratio (height to width) required by functional area

Area No.	Required area (m^2)	Aspect ratio	Area No.	Required area (m^2)	Aspect ratio
1	6000	0.60	6	9000	0.25
2	400	0.57	7	3500	0.33
3	4000	0.50	8	2000	0.625
4	600	0.40	9	1000	0.5
5	2000	0.33	10	1000	0.5

Table 4. Transfer passenger flow between functional areas

Area No.	1	2	3	4	5	6	7	8	9	10
1	0	1800	0	0	0	0	0	0	0	0
2	0	0	0	0	0	350	500	350	0	600
3	2150	0	0	0	0	0	0	0	0	0
4	0	0	850	0	0	0	0	0	0	0
5	0	0	200	200	0	0	0	0	0	0
6	0	0	200	100	0	0	0	0	0	50
7	0	0	300	200	0	0	0	0	0	50
8	0	0	0	0	0	0	0	0	0	0
9	0	0	200	150	0	0	0	0	0	0
10	0	0	400	200	0	0	0	0	0	0

The initial coding is conducted by genetic manipulation. The initial population of the genetic algorithm is 60 members, the number of iterations is 150, the cross-rate is 0.9 and the mutation rate is 0.1.

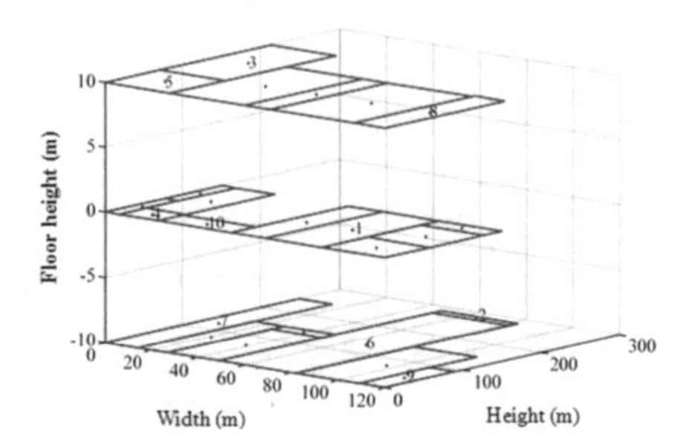

Fig. 1. Original functional area layout

Figure 1 shows a randomly generated initial encoding of a corresponding functional area layout scheme, with average expected traveling time 6.57 min, average cross-collision delay 3.51 s, total cost of hub 194 million yuan, and average deviation of aspect ratio 3.63. As the figure illustrates, the design scheme can be optimized from the aspects of the functional area layout, functional area shape and the total cost, etc.

Figures 1 and 2 shows two typical optimized layout solutions after 150 genetic iterations resulting in a Pareto-optimal solution set. The average expected traveling time is 4.24 min, the average cross-collision delay is 0.53 s, the traveling service level is D, the total cost of the hub is 135.1 million yuan, and the average deviation of the aspect ratio is 0.57.

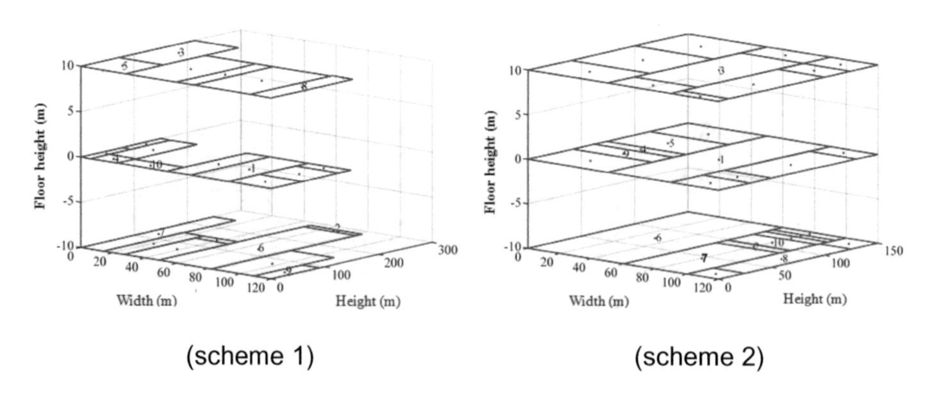

Fig. 2. Optimization of functional area layout

Comparing scheme 1 and scheme 2, the main differences between the two layouts are the total cost of the hub and the average expected traveling time. In scheme 1, the ticketing area, the social vehicles' visitor area and the taxi drop-off area are arranged on the 2nd floor, while the bus transfer area is arranged on the ground floor. In scheme 2, the ticketing area, the social vehicles' visitor area and the taxi drop-off area are arranged on the ground floor, while the bus transfer area is arranged on the 1st floor. From the cost point of view, scheme 1 is more expensive than scheme 2. However, in scheme 1, the social vehicles' visitor area and the taxi drop-off area are arranged near to the large passenger waiting area, making the passengers' traveling time shorter. Thus, when the construction costs are limited, scheme 2 is preferable, but when more attention is paid to convenience for passengers and the construction costs are less restrained, scheme 1 is preferable.

5 Summary

Based on the average expected passenger walking time, average cross-collision delays for passengers, and the cost of the hub, a model of macroscopical layout optimization of large passenger hub facilities is constructed in this paper, with constraint conditions on the functional areas' acreage, shape and position. Based on the features of this model and the methods used to tackle multi-objective optimization problems, the genetic algorithm NSGA-II was chosen to solve the model. Taking a three-storey high-speed rail hub as an example, the model was used to optimize the layout of the hub's functional areas. Through an evaluation, the model's validity and feasibility are proved. Therefore, the research results can be used to evaluate layout plans for the passenger transfer facilities of large hubs, and provide a theoretical foundation for perfecting hub transfer facilities.

Acknowledgement. This research was financed by the Natural Science Foundation of Zhejiang (No. LY18G030021 & No. LY18E080021).

References

Bie, Y., Wang, D., Qi, H.: Prediction model of bus arrival time at signalized intersection using GPS data. J. Transp. Eng. **138**(1), 12–20 (2012)

Mao, C.Y., Bie, Y.M., Pei, Y.L., Qi, W.W.: Research on safety and static-dynamic legibility of distressed pavement. J. Discrete Dyn. Nat. Soc., 1–7 (2015)

Zhang, L.: Analysis on the Change and Forecast of Passenger Flow of Wuhan-guangzhou High-speed Railway. MS. Central South University, China (2012)

Qu, X., Wang, S.: Long distance commuter lane: a new concept for freeway traffic management. Comput. Aided Civ. Infrastruct. Eng. **30**(10), 815–823 (2015)

Qu, X., Ren, L., Wang, S., Oh, E.: Estimation of entry capacity for single-lane modern roundabouts: a case study in Queensland, Australia. J. Transp. Eng. **140**(7), 05014002 (2014)

Feng, W.: Research on Key Problems of City's External Traffic Comprehensive Transfer Hubs System. PD.Southwest Jiaotong University, China (2010)

Qiu, L.L., Gu, B.N.: Analysis of Typical Layout Design in Foreign Integrated Transport Hub. PD. J. Urban Mass Transit, (3), 55–59 (2006)

Kusiak, A., Heragu, S.S.: The facility layout problem. Eur. J. Oper. Res. **29**(3), 229–251 (1987)

Lee, K.Y., Roh, M., Jeong, H.S.: An improved genetic algorithm for multi-floor facility layout problems having inner structure walls and passages. Comput. Oper. Res. **32**(4), 879–899 (2005)

Aiello, G., Scalia, G.L., Enea, M.: A multi objective genetic algorithm for the facility layout problem based upon slicing structure encoding. Expert Syst. Appl. **39**(10), 352–358 (2012)

McKendall, A.R., Shang, J., Kuppusamy, S.: Simulated annealing heuristics for the dynamic facility layout problem. Comput. Oper. Res. **33**(8), 2431–2444 (2006)

Traffic Impact Analysis of Gold Coast Light Rail Stage 2

Yan Kuang(✉), Barbara Yen, and Kate Barry

Griffith School of Engineering,
Griffith University, Gold Coast, QLD 4222, Australia
{y.kuang, b.yen}@griffith.edu.au,
kate.barry@griffithuni.edu.au

Abstract. The Gold Coast Light Rail project represents a significant investment in public transportation infrastructure on the Gold Coast and is one of the key initiatives set out in City of Gold Coast Council's Transport Strategy 2031. This paper conducts an impact analysis of Stage 2 on the peripheral transportation network. To assess the anticipated impacts, a traffic simulation was conducted in PTV Visum 16 for the base year of 2018 and forecasted for year 2031. For the base and forecasted models, a scenario was prepared both with and without Stage 2 of the system to allow a comparison of results between the variants. According to the results, it is found that Stage 2 has a positive impact on the transportation network by successfully implementing an integrated transport solution encouraging users to switch to public transport.

1 Introduction

Transportation plays an important role in urban planning. An increasing population is driving an increasing travel demand, resulting in a greater traffic demand on the network. With increasing traffic demand, a series of mobility-related problems occur such as congestion, air pollution, noise pollution and accidents etc. [1]. These issues become particularly prevalent in busy urban centres such as the Gold Coast. As the population in Australia continues to increase, shortcomings in Australia's national transport infrastructure have been identified [2]. These issues include, increasing traffic congestion, antiquated public transport networks and inadequate airport facilities [3].

Rapid population growth, along with a bias towards private travel has resulted in congestion issues throughout the region [4]. Household Travel Surveys conducted by DTMR in 2011 revealed that 85% of all travel on the Gold Coast is by car, with only 4% on public transport [4]. Cycling (1%), Walking (8%) and other modes (2%) made up the remaining shares [4]. This dependence on car travel causes problematic congestion issues, and many urbanised areas of the Gold Coast already fully utilise the road reserve, leaving little scope for future expansion of the roadway [4]. Congestion is problematic both environmentally and economically, and it slows the movement of freight as well as causing increased noise and pollution emissions [5]. Over time congestion leads to the progressive degradation of the environment, and decreased

G. De Pietro et al. (Eds.): KES-IIMSS-18 2018, SIST 98, pp. 397–406, 2019.
https://doi.org/10.1007/978-3-319-92231-7_41

urban amenity [5]. These adverse consequences are heightened in cities with widespread low density suburbs, such as the Gold Coast, where users habitually travel by car to the city centre [5, 6]. To reduce congestion and negate its effects, public transport must become an appealing choice, with the capacity to take on a substantial portion of previously private travel [5]. Only an efficiently functioning public transportation system can be competitive with private transport [7].

The Gold Coast Rapid Transit (GCRT) system is a dedicated transit corridor that houses a light rail system along a coastal stretch of the city. The first stage of the project is 13 km connecting Parkwood to Broadbeach, and the second stage will add an additional 7.3 km of track to connect from Parkwood to Helensvale [8]. The Gold Coast is one of Australia's fastest growing cities [9], and the corridor containing Stage 2 has undergone significant growth in the last few years with the opening of the Gold Coast University Hospital and Gold Coast Private Hospital, as well as being adjacent to Griffith University, the largest university campus on the Gold Coast. Additionally, as host to the 2018 Commonwealth Games, the athletes village is under construction within the GCRT corridor in Parkwood. After the completion of the Games, the athletes village will be converted to a residential and retail precinct [10], generating a significant increase in travel demand.

To improve the traffic safety and efficiency, massive research has been conducted by many researchers [11–15]. Traffic Impact Analysis (TIA) is an essential step for engineers and planners to undertake on any project that proposes changes to, or that will affect, the transportation system in the area [16]. A study by Xiao [17] indicates that a TIA is an essential step in planning to aid the prevention of conflicts between new developments and the resulting traffic generation. Where a new development or land use type change is proposed a TIA can be performed to assess the adequacy of existing or future transportation infrastructure to accommodate the additional trips generated by the changes [16]. The purpose of a TIA is to analyse both the intensity and spread of impact caused by newly increased traffic demand and use this information to propose solutions to negate this effect [17]. The model was created in Visum, macroscopic traffic simulation software that allows the modelling of private and public transport modes in a single integrated model [18]. This research offers a comparison of the network upon opening in early 2018 and in 2031 both with and without the light rail system included to demonstrate the impact of the system on the peripheral transport network.

This project aims to conduct an impact analysis on GCLR Stage 2 on the surrounding transportation network in 2018 and 2031, in terms of the usership of the system. To this end, this project investigates the changing modal share between private and public transportation modes by considering the introduction of Stage 2 and increase of travel demand. Furthermore, this project also provides an overview of the application of transportation planning procedures and traffic simulation at the macroscopic level and its applicability to city planning, particularly for the proposed future stages of the GCLR. Additionally, the sufficiency of the Gold Coast transportation network is to be studied by examining whether the current roads, bus network and light rail system will have sufficient capacities to accommodate the future travel demand. Lastly, this study can provide recommendations on the travel demand forecasts and future infrastructure upgrades in the area.

2 Literature Review

2.1 Integrated Transportation System

Traffic congestion has a notably negative affect on both society and the environment. As a result of this, methods of congestion mitigation are employed during the planning of urban areas to address the potential effects. All stakeholders are adversely affected by traffic congestion, most notably road users, public transport operators and local residents [5]. While usually seen simply as a frustrating inconvenience, congestion also results in increased travels costs and a higher noise and environmental pollution [5, 19]. A commonly proposed solution to alleviate congestion, and the associated negative effects, is the introduction of public transportation systems to minimise the usership of private cars [5, 19]. Previous studies indicate that in order to effectively compete with private transportation, public transportation systems must function efficiently and not disadvantage users [5, 19]. Previous research indicates that the introduction of an IUTS is an effective approach to increase the effectiveness and appeal of public transport [5, 6, 19].

An integrated transportation system is one in which the public and private transportation within an urban area are well linked [5]. Transport integration focuses on connecting the multiple transport modes that operate in a system in order to facilitate passenger transfer between different modes and ensure efficient movement of travellers from origin to destination [19]. Within an urban space, integration is focussed on linking the well-developed road infrastructure of a suburban system to the dense public transport system within a city [5]. Integration can be achieved by designing systems that allow door-to-door transfers between modes, such as park and ride facilities and the assimilation of fare and ticketing systems between various modes and carriers [19]. An IUTS can be implemented to generate an overall improvement in the performance of the system, increased service quality level, improved accessibility and better environmental protection [5, 19]. One such example of an integrated transport system is the GCRT system. Additionally, Stage 2 incorporates two large capacity carparks for travellers to park and ride on the system. If the future stages of the GCLR system are constructed the light rail will also connect to the Gold Coast Airport at Coolangatta, furthering the integration [20].

To assess the effectiveness of Stage 2 of the GCLR in providing an integrated transport network that supports an efficient roadway network, this project conducted an impact analysis of the surrounding network focusing on modal split. As part of the 2031 Transport Strategy CGC has put forward the mode share targets. Their 2018 target is in line with current modal split determined through DTMR's household surveys in 2011. This impact analysis aims to reveal if implementing Stage 2 of the GCLR will be a significant contributor in reaching the 2031 modal split targets and as a result help relieve the demand on the private roadway network. TIA is a specialised engineering study employed to determine the potential traffic impacts resulting from a proposed traffic generator [21]. A number of studies have been undertaken to demonstrate the usefulness of undertaking a TIA in order to prevent conflict resulting from additional traffic generation [16, 17]. The application of TIA studies is primarily directed towards assessing the effects of proposed developments, redevelopments and land rezoning [16]. The primary focus of this study is assessing the effect of Stage 2 on the ability of the transportation network to support the forecasted travel demand.

2.2 Standard 4-Step Model

With the network model finalised, the subsequent process was modelling the transport demand and associated impacts. PTV Visum includes the standard Four-Step Model (FSM) for modelling forecasted travel demand, a robust method for the simulation of traffic flows [7, 18, 22]. This FSM was employed to simulate the traffic scenario for this study. While simplistic models exist for smaller networks, the complexity of large-scale regional models necessitated the development of the sequential FSM to determine network flows [23]. The FSM approach includes the steps of trip generation, trip distribution, mode choice and route choice, as shown in **Error! Reference source not found.** [18]. The 4 stages of the model can be calculated within Visum as part of the demand modelling procedure [18]. In order to discuss each step of the FSM these systems must first be defined. The study area is separated into traffic analysis zones that represent the origin and destination of travel trips. Zoning is based on land-use types and is the basis of the FSM. In addition to internal traffic analysis zones, external zones must also be setup to represent the trips into, out of and through the study area [22].

3 Model Validation

To confidently forecast the future traffic scenario it was important to first establish the accuracy of the base model. Due to the scope of the project, several assumptions were made in order to complete the modelling efficiently and within imposed constraints. The investigated transportation modes were limited to travel by car, bus, and light rail and excluded active transport modes such as walking and cycling. To include these additional modes would require a more detailed network to be constructed in Visum, exceeding the capacity of the trial version. Additionally, the lack of available traffic count or survey data necessitated calculating the trip volumes in the study area from a Trip Generation Rate Table, resulting in a lesser degree of accuracy. In order to minimise the introduction of errors resulting from these assumptions, a model validation process was undertaken. To ensure the 2018 base model was correctly calibrated to reflect the existing travel demand situation, the model results were verified against published traffic count and mode share statistics. The link traffic volumes were assessed against traffic census data published by DTMR and the modal share was verified against DTMR survey information regarding the current shares.

Several of the link volumes were verified against traffic count data published for the SCR network in 2016. The traffic count data released by DTMR is in the form of Annual Average Daily Traffic (AADT) along particular state-controlled road sections. In order to compare the AADT traffic census counts to the analysis period for the project, the AADT was evaluated for the peak hour period. DTMR assumes that between 10% and 12.5% of AADT occurs during the peak period for highways and urban dual carriageways [24]. Table 1 displays the verification of link traffic volumes within the study area at several key locations. It can be seen in Table 1 that the

Table 1. Trip Volume Validation 2018-1 Model [23]

Road	AADT	% Heavy vehicle	Adjusted AADT	Model trip volume	% of Adjusted AADT
Southport-Nerang Rd	35,285	8.63%	32,240	3,529	10.9%
Olsen Avenue	26,741	4.88%	25,436	3,035	11.9%
Brisbane Rd	33,582	6.69%	31,335	3,275	10.5%
Smith Street Mwy	67,307	6.06%	63,228	4,458	7.1%

modelled trip volumes are within 10% to 12.5% AADT for three of the four chosen locations. The traffic volumes for Smith Street Motorway are slightly underestimated in comparison with the adjusted AADT. This can be attributed to the manual trip generation process having a lesser degree of accuracy than the more robust method involving survey data.

In addition to verifying the trip volumes, the mode share for the base model was also assessed to ensure the model correctly reflected the current situation. Table 2 shows that the base 2018-1 model closely reflects the most recent mode share information available for the Gold Coast region. The modelled 3.8% is slightly lesser than the known 4% however it should be noted that the published mode share information relates to the entire Gold Coast region and not only the study area investigated in this project. This has important implications as the chosen study area is not located within the dense coastal city strip of the Gold Coast which attracts a greater amount of public transport use. This fact accounts for the marginally lower estimated mode share in the 2018-1 model and as such the mode share was determined to closely reflect the existing situation. As the base 2018-1 model results closely align to the real situation for both traffic volumes and mode share, the forecasted models can be assumed to also closely reflect the future situation.

Table 2. Mode Share Validation 2018-1 Model [4]

Source	Public transport mode share
2011 Household Travel Surveys	4%
2018 CGC Mode Share Target	4.1%
2018-1 Model	3.8%

4 Results and Analysis

To assess the impact of Stage 2 of the GCLR the results of the traffic simulation models were analysed. The impact of GCLR Stage 2 was assessed by comparing the model variants for 2018 and 2031 with and without the inclusion of Stage 2. Consideration was given to the modal share, trip volumes, congestion issues and the contribution of Stage 2 in achieving the target mode shares given by CGC.

4.1 2018 Travel Demand

In this paper, model 2018-1 and 2018-2 were produced to simulate the traffic situation before and after the introduction of GCLR Stage 2. The FSM procedure was applied in Visum to produce the modal OD matrices and resulting trip volumes on each link of the network. The modal OD matrices for each of the 2018 models were analysed to calculate the modal split. It should be noted that for the purpose of calculating the modal split internal zone trips have been excluded (Table 3).

Table 3. 2018 Modal split

Mode	Model 2018-1	Model 2018-2
Private Transport (PrT)	96.2%	92.6%
Public Transport (PuT)	3.8%	7.4%

For the base model 2018-1, where GCLR Stage 2 is not present, the market share of public transport is relatively low at 3.8%. With the addition of Stage 2 in model 2018-2, the PuT market share increases to 7.4%. This shows that the inclusion of GCLR Stage 2 increases the usage of public transport within the study area. The 2018 models indicate that Stage 2 successfully implements an integrated transport system that increases public transport use and decreases the demand of network.

To further investigate the effect of GCLR Stage 2 in decreasing the demand on the roadway network several areas of congestion were compared between model 2018-1 and 2018-2. Table 4 shows the percentage reduction of private transport usage of certain major roads within the study area. It can be seen that the capacity of several of these roads is exceeded by the estimated trip volumes for both models, showing there will be congestion issues. It can be seen that introducing GCLR Stage 2 has reduced the PrT trip volumes on these roads by between 3.6% and 13.2%. Despite the reduction in trip volumes the capacity of several of these roads is still exceeded with the introduction of Stage 2. This indicates that infrastructure upgrades will be required to mitigate congestion issues on the Gold Coast road network.

Table 4. 2018 PrT Trip Volume Reduction

Road Link	Capacity (trips/h)	2018-1 Trip volume (trips/h)	2018-2 Trip volume (trips/h)	Trip volume decrease
Southport-Nerang Rd	2,400	3,529	3,375	4.4%
Olsen Ave 1	3,300	2,530	2,434	3.8%
Olsen Ave 2	2,400	2,267	2,171	4.2%
Brisbane Rd	2,600	3,275	3,032	7.4%
Smith St	2,600	3,456	3,332	3.6%
Napper Rd	2,400	1,619	1,406	13.2%
Arundel Dr	1,100	635	568	10.6%
Wardoo St	2,400	2,957	2,792	5.6%

To further investigate the effect of GCLR Stage 2 in decreasing the demand on the roadway network several areas of congestion were compared between model 2018-1 and 2018-2. Table 4 shows the percentage reduction of private transport usage of certain major roads within the study area. It can be seen that the capacity of several of these roads is exceeded by the estimated trip volumes for both models, showing there will be congestion issues. It can be seen that introducing GCLR Stage 2 has reduced the PrT trip volumes on these roads by between 3.6% and 13.2%. Despite the reduction in trip volumes the capacity of several of these roads is still exceeded with the introduction of Stage 2. This indicates that infrastructure upgrades will be required to mitigate congestion issues on the Gold Coast road network.

4.2 2031 Travel Demand

Further to the investigation of the current travel demand, the model variants 2031-1 and 2031-2 were produced to simulate the forecasted traffic situation in 2031 before and after the introduction of GCLR Stage 2, respectively. These models replicate the physical network of 2018 and employ the increased 2031 travel demand.

Table 5. 2031 Modal Split

Mode	Model 2031-1	Model 2031-2
Private Transport (PrT)	85.4%	75.6%
Public Transport (PuT)	14.6%	24.4%

The modal OD matrices for each of the 2031 models were analysed to calculate the modal split shown in Table 5. Again, the internal zone trips were excluded for the purpose of calculating this modal split. Table 5 shows that with the increased travel demand the public transport mode share increases to 14.6%, and further to 24.4% with the addition of GCLR Stage 2. This is a substantial increase from 3.8% and 7.4% in the equivalent 2018 scenarios. This implies that the current roadway configuration would not be independently sufficient to cope with the travel demand anticipated for 2031. The increase in PuT market share between model 2031-1 and 2031-2 implies that GCLR Stage 2 provides an appealing travel choice in the loaded network.

The follow-on effect of increased PuT travel is a decrease in PrT travel volumes. As discussed throughout Sect. 2 of this report, road congestion is greatly decreased with a shift from private to public travel. To further investigate the effect of GCLR Stage 2 in decreasing the roadway demand in 2031, congested link sections were investigated between model 2031-1 and 2031-2. Table 6 shows the percentage reduction of private transport usage for the major roads investigated in Table 4 for the 2031 travel demand.

The added travel demand in 2031 resulted in the addition of GCLR Stage 2 having a greater effect on the decrease of PrT trip volumes. Introducing Stage 2 caused a decrease in PrT travel volumes between 7.6% and 61.2%. Despite this considerable reduction in trip volumes, the capacity of several of these roads is still exceeded after the introduction of Stage 2. However, comparing these results to 2018 shows an improvement in the volume of PrT travel even with the significant increase in travel

Table 6. 2031 PrT Trip Volume Reduction

Road Link	Capacity (trips/h)	2031-1 Trip volume (trips/h)	2031-2 Trip volume (trips/h)	Trip volume decrease
Southport-Nerang Rd	2,400	3,307	2,659	19.6%
Olsen Avenue	3,300	2,710	2,405	11.3%
Olsen Avenue	2,400	2,210	1,934	12.5%
Brisbane Rd	2,600	2,940	1,941	34.0%
Smith Street Mwy	2,600	3,524	2,941	16.5%
Napper Rd	2,400	1,229	477	61.2%
Arundel Dve	1,100	626	365	41.7%
Wardoo Street	2,400	2,916	2,693	7.6%

demand. This strongly indicates that GCLR Stage 2 provides a significant improvement to the transportation network and will reduce the need for roadway upgrades in the future.

4.3 Public Transport Mode Share Target

The total equivalent public mode share target for 2031 is 26% for public transport and remains at 74% for cars. Table 5 shows that a public transport mode share of 24.4% is possible for 2031 with the inclusion of GCLR Stage 2. According to the results, without Stage 2 a public transport mode share of only 14.6% would be possible. This shows that GCLR Stage 2 is a major contributor to reaching the target mode share set by CGC for 2031. This indicates that further expansion of the light rail system should assist in reaching the even greater public transport mode share for 2040.

5 Conclusions and Recommendations

A traffic impact analysis was conducted to assess the effects of introducing Stage 2 of the Gold Coast Light Rail on the current transportation network. To determine the anticipated impact a traffic simulation model was created to mimic the travel behaviour of system users. The traffic simulation was created in PTV Visum using the Four-Step Model to model the travel demand. The results of the simulations were analysed and compared to determine the impact of GCLR Stage 2 on the peripheral network. The impact was assessed by investigating the modal share, trip volumes and congestion issues of the four model variants.

The results show that the implementation of the GCLR supports the desired increase in public transport use. This indicates that further expansion of the light rail system will assist in reaching the modal split targets set for 2040. It showed that Stage 2 decreases the utility costs associated with public transport and affirms the effectiveness of GCLR in implementing an integrated transport network that decreases traffic impacts normally associated with increased travel demands. The trip volume results also revealed some significant issues with the roadway network capacity in both 2018 and 2031.

While Stage 2 was shown to decrease the usage of private travel and consequently the demand on the roadway network, several major roads within the study area showed trip volumes that exceed the current capacity and will experience congestion related issues. This was supported by the large shift in private to public transport between the 2018 and 2031, explained by the increasing utility of private travel associated with congestion delays.

References

1. Farahani, R.Z., Miandoabchi, E., Szeto, W.Y., Rashidi, H.: A review of urban transportation network design problems. Eur. J. Oper. Res. **229**, 281–302 (2013)
2. Infrastructure Australia 2016. Australian Infrastructure Plan
3. James, M.: Transport Infrastructure. Parliamentary Library (2013). http://www.aph.gov.au/About_Parliament/Parliamentary_Departments/Parliamentary_Library/pubs/BriefingBook44p/Transport. Accessed 1 Sept 2017
4. GCCC: Gold Coast City Transport Strategy 2031 Technical report (2013)
5. Fierek, S., Zak, J.: Planning of an integrated urban transportation system based on macro–simulation and MCDM/A methods. Procedia Soc. Behav. Sci. **54**, 567–579 (2012)
6. Meng, L., Taylor, M.A., Scrafton, D.: Combining latent class models and gis models for integrated transport and land use planning–a case study application. Urban Policy Res. **34**, 305–329 (2016)
7. Solecka, K., Żak, J.: Integration of the urban public transportation system with the application of traffic simulation. Transp. Res. Procedia **3**, 259–268 (2014)
8. DTMR: Gold Coast Light Rail Stage 2 Newsletter (2016)
9. GCCC: Gold Coast Light Rail Feasability Study Summary Report (2004)
10. Queensland Government Department of Infrastructure Local Government and Planning. Gold Coast Health and Knowledge Precinct (2017) http://www.edq.qld.gov.au/economic-development-queensland/parklands-health-and-knowledge-precinct.html. Accessed 13 Mar 2017
11. Qu, X., Zhang, J., Wang, S.: On the stochastic fundamental diagram for freeway traffic: model development, analytical properties, validation, and extensive applications. Transp. Res. Part B **104**, 256–271 (2017)
12. Kuang, Y., Qu, X., Wang, S.: A tree-structured crash surrogate measure for freeways. Accid. Anal. Prev. **77**, 137–148 (2015)
13. Zhou, M., Qu, X., Li, X.: A recurrent neural network based microscopic car following model to predict traffic oscillation. Transp. Res. Part C **84**, 245–264 (2017)
14. Wang, S., Qu, X.: Station choice for Australian commuter rail lines: equilibrium and optimal fare design. Eur. J. Oper. Res. **258**(1), 144–154 (2017)
15. Bie, Y., Cheng, S., Easa, S., Qu, X.: Stop Line set back at a signalized roundabout: a new concept for traffic operations. J. Transp. Eng. ASCE **142**(3), 05016001 (2016)
16. Ponnurangam, P., Umadevi, G.: Traffic Impact Analysis (TIA) for Chennai IT Corridor. Transp. Res. Procedia **17**, 234–243 (2016)
17. Xiao, X.Q.: A study on the major problems of urban traffic impact analysis. In: Applied Mechanics and Materials, pp. 2619–2622. Trans Tech Publications (2012)
18. PTV AG, PTV Visum 16 Manual. Karlsruhe (2016)
19. Li, L., Loo, B.P.: Towards people-centered integrated transport: a case study of Shanghai Hongqiao Comprehensive Transport Hub. Cities **58**, 50–58 (2016)
20. GCCC: Gold Coast Rapid Transit Corridor Study (2011)

21. Yulianto, B., Setiono: Web application and database modeling of traffic impact analysis using Google Maps. In: AIP Conference Proceedings, p. 060002. AIP Publishing (2017)
22. McNally, M.G.: The Four-Step Model. Handbook of Transport Modelling, 2nd edn. Emerald Group Publishing Limited (2007)
23. DTMR: 2016 Traffic Census Data - Google Earth (2017)
24. DTMR: Cost-benefit Analysis Manual Road Projects (2011)

A Crowdsourcing Matching and Pricing Strategy in Urban Distribution System

Xin Lin[1], Yu-hang Chen[2], Lu Zhen[1(✉)], Zhi-hong Jin[3], and Zhan Bian[4]

[1] School of Management, Shanghai University, Shanghai 200044, China
linx_dlmu@163.com, lzhen@shu.edu.cn
[2] School of Information Science and Technology, Dalian Maritime University, Dalian 116026, China
444658025@qq.com
[3] School of Transportation Engineering, Dalian Maritime University, Dalian 116026, China
jinzhihong@dlmu.edu.cn
[4] School of Business Administration, Capital University of Economics and Business, Beijing 100070, China
bianzhan1990@163.com

Abstract. The vigorous development of O2O e-commerce promote the appearance of many small orders, increasing the stresses on logistics operators to carry out city distribution. However, a crowdsourcing joined to release these stresses is a new try and become more popular. This paper focuses on matching of the crowds and tasks from crowdsourcing platform for city distribution. To address exploring the impact of time, space and efficiency on the task matching in crowdsourcing platform, a bi-objective matching and differentiated pricing model creates to achieve the highest efficiency and the lowest total cost in urban distribution system. For solving the model, a two-dimensional and multi-stage roulette algorithm has been designed, with combining modeling and simulation method. The proposed method takes full use of economic development of the region where the task is located and the space-time efficient distance and space-time reachable distance. To illustrate the effectiveness and validity of the proposed method, a sample test is conducted with the actual operating data of a company in the PRD region, and the results show that 719 tasks out of 746 matching pairs are executed, the task matching rate is 89.4%, the completion rate is 86.1% and the total task price 39140 RMBs. Compared with the original matching situation, the total price increases at 3.90%, while the task completion rate is improved at 37.7%, which greatly enhance the efficiency of crowdsourcing platform member matching. The matching of the participants and tasks of the city distribution crowdsourcing platform, which combines with the measurement of differential pricing and crowds' credibility, can be applied in this problem successfully.

Keywords: City distribution · Crowdsourcing match
Space-time efficiency pricing · Two-dimensional multi-stage roulette algorithm

G. De Pietro et al. (Eds.): KES-IIMSS-18 2018, SIST 98, pp. 407–417, 2019.
https://doi.org/10.1007/978-3-319-92231-7_42

1 Introduction

With the continuous development of e-commerce, O2O transactions are gradually accepted by citizens. Based on it, city distribution service providers usually busy in more and more online orders with working stresses increasing. Therefore, many logistics operators are trying to find outsourcing services to replace them to solve this problem, so as to deal with the cost of spill losses (caused by insufficient loaded, backhaul, time window broken, etc.) and uncertainty of customers' requirements [1].

Crowd logistics, a crowdsourcing-based processing model, may be a direction to alleviate the problem of fewer batches in bulks in cities [2]. The crowdsourcing platform is able to release the discrete distribution tasks to the public via mobile APPs and to promote the competition among task receivers by differentiated pricing, thereby ensuring high-quality city distribution services as much as possible [3].

At present, the concept of crowdsourcing has been applied to many fields of transportation and logistics industry. For example, in some areas, crowds of taxis for last mile delivery has been launched [4]; large e-commerce companies will put forward some reward policy to encourage customers to take some additional packages for their neighbors when picking up their own packages [5]. But to the best of our knowledge, the research results of systematic modeling and simulation optimization are still lacking [6].

Therefore, this paper starts with designing a pricing mechanism that can motivate the crowd to participate enthusiastically, enabling them to competitively select city distribution tasks and finalize the task allocation plan after measuring the reliability of their tasks. Thereby achieving the effect of quality and quantity assurance while reducing the delivery cost as much as possible so as to promote a win-win situation for all parties. Combined with the concept of modeling and simulation, this paper constructs a model of city distribution task pricing based on space-time efficiency and designs a two-dimensional Multi-Stage Heuristic Roulette algorithm for numerical solution of real cases from three aspects of time, space and efficiency.

2 Problem Description

The operation of crowdsourcing platform for city distribution in O2O mode needs to consider the mutual benefits of logistics operators, crowds and customers [7], and they belong to a two-phase optimization problem that includes task matching and vehicle routing planning, as shown in Fig. 1.

After downloading an APP, every citizen could join to select their interest tasks. By the side, the platform working to matching as Fig. 1(a) shows, which enables the operation of the platform by pushing distribution tasks that are geographically appropriate to the crowds and paying their gratuities after they have completed the delivery tasks. This pricing match presents a "many-to-one" relationship, and the popularity of each distribution task depends on its distance from each of the participants and the pricing of tasks. It is the first phase in urban distribution system.

The second phase requires the planning of distribution route for each crowd. Taking a private delivery route as an example, as Fig. 1(b) shows, the participant firstly arrives at the nearest city distribution center to pick up the goods and issues the goods to the

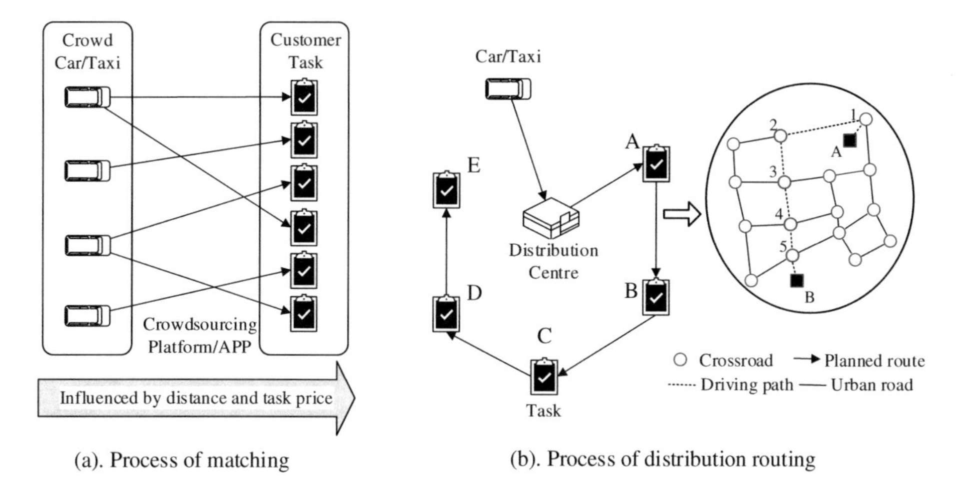

(a). Process of matching (b). Process of distribution routing

Fig. 1. Two stages of urban distribution crowdsourcing

customer successively on the way, which forming an "Origin-DC-ABCDE" open route. In each route, we need to consider the structural characteristics of urban road network and traffic environment, and find the most economical driving route between single source points. Such as the participants from customer A to customer B, go through the crossroads ("1-2-3-4-5") in turn. At present, corresponding solutions have been found for this stage of problem [8]. Therefore, this paper aims to explore the first phase of the task matching solution strategy.

The following three concepts need to be clarified in the matching process of distribution tasks: (1) Participant's credibility: Because of the large scale of the crowds, the customers may be worried that their distribution information will be leaked by the relevant people [9], therefore, this concept is introduced to measure the behavior reliability of the participants; (2) Participant's capacity: Consistent with the process of

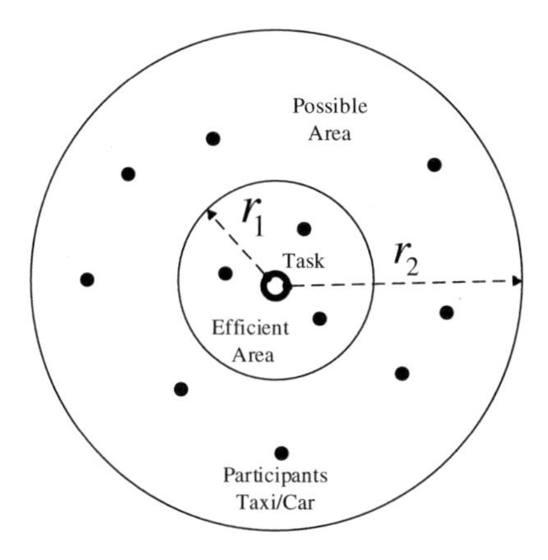

Fig. 2. Analysis of a single task surroundings

appointment and matching of other industries [10], crowdsourcing will also investigate the distribution capacity of every participant, and set the appropriate quota; (3) Differential pricing: In order to stimulate the participation of encouraging the crowds to complete the task and never give up when meeting some difficulties, the differentiated pricing should be designed [11]. Therefore, during the process of bilateral selection of the crowds and tasks, two trends can be found: (1) In the process of the participants choosing the tasks, the higher the price and the closer the tasks are, the more popular the tasks are; (2) In the process of the tasks matching, the successful of delivery depends on the credibility of each participant and scale of crowds in the task surrounding. Better credibility and more selections, Higher quality of delivery.

After the analysis of a single task, we can find that distribution of participants shown in Fig. 2 around it, which will greatly affect the efficiency of delivery tasks after the task is selected and matched by the participants. In general, the higher the number of participants with high credit and high quotas gathered around the task, the greater the probability that they will be successfully executed; otherwise, the smaller the probability. Refer to the concept of matching space-time normalized distance [12], in order for the task to be successfully implemented by participants as much as possible, following the space-time efficient distance r_1 and space-time reachable distance r_2, $r_1 \leq r_2$, a range of participants can be categorized and discussed. In the formulation of differentiated pricing scheme, the comprehensive effect on delivery effectiveness by the following three dimensions need considering: (1) Time dimension: the distance between the participants and the task influences the time spent by participants in reaching the task point and affects the willingness of the matched participants to implement; (2) Spatial dimension: the more participants with high credibility gathered in the space near the task, the greater the probability that the task will be executed successfully. (3) Efficiency dimension: When making system differentiation pricing, we should give full consideration to the reputation of participants affected by the start time and quota quantification of tasks in order to obtain high-quality matching pairs. In addition, the pricing of each task should take into account the economic development in the region (in this case, the economic development data of the region is shown in Table 1), we define the economic development coefficient K_i, $i = 1, 2, \ldots, 6$, which can be obtained by weighted average of Per capita GDP Q_i by Eq. (1).

$$K_i = 6Q_i \Big/ \sum_{i=1}^{6} Q_i \quad (1)$$

Table 1. Coefficient of regional economic development

Cluster (i)	1-Foshan	2-Guangzhou	3-Panyu	4-Dongguan	5-Bao'an	6-Longhua
Per capita GDP(Q_i)	108888	106480	104189	75213	96507	114018
Economical coefficient (K_i)	1.08	1.06	1.03	0.75	0.96	1.13

Data Sources: "Statistical Yearbook of Guangdong Province in 2015"

3 Model of Matching and Pricing

3.1 Hypothesis

(1) After pricing minded differentiation rules, each task price should keep constant in the process of matching.
(2) Only does task completion rate calculates based on current decision result. Re-matching or multi-matching result will not be considered in it.
(3) As a body of matching, the quality of each participant will not be disrupted by any other external factors except the inners, such as credibility, start booking time and capacity.

3.2 Variables and Parameters

Variables:

$$x_{ij} = \begin{cases} 1, & \text{task } i \text{ is matched to participant } j \\ 0, & \text{others} \end{cases}$$

$$y_{ij} = \begin{cases} 1, & \text{task } i \text{ is performed by participant } j \\ 0, & \text{others} \end{cases}$$

Parameters:

A: Set of tasks. Task i contains task serial number a_i and latitude and longitude coordinates $\left(\alpha_i^A, \beta_i^A\right)$, $i = 1, 2, \ldots, |A|$.

B: Set of crowds. Participant j contains participant serial number b_j and latitude and longitude coordinates $\left(\alpha_j^B, \beta_j^B\right)$, $j = 1, 2, \ldots, |B|$.

P_i: Differentiation price of task i, $i = 1, 2, \ldots, |A|$.

Po_i: Original price of task i, $i = 1, 2, \ldots, |A|$.

K_i: Economic development coefficient of task i. It is clustered by the geographical location of task i and its surroundings, $i = 1, 2, \ldots, |A|$.

γ_1: Comprehensive matching rate. It is the proportion of matched tasks in all.

γ_2: Performed rate. It is the proportion of successful performed tasks in matched group.

C_j: Capacity of participant j, $j = 1, 2, \ldots, |B|$.

δ_j: Reliability of participant j, $j = 1, 2, \ldots, |B|$.

R_i^1: Set of time-space efficient participants, where

$$D_{ij} = \sqrt{\left(\alpha_i^A - \alpha_j^B\right)^2 + \left(\beta_i^A - \beta_j^B\right)^2} \leq r_1, j \in R_i^1$$

R_i^2: Set of time-space possible arrived participants, where $r_1 < D_{ij} \leq r_2$, $j \in R_i^2$;

$\bar{\delta}$: Average reliability in crowds. It is weighted average by capacity and reliability of each participant, where $\bar{\delta} = \sum_{j=1}^{|B|} \delta_j C_j \Big/ \sum_{j=1}^{|B|} C_j$

3.3 Mathematical Model

$$\min \frac{1}{\gamma_1 \gamma_2} \sum_{i=1}^{|A|} \sum_{j=1}^{|B|} P_i x_{ij} \tag{2}$$

$$\gamma_1 = \frac{1}{|A|} \sum_{i=1}^{|A|} \sum_{j=1}^{|B|} x_{ij} \tag{3}$$

$$\gamma_2 = \sum_{i=1}^{|A|} \sum_{j=1}^{|B|} y_{ij} \Big/ \sum_{i=1}^{|A|} \sum_{j=1}^{|B|} x_{ij} \tag{4}$$

$$P_i = \frac{K_i Po_i}{r_2} \left(\frac{r_1 |R_i^1|}{|A|} - \frac{r_2 |R_i^2|}{|A|} \right), \ i = 1, 2, \ldots, |A| \tag{5}$$

$$s.t. \sum_{j=1}^{j=|B|} x_{ij} \leq 1 \ i = 1, 2, \ldots, |A| \tag{6}$$

$$\sum_{i=1}^{i=|A|} x_{ij} \leq C_j j{=}1, 2, \ldots, |B| \tag{7}$$

$$y_{ij} = \begin{cases} x_{ij}, & \delta_j \geq \bar{\delta} \\ x_{ij} \left\lceil \frac{\bar{\delta} Rnd(0,1)}{\delta_j} \right\rceil, & \delta_j < \bar{\delta} \end{cases} \ i{=}1, 2, \ldots, |A|, j = 1, 2, \ldots, |B| \tag{8}$$

Equation (2) is the objective to minimize generalized cost, which contains the maximize of multiplication of comprehensive matching rate and the bottom of performed rate and minimize of the total cost in urban distribution. Equation (3) calculates comprehensive matching rate; Eq. (4) calculates performed rate; Eq. (5) is task differentiated pricing function, where is influenced by economic development coefficient and estimated service distance (assessed by efficient distance and possible arrived distance) of each task. Eqs. (6) to (9) are all constraints. Equation (6) express a task should only be matched to a participant; Eq. (7) express the scale of matched tasks should not over its participant's capacity; Eq. (7) express the logical relationship between matched variable and performed variable. If reliability of such matched participant is exceed average reliability in crowds, its task could be recognized as performed; However, to create a random between 0 and 1 and then compare it with the proportion of reliability and the average, when over it unperformed, otherwise performed.

4 Two-Dimensional Multi-stage Heuristic Roulette Algorithm

To find a participant for each task, we design the MHR algorithm which based on the space-time efficient distance r_1 and space-time reachable distance r_2. Since the crowdsourcing system gives the beginning booking time t_j^B and capacity C_j which referred to the credibility of each participant when the tasks start, therefore, the design of MHR algorithm can be reduced to a time-step knapsack problem. Since the distance constraint are limited by r_1 and r_2, the scale of the alternative crowd gatherings will be effectively reduced, where $R_i^1 \cap R_i^2 = \emptyset$ and $R_i^1 \cup R_i^2 = R_i$. Since the elements in R_i are obtained through traversal from all the participants, the sum of the probabilities of all participants in the final decision is 1 during the multi-stage roulette process. Therefore, in large-scale calculation, this algorithm can effectively obtain feasible solution, and compared with the global traversal algorithm, this algorithm can reduce the time complexity more significantly [13, 19].

(1) **The main algorithm implementation steps**

To match the task set $A\{a_i, (\alpha_i^A, \beta_i^A), P_i\}$ and the crowd set $B\{b_j, (\alpha_j^B, \beta_j^B), C_j, t_j^B, q_j^B\}$, we can get the matching solution $F_i\{a_i, b_j\}$ for each task. If a task b_j=*null*, the decision variables x_{ij}=0; otherwise, x_{ij}= 1.

In order to make the alternative crowd gatherings have at least one high-degree-credibility participant, we calculate the space-time efficient distance:

$$r_1(i) = \min_{j \in B}\{D_{ij}\}, j = 1, 2, \ldots, |B| \tag{9}$$

To ensure that at least one alternative participates in each task, pessimistic criteria are used to calculate the space-time reachable distance:

$$r_2(i) = \max_{i \in A} \min_{j \in B}\{D_{ij}\} \tag{10}$$

The specific steps of the algorithm are as follows:

Step 1: Initialize $i = 1$;
Step 2: Select a_i, initialize $j = 1$;
Step 3: Select b_j, if $D_{ij} \le \min\{r_1(i), r_2(i)\}$, then save b_j into R_i^1, $j = j + 1$; Otherwise, $j = j + 1$;
Step 4: If $j > |B|$? If so, go to **Step 5**; Otherwise, return to **Step 3**;
Step 5: Call roulette, if x_{ij}= 1 then $R_i = R_i^1, C_j = C_j - 1, i = i + 1$, go to **Step 9**; Otherwise, go to **Step 6**;
Step 6: Select b_j, if $D_{ij} \le \max\{r_1(i), r_2(i)\}$, then save b_j into R_i^2, $j = j + 1$, Otherwise, $j = j + 1$;
Step 7: If $j > |B|$? If so, go to **Step 8**; Otherwise, return to **Step 6**;
Step 8: Call roulette and get R_i^2, $R_i = R_i^2$, $i = i + 1$;
Step 9: If $i > |A|$?If so, End the match and export all R_i; Otherwise, return to **Step 2**.

(2) **Multi-stage roulette rules**

In the case of full consideration of "each participant with the beginning booking time of the same task belong to the same priority batch" and "the probability of each participant in the same batch selects mission depends on the proportion of reservation capacity", we design the roulette rules with complete information sharing in the batch. Analyzing each element $r_{jn}\left(b_{jn}, \alpha^B_{b_{jn}}, \beta^B_{b_{jn}}, C_{b_{jn}}, t^B_{b_{jn}}\right)$ in R_i, $n = 1, 2, \cdots, |R_i|$, we divide participants into $R_i^{T(1)}$, $R_i^{T(2)}$,..., $R_i^{T(m)}$, $\bigcup\limits_{j=1}^{j=m} R_i^{T(j)} = R_i$. The crowds in the same set have the same t_j^B, Which is $\forall j \in R_i^{T(m)}, T(m) = t_j^B$. Based on the proposed differentiated pricing criteria, the probability of each participant selecting a task is proportional to the ratio of differentiated prices P_i to the original platform price Po_i.The probability $\Pr(b_{jn})$ of the participant b_{jn} being selected for the task and the unselected probability $\Delta\Pr(m)$ for each batch can be calculated:

$$\Pr(b_{jn}) = \frac{C_{b_{jn}}}{\sum\limits_{n=1}^{n=|R_i|} C_{b_{jn}}} \times \frac{P_i}{Po_i}, \Delta\Pr(m) = 1 - \sum_{k=1}^{k=m}\sum_{n=1}^{n=|T_m|} \Pr\left(b_{jn}^k\right)$$

The flow chart of MHR algorithm and its probability expressions are shown in Fig. 3.

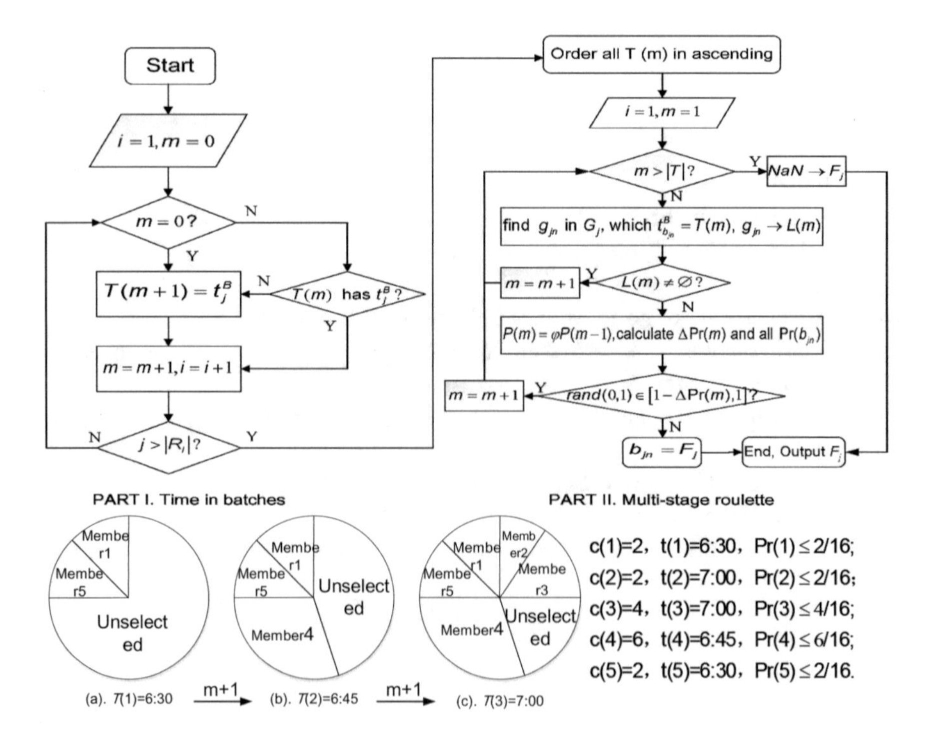

Fig. 3. Algorithm progress and rule of roulette

5 Numerical Result

Collecting a real operating data form a urban distribution company located in Pearl River Delta as case study, we code the algorithm in Matlab2015b and test data in a desktop (Inter Core i7-6700, 4 GB, Windows 7). Through a sensitivity analysis of distance parameters (Fig. 4), we found that the max possible arrived distance is 1 unit of latitude and longitude distance. To observe the differentiation of tasks' price, we changed the value of r_1 and r_2 between 0 and 1 and found when $r_1 = 0.50$ and $r_2 = 0.96$, the total cost of urban distribution could reach the lowest (Fig. 5).

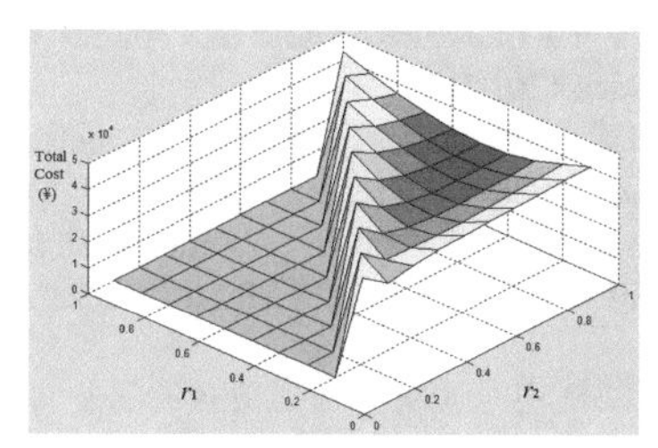

Fig. 4. Sensitivity of distance parameters

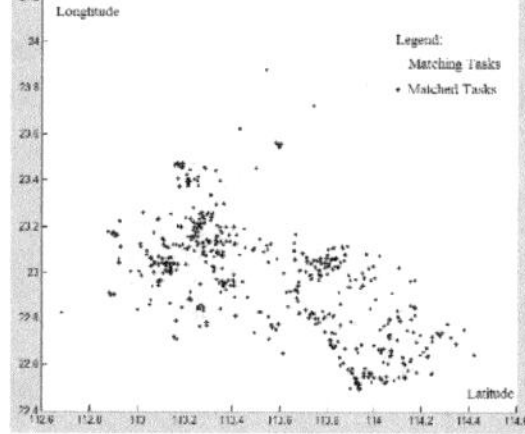

Fig. 5. Result of MHR algorithm

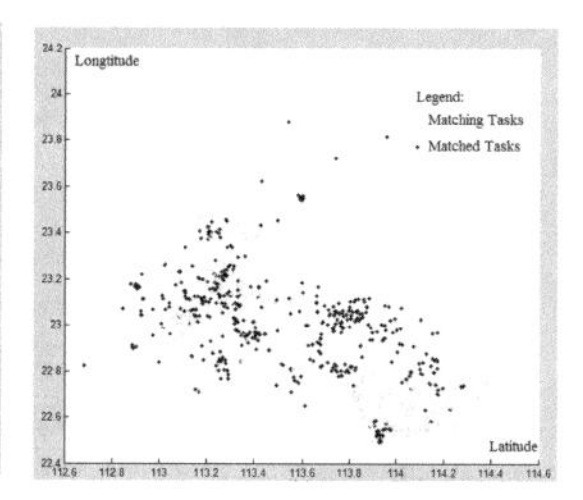

Fig. 6. Result of reality matches

The result of MHR algorithm calculated shows there has 746 matched groups, its total cost reached to 43591 RMBs under 89.4% comprehensive matching rate. Through a further calculate of Eq. (3), there has 719 performed matched groups, its total cost reached to 39140 RMBs under 86.1% performed rate. To compared with the usual operation result from the urban distribution company (522 matched groups in total cost of 37672 RMBs), MHR algorithm improves 37.7% performed rate with the limited increasing of 3.90% total cost. It could be a practical strategy in solving the urban distribution crowdsourcing problem with balancing the profit between crowds, customers and company (Fig. 6).

6 Conclusions and Future Research

The research work in this paper shows that with the help of MHR algorithm, by the method of integrating the measurement of differential pricing and crowds' credibility into the matching process between the tasks and participants in the crowdsourcing platform of city distribution, we can obtain matching programs which have higher quality and lower cost than the completely free matching. Future research will subdivide various factors affecting pricing, put forward more task matching strategies (such as packaging, discount, etc.), and make it link with the second-phase route optimization problems and applied to the software design of a type of APP using in city distribution crowdsourcing.

Acknowledgements. This study was supported by Research Fund for the National Natural Science of China (Grant Nos. 71572023, 71302044, 71431001, 71602130), EC-China Research Network on Integrated Container Supply Chains (Grant No. 612546), and the Fundamental Research Funds for the Central Universities (Grant No. 3132016301).

If you need the original data, please email 'linx_dlmu @ 163.com'

References

1. Slabinac, M.: Innovative solutions for a "Last-Mile" delivery–a European experience. In: Business Logistics in Modern Management, p. 15 (2014)
2. Mehmann, J., Frehe, V., Teuteberg, F.: Crowd logistics – a literature review and maturity model. In: Hamburg International Conference of Logistics (2015)
3. Lu, X.Y., Long, D.Z., Chen, Y.: The analysis and empirical research of influencing factors on users' loyalty in the model of crowdsourcing. Chin. J. Manage. **13**(7), 1038–1044 (2016)
4. Chen, C., Pan, S.: Using the crowd of taxis to last mile delivery in e-commerce: a methodological research. In: Service Orientation in Holonic and Multi-agent Manufacturing. Springer International Publishing, pp. 1–9 (2016)
5. Devari, A., Nikolaev, A.G., He, Q.: Crowdsourcing the last mile delivery of online orders by exploiting the social networks of retail store customers. Transp. Res. Part E Logist. Transp. Rev. **105**, 105–122 (2017)
6. Suh, K., Smith, T., Linhoff, M.: Leveraging socially networked mobile ICT platforms for the last-mile delivery problem. Environ. Sci. Technol. **46**(17), 9481–9490 (2012)
7. Song, T.S., Tong, Y.X., Wang, L.B., Xu, K.: Online task assignment for three types of objects under spatial crowdsourcing environment. Ruan Jian Xue Bao/J. Softw. **28**(3), 611–630 (2017). (in Chinese)
8. Lin, X., Geng, F., Jin, Z., et al.: The optimization of delivery vehicle scheduling considering the actual road network factors. In: International Conference on Logistics, Informatics and Service Sciences, pp. 1–7. IEEE (2017)
9. Varshney, L.R., Vempaty, A., Varshney, P.K.: Assuring privacy and reliability in crowdsourcing with coding. In: Information Theory and Applications Workshop, pp. 1–6. IEEE (2014)
10. Li, N., Chen, G., Govindan, K., et al.: Disruption management for truck appointment system at a container terminal: a green initiative. Transp. Res. Part D (2015)
11. Varshney, L.R.: Privacy and reliability in crowdsourcing service delivery, pp. 55–60 (2012)
12. Yue, Q., Fan, Z.P.: Decision method for two-sided matching based on cumulative prospect theory. J. Syst. Eng. **28**(1), 38–46 (2013). (in Chinese)
13. Jin, Z.H., Ji, M.J.: Practical Optimization Techniques for Logistics. China Material Press (2008). (in Chinese)
14. Zhen, L., Xu, Z., Wang, K., Ding, Y.: Multi-period yard template planning in container terminals. Transp. Res. Part B **93**, 700–719 (2016)
15. Zhen, L.: Modeling of yard congestion and optimization of yard template in container ports. Transp. Res. Part B **90**, 83–104 (2016)
16. Zhen, L., Wang, K.: A stochastic programming model for multi-product oriented multi-channel component replenishment. Comput. Oper. Res. **60**, 79–90 (2015)

17. Qu, X., Zhang, J., Wang, S.: On the stochastic fundamental diagram for freeway traffic: model development, analytical properties, validation, and extensive applications. Transp. Res. Part B **104**, 256–271 (2017)
18. Liu, Z., Yan, Y., Qu, X., Zhang, Y.: Bus stop-skipping scheme with random travel time. Transp. Res. Part C **35**, 46–56 (2013)
19. Qu, X., Wang, S., Zhang, J.: On the fundamental diagram for freeway traffic: a novel calibration approach for single-regime models. Transp. Res. Part B **73**, 91–102 (2015)

Initial Classification Algorithm for Pavement Distress Images Using Features Fusion

Zhigang Xu[1], Yanli Che[1], Haigen Min[1], Zhongren Wang[2], and Xiangmo Zhao[1](✉)

[1] School of Information Engineering, Chang'an University, Xi'an, Shaanxi, People's Republic of China
xmzhao@chd.edu.cn
[2] California Department of Transportation, Sacramento, CA 958833, USA

Abstract. In this paper, a novel two-staged pavement image processing framework is presented. The pavement images are classified into four general categories in the first stage, so that the images can be processed using category-specific algorithms in the 2nd stage. The proposed algorithm first fuses a local contrast enhanced image with a global grayscale corrected image to obtain an enhanced distressed pavement image. The enhanced image is then decomposed with a three-layer wavelet transform to obtain three texture features of the entire image including High-Amplitude Wavelet Coefficient Percentage (HAWCP), the High-Frequency Energy Percentage (HFEP), and the Standard Deviation (STD). In the meantime, an improved P-tile method is used to obtain the binary image. From the binary image, three additional shape features are extracted including the Average Area of all Connected Components (AA), the Area of the Maximum Connected Component (AM), and the Equivalent Length of the longest Connected Component (EL). Finally, a BP neural network is used to fuse both the texture and shape features sequentially to achieve the initial classification. Experimental results show that for the four types of pavement images, the proposed algorithm achieves an effective classification of the pavement distress image with the accuracy rates of 96.5%, 91.4%, 95.2% and 98.1% respectively, which are higher than those of the classification algorithm with a single-type feature.

1 Introduction

A complete and accurate roadway asset inventory is vital for data reporting, data analysis and decision support of daily management of road maintenance departments [1]. The inventory generally includes the status of the roadway assets, road geometry, pavement diseases and road roughness.

In recent years, the automated acquisition of pavement images has been improved rapidly, along with LASER and the high-speed and high resolution imaging techniques being applied. Currently, most commercially available APCS can collect shadow-free images with uniform illumination [2]. However, due to the complexity of pavement texture, the diverse characteristics and the feebleness of distress objects and the

G. De Pietro et al. (Eds.): KES-IIMSS-18 2018, SIST 98, pp. 418–427, 2019.
https://doi.org/10.1007/978-3-319-92231-7_43

presence of other external objects, professionals face huge challenges to develop a pavement distresses identification algorithm with high recognition accuracy, high adaptability and processing speed. In practice, most pavement image identification work is done by a combined method of artificial aid and automated recognition, which hinders the APCS being used widely. In California Department of Transportation (Caltrans), this problem also reduces the efficiency of pavement survey.

Generally, a complete identification process for pavement distresses includes four steps: 1. preprocessing of the pavement image; 2. segmentation of distress objects; 3. geometry measure for distress objects; 4. classification and evaluation for distress objects.

The segmentation of the distress objects is the critical step which directly determines the quality and efficiency of the classification and evaluation of the pavement images. In the past 30 years, intensive literatures focused on different image segmentation methods, which could be classified into five categories: 1. Thresholding-based segmentation [3]; 2. Segmentation based on edge detection [4]; 3. Segmentation based on multi-scale analysis [5–8]; 4. Segmentation based on graph theory [9–12]; 5. Segmentation based on Fractal geometric theory [13].

All the algorithms discussed above have a common assumption that there must be distress objects in the pavement images to be processed and the type of the distresses is known. Hence, plenty of the prior knowledge can be used to detect the distress objects in these algorithms.

Figure 1 shows eight groups of pavement images with the most common types: (a–b) Intact pavement; (c–d) Pothole distress; (e–f) Cracks with strong contrast; (g–h) Cracks with weak contras; (i–j) Alligator cracks; (k–l) Pavement seams; (m–n) Patches; (o–p) Road marking. These images are all collected by APCS. The resolution of each images are 1024×1024 pixels corresponds to a pavement area of about 40×40 in^2.

Through the experiments on pavement image collecting and statistical analysis, it can be found that most images captured by APCS are intact with good condition. Only a small portion of pavement images have distress objects. Generally, this portion is less than 20% and with diverse distress types. It is difficult to detect all types of distress objects with one single algorithm. For the pavement images with unknown distress type, if all the prepared algorithms are applied on the input image in a serial sequence, it will consume huge processing time [14].

To improve the efficiency of pavement image processing, this paper proposes a novel method of "classification before recognition". Firstly the pavement images are qualitatively analysed with a low-complexity algorithm. The input pavement images will be classified into the mentioned four categories before the detection of the distresses. After the classification, the specific algorithm will be applied to extract the distress objects. Finally, the distress objects will be measured and evaluated according to the criteria in the PMS protocols. This new procedure will convert the conventional serial process flow to a parallel flow and shorten the processing time.

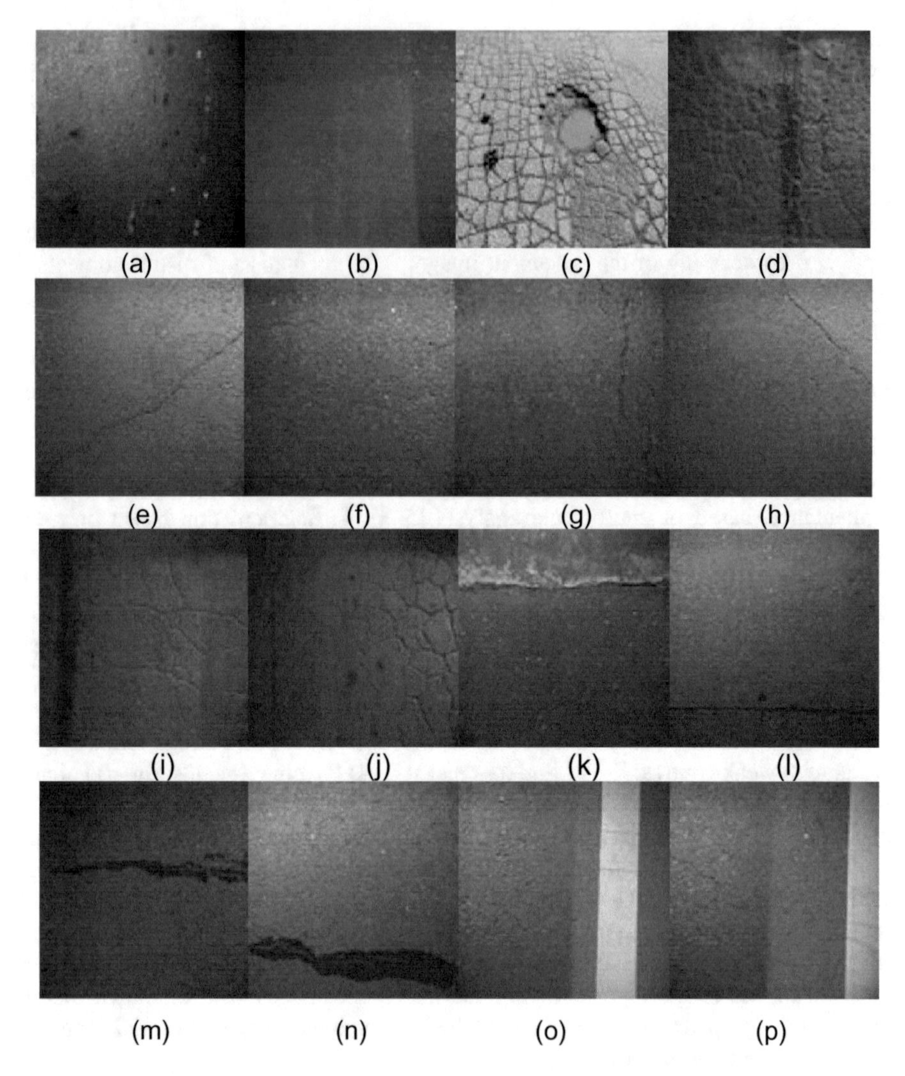

Fig. 1. 8 groups of pavement images with the most common distress types.

2 Framework of the Two-Staged Pavement Image Processing

In this paper, we present a novel two-staged pavement image processing framework as shown in Fig. 2. In the first stage, the pavement images are preprocessed and classified into four general categories, so that the images can be processed using category-specific algorithms in the 2nd stage. Then the geometry parameters of the distress objects will be measured. For instance, the parameters of the cracking-type distress including the width, length, loops number, and block size will be calculated. Finally, the pavement distress will be evaluated according to different protocols [14, 15].

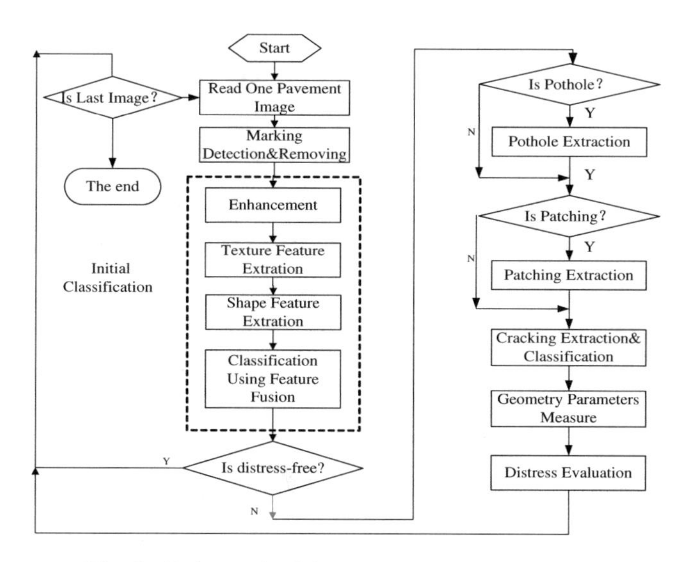

Fig. 2. Framework of the pavement image processing

This paper focuses the initial classification algorithm (See the dash box in Fig. 2), which includes 4 steps: (1) Enhancement of the input image, (2) Texture feature extraction, (3) Shape feature extraction, and (4) Classification using feature fusion. The software is developed with Visual C++ 2003 platform and runs on a DELL T630 Tower Server. The processing time of each input image is less than 100 ms.

3 Enhancement of the Pavement Images

Gao [16] proposed a fast pavement enhancement algorithm. We found when it is applied to enhance the pavement images, it has two deficiencies: 1. as shown in Fig. 3(d), the strong edges of cracks and patches are blurred, which is harmful for the detection of distress objects; 2. Though it can normalize the global grayscale of the entire pavement image, however the details of the images are lost.

This paper proposes an improved enhancement algorithm which fuses the local enhancement image P_t with the global enhancement image $I_t^{'}$ by the Eq. (1), where w is the fusion weight. Finally the enhanced image Y will be obtained, which is shown in Fig. 3(e), it can be found that the images enhanced by the proposed algorithm not only has a equalized background, but also reserves clear details compared with the Gao's algorithm.

$$Y = wP_t + (1 - w)I_t^{'} \tag{1}$$

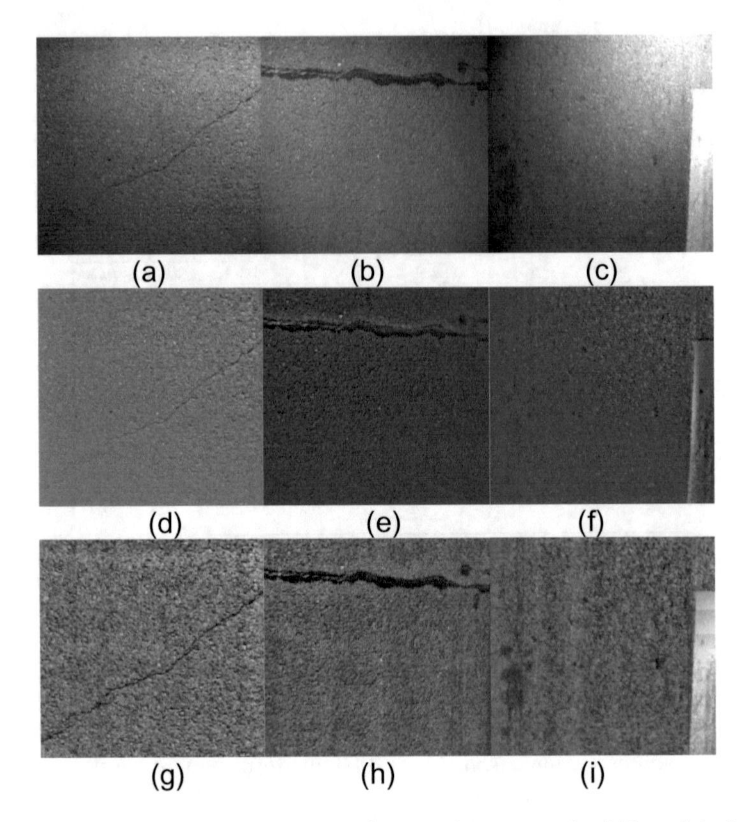

Fig. 3. Comparison experiments on pavement image enhancement: (a–c) The original pavement images; (d–f) The pavement images enhanced by Gao's algorithm; (g–i) The pavement images enhanced by the proposed algorithm

4 Feature Extraction of the Pavement Images

4.1 Texture Features Extraction Based on Wavelet Transform

With Wavelet transform, two sets of coefficients, the details and approximation components of an image can be obtained by taking inner product of the original image with a high-pass filter and a low-pass filter. These coefficients are condensed information which can used to express the global grayscale distribution, local details and phases for the description of image texture.

Zhou [6] proposed a pavement classification algorithm based on Wavelet texture, four Wavelet-based texture parameters are designed, including High Amplitude Wavelet Coefficient Percentage (HAWCP), High-Frequency Energy Percentage (HFEP), Moment of Wavelet Coefficients (MWC) and Standard deviation (STD). In order to reduce the complexity of the proposed algorithm, we use Linear Discriminated Analysis (LDA) to make feature selection on the four Wavelet-based texture features above. Finally, three features are remained, which are HAWCP, HFEP, and STD.

4.2 Shape Features Extraction Based on Shape Analysis

From the pavement images captured by the APCS, it can be observed that the distress object has significant shape features, which have distinctive ability to identify the pavement distresses compared with the texture features. So this paper selects some shape features as parts of criteria to classify the pavement images.

Generally, each pavement image needs to be converted into a binary image before shape extraction. As we know, the distress objects always occupy a small proportion of the entire pavement image with the darker intensity. So this paper use P-tile thresholding method to get the binary images of pavement distresses, which is interpreted by Eq. (2), where the threshold T satisfies Eq. (3).

$$I(x,y) = \begin{cases} 0, I(x,y) > T \\ 255, I(x,y) \leq T \end{cases} \tag{2}$$

$$\frac{1}{M \times N}\sum_{i=0}^{T} h(i) \leq P < \frac{1}{M \times N}\sum_{i=0}^{T+1} h(i) \tag{3}$$

In Eq. (3), M and N represent the width and height of the image respectively, $h(i)$ is the histogram function of I, and P is the prior knowledge, which represents the proportion of the distress objects in the pavement image. According to experimental experiences, P is often set with an initial value from 10% to 20% and adjusted with the illumination condition. In this paper, P is set to 15%. In order to reserve the details of the distress objects, we improve P-tile thresholding method before its application. In the first stage, the pavement image is divided into the non-overlapping sub-blocks with three kinds of block size (32 × 32 pixels, 64 × 64 pixels and 128 × 128 pixels). In the 2nd stage, the P-tile thresholding method is applied on each block. At last, all the blocks are overlaid into the final binary image as shown in Fig. 4. Note that the road markings produce some noises in the last two images above,we can use the mentioned method in Sect. 2 of this paper to remove them.

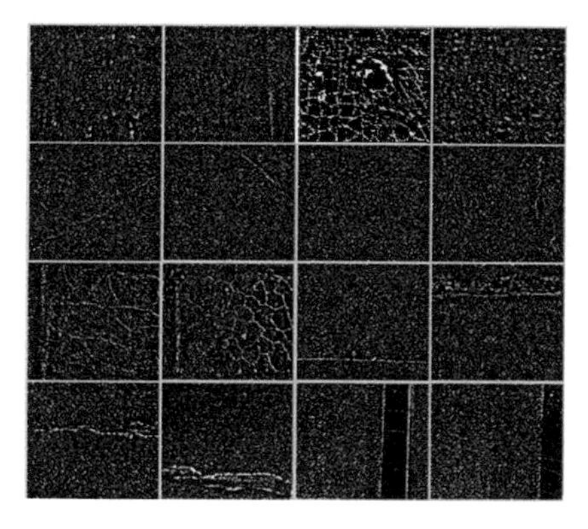

Fig. 4. Binary images resulting from P-tile thresholding method

From the binary pavement images shown in Fig. 4, we can find that these images have the shape features as follows:

(1) To the distress-free pavements, the area of each connected component is relatively small. The connected components have similar sizes due to the uniform pavement textures.
(2) The connected component corresponding to the crack has a curve-like shape with narrow width and long length.
(3) The connected component corresponding to the pothole has a big size and circular edges.
(4) The connected component corresponding to the patch has a big size, thick width and long length.

Considering the information above, we design three descriptors to qualify the shape features of the binary pavement images as follows.

(1) Average Area of All Connected Components (AA)

The definition of AA is described in Eq. (4), where N is the total number of the connected components in the image, and A_i is the area of each connected component. When AA is smaller, the probability is bigger for the pavement is distress-free. This descriptor reflects the macroscopic shape characteristics of the binary pavement image.

$$AA = \frac{1}{N}\sum_{i=1}^{N} A_i \tag{4}$$

(2) Area of the Maximal Connected Component (AM)

AM is the area of the maximal connected component in the image, when AM is bigger, the more likely the pavement image contains a distress. It reflects the microscopic shape characteristics of the binary pavement image.

(3) Equivalent Length of the Longest Connected Component (EL)

The value of EL is the total number of the pixels on the medial axis. Furthermore, in order to improve the distinctiveness of the proposed EL on distinguishing the cracking-types from other distress, we correct the EL with roundness by Eq. (5), where L is the medial axis length and f is the roundness of the shape. Note that the roundness of a circle is 1, but to a rectangular with length-width ration of 4, its roundness is 0.074, so it is good to amplify EL for curve-like objects. From the results in Table 1, it can be found that EL has significant distinctiveness compared with conventional descriptors.

$$EL = \frac{L}{f} \tag{5}$$

Table 1. Comparison results of different descriptors for distress length

Connected components #	Principal axis of the minimal surrounding ellipse	The longest cord	Medial axis length	Proposed EL
1	272	247	243	10942
2	252	266	354	4960
3	137	147	208	5118
4	71	76	74	184

5 Experimental Results and Discussion

In this section, 48 pavement images with four types (0: Distress-free, 1: Cracking-type, 2: Pothole-type, 3: Patching-type) are chosen as training samples. The texture features (HAWCP, HFEP and STD) and the shape features (AA, MA and STD) of these 48 images are calculated respectively.

Based on the texture and shape features, we design a BP neural network, and three feature scenarios are input to train the network, i.e., the pure texture features {HAWCP, HFEP, STD}, the pure shape features {AA, MA, EL} and the feature-fusion {HAWCP, HFEP, STD, AA, MA, EL}. The training results show that the network with fused features as input converges fast, its training iterations is 813 (objective error = 0.001), while those of the first two scenarios are 181,212 and 123,356 respectively with the same objective error.

Accordingly, the feature-fusion combination is the most suitable to describe the characteristic of the pavement image. Subsequently, we select 800 pavement images as

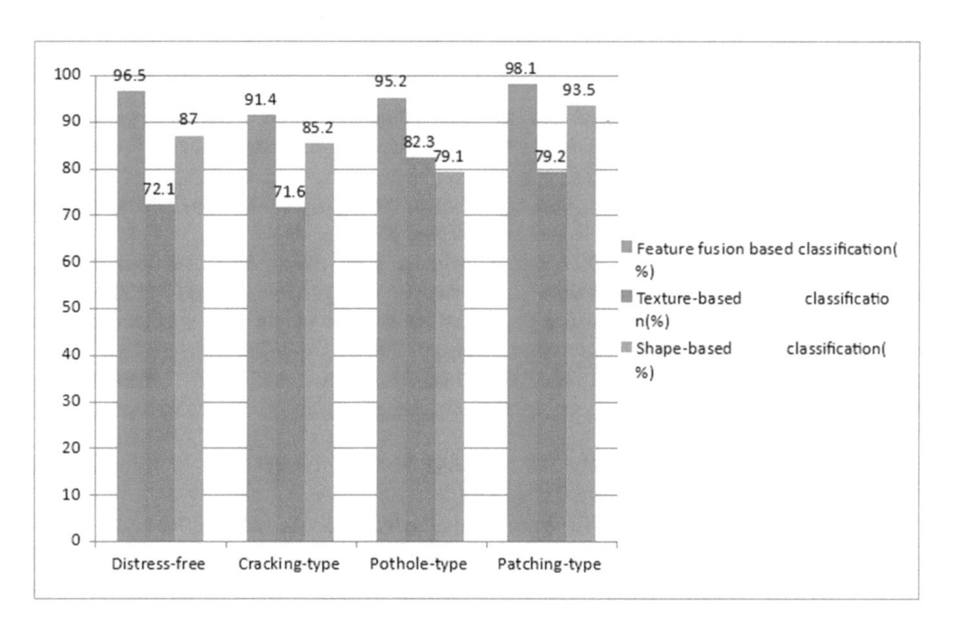

Fig. 5. The classification accuracy rates of the different classification modes

testing samples, where 400 are distress-free, 200 are cracking-type, 50 are pothole-type, and 150 are images with patches. Both the texture and the shape features are extracted from these 800 images. Figure 5 gives the classification accuracy rates which shows that for four types of pavement distress images the feature fusion has the highest classification accuracy rates (which are 96.5%, 91.4%, 95.2% and 98.1%, respectively) among the three classifications.

6 Conclusion

This paper proposes an efficient initial classification algorithm for the processing of the pavement distress images. The proposed algorithm fuses the local contrast enhanced image with the corrected global grayscale image to get an enhanced pavement distress image. With the enhanced pavement images, six distinguishing features of HAWCP, HFEP, STD, AA, MA, and EL are extracted using Wavelet transform and shape analysis, respectively. Finally, the neural network is used to fuse the texture and shape features serially and implement the initial classification. Experimental results show that the proposed algorithm achieves an effective classification of the pavement distress image with the accuracy rates of 96.5%, 91.4%, 95.2% and 98.1% for the four types of pavement images respectively, which are higher than those of the conventional classification algorithm with a single-type feature.

Acknowledgements. The authors appreciate the pavement images provided by UCPRC. The authors would also acknowledge financial support from China government under the NFSC and SNFSC programs (Grant no. 51278058, 2013JC9397).

References

1. Tsai, Y., Wu, M., Adams, E.D.: GPS/GIS enhanced road inventory system. In: Transportation Research Board 82nd Annual Meeting, 11–15 January 2003, Washington, DC (2003)
2. Zhang, A., Li, Q., Wang, K.C.P., Qiu, S.: Matched filtering algorithm for pavement cracking detection. Transp. Res. Rec. J. Transp. Res. Board **2367**(1), 30–42 (2013)
3. Cheng, H., Shi, X., Glazier, C.: Real-time image thresholding based on sample space reduction and interpolation approach. J. Comput. Civil Eng. **17**(4), 264–272 (2003)
4. Albert, A.P., Nii, A.O.: Evaluating pavement cracks with bidimensional empirical mode decomposition. EURASIP J. Adv. Signal Process. **2008**(1), 1–7 (2008)
5. Xu, Z., Zhao, X., Yang, L., Wei, N., Zhang, L.: Quick and precise road marking segmentation algorithm based on beamlet. J. Chang'an Univ. (Natural Science Edition) **33**(5), 101–108 (2013)
6. Zhou, J., Huang, P.S., Chiang, F.P.: Wavelet-based pavement distress detection and evaluation. Opt. Eng. **45**(2), 409–411 (2006)
7. Wang, K.C.P., Li, Q., Gong, W.: Wavelet-based edge detection with à trous algorithm for pavement distress survey. J. Transp. Res. Board **2024**(2024), 73–81 (2007)
8. Ma, C., Zhao, C., Hou, Y.: Pavement distress detection based on nonsubsampled contourlet transform. In: International Conference on Computer Science and Software Engineering, vol. 1, pp. 28–31 (2008)

9. Alekseychuk, D.I.: Detection of crack-like indications in digital radiography by global optimization of a probabilistic estimation function (2006)
10. Zou, Q., Cao, Y., Li, Q., Mao, Q., Wang, S.: CrackTree: automatic crack detection from pavement images. Pattern Recogn. Lett. **33**(3), 227–238 (2012)
11. Li, Q., Zou, Q., Zhang, D., Mao, Q.: FoSA: f* seed-growing approach for crack-line detection from pavement images. Image Vis. Comput. **29**(12), 861–872 (2011)
12. Kaul, V., Yezzi, A., Tsai, Y.: Detecting curves with unknown endpoints and arbitrary topology using minimal paths. IEEE Comput. Soc. **34**(10), 1952–1965 (2012)
13. Wang, H., Zhu, N., Wang, Q.: Fractal features analysis and classification for texture of pavement surfaces. J. Harbin Inst. Technol. **37**(6), 816–818 (2005)
14. California Department of Transportation, Caltrans Pavement Management System, Automated Pavement Condition Survey Manual, Interim version, October 2009
15. Xu, Z., Zhao, X., Zhang, L.: Asphalt pavement crack recognition algorithm using shape analysis. ICIC Expr. Lett. Int. J. Res. Surv. part B Appl. **2**, 671–678 (2011)
16. Gao, J., Ren, M., Yang, J.: A practical and fast method for non-uniform illumination correction. J. Image Graph. **7**(6), 548–552 (2002)

Author Index

G. De Pietro et al. (Eds.): KES-IIMSS-18 2018, SIST 98, pp. 429–431, 2019.
https://doi.org/10.1007/978-3-319-92231-7

Printed in the United States
By Bookmasters